Waste as a Resource

Series Editor

Cristina Trois, Civil Engineering, Stellenbosch University, Stellenbosch, South Africa

The Waste as a Resource series examines cutting-edge knowledge and innovations in waste management. It focuses on methods and technologies that maximize value recovery from waste through materials recycling and energy production. The series aims to facilitate the transition to a circular economy by presenting techniques and strategies to minimize waste and divert it from landfills. Most importantly, it explores potential pathways for treating waste as a valuable resource. The series has a global perspective, with a particular emphasis on developing countries and emerging economies. It aims to address a gap in the existing literature by adequately exploring valorization technologies suitable for and adaptable to low-income countries or informal contexts.

Himanshu Gupta · Soniya Dhiman ·
Ram Kishore Varshney
Editors

Advanced Technologies for Assessment, Recycling, and Management of Solid Waste

Springer

Editors
Himanshu Gupta
Department of Chemistry, School of Sciences
IFTM University
Moradabad, Uttar Pradesh, India

Soniya Dhiman
Helmholtz Institute Freiberg for Resource Technology
Helmholtz-Zentrum Dresden-Rossendorf
Dresden, Germany

Ram Kishore Varshney
Chandausi Inter College, Chandausi
Sambhal, Uttar Pradesh, India

ISSN 2731-8222 ISSN 2731-8230 (electronic)
Waste as a Resource
ISBN 978-3-032-24787-2 ISBN 978-3-032-24788-9 (eBook)
https://doi.org/10.1007/978-3-032-24788-9

This Springer imprint is published by the registered company Springer Nature Switzerland AG
The registered company address is: Gewerbestrasse 11, 6330 Cham, Switzerland

If disposing of this product, please recycle the paper.

Contents

An Overview of Advanced Technologies in Solid Waste Assessment, Recycling, and Management

Himanshu Gupta, Soniya Dhiman, and Ram Kishore Varshney

1 Introduction

Waste is referred as a material or substance which is no longer in use or it is replaced by any other material for better handling of the products. The leftover or used old material, substance or device becomes waste (Saleh & Rahman, 2016). The rapid urbanization, industrial expansion as well as increased consumption have resulted in the global escalation of the amounts of solid waste, which consists of hazardous substances. By 2050, it is estimated that global solid waste will reach around 3.4 billion metric tons per annum (Kaza et al., 2018). By 2025, per year around 2.2 billion tons municipal solid waste is generated, which is 1.42 kg/capita/day (Hoornweg & Bhada-Tata, 2012). Due to the increase in urban population and diverse industrial outputs, the amounts and complexity of solid wastes, the traditional landfilling and incineration of solid wastes are rendered obsolete. It is also reported that controlled incinerators utilize 3 times or higher cost than the landfilling cost per ton. However, landfilling is also problematic due to the concerns of residents in the neighboring areas (Hoornweg & Bhada-Tata, 2012). Waste management objectives generally include the minimization of toxic effects of wastes, environmental protection, ecosystem conservation, and social welfare (Seyyedi et al., 2024). Greenhouse gases such as ammonia, carbon dioxide, and methane are emitted by solid waste containing organic

H. Gupta (✉)
Department of Chemistry, School of Sciences, IFTM University, Moradabad, Uttar Pradesh, India
e-mail: hims.research@gmail.com

S. Dhiman
Helmholtz Institute Freiberg for Resource Technology, Helmholtz-Zentrum Dresden-Rossendorf, Dresden, Germany

R. K. Varshney
Chandausi Inter College, Chandausi, Sambhal, Uttar Pradesh, India

H. Gupta et al. (eds.), *Advanced Technologies for Assessment, Recycling, and Management of Solid Waste*, Waste as a Resource,
https://doi.org/10.1007/978-3-032-24788-9_1

matter. Landfilling of such wastes attracts rodents and insects as well as poses significant threat to the environment and public health. Potential health risks due to exposure to bacterial aerosols were observed in adult landfill workers during health evaluation (Abubakar et al., 2022; Kamdi et al., 2024; Louzizi et al., 2025). Waste burning and open dumping are commonly used in many countries, which release hazardous pollutants, causing exacerbating health issues and respiratory illness in the nearby areas (Lakhouit, 2025). There are different types of solid wastes which include paper, plastic, glass, metal, and organic wastes. It is reported that organic waste has its higher amounts in low-income countries, whereas plastic, paper, glass, and metal wastes are greater in high-income countries. In Morocco, in 2015, solid wastes were around 27 million tons, which are estimated to enhance by 2030 to 39 million tons (Louzizi et al., 2025). Waste management is important for the environment and human health. Production of waste is the major problem in different European countries, which creates pressure for the sustainable environment. In 2016, approx. 2.5 billion tons of waste was generated in European member states. It is also reported that landfilled municipal waste (24%) impacts groundwater. The generation of high volumes of solid wastes has enhanced extraction and mining processes greater than earlier (Sandhi & Rosenlund, 2024; Schaffartzik et al., 2014). In the present times, the model of circular economy is under consideration, providing more resource utilization (Kirchherr et al., 2017; Sandhi & Rosenlund, 2024). The current chapter gives an overview of the assessment, recycling, and management of solid wastes for different applications in order to protect the environment and maintain a circular economy. This book provides a comprehensive roadmap of these innovations, offering a technical and strategic guide for researchers, policymakers, and engineers dedicated to building a sustainable, waste-free future through the rigorous application of science and technology in the fields of assessment, recovery, and holistic management of solid wastes.

2 Types and Origin of Solid Wastes

A waste is a mandatory by-product of different human activities, industrial, and natural processes. The wastes are categorized further on the basis of source, composition, and environmental effects. Municipal solid wastes are reported to be maximum in countries such as India, China, and United States. Organic wastes are more produced in low-income countries, whereas other wastes such as paper, metal are produced more in high-income countries. Solid wastes are also produced by domestic, household, and industrial activities (Gupta, 2018; Gupta & Gupta, 2016; Nanda & Berruti, 2021; Singh et al., 2018). Based on the source, the waste is categorized as domestic, commercial, municipal, biomedical, mining, nuclear, marine, industrial, and e-waste. Based on physical state, waste can be solid, liquid, and gaseous waste. Based on toxic nature, wastes can be hazardous and non-hazardous. Solid waste can be originated at different places such as factories, hospitals, household, as well as laboratories (Cayumil et al., 2021; Song et al., 2015). The type as well as origin of solid wastes are diverse and complex in the modern societies. A better understanding of the type of waste and their separation practices can help to implement waste management strategies and reduce solid wastes at their sources.

3 Solid Waste Landfilling Contaminating Groundwater

Globally, landfilling is the mainly employed disposal method for solid waste. It is generally done in established locations or pits in a controlled manner. In particular cases, landfills are equipped with gas venting systems, leachate collection, impermeable liners, and environmental monitoring systems. Such systems safeguard groundwater and nearby regions (Cossu & Stegmann, 2018; Ma et al., 2022). But landfills are poorly managed in a major part of the world, particularly in developing countries, which results in the release of leachate to the groundwater, contaminating it. In order to mitigate groundwater contamination due to landfilling of solid wastes, the operational practices, leaching information, and landfill type and capacity should be mandatorily examined (Siddiqua et al., 2022). The toxic leachate consists of persistent organic pollutants, heavy metals, radioactive contaminants, etc. The leachate can percolate and reach the groundwater slowly and collect to cause higher contamination (Dagwar & Dutta, 2024; Gunarathne et al., 2024). With the continuous and widespread landfilling at a large scale, groundwater pollution has emerged as an environmental, public health, and socio-economic concern, which requires a proper solution or alternative to safeguard water for sustainability.

4 Applications of Solid Wastes for Environment Protection

A major portion of construction and demolition waste (60%), fly ash waste (47%), glass (75%), plastics (55%), and tires (75%) dispose to the landfills. Recycling of such wastes can be a sustainable solution. Also, 2.8 jobs are created upon waste disposal of 10,000 tons, whereas 9.2 jobs will be created upon recycling of this amount of waste (Ferdous et al., 2021). A huge quantity of vehicular tires are landfilled every year (Gupta & Gupta, 2018). An e-waste of around 60,000 tons/year is produced in Singapore alone, which comprises of around 1620 tons electric components. Approximately 6 wt% of global e-waste comes from printed circuit boards (Charpentier et al., 2023). Among all kinds of e-waste, printed circuit boards (PCBs) and lithium-ion batteries (LIBs) have significant potential for recycling both from an environmental and economic perspective. Around 1.36 million tons will be the global output for waste lithium-ion batteries by 2025 (Cunha et al., 2025; Gupta & Shiju, 2025; Gupta et al., 2025). The different kinds of solid wastes can be utilized for environment-friendly processes like composting of the organic waste using microorganisms (Potysz et al., 2019). Similarly, vermicomposting employs earthworms to break down the organic waste into granular and nitrogen-rich material

(Patwa et al., 2020). Pyrolysis, hydrothermal liquefaction, and gasification converts different wastes including plastic waste, polyurethane waste into value-added products (Poravou et al., 2022; Yek et al., 2024). Various precious metals can be separated from printed circuit boards using different techniques like pyrometallurgy, smelting, hydrometallurgy, and bioleaching (Udayakumar et al., 2022; Dong et al., 2023; Zhang & Xu, 2016). The waste LIBs can be recycled in a number of ways to recover valuable metals, which are toxic for the environment and public health, if processed in the waste streams (Agarwal et al., 2024; Dhiman & Gupta, 2019; Gupta et al., 2025). Solid waste can also be utilized for the preparation of different nanoparticles employing the carbothermal (Kim et al., 2018), microwave-aided (Al-Ghouti et al., 2021), co-precipitation, solvothermal, and hydrothermal techniques (Siddiqui et al., 2020). Green mixed nano-composite have also been synthesized utilizing solid wastes for degradation of toxic dyes (Singh et al., 2016). Municipal solid wastes can also be upcycled to synthesize carbon-based nanomaterials (Zhou et al., 2012). Different solid waste materials are used in the construction materials for the maintenance of circular economy. Many of the solid waste materials such as blast furnace slag, red mud are utilized for the preparation of adsorbents to remove environmental contaminants (Dhmees et al., 2018; Wang et al., 2019). An overview of the solid waste in the environment and its fate is given as Fig. 1.

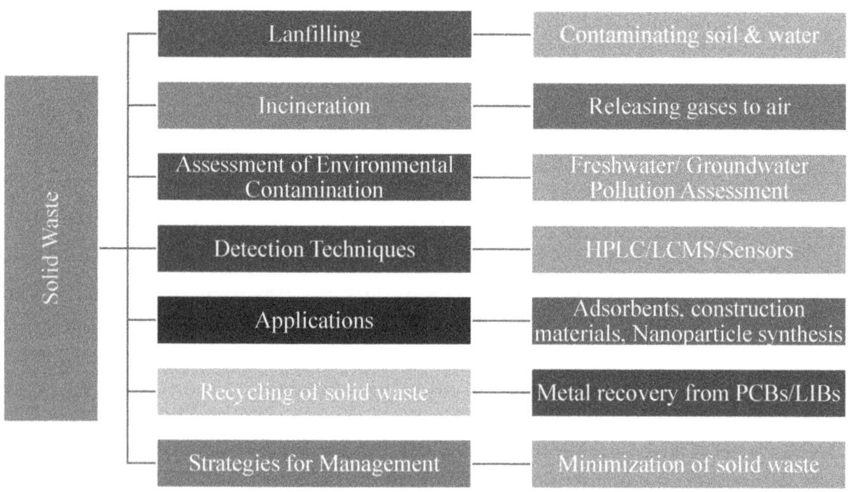

Fig. 1 An overview of solid waste and its fate in the environment

5 Conclusion

The rapid urbanization and industrial expansion have resulted in the global escalation of the amounts of solid waste, which necessitates its assessment, disposal, recycling, and proper management. By 2050, it is estimated that global solid waste will reach around 3.4 billion metric tons per annum. Greenhouse gases such as ammonia, carbon dioxide, and methane are emitted by solid waste containing organic matter. Potential health risks due to exposure to bacterial aerosols were observed in adult landfill workers during health evaluation. Waste burning and open dumping are commonly used in many countries, which release hazardous pollutants causing exacerbated health issues and respiratory illness. With the continuous and widespread landfilling at a large scale, groundwater pollution has emerged as an environmental, public health, and socio-economic concern. A major portion of construction and demolition waste (60%), fly ash waste (47%), glass (75%), plastics (55%), and tires (75%) dispose to the landfills. Recycling of such wastes can be a sustainable solution. Also, 2.8 jobs are created upon waste disposal of 10,000 tons, whereas 9.2 jobs will be created upon recycling of this amount of waste. Around 1.36 million tons will be the global output for waste lithium-ion batteries by 2025. Various precious metals can be separated from printed circuit boards and LIBs using different techniques like pyrometallurgy, smelting, hydrometallurgy, and bioleaching. The current chapter gives an overview of the assessment, recycling, and management of solid wastes for different applications in order to protect the environment and maintain a circular economy. This book provides a comprehensive roadmap of these innovations, offering a technical and strategic guide for researchers, policymakers, and engineers dedicated to building a sustainable, waste-free future.

References

Abubakar, I. R., Maniruzzaman, K. M., Dano, U. L., AlShihri, F. S., AlShammari, M. S., Ahmed, S. M. S., Al-Gehlani, W. A. G., & Alrawaf, T.I. (2022). Environmental sustainability impacts of solid waste management practices in the global South. *IJERPH 19*(19), 12717. https://doi.org/10.3390/ijerph191912717

Agarwal, S., Dhiman, S., & Gupta, H. (2024). Recovery of lithium and copper from anode electrode materials of spent LIBs by acidic leaching. *Environmental Science and Pollution Research, 31*, 34249–34257. https://doi.org/10.1007/s11356-024-33537-x

Al-Ghouti, M. A., Khan, M., Nasser, M. S., Saad, K. A., & Heng, O. E. (2021). A novel method for metals extraction from municipal solid waste using a microwave-assisted acid extraction. *Journal of Cleaner Production, 287*, Article 125039. https://doi.org/10.1016/j.jclepro.2020.125039

Cayumil, R., Khanna, R., Konyukhov, Y., Burmistrov, I., Kargin, J. B., & Mukherjee, P. S. (2021). An overview on solid waste generation and management: Current status in Chile. *Sustainability, 13*(21), 11644. https://doi.org/10.3390/su132111644

Charpentier, N. M., Maurice, A. A., Xia, D., Li, W. J., Chua, C. S., Brambilla, A., & Gabriel, J. C. P. (2023). Urban mining of unexploited spent critical metals from E-waste made possible using advanced sorting. *Resources, Conservation and Recycling, 196*, Article 107033.

Cossu, R., & Stegmann, R. (2018). *Solid waste landfilling: Concepts, processes, technology* (1st ed.). Elsevier.

Cunha, L., Monteiro, J., Futuro, A., Regufe, M. J., Soeiro, J., & Sousa, R. (2025). Recycling PCBs for nanoparticles production with potential applications in cosmetics, cement manufacturing, and CO_2 capture. *Waste Management 191*, 308–323. https://doi.org/10.1016/j.wasman.2024.11.018

Dagwar, P. P., & Dutta, D. (2024). Landfill leachate a potential challenge towards sustainable environmental management. *Science of the Total Environment, 926*, Article 171668.

Dhiman, S., & Gupta, B. (2019). Partition studies on cobalt and recycling of valuable metals from waste Li-ion batteries via solvent extraction and chemical precipitation. *Journal of Cleaner Production, 225*, 820–832. https://doi.org/10.1016/j.jclepro.2019.04.004

Dhmees, A. S., Khaleel, N. M., & Mahmoud, S. A. (2018). Synthesis of silica nanoparticles from blast furnace slag as cost-effective adsorbent for efficient azo-dye removal. *Egyptian Journal of Petroleum, 27*(4), 1113–1121. https://doi.org/10.1016/j.ejpe.2018.03.012

Dong, Y., Zan, J., & Lin, H. (2023). Recovery of precious metals from waste printed circuit boards though bioleaching route: A review of the recent progress and perspective. *Journal of Environmental Management, 348*, Article 119354.

Ferdous, W., et al. (2021). Recycling of landfill wastes (tyres, plastics and glass) in construction—A review on global waste generation, performance, application and future opportunities. *Resources, Conservation and Recycling, 173*, Article 105745. https://doi.org/10.1016/j.resconrec.2021.105745

Gunarathne, V., Phillips, A. J., Zanoletti, A., Rajapaksha, A. U., Vithanage, M., Di Maria, F., Pivato, A., Korzeniewska, E., & Bontempi, E. (2024). Environmental pitfalls and associated human health risks and ecological impacts from landfill leachate contaminants: Current evidence, recommended interventions and future directions. *Science of the Total Environment, 912*, Article 169026.

Gupta, H. (2018). PAH determination in effluent and sludge samples of paper industry. *Environmental Technology and Innovation, 9*, 115–121.

Gupta, H., Dhiman, S., Stam, P. G., Chaudhary, S., Murugavel, R., & Shiju, N. R. (2025). Sustainable approach to cobalt recycling: Levulinic acid leaching of waste lithium-ion battery cathode materials. *Environmental Technology & Innovation*, 104675. https://doi.org/10.1016/j.eti.2025.104675

Gupta, H., & Gupta, B. (2016). Adsorption of polycyclic aromatic hydrocarbons on banana peel activated carbon. *Desalination and Water Treatment, 57*(20), 9498–9509.

Gupta, H., & Gupta, B. (2018). Vehicular tire as potential adsorbent for the removal of polycyclic aromatic hydrocarbons. *Polycyclic Aromatic Compounds, 38*(4), 354–368.

Gupta, H., & Shiju, N. R. (2025). Recycling and reuse of olefinic separators in lithium-ion batteries: A mini review. *Current Opinion in Green and Sustainable Chemistry, 56*, Article 101054. https://doi.org/10.1016/j.cogsc.2025.101054

Hoornweg, D., & Bhada-Tata, P. (2012). What a waste: A global review of solid waste management. Urban Development Series, No. 15. World Bank. https://doi.org/10.1596/17388

Kamdi, P., Patil, S., Bafana, A., Lalwani, A., Middey, A., Kannan, K., & Sivanesan, S. (2024). Health risk assessment and characterization of PM2.5 bound bioaerosols at the municipal solid waste landfill site of Nagpur, India. *Journal of Aerosol Science, 178*, 106359. https://doi.org/10.1016/j.jaerosci.2024.106359

Kaza, S., Yao, L., Bhada-Tata, P., & Van Woerden, F. (2018). *What a waste 2.0: A global snapshot of solid waste management to 2050.* World Bank Publications.

Kim, C. H., Kim, J.-H., & Lee, S.-M. (2018). Transformation of solid waste into nanoporous carbon via carbothermic reduction. *ACS Omega, 3*(7), 7904–7910. https://doi.org/10.1021/acsomega.8b01280

Kirchherr, J., Reike, D., & Hekkert, M. (2017). Conceptualizing the circular economy: An analysis of 114 definitions. *Resources, Conservation & Recycling, 127*, 221–232.

Lakhouit, A. (2025). Revolutionizing urban solid waste management with AI and IoT: A review of smart solutions for waste collection, sorting, and recycling. *Results in Engineering, 25*, Article 104018. https://doi.org/10.1016/j.rineng.2025.104018

Louzizi, T., Chakir, E., & Sadoune, Z. (2025). A comprehensive review on solid waste management in Morocco: Assessment, challenges and potential transition to a circular economy. *Euro-Mediterranean Journal for Environmental Integration, 10*, 1281–1296. https://doi.org/10.1007/s41207-024-00662-5

Ma, S., Zhou, C., Pan, J., Yang, G., Sun, C., Liu, Y., Chen, X., & Zhao, Z. (2022). Leachate from municipal solid waste landfills in a global perspective: Characteristics, influential factors and environmental risks. *Journal of Cleaner Production, 333*, Article 130234.

Nanda, S., & Berruti, F. (2021). Municipal solid waste management and landfilling technologies: A review. *Environmental Chemistry Letters, 19*(2), 1433–1456. https://doi.org/10.1007/s10311-020-01100-y

Patwa, A., Parde, D., Dohare, D., Vijay, R., & Kumar, R. (2020). Solid waste characterization and treatment technologies in rural areas: An Indian and international review. *Environmental Technology & Innovation, 20*, 101066. https://doi.org/10.1016/j.eti.2020.101066.

Poravou, C. A., Tsongidis, N. I., Lekkos, C., et al. (2022). Valorization of plastic waste: A lab-scale approach with the aid of solar hydrothermal liquefaction technology. *Waste & Biomass Valorization, 13*, 3835–3844. https://doi.org/10.1007/s12649-022-01837-3

Potysz, A., Pędziwiatr, A., Hedwig, S., & Lenz, M. (2019). Rapid metal mobilisation through litter water and bioweathering as the legacy of historical copper smelting. *Journal of Geochemical Exploration, 206*, 106364. https://doi.org/10.1016/j.gexplo.2019.106364.

Saleh, H. E.-D. M., & Rahman, R. O. A. (Eds.). (2016). *Management of hazardous wastes*. InTech. https://doi.org/10.5772/61668

Sandhi, A., & Rosenlund, J. (2024). Municipal solid waste management in Scandinavia and key factors for improved waste segregation: A review. *Cleaner Waste Systems, 8*, Article 100144. https://doi.org/10.1016/j.clwas.2024.100144

Schaffartzik, A., Mayer, A., Gingrich, S., Eisenmenger, N., Loy, C., & Krausmann, F. (2014). The global metabolic transition: Regional patterns and trends of global material flows, 1950–2010. *Global Environmental Change, 26*, 87–97.

Seyyedi, S. R., Kowsari, E., Gheibi, M., Chinnappan, A., & Ramakrishna, S. (2024). A comprehensive review integration of digitalization and circular economy in waste management by adopting artificial intelligence approaches: Towards a simulation model. *Journal of Cleaner Production, 460*, Article 142584. https://doi.org/10.1016/j.jclepro.2024.142584

Siddiqua, A., Hahladakis, J. N., & Al-Attiya, W. A. K. (2022). An overview of the environmental pollution and health effects associated with waste landfilling and open dumping. *Environmental Science and Pollution Research, 29*, 58514–58536.

Siddiqui, M. T. H., Baloch, H. A., Nizamuddin, S., Haris, M., Mubarak, N. M., Czajka, M., Khalid, M., Griffin, G. J., & Srinivasan, M. (2020). Synthesis of novel magnetic carbon nano-composite from waste biomass: A comparative study of industrially adoptable hydro/solvothermal co-precipitation route. *Journal of Environmental Chemical Engineering, 8*(2), Article 103519. https://doi.org/10.1016/j.jece.2019.103519

Singh, S., Parveen, N., & Gupta, H. (2018). Adsorptive decontamination of rhodamine-B from water using banana peel powder: A biosorbent. *Environmental Technology & Innovation, 12*, 189–195.

Singh, S., Srivastava, V. C., Mandal, T. K., Mall, I. D., & Lo, S. L. (2016). Synthesis and application of green mixed-metal oxide nano-composite materials from solid waste for dye degradation. *Journal of Environmental Management, 181*, 146–156. https://doi.org/10.1016/j.jenvman.2016.06.007

Song, Q., Li, J., & Zeng, X. (2015). Minimizing the increasing solid waste through zero waste strategy. *Journal of Cleaner Production, 104*, 199–210. https://doi.org/10.1016/j.jclepro.2014.08.027

Udayakumar, S., Razak, M. I. B. A., & Ismail, S. (2022). Recovering valuable metals from Waste Printed Circuit Boards (WPCB): A short review. *Materials Today: Proceedings, 66*(5), 3062–3070. https://doi.org/10.1016/j.matpr.2022.07.364.

Wang, L., Sun, N., Tang, H., & Sun, W. (2019). A review on comprehensive utilization of red mud and prospect analysis. *Minerals, 9*(6), 362. https://doi.org/10.3390/min9060362

Yek, P. N. Y., Chan, Y. H., Foong, S. Y., et al. (2024). Co-processing plastics waste and biomass by pyrolysis–gasification: A review. *Environmental Chemistry Letters, 22*, 171–188. https://doi.org/10.1007/s10311-023-01654-7

Zhang, L., & Xu, Z. (2016). A review of current progress of recycling technologies for metals from waste electrical and electronic equipment. *Journal of Cleaner Production, 127*, 19–36.

Zhou, C., Alves, J. O., Tenorio, J. A. S., & Levendis, Y. A. (2012). Synthesis of carbon nanomaterials through up-cycling agricultural and municipal solid wastes. *Industrial & Engineering Chemistry Research, 51*(7), 2922–2930. https://doi.org/10.1021/ie202711h

Domestic, Household, and Industrial Solid Wastes Management Strategies for Sustainable Development

Rashmi Verma, Richa Saxena, Shreya Kotnala, Shalini Tiwari, B. S. Rawat, and Latif Arifudin

1 Introduction

Waste is something, which is no longer utilised, or is removed from the action for which it was previously used. It can be generated by such sources as households, industries, agriculture, and healthcare facilities (Saleh & Abdel Rahman, 2016). Solid, liquid, organic, hazardous, and e-waste are some of the types of waste. Studies on discards have shown that garbage is not only a technical issue for public health officials and sanitary engineers, or a sign of an all-too-human need for purpose (Reno, 2015). With the increase in population, setting up of industries, there is an increase in waste that leads to harmful effect on human health and ecosystem (Olawade et al., 2024). Hence, in order to control and manage the waste produced, waste management techniques are used. The term waste management commonly applies to the collection and disposal of waste, including materials as a result of the

R. Verma
Department of Biotechnology, School of Basic & Applied Sciences, Shri Guru Ram Rai University, Dehradun, India

R. Saxena (✉)
Department of Physics, School of Basic & Applied Sciences, Shri Guru Ram Rai University, Dehradun, India
e-mail: saxena.richa23@gmail.com

S. Kotnala · S. Tiwari
Department of Chemistry, School of Basic & Applied Sciences, Shri Guru Ram Rai University, Dehradun, India

B. S. Rawat
Department of Physics, Uttaranchal Institute of Technology & School of Applied & Life Sciences, Uttaranchal University, Dehradun, India

L. Arifudin
Department of Mechanical Engineering, King Saud University, Riyadh, Saudi Arabia

H. Gupta et al. (eds.), *Advanced Technologies for Assessment, Recycling, and Management of Solid Waste*, Waste as a Resource,
https://doi.org/10.1007/978-3-032-24788-9_2

extraction, production, consumption, and other human activities. Waste may be seen. The main objectives of waste management are minimising the hazardous impacts resulting from waste, utilisation of resources, and conservation of human health and ecosystem (Gupta & Gupta, 2016, 2018; Seyyedi et al., 2024). Uncontrolled management of waste results in fires, explosions, air, soil, and water contamination, all of which pose major health risks. Similarly, inappropriate waste management and disposal can lead to expensive cleanups and endanger the lives of residents in surrounding towns (Amasuomo & Baird, 2016; Dijkgraaf & Vollebergh, 2004; Visvanathan, 1996). Effective waste management is critical to minimising greenhouse gas emissions from landfills, lowering pollution, conserving resources, protecting human health, and combating climate change. Composting, zero-waste initiatives, and circular economies are examples of sustainable practices that are increasingly important in addressing the world's waste issues. Collaboration between communities, businesses, and governments is necessary to achieve long-term environmental sustainability (Seadon, 2010). Yet even with strong advances in some areas, many barriers still remain to implementing more sustainable waste management worldwide. The challenges are a lack of adequate infrastructure, low financial capacity, ineffective regulations, and poor interdisciplinary connection with the public. In most of these developing countries, the informal sector prevails in waste management, which has low recycling rates, low labour conditions, and environmental contamination (Kumari & Raghubanshi, 2023). Solving these challenges is a joint effort between decision makers, industry, civil society, and individuals to encourage sustainable production and consumption, and invest in better preservation and processing of post-harvest fish and fishery products. The worsening state of its waste problem further emphasises the importance of immediate, integrated, and creative approaches that favour waste minimisation, resource recovery, and environmental promotion (Demirbas, 2011). With countries working to meet global sustainability targets and combating climate change, the management of waste will continue to be an important part of environmental policy and public health efforts. Transforming our relationship with waste requires coordinated, innovative policy, and behavioural solutions that create a cleaner, healthier, and more environmentally responsible world.

2 Types and Origin of Waste

Waste is a necessary by-product of industrial processes, human activity, and the natural cycles. Depending on the source of the waste, composition, and effect on the environment, they are further categorised. The most leading waste is the solid waste that is disposed of in solid form. United States, China, and India are the leading countries for creating municipal solid waste. Solid wastes are also affected by the income of the people. Organic wastes are created by low to middle class, and paper, metal, and glass wastes are created by high class people. Industries and households are responsible for these kind of wastes (Nanda & Berruti, 2021). Liquid waste is another type of waste consisting of liquid substances as discarded. Sewage, chemicals, oils, and grease are a few examples of liquid waste (Sravya et al., 2025). Dangerous substance is a category for special treatment/handling due to possible health hazards to humans and the environment. The origin of this kind of waste can be several places, including laboratories, households, hospitals, and factories that make chemicals (Song et al., 2015). The harmful by-products of industrial processes, such as heavy metal sludge, corrosive acids, and the harmful toxic solvents, need to be handled carefully before they can be disposed of (Gupta, 2018). Other waste categories may include electronic waste, agricultural waste, and plastic waste (Gupta & Shiju, 2025; Gupta et al., 2025; Singh et al., 2018). The nature and origins of wastes are at least as varied as the activities of humanity itself—from simple domestic dwellings to vast industrial installations, from farmland to high-tech hospitals. That ranges from food scraps and packaging materials to discarded furniture and electronic gadgets. Every home also adds to this stream with everyday activities from cooking, to cleaning, and shopping, and businesses add to it with packaging, old equipment, and day-to-day business. In the case of MSW, its composition is highly related to the level of development in a country, being richer countries generating a larger amount of packaging and electronic material waste, and poorer nations producing less of such materials and a greater percentage of organic content and non-recyclable materials (Panchal et al., 2021). Factories typically accumulate waste such as rejected products, scraps, and chemical waste during the manufacturing process. When structures are built, remodelled, or demolished, the construction industry contributes massive amounts of waste, such as concrete, wood, metals, and insulation materials. Miners create massive overburden and tailings that can be laced with toxic heavy metals, while oil refineries yield petrochemical sludge and doped water. Sources of different wastes are depicted in Fig. 1 (Hajam et al., 2023). All these industries have their specific waste profile, such as the textile industry discharges water containing dyes, excess fabric cut-pieces, the electronic industry discharges solvents and heavy metal products. Industrial waste is often orders of magnitude larger than municipal waste streams; its treatment is complicated by the widespread presence of hazardous substances that produce contamination of soil and water bodies (Kiyasudeen et al., 2016). Hence, the origin and type of wastes are diverse, which reflects the complexity of modern societies, and by understanding the different types and origins of waste, individuals can reduce waste, develop sustainable waste management practices, and promote recycling (Table 1).

3 Waste Management Strategies

Every time a simple edible food item is thrown in the trash, a damaged toy is thrown away, a smart device or other electronic gadget needs to be replaced, or anything else is disposed of, resources are being depleted and also adding to the environmental pressures in all places. Additionally, a whole new set of effects arises when a product

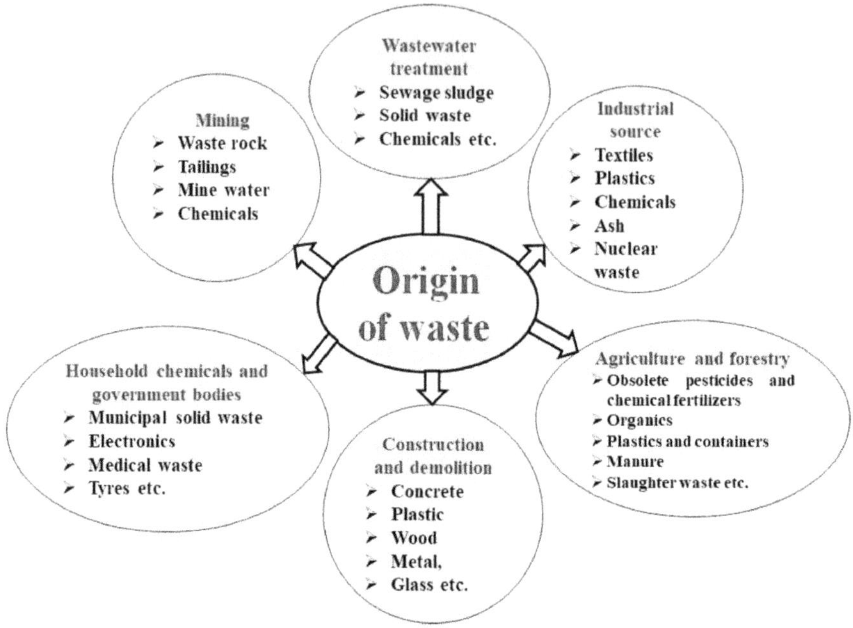

Fig. 1 The different sources of origin of waste (published in open-access article license (https://creativecommons.org/licenses/by/4.0/)) (Hajam et al., 2023)

is discarded and turns into garbage. Hence, for this, proper waste management strategies need to be adopted. Waste management strategies refer to a systematic approach that is used to manage wastes in a way that its disposal doesn't affect the environment (Anokye & Mohammed, 2024). Many of the reported pollution accidents linked to improper waste management techniques raised public concerns about a lack of regulations, insufficient laws, and the effects on the environment and human health (Schultz, 2002).

The goal of these strategies is to minimise the harmful effects of waste on the environment and human health. Hence, the most widely adopted framework for waste management is the Integrated Waste Management Hierarchy (Siddiqua et al., 2022).

The waste management strategies have evolved from simple garbage disposal to complex, integrated systems that maximise resource recovery while minimising the environmental damage (Romero-Hernández & Romero, 2018). Recycling reduces the repository usage and conserves materials, which just use 4% of primary energy when recycled. Hence, all these blended economic, technical, and institutional strategies underpinned by lifecycle, emission, and equity analyses point to a resilient pathway for efficient waste management (Oppelt, 1987). The waste management strategies involve traditional or conventional approaches and modern emerging methods of waste management.

Table 1 Classification of waste based on its types and origin

Categorisation	Type of waste	Example
Based on biological properties	Domestic waste	Kitchen waste, papers, and cloth,
	Commercial waste	Cardboard, plastics, and packaging
	Municipal waste	Market wastes, dead animals, and street sweepings
	Bio-medical waste	Bandages, syringes, needles
	E-waste	Mobiles, computers, batteries
	Mining waste	Tailings, waste rocks, overburden
	Nuclear waste	Radioactive isotopes
	Marine waste	Debris, plastic bottles
	Space waste	Non-functional satellites
	Industrial waste	Dyes, chemicals, metal residues
Based on physical properties	Solid waste	Plastics, metal scraps, broken glass
	Liquid waste	Industrial effluent, oil, chemicals
	Gaseous waste	Harmful gases like No_2, CO_2, SO_2
Based on decomposition time	Quickly decomposable	Fruits, vegetables
	Slow decomposable	Bones, cotton cloth
	Non-decomposable	Plastic, metals
Based on the recycling potential	Recyclable waste	Paper, cardboard, and metals
	Non-recyclable waste	Contaminated plastics, soiled diapers
	Upcyclable waste	Glass bottles, pallets
Based on hazard potential	Hazardous waste	Batteries, solvents, etc.
	Non-hazardous waste	Vegetable peels, biodegradable garden waste

3.1 Traditional Approaches to Waste Management

Management of waste is important for urban infrastructure and practicalities, urban sanitation and disease control, and urban development and planning (Wan et al., 2019). By this term is understood the regular parts of the removal from among men, the management, treatment, processing, and final disposition of refuse collected from, equipment of, or methods employed in, managing an operation (operation here refers to the overall process and equipment specifically required for management, treatment, storage, and final disposition of solid waste by methods that will preserve and protect human health and the environment; this does not include a waste pile, or an inactive or disposed of surface impoundment). Conventional waste management methods have been used for decades, and remain the most common practice, especially in developing countries and areas where limited access to modern waste treatment facilities is available (Osmani & Villoria-Sáez, 2019) Although such conventional approaches have contributed to adhering to basic hygiene and cleanliness, they are widely criticised because of their environmental, social, and health issues. This article reviews various common waste management practices, how they work, and their strengths and weaknesses (Fudala-Ksiazek et al., 2016) Open dumping is the most popular traditional way of waste management. It involves open defecation in areas such as bushes, vacant lots, roadside tracks, river banks, and open fields with no systematic control of excreta and little or no treatment of waste. Even though open dumping is a quick and inexpensive way to get rid of trash, it poses a serious threat to the environment and public health (Czekała et al., 2023). Uncontrolled dumps attract rats and stray dogs, which spread disease, and serve as flies and mosquitoes breeding grounds.

Another popular conventional waste management method is landfilling. Trash is dumped in landfills, which are specially constructed sites where it is occasionally covered in dirt, to reduce odour and prevent the spread of disease. A landfill is an engineered pit that is specifically made to hold compacted solid waste and has a specific covering to allow for waste disposal. To prevent the waste from contaminating subterranean water, a lining has been placed at the bottom of the landfill. In order to reduce volume, landfills are made to receive concentrated garbage in compacted layers.

A landfill's bottom is shielded to prevent contamination of the subsurface water. Essentially, at the conclusion of each day, soil should cover the dumped trash. This will guarantee that the rubbish cannot be dug up by animals or flies.

Even though modern sanitary landfills have engineering controls like impermeable liners, leachate collection systems, and gas capture devices, many conventional landfills lack these safeguards, especially in low-income areas. Therefore, conventional landfills might be harmful to the environment (Rubinov & Fetisov, 2021). Leachate, a poisonous liquid produced by the breakdown of organic waste, can leak into nearby soil and groundwater if it is not adequately contained. Drinking water supplies are contaminated, endangering ecosystems and human health. Large tracts of land that could be utilised for housing, green spaces, or agriculture are also taken up by the build-up of waste in landfills (Giusti, 2009) Concern over past inappropriate disposal methods has led to the enactment of numerous hazardous waste cleanup and control laws at the federal and state levels in the past ten years, which have an unparalleled breadth. Waste management procedures will be significantly altered as a result of these several statutes. Waste minimisation at the source of generation, waste reuse, physical, chemical, and biological treatment, incineration, and chemical stabilisation techniques will largely replace the more conventional and economical methods. Properly built incineration systems can manage the huge amount of waste, making them the most effective terminal treatment technology (Samui et al., 2022).

Composting is a biological waste management technique that creates a nutrient-rich substance that is used as a soil conditioner by breaking down organic waste materials. However, it is a long-standing technology; its widespread use and effectiveness have been diminished by certain issues (Dawar et al., 2025) Conventional composting techniques, which are widely used in both urban and rural settings, entail piling organic waste outdoors in pits or heaps and letting natural microbial processes break it down. Conventional composting methods are frequently unregulated and ineffective, even though composting is an environmentally friendly substitute for landfilling and incineration when it comes to managing organic waste (Sharma & Vaid, 2021). Figure 2 summarises various household strategies (Viljoen et al., 2021). Many people blame socioeconomic and institutional factors for the continued use of traditional waste management methods. Many developing countries have trouble managing their municipal waste because they don't have enough money, they don't know how to use technology, the rules aren't strict enough, and cities are growing too quickly. The problem is worse because most people don't know how to recycle, consume sustainably, or separate their trash correctly. So, even though they have their problems, traditional methods are still the most common way to do things in many parts of the world. People are starting to realise that traditional ways of dealing with waste are not going to work in the long run. They cause important resources to run out, put people's health at risk, and hurt the environment (Sithole et al., 2024).

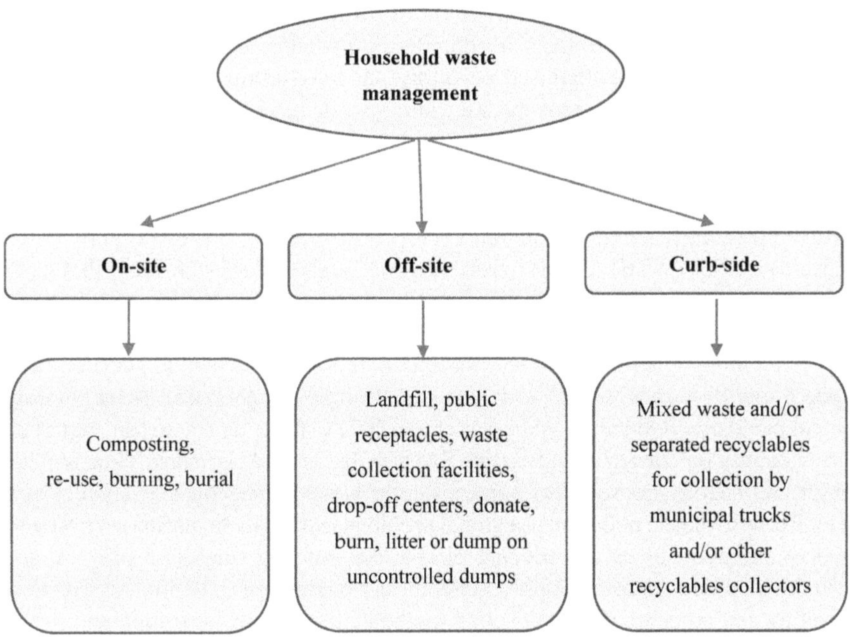

Fig. 2 Household waste management strategies (published in open-access article licence (https://creativecommons.org/licenses/by/4.0/)) (Viljoen et al., 2021)

3.2 Modern Waste Management Strategies

The old collect and dump methods of waste management have given way to more effective, sustainable, and ecologically friendly methods. Advanced waste management systems must be developed and put into place because of the growing amount and complexity of waste produced by urbanisation, industrialisation, and population growth worldwide (Shukla et al., 2024). Modern strategies seek to reduce waste generation, encourage resource recovery, guarantee environmental protection, and incorporate technological advancements with policy frameworks, in contrast to conventional methods that mainly concentrate on disposing of waste through landfilling or incineration. Waste minimisation, or lowering the amount and toxicity of waste at its source, is one of the core tenets of contemporary waste management. Product redesign, material substitution, enhanced manufacturing techniques, and consumer awareness campaigns are used to accomplish this. For example, businesses are embracing lean manufacturing practices, sustainable packaging, and Extended Producer Responsibility (EPR) for proper disposal returns the goods to the manufacturer (Hashim, 2024). The amount and complexity of trash have grown a lot. In the twenty-first century, because cities are growing quickly, industries are growing, and the population is growing. Open dumping and uncontrolled landfilling are no longer good ways to get rid of trash because they hurt the environment, the climate, and people's health. Because of this, modern waste management strategies use more organised, scientific, and long-lasting methods to deal with these problems. These methods include cutting waste at its source, improving recycling and resource recovery, coming up with new ways to treat waste, and putting together policy frameworks (Barati et al., 2022). They all help turn trash from a problem into a useful resource. The key pillars of waste management strategies involve Prevention, minimisation, reuse, recycling, recovery, and disposal of waste. Waste-to-energy (WTE) technologies are good ways to deal with trash that can't be recycled, especially in cities where land is hard to come by. People often use incineration, gasification, and pyrolysis to turn solid waste into electricity, heat, or fuel. Even though there are still worries about emissions and how to get rid of ash, modern incinerators with high-tech flue gas cleaning systems meet strict environmental standards (Goutam Mukherjee et al., 2021). WTE plants are an important part of energy policies in countries like Sweden, Japan, and Germany. They provide power and heat to thousands of homes. Plastic waste management is another new field. Because plastic pollution is a global problem, especially in marine ecosystems, modern strategies focus on reducing plastic use, banning single-use plastics, and promoting biodegradable alternatives. People are paying more attention to chemical recycling and other advanced recycling methods, like depolymerisation and pyrolysis, because they can turn used plastics back into raw materials or fuels (Hoang et al., 2022). Waste tracking, collection, and treatment are being revolutionised by management systems. Real-time waste bin monitoring, collection route optimisation, and data-driven decision-making are made possible by technologies like RFID tags, GIS mapping, IoT (Internet of Things), and mobile applications. These technological advancements lower expenses, boost citizen participation, and improve operational efficiency (Pandey et al., 2023; Table 2).

The modern emerging techniques of waste management have gained significant attention over a couple of years as the old conventional methods like open dumping have been proven harmful to the environment. The increasing waste problem worldwide, involving environmental, economic, and health implications, calls for emerging and promising methods to treat waste in a more sustainable way. One of the options in the recent technologies is the Waste-to-Energy (WtE) technologies. Waste reduction in the form of incineration, pyrolysis, gasification, or anaerobic digestion not only minimises waste volume, but also produces energy in the form of electricity or heat (Szpilko et al., 2023) WtE technologies have mainly concentrated on reduction. Notwithstanding the encouraging advancements, barrier reviews have also underlined barriers including a high initial investment cost, technological barriers in transition countries, low public awareness, and policy deficits that prevent a wide-scale deployment of these technologies. Nevertheless, with progressive research, public–private partnerships, and enabling policy environments, these nascent waste

Table 2 Comparison between the conventional waste management strategies and modern waste management strategies

Characteristics	Conventional waste management strategies	Modern waste management strategies
Segregation of waste	Absent and if present, rare	Segregation of waste through different colour coded bins and automated sorting facilities
Primary approach	Disposal focussed	
Environmental impact	High chances of pollution	Reduced pollution with emission control
Economic benefits	The economic benefits are minimal, with often costly disposal techniques	Generate revenue through recycling, energy recovery
Transportation	Through open trucks or basic vehicles	Through fully covered, even GPS-enabled vehicles
Recycling approach	Low efficiency and informal sector-driven	Formal material recovery facilities
Handling hazardous waste	Unsafe practice as it is often mixed with general waste	Proper hazardous waste treatment facilities
Plastic waste handling	Burning or dumping	Pyrolysis for fuel recovery, plastic recycling plants
Monitoring and data	Manual records	Data analytics, IoT-enabled monitoring
Smart city integration	Not integrated with the urban planning	Waste management embedded in Smart city frameworks
Involvement of technology	Minimal	Advanced technologies involved

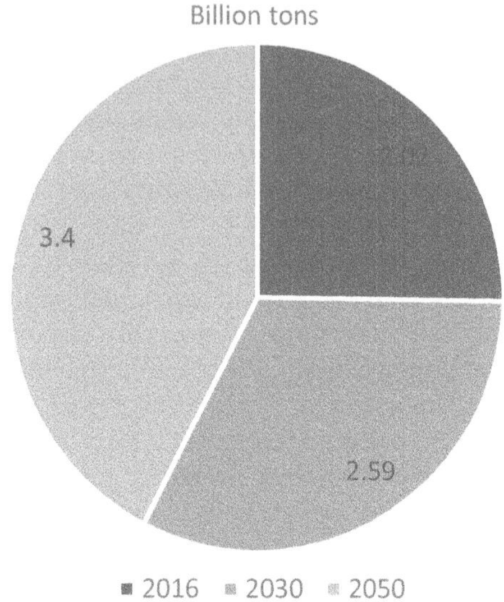

Fig. 3 MSW production trend from 2016–2050 (published in open-access article licence (https://creativecommons.org/licenses/by/4.0/)) (Bassey et al., 2024)

management strategies represent templates for sustainable waste management, anti-degradation of the natural environment, and improvements towards actualising global circular economy objectives. An estimated 2.01 billion tonnes of municipal solid waste are produced annually worldwide, of which roughly 33% is improperly managed. By 2050, the amount of garbage produced worldwide is expected to exceed 3.4 billion tonnes. Figure 3 shows the future waste generation at global level (Hoque & Rahman, 2020; Khosravani et al., 2023; Silpa et al., 2018).

Green Practices and Corporate Social Responsibility (CSR)

By fusing environmental care with business strategy, green practices and corporate social responsibility (CSR) are essential to sustainable waste management (Awewomom et al., 2024). These strategies emphasise reducing the negative effects on the environment, making the best use of available resources, and encouraging stakeholder participation in sustainability initiatives (Lee et al., 2018). Through the use of sustainable resource management, renewable energy sources, and eco-friendly technologies, green practices place a high priority on ecological preservation (Bei & Wang, 2023). Eco-design, sustainable sourcing, and energy-efficient manufacturing are important tactics that lower greenhouse gas emissions and resource waste (Hariyani et al., 2024; Khan et al., 2021). Furthermore, using recyclable and biodegradable materials reduces pollution and the need for landfill disposal (Flury & Narayan, 2021; Mazhandu et al., 2020). These actions improve stakeholder trust and the company's reputation in addition to improving environmental performance

(Baah et al., 2021; Sarkis et al., 2010). By incorporating green practices, enterprises can contribute to long-term sustainability by striking a balance between ecological responsibility and economic growth.

The important element of advanced waste management is CSR that motivates organisations for operational frameworks (Camilleri, 2017). Minimising the waste, recycling of wastes, role of community are the main focus of CSR so that their impact on environment is reduced (Mogha & Tebourbi, 2025; Rela et al., 2020). CSR is an essential instrument for fostering green innovation, obtaining a competitive edge, and raising shareholder value, claim Porter and Kramer (Porter & Kramer, 2006). According to Martinez-Conesa et al. CSR might boost the possibility of funding R&D, which can spur the creation of "green" or ecologically friendly technologies (Martinez-Conesa et al., 2017). Companies that participate in CSR programmes have the potential to foster ecologically sustainable innovation, (Hull & Rothenberg, 2008). Additionally, participating in CSR efforts might stimulate the advancement of green innovation (Hong et al., 2020).

Chemical Precipitation and Coagulation

In order to improve water quality and satisfy regulatory requirements, chemical precipitation and coagulation are crucial techniques for treating wastewater by eliminating suspended particles, heavy metals, and other impurities. By adding reagents such as alum, lime, or ferric chloride to wastewater, chemical precipitation creates insoluble chemicals that separate and can be eliminated by filtration or sedimentation. This technique works very well for (i) water softening and (ii) eliminating phosphates and heavy metals (Rela et al., 2020). Chemical coagulation, on the other hand, destabilises colloidal particles by adding coagulants such as ferric chloride or alum. Waste water treatment can benefit from enhancing quality of water, regular compliance, operations on downstreams, precipitation by chemical process, and coagulation (Dutta et al., 2021). They successfully (i) eliminate a wide variety of pollutants, (ii) enhance the safety, colour, and purity of water, and (iii) are adaptable enough to be used on different kinds of wastewater, such as municipal and industrial effluents. By lowering pollutant loads, these techniques also aid in lowering downstream treatment costs (Semerjian & Ayoub, 2003; Sher et al., 2013). Due to the higher surface area of the resultant flocs and the wider pH range, iron-based coagulants frequently outperform aluminium-based coagulants by a small margin (Wang et al., 2012; Chiang et al., 2012). Iron-based coagulants always produce flocs that are loose and open, which guarantee a large surface area and three-dimensional lattice size for heavy metal complexing and adsorption (Kang et al., 2003).

Incineration

One common thermal treatment technique is incineration, which is burning garbage at a high temperature. Ash, flue gas, and heat are the end products of this. Incineration is appropriate for all types of waste because it efficiently lowers waste volumes and recovers energy (Lindberg et al., 2015; Lombardi et al., 2015). Incineration technologies include mass burn incinerators that convert wastes into uniform fuels, increasing efficiency (Bosmans & Helsen, 2010), managing the wastes production of energy within control by refuse-derived fuel (RDF) systems (McDougall et al., 2001), by using high maintenance costly rotatory kiln incinerators for destroying toxic wastes (Jiang et al., 2019; Wey et al., 2006), and by highly expensive fluidised bed incinerators for highly efficient combustion (Van Caneghem et al., 2012). It can minimise the wastes by 90%. This significantly lowers the need for landfills (Silva et al., 2020). They also make it possible to recover energy, generate electricity from wastes, and efficiently destroy harmful elements, including poisonous chemicals and germs. The environmental impact is reduced as a result (Mulchandani & Westerhoff, 2016; Oladejo et al., 2018). The feedstock composition and the circumstances of the procedure determine the intricate reaction processes that take place during hydrothermal treatment (Zhang et al., 2023). Plastics go through distinct reaction paths, frequently involving hydrolysis, depolymerisation, and repolymerisation stages (Zhang et al., 2023). About 50.7% of all MSW disposed of in China in 2019 was handled by incinerating the material, which is the most often used method (Khan et al., 2022).

Pyrolysis

Pyrolysis is an important process in which wastes are broken down by the help of heat into oil, char, and syngas. It is a technique by which wastes are converted into useful products (Andooz et al., 2023; Sathiskumar & Karthikeyan, 2019). The process starts with the preparation of the feedstock, which involves drying and shredding the materials to provide even heating. To break down the feedstock into smaller molecules, it is heated in a pyrolysis reactor. By modifying the process settings, the percentage of these products can be maximised. For example, slow pyrolysis promotes the creation of char, whereas quick pyrolysis conditions (high heating rates, moderate temperatures about 500 °C, and short vapour residence durations) usually maximise the bio-oil output (Chiang et al., 2024). Bio-oil is a complex mixture of organic compounds, and char is a solid that is rich in carbon (Jahirul et al., 2012; Uddin et al., 2018).

Furthermore, pyrolysis tackles major waste issues and aids in recycling initiatives, especially for tyres and plastics. To maximise product yields and quality, the process must precisely manage operating conditions, which is one of its problems (Palos et al., 2021; Qureshi et al., 2020). Commercial applications of bio-oil and syngas may be impacted by variations in their composition caused by feedstock and process conditions (Jahirul et al., 2012; Ringer et al., 2006). Significant technological and financial challenges also exist enlarging the pyrolysis systems (Perkins et al., 2018; Richard, 2010; Roy & Dias, 2017; Siddique et al., 2022; Sorunmu et al., 2020). The pyrolysis of polystyrene was investigated by Albor and Mirkouei et al., who noted that styrene monomer was produced as a significant liquid product, highlighting the possibility of useful chemical feedstocks (Albor et al., 2023).

Gasification

Gasification requires a high temperature that transforms wastes (trash, biomass, coal) into syngas by utilising a small quantity of oxygen, air, or steam (Santos & Alencar, 2020; Speight, 2015). Syngas can be produced through this process, which takes place in an oxygen-deficient atmosphere (You et al., 2018). Many gasifiers are created on the basis of feedback and conscious resulted good (Krishnamoorthy & Pisupati, 2015; Warnecke, 2000). A wide category of materials can be controlled (Stiegel & Maxwell, 2001). High energy efficiency is attained by the process, particularly in integrated gasification combined cycle (IGCC) plants, where the creation of syngas helps to generate more electricity (Chen et al., 2015). Furthermore, compared to conventional burning, gasification produces fewer emissions. Gasification's regulated environment makes it easier to catch and remove impurities. Process stability and efficiency may be impacted by variations in feedstock quality (Li et al., 2016). Effective cleanup techniques are required since tar build-up can cause equipment fouling and clogging (Rakesh & Dasappa, 2018; Woolcock & Brown, 2013). The availability of feedstock, market prices, and competition from other technologies are the factors that affect sustainability (Emenike et al., 2020). Cai et al. used a fixed-bed reactor on a laboratory scale with temperatures between 600 and 900 °C to co-gasify fuel generated from waste and straw mixes. To investigate the co-gasification process's synergistic effects, the outcomes were contrasted with those of gasification of a single feedstock. According to the author, co-gasification outperformed conventional gasification in terms of carbon conversion, cold gas efficiency, and gaseous product output (Cai et al., 2021).

Plasma Arc Gasification

The thermal treatment techniques transform wastes into syngas and slag by using a high-temperature plasma arc that can reach temperatures of 3000 °C to 10,000 °C. The feedstock is first pre-processed, then an electrical discharge is used to create a plasma arc, and lastly, waste is uncovered to high temperature in gasification reactors. As a result, inorganic materials melt into a stable, glass-like vitrified slag, and organic components break down into syngas. The vitrified slag can subsequently be used in building, whereas syngas can be used as fuel for electricity generation and chemical production (Lemmens et al., 2007). Quintero-Coronel et al. investigated the co-gasification of coal and biomass as a potentially efficient method of combining the generation of syngas with diverse gasification feedstocks in a distinct study (Quintero-Coronel et al., 2022). This tactic could be employed in addition to natural gas in home and business burners. The performance of co-gasification between high-volatile bituminous coal and palm kernel shell was assessed utilising a top-lit updraft gasifier with a changing ignition front. The study found that the igniting front spread more uniformly and fast as biomass volume increased (Quintero-Coronel et al., 2022).

Sensor-Based Waste Monitoring

The method known as sensor-based trash monitoring makes use of sensors to quantify the amount of garbage produced, pinpoint its sources, and assess how well waste management plans are working in a given location. A wireless sensor network is a collection of wireless sensors that are organised by themselves in a way to monitor the environment and physical conditions (Gurram et al., 2022). Typically, ultrasonic sensors measure the distance between the trash in the compartment and the bin's lid to determine the fill level. To determine the weight of the waste, a load cell sensor is occasionally added at the bottom of the bin (Aleyadeh & Taha, 2018; Pardini et al., 2020). For instance, the authors of (Kaushik & Yadav, 2017) suggest a system that uses ultrasonic sensors and a Microcontroller Unit (MCU) to send an SMS message to the municipality in the event that the garbage level rises above a predetermined point. The routing and scheduling of the garbage collecting operations can be optimised by knowing the waste levels and the locations of the relevant bins; thus, the authors assert that the service cost can be reduced by 50%.

Waste-Sorting Robots

For the management of municipal solid waste, garbage classification is highly advised, and the use of robots can significantly increase rubbish classification efficiency. However, in order to classify rubbish in extremely complex, unexpected, and heterogeneous industrial contexts, robots need sophisticated visual and operational skills (Koskinopoulou et al., 2021). Recent studies have concentrated on increasing the precision and effectiveness of garbage classification robots, which calls for the creation of better cameras and sensors to distinguish between various waste categories, as well as enhanced artificial intelligence algorithms for waste classification. Finding the target region of interest using hyperspectral pictures is a promising strategy (Xiao et al., 2020).

Construction and demolition debris can be automatically collected by them (Wang et al., 2020). The contours of any object in an image, including rubbish from construction and demolition, can be precisely detected by deep learning technologies like instance segmentation (Chen et al., 2022). Manual collection and classification are sometimes ineffective and dangerous due to the complexity of building sites and the massive volumes of garbage produced. As a result, research is being focused on recovering building trash (Yang et al., 2023).

4 Challenges and Future Prospects in Waste Management

India's solid waste management (SWM) situation is currently dire since the most effective and suitable techniques for collecting and disposing of waste. SWM is less, and there are less management facilities in this field. At present, the accountability of SWM systems is less in India (Deepalakkshmi et al., 2023). In India, municipal authorities are having less budgets for making arrangements for garbage collection and disposal treatments. Numerous methods have been created for addressing the waste management issue, that is a pivotal element in sustainability of environment (Agboola et al., 2025). Technical and physical obstacles to sustainable MSW management include things like a lack of waste collection locations, irregularities in waste collection, insufficient vehicles for waste collection, restricted access to waste bins, alternatives to final waste disposal, inadequate waste separation facilities, waste volume, and space constraints. Sustainable MSW management is hampered by a lack of planning and strategy, a weak policy, low programme participation, poor public communication, a lack of knowledge, improper media, etc. The main political and legal obstacles to sustainable MSW management include population expansion, competing interests among parties, and inadequate and weak legislation.

Digital technologies like artificial intelligence (AI) and the Internet of Things (IoT) will improve tracking, sorting, and monitoring along the recycling chain, streamlining trash management operations (Fang et al., 2023). Accepting these advancements will promote resource conservation and sustainable economic growth while reducing the environmental impact of rubbish disposal (Ikponmwosa Aiguobarueghian et al., 2024). In order to create legislation, encourage innovation, and ease the transition to a more circular economy where waste is transformed into a valuable resource, businesses, governments, and educational institutions will need to collaborate.

5 Conclusion

Proper industrial solid waste treatment is at the basis of reducing a negative effect on the environment and human well-being, as well as promoting global sustainability objectives. The reuse of waste materials through recycling and energy recovery, and the production of by-products, illustrates how environmental issues could be used as a source of innovation. These practices not only make the industries less eco-footprint but also enhance the security of resources, economic resilience, and sustainable development. Circular economy principles represent an example of paradigmatic shift from linear "take-make-dispose" to regenerative systems, which aim to increase the use rate of material, the production with zero waste, and maintenance of system's health in a long-term perspective. This is an investment in the survival and health of our generations to come but also a common-sense environmental one. By implementing circular practices, industrial ecosystems can achieve environmental co-benefits (e.g. lowering GHG emissions, conservation of resources, and cleaner production processes) Key areas for future action: Future initiatives need to focus on technological breakthroughs, on strengthening the regulatory framework, and on fostering intersectoral collaboration if this vision is to be realised in his entirety. It will be largely contingent upon an integrated process to reach sustainable development goals and ensure a future of balance between the care of the environment and economic growth.

References

Agboola, S.O., Inetabor, G.M., Bello, O.O., & Bello, O.S. (2025). Waste pollution and management: Current challenges and future perspectives. In: Yatoo, A.M., Sillanpää, M. (Eds.), *Smart waste and wastewater management by biotechnological approaches.* Interdisciplinary Biotechnological Advances. Springer, Singapore. https://doi.org/10.1007/978-981-97-8673-2_1

Aiguobarueghian, I., Adanma, U. M., Ogunbiyi, E. O., & Solomon, N. O. (2024). Waste management and circular economy: A review of sustainable practices and economic benefits. *World Journal of Advanced Research and Reviews, 22*(2), 1708–1719. https://doi.org/10.30574/wjarr.2024.22.2.1517

Albor, G., Mirkouei, A., McDonald, A. G., Struhs, E., & Sotoudehnia, F. (2023). Fixed bed batch slow pyrolysis process for polystyrene waste recycling. *Processes, 11*(4), 1126. https://doi.org/10.3390/pr11041126

Aleyadeh, S., & Taha, A.-E. M. (2018). An IoT-based architecture for waste management. In *2018 IEEE International Conference on Communications Workshops (ICC Workshops)* (pp. 1–4). IEEE. https://doi.org/10.1109/ICCW.2018.8403750

Amasuomo, E., & Baird, J. (2016). The concept of waste and waste management. *Journal of Management and Sustainability, 6*(4), 88. https://doi.org/10.5539/jms.v6n4p88

Andooz, A., Eqbalpour, M., Kowsari, E., Ramakrishna, S., & Ansari Cheshmeh, Z. (2023). A comprehensive review on pyrolysis from the circular economy point of view and its environmental and social effects. *Journal of Cleaner Production, 388*, Article 136021. https://doi.org/10.1016/j.jclepro.2023.136021

Anokye, K., & Mohammed, A. S. (2024). Indigenous traditional knowledge for cleaner waste systems and sustainable waste management system in Ghana. *Cleaner Waste Systems, 9*, Article 100165. https://doi.org/10.1016/j.clwas.2024.100165

Awewomom, J., et al. (2024). Addressing global environmental pollution using environmental control techniques: A focus on environmental policy and preventive environmental management. *Discover Environment, 2*(1), 8. https://doi.org/10.1007/s44274-024-00033-5

Baah, C., Opoku-Agyeman, D., Acquah, I. S. K., Agyabeng-Mensah, Y., Afum, E., Faibil, D., & Abdoulaye, F. A. M. (2021). Examining the correlations between stakeholder pressures, green production practices, firm reputation, environmental and financial performance: Evidence from manufacturing SMEs. *Sustainable Production and Consumption, 27*, 100–114. https://doi.org/10.1016/j.spc.2020.10.015

Barati, B., Zafar, F. F., & Wang, S. (2022). Different waste management methods, applications, and limitations. In *Waste-to-energy* (pp. 21–58). Springer. https://doi.org/10.1007/978-3-030-91570-4_2

Bassey, U., et al. (2024). Characteristics and management of municipal solid waste in Uyo, Akwa Ibom state, Nigeria. *Scientific Reports, 14*(1), 10971. https://doi.org/10.1038/s41598-024-61108-0

Bei, J., & Wang, C. (2023). Renewable energy resources and sustainable development goals: Evidence based on green finance, clean energy and environmentally friendly investment. *Resources Policy, 80*, Article 103194. https://doi.org/10.1016/j.resourpol.2022.103194

Bosmans, A., & Helsen, L. (2010). Energy from waste: Review of thermochemical technologies for refuse derived fuel (RDF) treatment. In *Proceedings Venice 2010, Third International Symposium on Energy from Biomass and Waste Venice*, Italy, November 8–11, 2010.

Cai, J., et al. (2021). Synergistic effects of co-gasification of municipal solid waste and biomass in fixed-bed gasifier. *Process Safety and Environmental Protection, 148*, 1–12. https://doi.org/10.1016/j.psep.2020.09.063

Camilleri, M. A. (2017). *Corporate sustainability, social responsibility and environmental management*. Springer. https://doi.org/10.1007/978-3-319-46849-5

Chen, S., Lior, N., & Xiang, W. (2015). Coal gasification integration with solid oxide fuel cell and chemical looping combustion for high-efficiency power generation with inherent CO_2 capture. *Applied Energy, 146*, 298–312. https://doi.org/10.1016/j.apenergy.2015.01.100

Chen, X., Huang, H., Liu, Y., Li, J., & Liu, M. (2022). Robot for automatic waste sorting on construction sites. *Automation in Construction, 141*, Article 104387. https://doi.org/10.1016/j.autcon.2022.104387

Chiang, P. F., Han, S., Claire, M. J., Maurice, N. J., Vakili, M., & Giwa, A. S. (2024). Sustainable treatment of spent photovoltaic solar panels using plasma pyrolysis technology and its economic significance. *Clean Technologies, 6*(2), 432–452. https://doi.org/10.3390/cleantechnol6020022

Chiang, Y. W., Ghyselbrecht, K., Santos, R. M., Martens, J. A., Swennen, R., Cappuyns, V., & Meesschaert, B. (2012). Adsorption of multi-heavy metals onto water treatment residuals: Sorption capacities and applications. *Chemical Engineering Journal, 200–202*, 405–415. https://doi.org/10.1016/j.cej.2012.06.070

Czekała, W., Drozdowski, J., & Łabiak, P. (2023). Modern technologies for waste management: A review. *Applied Sciences, 13*(15), 8847. https://doi.org/10.3390/app13158847

Dawar, I., Srivastava, A., Singal, M., Dhyani, N., & Rastogi, S. (2025). A systematic literature review on municipal solid waste management using machine learning and deep learning. *Artificial Intelligence Review, 58*(6), 183. https://doi.org/10.1007/s10462-025-11196-9

Deepalakkshmi, B., Gowsalya, K., Poornima, J., & Elumalai, M. (2023). Solid waste management (SWM) in India and their resolving strategies. *International Journal of Innovative Science and Research Technology, 8*(9), 318–326. https://doi.org/10.5281/zenodo.8354874

Demirbas, A. (2011). Waste management, waste resource facilities and waste conversion processes. *Energy Conversion and Management, 52*(2), 1280–1287. https://doi.org/10.1016/j.enconman.2010.09.025

Dijkgraaf, E., & Vollebergh, H. R. J. (2004). Burn or bury? A social cost comparison of final waste disposal methods. *Ecological Economics, 50*(3–4), 233–247. https://doi.org/10.1016/j.ecolecon.2004.03.029

dos Santos, R. G., & Alencar, A. C. (2020). Biomass-derived syngas production via gasification process and its catalytic conversion into fuels by Fischer Tropsch synthesis: A review. *International Journal of Hydrogen Energy, 45*(36), 18114–18132. https://doi.org/10.1016/j.ijhydene.2019.07.133

Dutta, D., Arya, S., & Kumar, S. (2021). Industrial wastewater treatment: Current trends, bottlenecks, and best practices. *Chemosphere, 285*, Article 131245. https://doi.org/10.1016/j.chemosphere.2021.131245

Emenike, O., Michailos, S., Finney, K. N., Hughes, K. J., Ingham, D., & Pourkashanian, M. (2020). Initial techno-economic screening of BECCS technologies in power generation for a range of biomass feedstock. *Sustainable Energy Technologies and Assessments, 40*, Article 100743. https://doi.org/10.1016/j.seta.2020.100743

Fang, B., et al. (2023). Artificial intelligence for waste management in smart cities: A review. *Environmental Chemistry Letters, 21*(4), 1959–1989. https://doi.org/10.1007/s10311-023-016 04-3

Flury, M., & Narayan, R. (2021). Biodegradable plastic as an integral part of the solution to plastic waste pollution of the environment. *Current Opinion in Green and Sustainable Chemistry, 30*, Article 100490. https://doi.org/10.1016/j.cogsc.2021.100490

Fudala-Ksiazek, S., Pierpaoli, M., Kulbat, E., & Luczkiewicz, A. (2016). A modern solid waste management strategy—The generation of new by-products. *Waste Management, 49*, 516–529. https://doi.org/10.1016/j.wasman.2016.01.022

Giusti, L. (2009). A review of waste management practices and their impact on human health. *Waste Management, 29*(8), 2227–2239. https://doi.org/10.1016/j.wasman.2009.03.028

Goutam Mukherjee, A., et al. (2021). A review on modern and smart technologies for efficient waste disposal and management. *Journal of Environmental Management, 297*, Article 113347. https://doi.org/10.1016/j.jenvman.2021.113347

Gupta, H. (2018). PAH determination in effluent and sludge samples of paper industry. *Environmental Technology and Innovation, 9*, 115–121.

Gupta, H., Dhiman, S., Stam, P. G., Chaudhary, S., Murugavel, R., & Shiju, N. R. (2025). Sustainable approach to cobalt recycling: Levulinic acid leaching of waste lithium-ion battery cathode materials. *Environmental Technology & Innovation*, 104675. https://doi.org/10.1016/j.eti.2025.104675

Gupta, H., & Gupta, B. (2016). Adsorption of polycyclic aromatic hydrocarbons on banana peel activated carbon. *Desalination and Water Treatment, 57*(20), 9498–9509.

Gupta, H., & Gupta, B. (2018). Vehicular tire as potential adsorbent for the removal of polycyclic aromatic hydrocarbons. *Polycyclic Aromatic Compounds, 38*(4), 354–368.

Gupta, H., & Shiju, N. R. (2025). Recycling and reuse of olefinic separators in lithium-ion batteries: A mini review. *Current Opinion in Green and Sustainable Chemistry, 56*, Article 101054. https://doi.org/10.1016/j.cogsc.2025.101054

Gurram, G. V., Shariff, N. C., & Biradar, R. L. (2022). A Secure Energy Aware Meta-Heuristic Routing Protocol (SEAMHR) for sustainable IoT-Wireless Sensor Network (WSN). *Theoretical Computer Science, 930*, 63–76. https://doi.org/10.1016/j.tcs.2022.07.011

Hajam, Y. A., Kumar, R., & Kumar, A. (2023). Environmental waste management strategies and vermi transformation for sustainable development. *Environmental Challenges, 13*, Article 100747. https://doi.org/10.1016/j.envc.2023.100747

Hariyani, D., Hariyani, P., Mishra, S., & Sharma, M. K. (2024). A literature review on green supply chain management for sustainable sourcing and distribution. *Waste Management Bulletin, 2*(4), 231–248. https://doi.org/10.1016/j.wmb.2024.11.009

Hashim, S., Khan, A. A., Munir, F., & Bashir, R. (2024). Comparative evaluation of mechanized and conventional composting process. *Journal of Computing & Biomedical Informatics.*

Hoang, A. T., et al. (2022). Perspective review on Municipal Solid Waste-to-energy route: Characteristics, management strategy, and role in circular economy. *Journal of Cleaner Production, 359*, Article 131897. https://doi.org/10.1016/j.jclepro.2022.131897

Hong, M., Drakeford, B., & Zhang, K. (2020). The impact of mandatory CSR disclosure on green innovation: Evidence from China. *Green Finance, 2*(3), 302–322. https://doi.org/10.3934/GF.2020017

Hoque, Md. M., & Rahman, M. T. U. (2020). Landfill area estimation based on solid waste collection prediction using ANN model and final waste disposal options. *Journal of Cleaner Production, 256*, Article 120387. https://doi.org/10.1016/j.jclepro.2020.120387

Hull, C. E., & Rothenberg, S. (2008). Firm performance: The interactions of corporate social performance with innovation and industry differentiation. *Strategic Management Journal, 29*(7), 781–789. https://doi.org/10.1002/smj.675

Jahirul, M., Rasul, M., Chowdhury, A., & Ashwath, N. (2012). Biofuels production through biomass pyrolysis —A technological review. *Energies, 5*(12), 4952–5001. https://doi.org/10.3390/en5124952

Jiang, X., Li, Y., & Yan, J. (2019). Hazardous waste incineration in a rotary kiln: A review. *Waste Disposal & Sustainable Energy, 1*(1), 3–37. https://doi.org/10.1007/s42768-019-00001-3

Kang, M., Kamei, T., & Magara, Y. (2003). Comparing polyaluminum chloride and ferric chloride for antimony removal. *Water Research, 37*(17), 4171–4179. https://doi.org/10.1016/S0043-1354(03)00351-8

Kaushik, D., & Yadav, S. (2017). Multipurpose street smart garbage bin based on IoT. *International Journal of Advanced Research in Computer Science, 8*, 1145–1149.

Khan, N., Jhariya, M. K., Raj, A., Banerjee, A., & Meena, R. S. (2021). Eco-designing for sustainability. In *Ecological intensification of natural resources for sustainable agriculture* (pp. 565–595). Springer. https://doi.org/10.1007/978-981-33-4203-3_16

Khan, M. S., Mubeen, I., Caimeng, Y., Zhu, G., Khalid, A., & Yan, M. (2022). Waste to energy incineration technology: Recent development under climate change scenarios. *Waste Management & Research: THe Journal for a Sustainable Circular Economy, 40*(12), 1708–1729. https://doi.org/10.1177/0734242X221105411

Khosravani, F., Abbasi, E., Choobchian, S., & Jalili Ghazizade, M. (2023). A comprehensive study on criteria of sustainable urban waste management system: Using content analysis. *Scientific Reports, 13*(1), 22526. https://doi.org/10.1038/s41598-023-49187-x

Kiyasudeen, K., Ibrahim, M. H., Quaik, S., & Ismail, S. A. (2016). Composting: A traditional practice of waste treatment. In *Prospects of organic waste management and the significance of earthworms* (pp. 45–68). Springer. https://doi.org/10.1007/978-3-319-24708-3_3

Koskinopoulou, M., Raptopoulos, F., Papadopoulos, G., Mavrakis, N., & Maniadakis, M. (2021). Robotic waste sorting technology: Toward a vision-based categorization system for the industrial robotic separation of recyclable waste. *IEEE Robotics & Automation Magazine, 28*(2), 50–60. https://doi.org/10.1109/MRA.2021.3066040

Krishnamoorthy, V., & Pisupati, S. (2015). A critical review of mineral matter related issues during gasification of coal in fixed, fluidized, and entrained flow gasifiers. *Energies, 8*(9), 10430–10463. https://doi.org/10.3390/en80910430

Kumari, T., & Raghubanshi, A. S. (2023). Waste management practices in the developing nations: Challenges and opportunities. In *Waste management and resource recycling in the developing world* (pp. 773–797). Elsevier. https://doi.org/10.1016/B978-0-323-90463-6.00017-8

Lee, J. W., Kim, Y. M., & Kim, Y. E. (2018). Antecedents of adopting corporate environmental responsibility and green practices. *Journal of Business Ethics, 148*(2), 397–409. https://doi.org/10.1007/s10551-016-3024-y

Lemmens, B., Elslander, H., Vanderreydt, I., Peys, K., Diels, L., Oosterlinck, M., & Joos, M. (2007). Assessment of plasma gasification of high caloric waste streams. *Waste Management, 27*(11), 1562–1569. https://doi.org/10.1016/j.wasman.2006.07.027

Li, C., Aston, J. E., Lacey, J. A., Thompson, V. S., & Thompson, D. N. (2016). Impact of feedstock quality and variation on biochemical and thermochemical conversion. *Renewable and Sustainable Energy Reviews, 65*, 525–536. https://doi.org/10.1016/j.rser.2016.06.063

Lindberg, D., Molin, C., & Hupa, M. (2015). Thermal treatment of solid residues from WtE units: A review. *Waste Management, 37*, 82–94. https://doi.org/10.1016/j.wasman.2014.12.009

Lombardi, L., Carnevale, E., & Corti, A. (2015). A review of technologies and performances of thermal treatment systems for energy recovery from waste. *Waste Management, 37*, 26–44. https://doi.org/10.1016/j.wasman.2014.11.010

Martinez-Conesa, I., Soto-Acosta, P., & Palacios-Manzano, M. (2017). Corporate social responsibility and its effect on innovation and firm performance: An empirical research in SMEs. *Journal of Cleaner Production, 142*, 2374–2383. https://doi.org/10.1016/j.jclepro.2016.11.038

Mazhandu, Z. S., Muzenda, E., Mamvura, T. A., Belaid, M., & Nhubu, T. (2020). Integrated and consolidated review of plastic waste management and bio-based biodegradable plastics: Challenges and opportunities. *Sustainability, 12*(20), 8360. https://doi.org/10.3390/su12208360

McDougall, F. R., White, P. R., Franke, M., & Hindle, P. (2001). Integrated solid waste management: A Life Cycle Inventory. *The International Journal of Life Cycle Assessment, 6*(5), 320–320. https://doi.org/10.1007/BF02978794

Mogha, V., & Tebourbi, N. (2025). Digital platform to scale initiatives for community engagement and social impact: A case study in the Indian plastic recycling industry. *Journal of Asia Business Studies, 19*(3), 586–605. https://doi.org/10.1108/JABS-07-2024-0389

Mulchandani, A., & Westerhoff, P. (2016). Recovery opportunities for metals and energy from sewage sludges. *Bioresource Technology, 215*, 215–226. https://doi.org/10.1016/j.biortech.2016.03.075

Nanda, S., & Berruti, F. (2021). Municipal solid waste management and landfilling technologies: A review. *Environmental Chemistry Letters, 19*(2), 1433–1456. https://doi.org/10.1007/s10311-020-01100-y

Oladejo, J., Shi, K., Luo, X., Yang, G., & Wu, T. (2018). A review of sludge-to-energy recovery methods. *Energies, 12*(1), 60. https://doi.org/10.3390/en12010060

Olawade, D. B., Fapohunda, O., Wada, O. Z., Usman, S. O., Ige, A. O., Ajisafe, O., & Oladapo, B. I. (2024). Smart waste management: A paradigm shift enabled by artificial intelligence. *Waste Management Bulletin, 2*(2), 244–263. https://doi.org/10.1016/j.wmb.2024.05.001

Oppelt, E. T. (1987). Incineration of hazardous waste. *JAPCA, 37*(5), 558–586. https://doi.org/10.1080/08940630.1987.10466245

Osmani, M., & Villoria-Sáez, P. (2019). Current and emerging construction waste management status, trends and approaches. In *Waste* (pp. 365–380). Elsevier. https://doi.org/10.1016/B978-0-12-815060-3.00019-0

Palos, R., Gutiérrez, A., Vela, F. J., Olazar, M., Arandes, J. M., & Bilbao, J. (2021). Waste refinery: The valorization of waste plastics and end-of-life tires in refinery units. A review. *Energy & Fuels, 35*(5), 3529–3557. https://doi.org/10.1021/acs.energyfuels.0c03918

Panchal, R., Singh, A., & Diwan, H. (2021). Does circular economy performance lead to sustainable development?—A systematic literature review. *Journal of Environmental Management, 293*, Article 112811. https://doi.org/10.1016/j.jenvman.2021.112811

Pandey, P., Dhiman, M., Kansal, A., & Subudhi, S. P. (2023). Plastic waste management for sustainable environment: Techniques and approaches. *Waste Disposal & Sustainable Energy, 5*(2), 205–222. https://doi.org/10.1007/s42768-023-00134-6

Pardini, K., Rodrigues, J. J. P. C., Diallo, O., Das, A. K., de Albuquerque, V. H. C., & Kozlov, S. A. (2020). A smart waste management solution geared towards citizens. *Sensors, 20*(8), 2380. https://doi.org/10.3390/s20082380

Perkins, G., Bhaskar, T., & Konarova, M. (2018). Process development status of fast pyrolysis technologies for the manufacture of renewable transport fuels from biomass. *Renewable and Sustainable Energy Reviews, 90*, 292–315. https://doi.org/10.1016/j.rser.2018.03.048

Porter, M. E., & Kramer, M. R. (2006). Strategy and society: The link between competitive advantage and corporate social responsibility. *Harvard Business Review, 84*(12), 78–92, 163.

Quintero-Coronel, D. A., Lenis-Rodas, Y. A., Corredor, L., Perreault, P., Bula, A., & Gonzalez-Quiroga, A. (2022). Co-gasification of biomass and coal in a top-lit updraft fixed bed gasifier: Syngas composition and its interchangeability with natural gas for combustion applications. *Fuel, 316*, Article 123394. https://doi.org/10.1016/j.fuel.2022.123394

Qureshi, M. S., et al. (2020). Pyrolysis of plastic waste: Opportunities and challenges. *Journal of Analytical and Applied Pyrolysis, 152*, Article 104804. https://doi.org/10.1016/j.jaap.2020.104804

Rakesh, N., & Dasappa, S. (2018). A critical assessment of tar generated during biomass gasification—Formation, evaluation, issues and mitigation strategies. *Renewable and Sustainable Energy Reviews, 91*, 1045–1064. https://doi.org/10.1016/j.rser.2018.04.017

Rela, I. Z., Awang, A. H., Ramli, Z., Md Sum, S., & Meisanti, M. (2020). Effects of environmental corporate social responsibility on environmental well-being perception and the mediation role of community resilience. *Corporate Social Responsibility and Environmental Management, 27*(5), 2176–2187. https://doi.org/10.1002/csr.1956

Reno, J. (2015). Waste and waste management. *Annual Review of Anthropology, 44*(1), 557–572. https://doi.org/10.1146/annurev-anthro-102214-014146

Richard, T. L. (2010). Challenges in scaling up biofuels infrastructure. *Science, 329*(5993), 793–796. https://doi.org/10.1126/science.1189139

Ringer, M., Putsche, V., & Scahill, J. (2006). *Large-scale pyrolysis oil production: A technology assessment and economic analysis*. Technical Report.

Romero-Hernández, O., & Romero, S. (2018). Maximizing the value of waste: From waste management to the circular economy. *Thunderbird International Business Review, 60*(5), 757–764. https://doi.org/10.1002/tie.21968

Roy, P., & Dias, G. (2017). Prospects for pyrolysis technologies in the bioenergy sector: A review. *Renewable and Sustainable Energy Reviews, 77*, 59–69. https://doi.org/10.1016/j.rser.2017.03.136

Rubinov, V. V., & Fetisov, V. A. (2021). Problems of a modern approach to the technological process of waste management. In *2021 Wave Electronics and its Application in Information and Telecommunication Systems (WECONF)* (pp. 1–4). IEEE. https://doi.org/10.1109/WECONF51603.2021.9470744

Saleh, H. E.-D. M., & Abdel Rahman, R. O. (Eds.). (2016). *Management of hazardous wastes*. InTech. https://doi.org/10.5772/61668

Samui, P., GuhaRay, A., & Mahmoudi, E. (Eds.). (2022). *Facets of a smart city: Computational and experimental techniques for sustainable urban development*. Bentham Science Publishers. https://doi.org/10.2174/97898150490771220101

Sarkis, J., Gonzalez-Torre, P., & Adenso-Diaz, B. (2010). Stakeholder pressure and the adoption of environmental practices: The mediating effect of training. *Journal of Operations Management, 28*(2), 163–176. https://doi.org/10.1016/j.jom.2009.10.001

Sathiskumar, C., & Karthikeyan, S. (2019). Recycling of waste tires and its energy storage application of by-products—A review. *Sustainable Materials and Technologies, 22*, Article e00125. https://doi.org/10.1016/j.susmat.2019.e00125

Schultz, J. A. M. (2002). *A study of changing traditional urban solid waste management approaches to sustainable practices* (pp. 209–222). https://doi.org/10.1007/978-94-017-2170-7_10

Seadon, J. K. (2010). Sustainable waste management systems. *Journal of Cleaner Production, 18*(16–17), 1639–1651. https://doi.org/10.1016/j.jclepro.2010.07.009

Semerjian, L., & Ayoub, G. M. (2003). High-pH–magnesium coagulation–flocculation in wastewater treatment. *Advances in Environmental Research, 7*(2), 389–403. https://doi.org/10.1016/S1093-0191(02)00009-6

Seyyedi, S. R., Kowsari, E., Gheibi, M., Chinnappan, A., & Ramakrishna, S. (2024). A comprehensive review integration of digitalization and circular economy in waste management by adopting artificial intelligence approaches: Towards a simulation model. *Journal of Cleaner Production, 460*, Article 142584. https://doi.org/10.1016/j.jclepro.2024.142584

Sharma, P., & Vaid, U. (2021). Emerging role of artificial intelligence in waste management practices. *IOP Conference Series: Earth and Environmental Science, 889*(1), Article 012047. https://doi.org/10.1088/1755-1315/889/1/012047

Sher, F., Malik, A., & Liu, H. (2013). Industrial polymer effluent treatment by chemical coagulation and flocculation. *Journal of Environmental Chemical Engineering, 1*(4), 684–689. https://doi.org/10.1016/j.jece.2013.07.003

Shukla, K. A., Sofian, A. D. A. B. A., Singh, A., Chen, W. H., Show, P. L., & Chan, Y. J. (2024). Food waste management and sustainable waste to energy: Current efforts, anaerobic digestion, incinerator and hydrothermal carbonization with a focus in Malaysia. *Journal of Cleaner Production, 448*, 141457. https://doi.org/10.1016/j.jclepro.2024.141457

Siddiqua, A., Hahladakis, J. N., & Al-Attiya, W. A. K. A. (2022). An overview of the environmental pollution and health effects associated with waste landfilling and open dumping. *Environmental Science and Pollution Research, 29*(39), 58514–58536. https://doi.org/10.1007/s11356-022-21578-z

Siddique, I. J., Salema, A. A., Antunes, E., & Vinu, R. (2022). Technical challenges in scaling up the microwave technology for biomass processing. *Renewable and Sustainable Energy Reviews, 153*, Article 111767. https://doi.org/10.1016/j.rser.2021.111767

Silpa, K., Yao, L., Bhada-Tata, P. Van Woerden, F., & Levine, D. (2018). *What a Waste 2.0: A global snapshot of solid waste management to 2050*. World Bank Publications.

Silva, L. J. de V. B. da, Santos, I. F. S. dos, Mensah, J. H. R., Gonçalves, A. T. T., & Barros, R. M. (2020). Incineration of municipal solid waste in Brazil: An analysis of the economically viable energy potential. *Renewable Energy, 149*, 1386–1394. https://doi.org/10.1016/j.renene.2019.10.134

Singh, S., Parveen, N., & Gupta, H. (2018). Adsorptive decontamination of rhodamine-B from water using banana peel powder: a biosorbent. *Environmental Technology & Innovation, 12*, 189–195.

Sithole, T., Jobodwana, L., & Magedi, F. (2024). Revolutionizing waste management: Solidification of landfill leachates using alkali-activated slag. *Topics in Catalysis*. https://doi.org/10.1007/s11244-024-02006-7

Song, Q., Li, J., & Zeng, X. (2015). Minimizing the increasing solid waste through zero waste strategy. *Journal of Cleaner Production, 104*, 199–210. https://doi.org/10.1016/j.jclepro.2014.08.027

Sorunmu, Y., Billen, P., & Spatari, S. (2020). A review of thermochemical upgrading of pyrolysis bio-oil: Techno-economic analysis, life cycle assessment, and technology readiness. *GCB Bioenergy, 12*(1), 4–18. https://doi.org/10.1111/gcbb.12658

Speight, J. G. (2015). Gasification processes for syngas and hydrogen production. In *Gasification for synthetic fuel production* (pp. 119–146). Elsevier. https://doi.org/10.1016/B978-0-85709-802-3.00006-0

Sravya, B., Sowmika, S., Anjum, M. R., & Golla, N. (2025). *The ecological impact and treatments for the disposal of liquid waste* (pp. 161–186). https://doi.org/10.1007/978-981-97-7221-6_6

Stiegel, G. J., & Maxwell, R. C. (2001). Gasification technologies: The path to clean, affordable energy in the 21st century. *Fuel Processing Technology, 71*(1–3), 79–97. https://doi.org/10.1016/S0378-3820(01)00138-2

Szpilko, D., de la Torre Gallegos, A., Jimenez Naharro, F., Rzepka, A., & Remiszewska, A. (2023). Waste management in the smart city: Current practices and future directions. *Resources, 12*(10), 115. https://doi.org/10.3390/resources12100115

Uddin, M. N., Techato, K., Taweekun, J., Rahman, M. M., Rasul, M. G., Mahlia, T. M. I., & Ashrafur, S. M. (2018). An overview of recent developments in biomass pyrolysis technologies. *Energies, 11*(11), 3115. https://doi.org/10.3390/en11113115

Van Caneghem, J., et al. (2012). Fluidized bed waste incinerators: Design, operational and environmental issues. *Progress in Energy and Combustion Science, 38*(4), 551–582. https://doi.org/10.1016/j.pecs.2012.03.001

Viljoen, J. M. M., Schenck, C. J., Volschenk, L., Blaauw, P. F., & Grobler, L. (2021). Household waste management practices and challenges in a rural remote town in the Hantam Municipality in the Northern Cape, South Africa. *Sustainability, 13*(11), 5903. https://doi.org/10.3390/su13115903

Visvanathan, C. (1996). Hazardous waste disposal. *Resources, Conservation and Recycling, 16*(1–4), 201–212. https://doi.org/10.1016/0921-3449(95)00057-7

Wan, C., Shen, G. Q., & Choi, S. (2019). Waste management strategies for sustainable development. In *Encyclopedia of sustainability in higher education* (pp. 2020–2028). Springer. https://doi.org/10.1007/978-3-030-11352-0_194

Wang, W., Zhang, X., Wang, H., Wang, X., Zhou, L., Liu, R., & Liang, Y. (2012). Laboratory study on the adsorption of Mn^{2+} on suspended and deposited amorphous $Al(OH)_3$ in drinking water distribution systems. *Water Research, 46*(13), 4063–4070. https://doi.org/10.1016/j.watres.2012.05.017

Wang, Z., Li, H., & Yang, X. (2020). Vision-based robotic system for on-site construction and demolition waste sorting and recycling. *Journal of Building Engineering, 32*, Article 101769. https://doi.org/10.1016/j.jobe.2020.101769

Warnecke, R. (2000). Gasification of biomass: Comparison of fixed bed and fluidized bed gasifier. *Biomass and Bioenergy, 18*(6), 489–497. https://doi.org/10.1016/S0961-9534(00)00009-X

Wey, M.-Y., Liu, K.-Y., Tsai, T.-H., & Chou, J.-T. (2006). Thermal treatment of the fly ash from municipal solid waste incinerator with rotary kiln. *Journal of Hazardous Materials, 137*(2), 981–989. https://doi.org/10.1016/j.jhazmat.2006.03.024

Woolcock, P. J., & Brown, R. C. (2013). A review of cleaning technologies for biomass-derived syngas. *Biomass and Bioenergy, 52*, 54–84. https://doi.org/10.1016/j.biombioe.2013.02.036

Xiao, W., Yang, J., Fang, H., Zhuang, J., Ku, Y., & Zhang, X. (2020). Development of an automatic sorting robot for construction and demolition waste. *Clean Technologies and Environmental Policy, 22*(9), 1829–1841. https://doi.org/10.1007/s10098-020-01922-y

Yang, J., Jiang, P., Nassar, R.-U.-D., Suhail, S. A., Sufian, M., & Deifalla, A. F. (2023). Experimental investigation and AI prediction modelling of ceramic waste powder concrete—An approach towards sustainable construction. *Journal of Materials Research and Technology, 23*, 3676–3696. https://doi.org/10.1016/j.jmrt.2023.02.024

You, S., Ok, Y. S., Tsang, D. C. W., Kwon, E. E., & Wang, C.-H. (2018). Towards practical application of gasification: A critical review from syngas and biochar perspectives. *Critical Reviews in Environmental Science and Technology, 48*(22–24), 1165–1213. https://doi.org/10.1080/10643389.2018.1518860

Zhang, B., Biswal, B. K., Zhang, J., & Balasubramanian, R. (2023). Hydrothermal treatment of biomass feedstocks for sustainable production of chemicals, fuels, and materials: Progress and perspectives. *Chemical Reviews, 123*(11), 7193–7294. https://doi.org/10.1021/acs.chemrev.2c00673

Assessment of Contamination Levels in Freshwater Resources due to Solid Waste

Manish Pant, Sankar Chakma, Jabrinder Singh, Pushpa Kewlani, and Tarumay Ghoshal

1 Introduction

The rapid growth of urban areas and industrial activities has led to the generation of large amounts of solid waste and polluted water (Gupta & Dhiman, 2024; Saxena, 2025). Solid waste refers to discarded solid or semi-solid materials that have lost their original usefulness. Human activities have affected various aquatic ecosystems, in addition to contributing to the disposal of different types of solid waste. Based on their source, solid waste can be categorized into municipal, agricultural, and industrial types. Besides occupying substantial land space, solid waste (SW) pollutes soil, water, and air, posing significant health risks to humans (Madhavaraj & Karthikeyan, 2025). Due to its large volume, high resource consumption, and variety of forms, industrial solid waste resource exploitation has attracted a lot of attention as a result of continuous industrial development. Utilizing resources can increase the value of solid waste, which is advantageous to society as a whole (Hou et al., 2023). Some of the industrial waste has been used for mine backfill, soil reclamation, and other uses in the construction sector. For example, fly ash and slag are frequently utilized

M. Pant
Centre for Land and Water Resource Management (CLWRM), G.B. Pant National Institute of Himalayan Environment, Kosi-Katarmal, Almora 263643, Uttarakhand, India

M. Pant (✉) · S. Chakma
Department of Chemical Engineering, IISER Bhopal, Bhopal 462066, Madhya Pradesh, India

J. Singh · T. Ghoshal
Land Air and Water (LAW)- Centre of Excellence, DIT University, Dehradun 248009, Uttarakhand, India

P. Kewlani
Department of Biotechnology, Graphic Era (Deemed to be University), Dehradun 248002, Uttarakhand, India

H. Gupta et al. (eds.), *Advanced Technologies for Assessment, Recycling, and Management of Solid Waste*, Waste as a Resource,
https://doi.org/10.1007/978-3-032-24788-9_3

as cementitious materials, and crushed coal gangue as well as tailings can be utilized as aggregates (Behera et al., 2021).

Greenhouse gases like methane, which are released by solid waste, significantly worsen air pollution and accelerate climate change. In many developing countries, a significant portion of municipal solid waste (MSW) recycling is carried out in open landfills, which contribute to soil, water, and air pollution while also releasing greenhouse gases. Kristanto and Koven (2019) reported that in 2016, solid waste was responsible for 5% of global greenhouse gas emissions. It is vital to enhance solid waste disposal globally since two of the most important strategies to fight climate change are safe disposal and minimizing the open burning of trash. The World Bank estimates that over 33% of the 2.01 billion tons of municipal solid waste (MSW) generated each year globally is not adequately managed. Without improvements in the solid waste management (SWM) sector, emissions from solid waste are projected to exceed 2.6 billion tons of CO_2 equivalent by 2050 (Elsheekh et al., 2021; Kaza et al., 2018). Proper management of solid waste can significantly reduce the release of carbon dioxide and other greenhouse gases into the atmosphere. It is noteworthy that wastewater has been filtered using physical, chemical, and biological methods using materials derived from solid waste. Coal fly ash, a typical type of solid waste, has been used extensively in research on the development of adsorbents and catalysts. The created materials have proven to be incredibly effective at treating wastewater (Asl et al., 2019; Gupta, 2018; Wang et al., 2023).

Clean water is essential for both individual health and public safety. Assessment of water quality and investigation of contamination are necessary before fresh water resources can be accessed. Research on it has been quite comprehensive (Pant & Singh, 2024b; Sharma et al., 2022). Landfilling, incineration, composting, and other related practices are examples of these conventional treatment methods. However, more contamination would result from these remediation techniques. Most solid rubbish is made up of recyclable materials that can be turned into a variety of resources; therefore, trash can become treasure (Chilakamarry et al., 2022; Nallapaneni et al., 2023). Thus, the implementation of solid waste treatment and resource utilization is essential. Today, new possibilities for managing water quality are made possible by the convergence of artificial intelligence, machine learning, and nanotechnology (Pant & Singh, 2025).

The 2030 Sustainable Development Goals (SDGs) are among the challenges and intersecting issues that nations all over the world hope to resolve. While not required, achieving the SDGs will help reduce the numerous detrimental impacts on the economy, society, environment, and metropolitan areas that endanger cities, as well as the benefits that accompany that reduction. To improve SDG achievement, the study will examine and identify the connections between SWM strategies and programs, as well as the specific targets associated with each goal. The study also utilizes expert questionnaires to evaluate the effects of solid waste management (SWM) plans and programs on the 17 Sustainable Development Goals (SDGs), categorizing them according to the extent of their impact. This book chapter emphasizes pollution control, environmental materials, and the long-term chemistry of ecological interfaces, which may be particularly relevant for many low-income countries where SWM is often conducted unsustainably.

2 Types of Solid Waste and Their Sources

2.1 Sources and Classification of Solid Waste

Any undesirable material or garbage that is neither liquid nor gaseous is referred to as solid waste (SW). Families, businesses, and industries frequently dispose of a wide range of materials under this heading (Dehghani et al., 2021). The components of this diverse collection are classified by their source and characteristics. Waste generated from households, commercial establishments, and gardens is classified as municipal solid waste (MSW); construction debris and by-products from manufacturing are termed industrial waste; chemical and biomedical waste fall under hazardous waste; organic residues and inorganic farming materials are considered agricultural waste; discarded electronic devices and components are categorized as electronic waste; mining tailings and overburden are identified as mining waste; and food scraps along with other biodegradable materials are classified as organic waste. Additionally, recyclable materials include metals, plastics, glass, and paper (Nanda & Berruti, 2021). Solid wastes have also been utilized in various physical, chemical, and biological processes to produce fuels, adsorbents, electrodes, nanoparticles, and numerous other products (Agarwal et al., 2024; Ghumra et al., 2022; Gupta et al., 2025). Bioconversion technology is an eco-friendly, sustainable approach that transforms organic solid waste into valuable bioproducts—such as biogas, bioethanol, biohydrogen, and biofertilizer—through the metabolic activities of living organisms, including microorganisms, worms, and insects. Reducing the hazardous nature of trash through the use of particular chemicals, such as acids, bases, oxidizers, and reducers, is known as chemical treatment (El-Emam, 2024; Thomson et al., 2020). Figure 1 demonstrates the several types of SW generated around the world are classified into 3 categories based on types, origin, and their property, which are discussed in further depth in the ensuing sections.

Municipal Solid Waste (MSW)

MSW consists of common objects thrown by the general people. It is made up of both biodegradable and non-biodegradable elements, and its composition changes depending on urbanization, lifestyle, including waste management procedures. The majority of MSW originates from homes; however, municipal solid waste also includes waste from businesses, workplaces, and organizations. Living standards, consumer behavior, and the nation's economic development stage all influence the

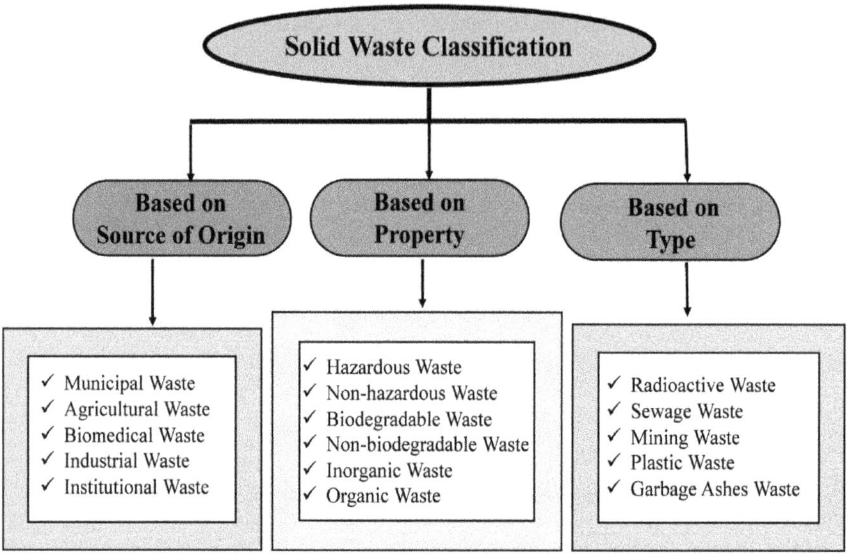

Fig. 1 Solid waste classification based on source of origin, types, and properties reprinted by original article (Prajapati et al., 2021) after taking permission from Elsevier

quantity and makeup of MSW. Municipal waste creation is quantified using the waste per person generation indicator, which is measured in kilograms per person per year (Zhang et al., 2024).

Industrial Waste

Industrial waste refers to materials generated before, during, or after manufacturing processes, as well as by-products from various industries, including animal slaughterhouses, fish processing, fruit and vegetable processing, dairy production, pulp and paper manufacturing, poultry processing, edible oil production, and sugar production (Abdel-Shafy & Mansour, 2018). Currently, industrial waste is a significant environmental polluter and a huge ecological hazard. On the other hand, the government and the public compelled industry to recycle or reduce trash in order to meet strict standards. Industries such as chemical, paper, food processing, cement, and others that utilize boiler feed water as a coolant primarily depend on water as their main supply. However, this water must be free from harmful chemicals, caustic substances, scale-forming compounds, and pathogens. Decline in performance, deterioration of products and contamination, as well as process decrease, and even process failure can result from improper treatment (Ogunwumi & Salami, 2023).

Biomedical Waste

Biomedical waste presents a serious risk to human health. A potential source of pathogens is untreated biomedical waste. Studies have identified more than 40 different kinds of harmful bacteria that can spread illness and cause harm to humans. The pandemic's etiological agent spreads quickly through a number of routes and is highly contagious (Chand et al., 2021). For hospital cleaning, hygiene, and maintenance procedures to be successful, biological waste management must be done well. One essential component of quality control in healthcare institutions is an efficient hospital waste management system. In Ayurvedic healthcare facilities, a significant number of medical staff demonstrated limited awareness about the correct handling of biological waste. The primary objectives of waste management are to prevent disease and protect the environment (Joshi & Diwan, 2015).

Agricultural and E-Waste

Agricultural wastes are farming by-products that are mostly composed of 25–30% lignin, 35–50% cellulose, and 25–30% hemicellulose. India's farming practices produce a significant amount of agri-residue (>30% of gross residue created) as excess. Additionally, these stubbles and straw are more frequently ablazed. After harvesting crops such as wheat, cotton, paddy, and other grains, the intentional practice of burning the remaining straw stubble is known as stubble burning (Singh, 2018). Bananas, pineapples, papayas, oranges, mangoes, and other fruits are used to make fruit peels. Animal feed is also made from the leftovers of fruits and vegetables. After being processed industrially, rice husk and sugarcane bagasse are also dumped straight into waterways. Therefore, it also affects the quality of water supplies (Pant et al., 2025a).

Plastic and Microplastic Waste

Microplastics (MPs) have been found in tea bags, bottled beverages, seafood, sea salt, and a number of other foods. They have also found their way into the atmosphere, waterways, and terrestrial habitats. Additionally, they have been found in marine organisms' digestive systems. They may cause an immune reaction because they are foreign substances within the tissues (Ragusa et al., 2021). Their broad surface area, in spite of their little size, improves chemical interactions with tissues and physiological fluids. Microplastics are particularly concerning due to their potential toxicity, tendency to bioaccumulate in the food chain of humans and animals, persistence in the environment, and capacity to act as carriers of pollutants and pathogens. The use of synthetic textiles in clothing and fabrics, such as polyester, nylon, and acrylic, could be a possible source of microplastic exposure during this time. Frequent washing and use of synthetic clothing leads to the release of microfibers into the environment. These microscopic fibers, sometimes imperceptible to the human eye, may accumulate indoors, such as in houses and offices, before they are released into the atmosphere. Furthermore, new microplastics might be created by the degradation of existing ones. Large plastic items, such as used bottles, packaging materials, and discarded toys, are exposed to environmental factors including sunlight, temperature fluctuations, and mechanical stress. According to Manzoor et al. (2022), these factors cause plastics to break down and fragment over time due to photooxidation, producing tiny particles that may eventually permit indirect exposure in pregnant women and newborns.

2.2 *Sources of Solid Waste*

The effectiveness of novel SWM systems depends on the identification of SW causes. Sources of solid waste (SW) include households, commercial establishments, and hotels; medical facilities such as hospitals and clinics; industries including rubber, foam, textiles, automotive, leather, plastics, and metal; as well as residues from farmers' markets and slaughterhouses (Chaturvedi & Khare, 2022). Table 1 lists the solid waste sources in a community.

3 Freshwater Contamination Pathways and Types due to Solid Waste

3.1 *Freshwater Contamination Pathways*

Freshwater contamination at solid waste disposal sites poses a significant threat to both the environment and human health, and can occur through multiple pathways. The pollutants originate from several significant pathways. The primary sources of contamination include leachate percolation, chemical reactions, microbial activity, physical processes, landfill design and management practices, hydrogeological conditions, and other related factors (Alao, 2025; Alessandrello, 2024). Pollutant distribution is facilitated by hydrogeological circumstances, waste material mobility, and leachate from rainwater infiltration. How pollutants move through the subsurface is determined by the aquifer's geological characteristics, including porosity, permeability, and heterogeneity (Satish Kumar et al., 2016). Hydrological parameters including groundwater flow rates, directions, and gradients are also important when determining the migration paths of contaminants. Furthermore, pollutants' behavior in groundwater is influenced by their physicochemical properties, including their solubility, sorption affinity, and degradation rates (Aderemi et al., 2011). The transportation and distribution of groundwater systems can be further impacted by human activities that introduce contaminants, such as trash disposal,

Table 1 The types and sources of solid trash in a community

Source	Typical waste generators	Types of solid waste
Residential and household waste	Individual residences, apartments, and apartment buildings	Paper waste, food leftovers textiles, leather, wood, plastics, paper and cardboard, garden debris, ashes, glass, cans and specific wastes such bulky goods, batteries, tires, oils, appliances, e-waste, and other hazardous elements found in homes
Manufacturing waste	Power stations, manufacturing facilities, building and fabrication sites, and chemical industries	Ash, building waste, demolition debris, food-related waste, housekeeping waste, packaging materials, and other hazardous by-products
Commercial waste	Stores, workplaces, dining establishments, lodging facilities, marketplaces, gas stations, and print and repair businesses	Food scraps, cardboard, plastics, paper, metals, and possibly hazardous trash
Institutional waste	Government buildings, medical facilities, research institutes, and educational institutions	Same as commercials
Building and destruction waste	Road repairs, renovations, building demolitions, and new construction	Steel, wood, bricks, tiles, concrete debris, as well as other structural remnants
Municipal waste	Wastewater treatment facilities, parks, beaches, street cleaning, and landscaping	Sewage sludge, beach waste, street sweepings, tree and grass cuttings, and litter from public areas
Agricultural waste	Dairies, farms, vineyards, orchards, crop fields, and animal feedlots	Animal manure, crop remnants, damaged products, and agrochemical waste, including pesticides and fertilizers

industrial operations, and agricultural practices (Economou-Eliopoulos & Megremi, 2021; Ravindiran et al., 2023).

Municipal landfill leachate is a highly concentrated and complex effluent containing inorganic compounds such as ammonium, magnesium, sodium, potassium, calcium, iron, sulfates, and chlorides, alongside heavy metals including cadmium, chromium, lead, copper, nickel, and zinc, as well as various xenobiotic organic substances (Carvajal-Flórez & Cardona-Gallo, 2019; Ma et al., 2022; Satish Kumar et al., 2016). Landfill leachate composition, amount, and removal of potential contaminants from trash are influenced by a number of parameters, including waste type, extent of compaction, absorptive potential, age, climate, landfill temperature, precipitation levels, geology, size, engineering, and operations. Pollution from human activities is the primary cause of water contamination; however, natural occurrences can also contribute. The nature and risk of contamination are frequently determined by water supply management and quality. The following are some significant sources of water contamination:

Agricultural Activities

It is widely recognized that agriculture accounts for approximately 70% of global surface water use, making it the largest consumer of freshwater resources. Except for the portion lost through evapotranspiration, agricultural water eventually returns to surface water or groundwater systems (Huang et al., 2019). However, water contamination is both a cause and an effect of agriculture. Poor agricultural methods produce net soil loss, irrigated land becomes salinized and waterlogged, and contaminants and silt are released into surface and/or groundwater. Wastewater and contaminated surface and groundwater are its victims, contaminating crops and spreading illness to customers and farm workers. Pesticides, fertilizers, and badly treated animal waste can leech into and contaminate water. Agricultural pesticides seep into groundwater, streams, and rivers during rainy seasons. If farm animal feces are not properly handled, they can contaminate water with germs, viruses, and parasites (Chauhan & Banerjee, 2025).

Industrial Waste

The term "industrial waste" refers to waste produced during manufacturing or other industrial operations. The discharge of industrial effluents into water bodies can cause pollution, posing serious risks to both the environment and human health. Industries release heavy metals and chemicals as by-products. When these toxins are illegally disposed of or stored, they have the potential to leach into nearby waters. Furthermore, underground oil or gasoline storage tanks might leak, contaminating groundwater (Chauhan & Banerjee, 2025).

Sewage

Sewage, which is often used to refer to any wastewater, is the portion of wastewater that contains feces or urine. Sewage refers to wastewater originating from residential, commercial, or municipal sources, typically conveyed through pipes or sewer systems. Untreated sewage can contain water, nutrients such as nitrogen and phosphorus, solids including organic matter, pathogens like bacteria, viruses, and protozoa, helminths (intestinal worms and similar parasites), oils and greases, as well as runoff from streets, parking areas, and rooftops. It may also carry heavy metals such as mercury, cadmium, lead, chromium, and copper, along with a variety of hazardous chemicals, including PCBs, PAHs, dioxins, furans, pesticides, phenols, and chlorinated organic compounds. Septic and sewage treatment plant discharges have the potential to contaminate water systems with human and industrial waste. When sewage is not properly handled, it releases harmful chemicals, bacteria, and viruses into the environment (Pant & Singh, 2024a; Singh et al., 2023).

Natural Events

Wastewater becomes contaminated by events of nature, including erosion, floods, and storms, that are made worse by human solid waste. Natural events such as volcanic eruptions, earthquakes, and storms can disturb water systems, allowing pollutants and sediments to enter water sources. These impacts are often worsened by poor land use practices and climate change. Furthermore, such events can damage water treatment facilities, reducing access to safe and clean water (Akhtar et al., 2021).

Mining

Seepage from tailings as well as waste rock impoundments, contaminating water from mine effluent discharge, and the substantial water consumption throughout ore processing are all ways that mining affects fresh water. The water supplies that we all rely on are increasingly in danger due to human activities like mining. The release of heavy metals and other contaminants into surrounding water bodies during extraction processes can present significant environmental and health hazards (Mohapatra & Kirpalani, 2017).

3.2 *Water Contamination Types*

Physical Contaminants

Solid particles can alter the color, purity, and appearance of water. These contaminants can be caused by human activities, like as construction and land development, as well as natural events such as erosion and storms. They are composed of dust, silt, and other suspended particles (Gupta & Gupta, 2018; Singh et al., 2025).

Chemical Contaminants

The environment and human health may be negatively impacted by substances that are not typically found in water. Among these include fluoride, heavy metals, salts, minerals, fertilizers, pesticides, and hazardous compounds. Exposure to certain chemicals can lead to severe health issues, including cancer, birth defects, and cognitive impairments. Besides posing risks to aquatic organisms and wildlife, these pollutants can also degrade water quality, impacting activities such as swimming and fishing (Singh et al., 2020).

Biological Contaminants

Organic materials, whether or not living. Animal or human excrement can contaminate water supplies, allowing bacteria, viruses, algae, fungi, parasites, and pathogens to thrive. Exposure to these pollutants can lead to significant health issues, including respiratory diseases, skin disorders, and gastrointestinal problems. The purpose of water treatment methods is to eradicate microbes in order to prevent waterborne infections (Pant et al., 2025b; Pant et al., 2023).

Disinfection By-products

Substances generated during the process of disinfecting drinking water. These substances may have detrimental effects on health, including cancer, birth defects, and reproductive issues. Despite being officially classified as chemical pollutants, they should be classified separately due to their distinct source (Fig. 2).

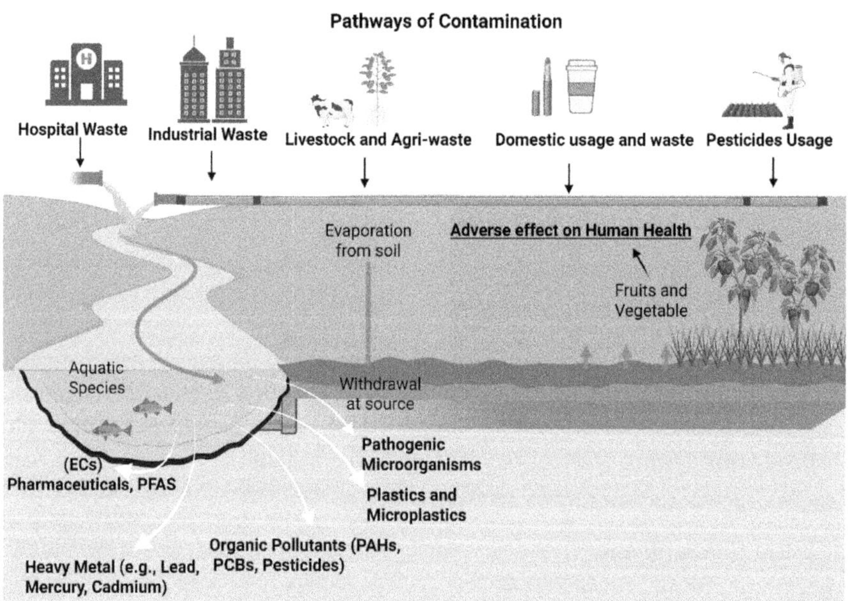

Fig. 2 Illustrated freshwater contamination caused by solid waste

4 Case Studies

The buildup of plastic waste represents one of the most visible human-induced pressures on aquatic ecosystems, yet it remains among the least well understood. Plastics "produce significant advantages to human society," but because of their longevity, unsustainable use, and improper waste management, they also "accumulate extensively in the natural habitats". According to Zhou et al. (2024) study claims that it provides significant new data regarding the state of microplastics in China's Northeast Plain. Microplastic (MP) concentrations varied from 300 to 12,800 items/kg, with a median of 1550 items/kg and an average of 1994 items/kg. Smaller MPs (< 500 μm) were less abundant than medium-sized MPs (500–5000 μm). Textiles and packaging were the main sources of MPs. The two main polymers that are present in textiles were rayon and polyethylene terephthalate (PET).

Groundwater at Delhi's Bhalaswa landfill has been found to be significantly contaminated by leachate from the site. A thorough analysis revealed that pollutants such as heavy metals and chlorides were identified in considerable amounts in groundwater and leachate samples taken at different separations from the landfill. Delhi has trouble preventing pollution because of inadequate waste management; as a result, each person produces 500 g of rubbish each day, which is about five times the national average. Delhi's waste is mostly generated by industries, hospitals, slaughterhouses, retail and commercial marketplaces, construction and demolition sites, and agricultural markets. Improper waste disposal leads to large amounts of trash entering the environment on a daily basis (Majumder et al., 2024).

One study looks into how a possible leachate leak from a MSW landfill in Mavallipura, Bangalore, India, would affect neighboring water bodies. Waste began to arrive at the more than 100-acre dump in 2005. Leachate accumulated inside the MSW mass, seeped into adjacent water bodies including ponds and open wells, and created steep and unstable slopes as a result of the uncontrolled disposal of MSW. Significant contamination of both organic and inorganic chemicals was found by physicochemical analysis. There were only trace levels of heavy metals, suggesting that the majority of the trash released was municipal waste. The water quality was deemed inadequate for human consumption based on physicochemical analyses (Naveen et al., 2018).

A study of 16 monitoring wells and two ponds around the Diaoyushan municipal solid waste (MSW) landfill in China found that leachate leakage was the leading source of groundwater contamination, accounting for 49.18% of the observed pollution. The quantity of water that reaches crops in Dongdasha Village is increased by rainwater and sewage ditches and canals, but the impact of pollution is also increased. Total hardness, Cl^-, SO_4^{2-}, TDS, and Pb levels in the contaminated groundwater area were high and were considerably over the drinking water quality limit (Chen et al., 2019).

Groundwater samples collected from northeastern Poland to assess solid waste pollution showed elevated levels of COD and BOD, measuring several hundred mg/L and around 100 mg/L, respectively. For groundwater, significantly lower levels of

C, N, and P were found in both examined forms. All groundwater samples under examination had an average total nitrogen content over 10 mg/L, which is the value associated with pure natural waters (Singh et al., 2006). This demonstrates the significant human influence on the groundwater quality under study. Regardless of wherever these samples came from, it was found that the values for N were nearly an order of magnitude greater than those for C. This finding demonstrates how leachates affect groundwater near a landfill. According to Kapelewska et al. (2019), this generally aligns with the findings made above using pollution indices and suggests that there may be a groundwater contamination risk due to a leak from the specific landfill. A study conducted in Southwest Ethiopia revealed that the BOD5 and COD levels in both leachate and stream water samples were greater than the recommended levels. According to the findings, the research area's soil and water quality are negatively impacted by solid waste open dump sites, and they are likely to pose a risk to human health through the food chain (Mekonnen et al., 2020).

The impact of unregulated disposal of 3800 tons of MSW per day on groundwater and surface water downstream of the Jawaharnagar, India, landfill was investigated. Surface water samples showed average concentrations of total hardness (TH) at 753 mg/L, Na^+ at 813 mg/L, K^+ at 530 mg/L, HCO_3^- at 978 mg/L, and NO_3^- at 262 mg/L, indicating that tanks and streams near the landfill acted as leachate collection points. Among the four sampling sessions, polluted groundwater samples exhibited elevated levels of EC, total hardness (TH, 1624 mg/L), and SO_4^{2-} (284 mg/L), all significantly exceeding permissible thresholds. Both surface water (SW) and groundwater (GW) contained very high concentrations of total organic carbon (TOC; SW average 241 mg/L, GW 154 mg/L), BOD (SW 5410 mg/L, GW 117 mg/L), and COD (SW 6427 mg/L, GW 176 mg/L), indicating substantial organic pollution originating from household waste dumps.

The study found that once solid waste dumps reached the methanogenic phase, they acted as point sources of pollution, generating leachate and releasing contaminants into aquatic environments through preferential pathways influenced by factors such as topography, soil characteristics, aquifer hydraulics, and contaminant kinetics (Rao et al., 2022). Key parameters, including COD, BOD_5, total solids (TS), nitrogen, and phosphorus, must be closely monitored, as they significantly affect the efficiency of leachate treatment in wastewater treatment facilities.

The utilization of municipal organic waste for soil enhancement, nutrient recycling, and sustainable waste management is demonstrated by numerous case studies. For a variety of uses, such as garden design, farming, and soil modification, compost-derived nutrients provide an eco-friendly way to improve soil fertility and quality. Composting biodegradable organic waste improves the chemical, physical, and biological characteristics of soils, hence boosting agricultural productivity and soil health (de Silva Gaspar et al., 2023). It also recycles nutrients. Composting diverts organic waste from agricultural landfills, which has been demonstrated to lower methane emissions. Composting or vermicomposting of soils has been shown to improve both tomato yield and nutritional quality, suggesting that these organic amendments could potentially replace mineral fertilizers in intensive agriculture (Kraut-Cohen et al., 2023). Additionally, the application of organic amendments such as compost has been found to influence soil microbial communities, which play a critical role in nutrient cycling and maintaining soil health. There are many uses for compost, as evidenced by the high-quality composts that may be produced by composting agricultural waste with urban trash to restore agricultural soils. MSW compost serves as a sustainable source of phosphorus for landscaping and soil restoration, enhancing soil P levels without causing significant environmental risks (Li et al., 2023). Following numerous MSW compost treatments, the microbial community dynamics, enzymatic activity, and soil C and N content all changed. These findings can be used to agricultural and landscape applications (Nordahl et al., 2023). According to a different study, incorporating composted garden waste into coastal saline soils enhances desalination effectiveness, nitrogen availability, and microbial diversity, guaranteeing an ecological restoration technique (Luo et al., 2023). Composting vineyards have recently drawn attention for their capacity to improve soil health while lowering greenhouse gas emissions, the most significant environmental trade-off (Wong et al., 2023).

Technological advancements, such as improved composting techniques and bioaerosol control in composting facilities, have helped mitigate the environmental and human health impacts associated with organic waste processing. The environmental performance of composting facilities improved as bioaerosol management progressed, reducing exposure levels and enhancing risk evaluations (Pascale et al., 2022). The environmental effects of waste management facilities are examined using life cycle assessment (LCA) techniques, which also show how renewable energy solutions may improve energy efficiency while having a less environmental impact (Daskal et al., 2022).

Local initiatives play a crucial role in collecting and composting various types of organic waste. For instance, Maine's Organics Recycling Law (2014) offers technical support to food waste producers to divert waste toward composting instead of landfilling. Similarly, San Francisco, California's Mandatory Organics Recycling Ordinance (2009) requires separate collection of organic waste from all residents and businesses. Both programs have successfully increased compost production while reducing the volume of waste sent to landfills.

5 Ecosystem Impacts and Human Health

One of the primary causes of a variety of environmental issues these days is plastic. Plastic pollution is man-made, non-biodegradable, and causes long-lasting environmental impact (Cole et al., 2013). Microplastics are formed when plastic, which is ubiquitous in nearly all environments, breaks down into smaller particles. The extensive use of plastic is increasing the amount of microplastic in aquatic environments across the globe. As per Prata (2018) and De Sá et al. (2018) state that sewage facilities, cities, industry, and household activities are the primary producers of microplastic (Andrady, 2017). The increasing amount of microplastics in different water compartments is causing the scientific community to grow more concerned.

Despite the fast expansion of scientific studies on microplastics, little is known about how they affect many different species. Both marine and terrestrial creatures may suffer when plastic garbage decomposes into tiny pieces. Microplastic is a novel issue that can have serious effects on the health of numerous animals and is beyond the scope of present therapeutic methods. Most organisms mistake microplastics for food, resulting in their accumulation within the body.

Global plastic production and consumption are driving a sharp rise in the enormous amount of microplastic (Pittura et al., 2022; Plastics Europe, 2020). The birds, fish, zooplankton, and mammals are among the aquatic animals that can be harmed by the tiny size of microplastic (Germanov et al., 2018). Marine megafauna, such as sharks and baleen whales, are at risk of ingesting large quantities of microplastics because they filter vast volumes of water daily to obtain sufficient nutrients. These organisms consume microplastics both directly from the ocean and indirectly through contaminated plankton (Germanov et al., 2019; Setälä et al., 2016). The rate of absorption of microplastic is influenced not only by dietary choices but also by the shape of the digestive system and the absorption process. Because aquatic species obtain less sustenance but ingest the same amount of particulate matter, their nutritional condition may be inferior. Microplastic accumulation can severely impact freshwater fish by concentrating in vital organs such as the stomach and intestines (Jabeen et al., 2017). The environmental and human health effects of solid waste contamination are illustrated in Fig. 3.

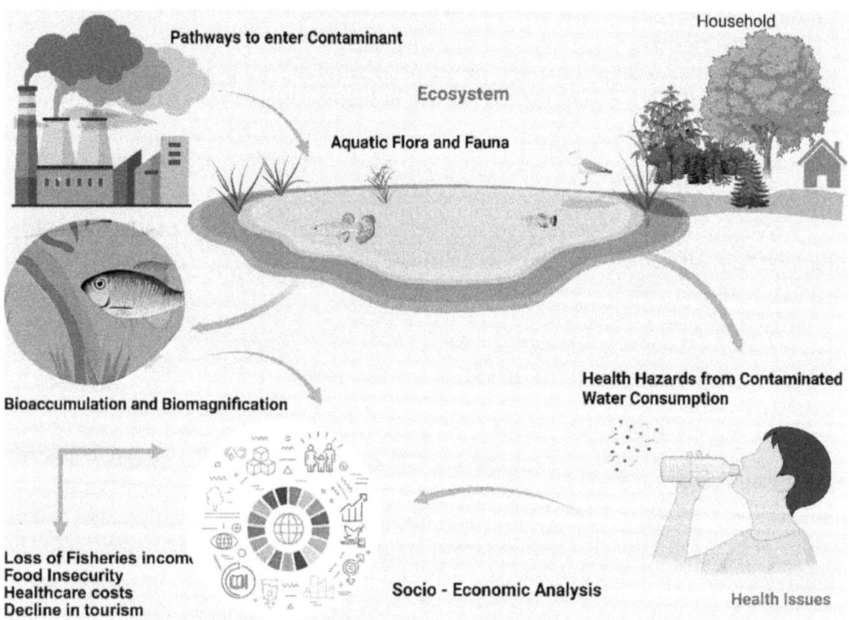

Fig. 3 Impacts of solid waste contamination on human health and the environment

6 Management Strategies and Mitigation Measures

6.1 *Water Environment*

Groundwater may become contaminated during landfill operations if surface runoff from wet seasons enters both active and closed landfill cells. Over time, gravity can cause the water accumulated in these cells to infiltrate and reach the subsurface water table. If not adequately managed, the effluent from residential activities on the job site could have an impact on groundwater supplies. As long as the project is in operation, the management plan ensures that the quality of the groundwater is regularly inspected for impurities (Datta et al., 2024; Ishaq et al., 2022).

Mitigation Measures

The leachate treatment facility should effectively handle leachate before using it for floor, vehicle, and construction cleaning. A geosynthetic clay liner (GCL) system should be installed to effectively collect and contain leachate generated from the existing landfill. A geotextile is a sheet material, either woven or nonwoven, designed to resist liquid penetration and provide enhanced durability against puncture damage. Additionally, a final cover system should be implemented at the landfill to allow controlled water infiltration, improve surface drainage, and support vegetation growth. Surface runoff is collected and removed from the waste site via a drainage pipe. An environmental monitoring system that collects and analyzes samples from the air, surface waters, soil gas, as well as groundwater around the dumping site on a regular basis.

6.2 *Land Environment*

The present landfill was constructed in compliance with the MoEF & CC standards, the CPCB, and the SWM Rules (2016). The landfill has minimal impact on groundwater and surface water habitats. The liner system prevents leachate from infiltrating and contaminating the groundwater. To minimize dust generation around the leachate holding tank, proper collection and recycling of landfill leachate are essential. For example, excess leachate from a spray drier must be appropriately managed (Hossain et al., 2014).

Mitigation Measures

The industrial complex must properly collect, store, and dispose of all solid and hazardous waste. In the event of a spill, the entire dumping site is totally drained, enclosed, and decontaminated, preventing toxins from seeping in. As a result, no changes were made to the soil quality at the dump.

6.3 Air Environment

The primary sources of fugitive dust during this period are the movement of trucks and tractors transporting waste to the landfill. Significant amounts of suspended particulates are generated during the unloading of garbage trucks and tractors. Additionally, further dust emissions occur when the deposited solid waste is leveled using a wheel loader, contributing to the overall release of fugitive materials. The vehicles and tractors that transport trash on a regular basis emit gaseous pollutants. It is anticipated that the sanitary landfill will emit methane and ammonia. According to the most recent data, the proposed gas collection system has been shown to have methane and carbon dioxide concentrations (Farzadkia et al., 2021).

Mitigation Measures

Tarpaulins must completely cover tractors and haulage vehicles used for rubbish hauling in order to stop dirt leaks while in motion. The CPCB recommends a dense green belt of rapidly developing and climatologically adapted flower species to encircle the dump site.

6.4 Impact on the Biological Environment: Biota and Habitat Degradation

Although the negative consequences on the ecosystem are severe, they are not always irreversible. The volume of solid waste generated—estimated at 1800 TPD—would require the equivalent ecological services of hundreds of mature trees to offset its impacts. Local wildlife, including rhesus macaques, palm civets, nilgais, and leopards, is severely affected, with frequent reports of animals being killed in traffic accidents near the disposal site as they search for food or migrate through disturbed habitats. During the field study, several residents also reported their household pets falling ill or dying, likely due to exposure to contaminated surroundings. In addition, many people living nearby suffer from chronic health conditions such as skin disorders, asthma, and cancer, attributed to the hazardous environmental conditions created by improper solid waste management (Kolawole & Iyiola, 2023).

Mitigation Measures

The disposal site should be surrounded by a green belt to reduce noise that penetrates beyond the walls of the facility and to provide habitat for birds and small animals.

Only activities on approach/service roads should be carried out inside the green buffer. Employees must be considerate to wildlife, and poaching and illegal hunting should be outlawed in the region.

7 Limitation and Future Prospects

SWM is a complex issue with many facets, including technical, institutional, social, economic, and environmental aspects. Pal and Bhatia (2022) state that the primary objective of MSWM is to minimize the adverse impacts of waste generation and disposal on human health and the environment while maximizing the recovery and efficient utilization of valuable resources. Nonetheless, there are certain problems with MSWM's prospects and constraints that need to be resolved. These consist of, among other things:

1. Insufficient data and information about the production, composition, collection and treatment as well as disposal of waste make it difficult to plan, monitor, and assess MSWM systems (Cayumil et al., 2021).
2. There is less opportunity for waste avoidance and resource recovery when the public is unaware of and does not participate in reduction of waste, reuse, and recycling.
3. The inability of organizations and financial resources, especially in low-income and developing countries, to set up and maintain effective MSWM systems.
4. Rapid urbanization and population increase strain existing infrastructure and MSWM services, making waste management in rural and peri-urban areas increasingly challenging.
5. New classification of waste, including hazardous, medical, and electronic wastes, is emerging that requires specific technology and management strategies to ensure their safe processing and disposal.
6. Infrastructure, equipment, and facilities that are insufficient for trash collection, transportation and treatment, as well as disposal.
7. Water quality control and solid waste management are made possible by the integration of AI, ML, and nanotechnology.
8. It is challenging to apply universal or conventional management and treatment strategies due to the very variable and unexpected composition and features of MSW and landfill leachate.
9. To overcome current limitations and envision sustainable future pathways, MSWM requires a comprehensive and integrated approach that considers the entire waste life cycle—from generation to final disposal.

The following components should form the core of this strategy:

1. Developing and carrying out all-encompassing, integrated waste management plans and strategies that take local context, requirements, and goals into consideration.
2. Obtaining adequate funds and establishing monetary rewards to promote waste reduction, avoidance, recycling, recovery, and reuse.
3. Educating the public and raising knowledge of the advantages of trash management and the obligations of rubbish producers and handlers.

4. Enhancing waste management requires strengthening data collection, monitoring, and reporting systems through the use of intelligent sensors and advanced technologies such as remote sensing and Geographic Information Systems (GIS).

8 Conclusion

SW is a major global problem. Significant urban pollution is caused by poor garbage collection and management techniques, which lowers human well-being, especially for the impoverished, and has long-term ecological effects. Traditional practices like burning, landfilling, and negligent trash disposal cause a host of ecological problems, including tainted water, global warming, and other detrimental health effects. In order to guarantee sustainable development, SW needs to be well planned and executed. This can be achieved by adopting integrated waste management approaches that emphasize efficient resource utilization and address every stage of waste generation, including resource recovery, treatment, transportation, segregation, and secure disposal in engineered landfills. The escalating generation of municipal solid waste (MSW), along with increasing concerns about rising heavy metal concentrations in groundwater—which can cause severe health issues and environmental impacts such as toxicity and groundwater pollution—further contributes to the progressive worsening of the leachate problem. The health of the general public, especially those who live close to trash disposal sites, has been the primary subject of research studies. Concern has grown over the prevalence of heavy metals as Cd, As, Cr, and Ni, which are thought to be carcinogenic and also have additional harmful health effects on the liver, kidneys, heart, lungs, skin, reproductive system, and central nervous system, depending on the quantity and length of exposure. Moreover, waste-to-energy technologies—such as incineration, pyrolysis, gasification, and anaerobic digestion—are increasingly recognized as essential components of solid waste management. This chapter suggests that, with the adoption of improved waste-to-energy methods, solid waste management could serve as a substantial source of renewable energy. Such approaches would not only ease dependence on conventional fuels amid rising electricity demand but also help address waste management challenges. The chapter concludes that integrating energy and material recovery within a comprehensive waste management framework is likely the most sustainable strategy. This approach would support the achievement of the Sustainable Development Goals (SDGs) while reducing the adverse impacts associated with solid waste.

References

Abdel-Shafy, H. I., & Mansour, M. S. (2018). Solid waste issue: Sources, composition, disposal, recycling, and valorization. *Egyptian Journal of Petroleum, 27*(4), 1275–1290.

Aderemi, A. O., Oriaku, A. V., Adewumi, G. A., & Otitoloju, A. A. (2011). Assessment of groundwater contamination by leachate near a municipal solid waste landfill. *African Journal of Environmental Science and Technology, 5*(11), 933–940.

Agarwal, S., Dhiman, S., & Gupta, H. (2024). Recovery of lithium and copper from anode electrode materials of spent LIBs by acidic leaching. *Environmental Science and Pollution Research, 31*, 34249–34257. https://doi.org/10.1007/s11356-024-33537-x

Akhtar, N., Syakir Ishak, M. I., Bhawani, S. A., & Umar, K. (2021). Various natural and anthropogenic factors responsible for water quality degradation: A review. *Water, 13*(19), 2660. https://doi.org/10.3390/w13192660

Alao, J. O. (2025). The factors influencing the landfill leachate plume contaminants in soils, surface and groundwater and associated health risks: A geophysical and geochemical view. *Public Health and Environment, 1*(1), 20–43. https://doi.org/10.70737/7ejde223

Alessandrello, C. (2024). Characterization plan of Bortolotto large area and multi-technique approach for the study of the thermodynamic and sequestering properties towards some toxic metals of leachate from municipal solid waste. https://hdl.handle.net/20.500.14242/101558

Andrady, A. L. (2017). The plastic in microplastics: A review. *Marine Pollution Bulletin, 119*(1), 12–22. https://doi.org/10.1016/j.marpolbul.2017.01.082

Asl, S. M. H., Javadian, H., Khavarpour, M., Belviso, C., Taghavi, M., & Maghsudi, M. (2019). Porous adsorbents derived from coal fly ash as cost-effective and environmentally-friendly sources of aluminosilicate for sequestration of aqueous and gaseous pollutants: A review. *Journal of Cleaner Production, 208*, 1131–1147. https://doi.org/10.1016/j.jclepro.2018.10.186

Behera, S. K., Mishra, D. P., Singh, P., Mishra, K., Mandal, S. K., Ghosh, C. N., Kumar, R., & Mandal, P. K. (2021). Utilization of mill tailings, fly ash and slag as mine paste backfill material: Review and future perspective. *Construction and Building Materials, 309*, Article 125120. https://doi.org/10.1016/j.conbuildmat.2021.125120

Carvajal-Flórez, E., & Cardona-Gallo, S. A. (2019). Technologies applicable to the removal of heavy metals from landfill leachate. *Environmental Science and Pollution Research, 26*(16), 15725–15753. https://doi.org/10.1007/s11356-019-04888-7

Cayumil, R., Khanna, R., Konyukhov, Y., Burmistrov, I., Kargin, J. B., & Mukherjee, P. S. (2021). An overview on solid waste generation and management: Current status in Chile. *Sustainability, 13*(21), 11644. https://doi.org/10.3390/su132111644

Chand, S., Shastry, C. S., Hiremath, S., Joel, J. J., Krishnabhat, C. H., & Mateti, U. V. (2021). Updates on biomedical waste management during COVID-19: The Indian scenario. *Clinical Epidemiology and Global Health, 11*, Article 100715.

Chaturvedi, S., & Khare, A. (2022). Solid wastes: Characteristics, composition and adverse effects on environment and public health. *Asian Journal of Advanced Research and Reports, 16*(7), 9–30. https://doi.org/10.9734/AJARR/2022/v16i730483

Chauhan, A., & Banerjee, R. (2025). A critical review of microplastic contamination in high altitude ecosystems, transport pathways, ecological risks, and imminent solutions. *Water, Air, & Soil Pollution, 236*(12), 1–19.

Chen, G., Sun, Y., Xu, Z., Shan, X., & Chen, Z. (2019). Assessment of shallow groundwater contamination resulting from a municipal solid waste landfill—A case study in Lianyungang, China. *Water, 11*(12), 2496. https://doi.org/10.3390/w11122496

Chilakamarry, C. R., Sakinah, A. M., Zularisam, A. W., Sirohi, R., Khilji, I. A., Ahmad, N., & Pandey, A. (2022). Advances in solid-state fermentation for bioconversion of agricultural wastes to value-added products: Opportunities and challenges. *Bioresource Technology, 343*, Article 126065. https://doi.org/10.1016/j.biortech.2021.126065

Cole, M., Lindeque, P., Fileman, E., Halsband, C., Goodhead, R., Moger, J., & Galloway, T. S. (2013). Microplastic ingestion by zooplankton. *Environmental Science & Technology, 47*(12), 6646–6655. https://doi.org/10.1021/es400663f

da Silva Gaspar, S., Assis, L. L. R. D., Prado, M. P. R. D., Pedroso Miguel, M. G., dos Reis, M., Ferreira, G., Schwan, R. F., Pasqual, M., Rigobelo, E. C., Castro, R. P., Buttrós, V. H., & Dória, J. (2023). Diversity and enzymatic activity of the microbiota isolated from composted on restaurant waste and yard trimmings. *Frontiers in Sustainable Food Systems, 7*, 1013361. https://doi.org/10.3389/fsufs.2023.1013361

Daskal, S., Asi, O., Sabbah, I., Ayalon, O., & Baransi-Karkaby, K. (2022). Decentralized composting analysis model—Benefit/cost decision-making methodology. *Sustainability, 14*(24), 16397. https://doi.org/10.3390/su142416397

Datta, S., Mitra, M., & Roy, S. (2024). Sustainable solid waste disposal to mitigate water pollution problem and its social environmental impact. In *Solid waste management in delta region for SDGs fulfillment: Delta sustainability by waste management* (pp. 47–65). Springer.

De Sá, L. C., Oliveira, M., Ribeiro, F., Rocha, T. L., & Futter, M. N. (2018). Studies of the effects of microplastics on aquatic organisms: What do we know and where should we focus our efforts in the future? *Science of the Total Environment, 645*, 1029–1039. https://doi.org/10.1016/j.scitotenv.2018.07.207

Dehghani, M. H., Omrani, G. A., & Karri, R. R. (2021). Solid waste sources, toxicity, and their consequences to human health. In *Soft computing techniques in solid waste and wastewater management* (pp. 205–213). Elsevier. https://doi.org/10.1016/B978-0-12-824463-0.00013-6

Economou-Eliopoulos, M., & Megremi, I. (2021). Contamination of the soil–groundwater–crop system: Environmental risk and opportunities. *Minerals, 11*(7), 775. https://doi.org/10.3390/min11070775

El-Emam, D. A. (2024). Valorization of solid waste products from a variety of sectors in wastewater treatment processes. In *Technical landfills and waste management: Volume 2: Municipal solid waste management* (pp. 207–257). Springer.

Elsheekh, K. M., Kamel, R. R., Elsherif, D. M., & Shalaby, A. M. (2021). Achieving sustainable development goals from the perspective of solid waste management plans. *Journal of Engineering and Applied Science, 68*(1), 9. https://doi.org/10.1186/s44147-021-00009-9

Farzadkia, M., Mahvi, A. H., Norouzian Baghani, A., Sorooshian, A., Delikhoon, M., Sheikhi, R., & Ashournejad, Q. (2021). Municipal solid waste recycling: Impacts on energy savings and air pollution. *Journal of the Air & Waste Management Association, 71*(6), 737–753.

Germanov, E. S., Marshall, A. D., Bejder, L., Fossi, M. C., & Loneragan, N. R. (2018). Microplastics: No small problem for filter-feeding megafauna. *Trends in Ecology & Evolution, 33*(4), 227–232.

Germanov, E. S., Marshall, A. D., Hendrawan, I. G., Admiraal, R., Rohner, C. A., Argeswara, J., Wulandari, R., Himawan, M. R., & Loneragan, N. R. (2019). Microplastics on the menu: Plastics pollute Indonesian manta ray and whale shark feeding grounds. *Frontiers in Marine Science, 6*, Article 487857. https://doi.org/10.3389/fmars.2019.00679

Ghumra, D. P., Rathi, O., Mule, T. A., Khadye, V. S., Chavan, A., Barba, F. C., Main, S., Odaneth, A., & Thorat, B. N. (2022). Technologies for valorization of municipal solid wastes. *Biofuels, Bioproducts and Biorefining, 16*(3), 877–890. https://doi.org/10.1002/bbb.2340

Gupta, H., & Dhiman, S. (2024). Emerging contaminants in the environment introduction. In *Occurrence, distribution and toxic effects of emerging contaminants* (pp. 1–16). CRC Press. https://doi.org/10.1201/9781003335757

Gupta, H., Dhiman, S., Stam, P. G., Chaudhary, S., Murugavel, R., & Shiju, N. R. (2025). Sustainable approach to cobalt recycling: Levulinic acid leaching of waste lithium-ion battery cathode materials. *Environmental Technology & Innovation*, 104675. https://doi.org/10.1016/j.eti.2025.104675

Gupta, H. (2018). Anthracene removal from water onto activated carbon derived from vehicular tyre. *Separation Science and Technology, 53*(4), 613–625. https://doi.org/10.1080/01496395.2017.1405038

Gupta, H., & Gupta, B. (2018). Vehicular tire as potential adsorbent for the removal of polycyclic aromatic hydrocarbons. *Polycyclic Aromatic Compounds, 38*(4), 354–368. https://doi.org/10.1080/10406638.2016.1217890

Hossain, M. L., Das, S. R., & Hossain, M. K. (2014). Impact of landfill leachate on surface and ground water quality.

Hou, H., Su, L., Guo, D., & Xu, H. (2023). Resource utilization of solid waste for the collaborative reduction of pollution and carbon emissions: Case study of fly ash. *Journal of Cleaner Production, 383*, Article 135449. https://doi.org/10.1016/j.jclepro.2022.135449

Huang, Z., Hejazi, M., Tang, Q., Vernon, C. R., Liu, Y., Chen, M., & Calvin, K. (2019). Global agricultural green and blue water consumption under future climate and land use changes. *Journal of Hydrology, 574*, 242–256. https://doi.org/10.1016/j.jhydrol.2019.04.046

Ishaq, A., Said, M. I. M., Azman, S., Abdulwahab, M. F., & Alfa, M. I. (2022). Impact, mitigation strategies, and future possibilities of Nigerian municipal solid waste leachate management practices: A review. *Nigerian Journal of Technological Development, 19*(3), 181–194.

Jabeen, K., Su, L., Li, J., Yang, D., Tong, C., Mu, J., & Shi, H. (2017). Microplastics and mesoplastics in fish from coastal and fresh waters of China. *Environmental Pollution, 221*, 141–149. https://doi.org/10.1016/j.envpol.2016.11.055

Joshi, S. C., & Diwan, V. (2015). Staff perception on biomedical or health care waste management: A qualitative study in a rural tertiary care hospital in India. *PLoS ONE, 10*(5), Article e0128383.

Kapelewska, J., Kotowska, U., Karpińska, J., Astel, A., Zieliński, P., Suchta, J., & Algrzym, K. (2019). Water pollution indicators and chemometric expertise for the assessment of the impact of municipal solid waste landfills on groundwater located in their area. *Chemical Engineering Journal, 359*, 790–800. https://doi.org/10.1016/j.cej.2018.11.137

Kaza, S., Yao, L., Bhada-Tata, P., & Van Woerden, F. (2018). *What a waste 2.0: A global snapshot of solid waste management to 2050*. World Bank Publications.

Kolawole, A. S., & Iyiola, A. O. (2023). Environmental pollution: Threats, impact on biodiversity, and protection strategies. In *Sustainable utilization and conservation of Africa's biological resources and environment* (pp. 377–409). Springer.

Kraut-Cohen, J., Zolti, A., Rotbart, N., Bar-Tal, A., Laor, Y., Medina, S., Shawahna, R., Saadi, I., Raviv, M., Green, S. J., & Yermiyahu, U. (2023). Short-and long-term effects of continuous compost amendment on soil microbiome community. *Computational and Structural Biotechnology Journal, 21*, 3280–3292.

Kristanto, G. A., & Koven, W. (2019). Estimating greenhouse gas emissions from municipal solid waste management in Depok, Indonesia. *City and Environment Interactions, 4*, Article 100027. https://doi.org/10.1016/j.cacint.2020.100027

Li, J., Zhang, H., & Zheng, L. (2023). Influence of organic amendments based on garden waste for microbial community growth in coastal saline soil. *Sustainability, 15*(6), 5038. https://doi.org/10.3390/su15065038

Luo, Y., Chavez-Rico, V. S., Sechi, V., Bezemer, T. M., Buisman, C. J., & Ter Heijne, A. (2023). Effect of organic amendments obtained from different pretreatment technologies on soil microbial community. *Environmental Research, 232*, Article 116346. https://doi.org/10.1016/j.envres.2023.116346

Ma, S., Zhou, C., Pan, J., Yang, G., Sun, C., Liu, Y., Chen, X., & Zhao, Z. (2022). Leachate from municipal solid waste landfills in a global perspective: Characteristics, influential factors and environmental risks. *Journal of Cleaner Production, 333*, Article 130234. https://doi.org/10.1016/j.jclepro.2021.130234

Madhavaraj, M., & Karthikeyan, K. (2025). A review of municipal solid waste management and landfills in India: Environmental impacts, sustainable strategies and policy insights. *Journal of Material Cycles and Waste Management*, 1–24. https://doi.org/10.1007/s10163-025-02287-4

Majumder, N., Ray, A., Goswami C., & Hembram, M. (2024). The case study of improper waste management and microplastics uses: Impacts on the natural environment in India. *IJNRD—International Journal of Novel Research and Development, 9*(1), b71–b79. ISSN: 2456-4184. Available at: https://ijnrd.org/papers/IJNRD2401111.pdf

Manzoor, S., et al. (2022). Plastic material degradation and formation of MP in the environment: A review. *Materials Today: Proceedings, 56*, 3254–3260. https://doi.org/10.1016/j.matpr.2021.09.379

Mekonnen, B., Haddis, A., & Zeine, W. (2020). Assessment of the effect of solid waste dump site on surrounding soil and river water quality in Tepi town, Southwest Ethiopia. *Journal of Environmental and Public Health, 2020*(1), 5157046. https://doi.org/10.1155/2020/5157046

Mohapatra, D. P., & Kirpalani, D. M. (2017). Process effluents and mine tailings: Sources, effects and management and role of nanotechnology. *Nanotechnology for Environmental Engineering, 2*(1), 1. https://doi.org/10.1007/s41204-016-0011-6

Nallapaneni, M. K., Hait, S., Priya, A., & Bohra, V. (2023). From trash to treasure: Unlocking the power of resource conservation, recycling, and waste management practices. *Sustainability, 15*(18), 13863. https://doi.org/10.3390/su151813863

Nanda, S., & Berruti, F. (2021). Municipal solid waste management and landfilling technologies: A review. *Environmental Chemistry Letters, 19*(2), 1433–1456. https://doi.org/10.1007/s10311-020-01100-y

Naveen, B. P., Sumalatha, J., & Malik, R. K. (2018). A study on contamination of ground and surface water bodies by leachate leakage from a landfill in Bangalore, India. *International Journal of Geo-Engineering, 9*(1), 27. https://doi.org/10.1186/s40703-018-0095-x

Nordahl, S. L., Preble, C. V., Kirchstetter, T. W., & Scown, C. D. (2023). Greenhouse gas and air pollutant emissions from composting. *Environmental Science & Technology, 57*(6), 2235–2247. https://doi.org/10.1021/acs.est.2c05846

Ogunwumi, O. T., & Salami, L. (2023). Perspective chapter: Industrial waste landfills. In *Solid waste and landfills management—Recent advances*. IntechOpen. https://doi.org/10.5772/intechopen.108787

Pal, M. S., & Bhatia, M. (2022, August). Smart solid waste management system using IoT technology: Comparative analysis, gaps, and challenges. In *International Conference on Intelligent Cyber Physical Systems and Internet of Things* (pp. 795–811). Springer. https://doi.org/10.1007/978-3-031-18497-0_58

Pant, M., Singh, P., Landge, A.S., Bisen, D., Singh, J., Manori, S. and Shukla, R.K. (2025b). Graphene oxide-p-phenylenediamine nanocomposite for efficient dye degradation and bacterial eradication in wastewater treatment. *Journal of Water Process Engineering, 75*, p.108009. https://doi.org/10.1016/j.jwpe.2025.108009

Pant, M., & Singh, J. (2024a). Distribution of heavy metals contamination in the environment and their toxicological impacts. In *Heavy metal contamination in the environment* (pp. 95–113). CRC Press. https://doi.org/10.1201/9781032685793

Pant, M., Bisen, D., Kewlani, P., Srivastav, A. L., Bhatt, I. D., & Chakma, S. (2025a). Review of food waste valorization technologies: A sustainable approach to resource recovery and utilization. *Biomass Futures*, 100001. https://doi.org/10.1016/j.bmf.2025.100001

Pant, M., & Singh, J. (2025). Integrating AI, machine learning, and nanotechnology: Shaping the future of water quality management. In *Computational automation for water security*. Elsevier Inc. https://doi.org/10.1016/B978-0-443-33321-7.00006-8

Pant, M., & Singh, J. (2024b). Seasonal variation of surface water quality and streamflow in Rispana: A tributary of Ganges River, India. *Environmental Quality Management, 34*(1), Article e22199. https://doi.org/10.1002/tqem.22199

Pant, M., Singhal, N., & Singh, J. (2023). Spatio-temporal variations in water quality of Rispana river in Dehradun, India. *Sustainable Water Resources Management, 9*(4), 123. https://doi.org/10.1007/s40899-023-00906-2

Pascale, E., Franchitti, E., Zanchi, N., Anedda, E., Bonetta, S., & Traversi, D. (2022). Size-fractionated PM10 and bioaerosol indicator development by different methods in composting plants for risk assessment. *Frontiers in Environmental Science, 10*, Article 777598. https://doi.org/10.3389/fenvs.2022.777598

Pittura, L., Gorbi, S., Mazzoli, C., Nardi, A., Benedetti, M., & Regoli, F. (2022). Microplastics and nanoplastics. In *Marine analytical chemistry* (pp. 349–388). Springer. https://doi.org/10.1007/978-3-031-14486-8

Plastic Soup Foundation. (2020). Beat the microbead. Available at: https://www.beatthemicrobead.org/about-us/. Accessed February 7, 2022.

Prajapati, P., Varjani, S., Singhania, R. R., Patel, A. K., Awasthi, M. K., Sindhu, R., Zhang, Z., Binod, P., Awasthi, S. K., & Chaturvedi, P. (2021). Critical review on technological advancements for effective waste management of municipal solid waste—Updates and way forward. *Environmental Technology & Innovation, 23*, 101749. https://doi.org/10.1016/j.eti.2021.101749

Prata, J. C. (2018). Microplastics in wastewater: State of the knowledge on sources, fate and solutions. *Marine Pollution Bulletin, 129*(1), 262–265. https://doi.org/10.1016/j.marpolbul.2018.02.046

Ragusa, A., Svelato, A., Santacroce, C., Catalano, P., Notarstefano, V., Carnevali, O., Papa, F., Rongioletti, M. C. A., Baiocco, F., Draghi, S., D'Amore, E., Rinaldo, D., Matta, M., & Giorgini, E. (2021). Plasticenta: First evidence of microplastics in human placenta. *Environment International, 146*, Article 106274. https://doi.org/10.1016/J.ENVINT.2020.106274

Rao, P. N., Reddy, A. G. S., Kumar, G. R., Babu, T. R., Prasad, K. M., & Madhusudhan, B. J. (2022). Assessment of water contamination at municipal solid waste disposal site, Jawaharnagar, Hyderabad, Telangana, India. *International Journal of Environment and Climate Change, 12*(10), 194–213. https://doi.org/10.9734/IJECC/2022/v12i1030786

Ravindiran, G., Rajamanickam, S., Sivarethinamohan, S., Karupaiya Sathaiah, B., Ravindran, G., Muniasamy, S. K., & Hayder, G. (2023). A review of the status, effects, prevention, and remediation of groundwater contamination for sustainable environment. *Water, 15*(20), 3662. https://doi.org/10.3390/w15203662

Satish Kumar, V., Amarender, B., Dhakate, R., Sankaran, S., & Raj Kumar, K. (2016). Assessment of groundwater quality for drinking and irrigation use in shallow hard rock aquifer of Pudunagaram, Palakkad District Kerala. *Applied Water Science, 6*(2), 149–167. https://doi.org/10.1007/s13201-014-0214-6

Saxena, V. (2025). Water quality, air pollution, and climate change: Investigating the environmental impacts of industrialization and urbanization. *Water, Air, & Soil Pollution, 236*(2), 73. https://doi.org/10.1007/s11270-024-07702-4

Setälä, O., Norkko, J., & Lehtiniemi, M. (2016). Feeding type affects microplastic ingestion in a coastal invertebrate community. *Marine Pollution Bulletin, 102*(1), 95–101. https://doi.org/10.1016/j.marpolbul.2015.11.053

Sharma, R., Kumar, R., Sharma, D. K., Sarkar, M., Mishra, B. K., Puri, V., Priyadarshini, I., Thong, P. H., Ngo, P. T. T., & Nhu, V. H. (2022). Water pollution examination through quality analysis of different rivers: A case study in India. *Environment, Development and Sustainability, 24*(6), 7471–7492. https://doi.org/10.1007/s10668-021-01777-3

Singh, J. (2018). Paddy and wheat stubble blazing in Haryana and Punjab states of India: A menace for environmental health. *Environmental Quality Management*. https://doi.org/10.1002/tqem.21598

Singh, J., Pant, M., Lakhera, A., & Singhal, N. (2023). Water quality assessment of Rispana River at Dehradun (Uttarakhand) India with reference to WQI and microbial contamination. https://doi.org/10.21203/rs.3.rs-3321030/v1

Singh, J., Pant, M., Ghoshal, T., Srivastava, A. L., & Singh, S. (2025). Spatio-seasonal observations of water quality index, heavy-metal pollution index and microbial contagion in springs of lower Indian Himalayan zone. *Environmental Claims Journal*, 1–36. https://doi.org/10.1080/10406026.2025.2482232

Singh, K. P., Singh, V. K., Malik, A., & Basant, N. (2006). Distribution of nitrogen species in groundwater aquifers of an industrial area in alluvial Indo-Gangetic Plains—A case study. *Environmental Geochemistry and Health, 28*(5), 473–485. https://doi.org/10.1007/s10653-006-9053-1

Singh, S., Kumar, A., & Gupta, H. (2020). Activated banana peel carbon: A potential adsorbent for Rhodamine B decontamination from aqueous system. *Applied Water Science, 10*(8), 1–8. https://doi.org/10.1007/s13201-020-01274-4

Thompson, V. S., Ray, A. E., Hoover, A., Emerson, R., Hartley, D., Lacey, J. A., Yancey, N., & Thompson, D. N. (2020). Assessment of municipal solid waste for valorization into biofuels.

Environmental Progress & Sustainable Energy, 39(4), Article e13290. https://doi.org/10.1002/ep.13290

Wang, G., Xiang, J., Liang, G., Wang, J., Ma, S., & He, C. (2023). Application of common industrial solid waste in water treatment: A review. *Environmental Science and Pollution Research, 30*(52), 111766–111801. https://doi.org/10.1007/s11356-023-30142-2

Wong, C. T. F., Falcone, M., Rich, G., Stubler, C., Malama, B., Lazcano, C., & Decock, C. (2023). Short-term effects of increasing compost application rates on soil C and greenhouse gas (N_2O and CO_2) emissions in a California central coast vineyard. *Frontiers in Environmental Science, 11*, 1123510. https://doi.org/10.3389/fenvs.2023.1123510

Zhang, Z., Chen, Z., Zhang, J., Liu, Y., Chen, L., Yang, M., Osman, A. I., Farghali, M., Liu, E., Hassan, D., & Ihara, I. (2024). Municipal solid waste management challenges in developing regions: A comprehensive review and future perspectives for Asia and Africa. *Science of the Total Environment, 930*, Article 172794. https://doi.org/10.1016/j.scitotenv.2024.172794

Zhou, Y., Zhang, Z., Bao, F., Du, Y., Dong, H., Wan, C., Huang, Y., & Zhang, H. (2024). Considering microplastic characteristics in ecological risk assessment: A case study for China. *Journal of Hazardous Materials, 470*, Article 134111. https://doi.org/10.1016/j.jhazmat.2024.134111

Groundwater Contamination Due to Landfilling of Toxic Wastes

Diksha, Deepak Chandra Joshi, and Priyanka Uniyal

1 Introduction

1.1 Overview of Groundwater Resources

Groundwater is a key part of the Earth's hydrologic system, filling the immense pore spaces within soil and rock cracks below the Earth's surface. It is one of the most essential and reliable sources of freshwater on our planet, ranking second to polar ice caps and glaciers in volume, but the most readily available for human consumption (Al-Hashimi et al., 2021). In contrast to surface water, which is susceptible to seasonal variation and quick pollution, groundwater is naturally filtered as it moves through rock and soil strata, making it relatively pathogen- and suspended particle-free. This natural purging makes groundwater a favored source of water for consumption, irrigation, and industrial use in both rural and urban settings (Nguyen & Huynh, 2023). Worldwide, almost 2.5 billion individuals rely on groundwater for drinking water, with more than 40% of the globe's irrigation and nearly one-third of industrial water from it (Khondoker et al., 2023). Aquifers, the subsurface layer of water-permeable rock or sediment, are mainly classified as unconfined and confined. Unconfined aquifers, being nearer the land surface and open to infiltration from the top, are more prone to contamination from activities on the surface (Tom & Adama, 2023). Confined aquifers are trapped between layers of impermeability, presenting some

Diksha
Department of Pharmaceutics, Dasmesh College of Pharmacy, Faridkot, Punjab, India

D. C. Joshi (✉)
Department of Pharmacy, Faculty of Pharmaceutical Science & Nursing, Vivekananda Global University, Jaipur, Rajasthan, India

P. Uniyal
Department of Pharmaceutical Sciences, School of Health Sciences and Technology, UPES, Dehradun, Uttarakhand, India

H. Gupta et al. (eds.), *Advanced Technologies for Assessment, Recycling, and Management of Solid Waste*, Waste as a Resource,
https://doi.org/10.1007/978-3-032-24788-9_4

level of protection against contaminants but harder to replenish and clean up when contaminated (Sharma et al., 2020). The importance of groundwater is all the more significant in developing nations like India, where it supplies almost 85% of rural drinking water demand and around 60% of irrigation needs for agriculture. With mounting surface water contamination and swift urbanization, reliance on groundwater has been growing, and it has become the only lifeline for millions of people, especially in arid and drought-hit areas (Abanyie et al., 2023; Banerjee & Ganguly, 2023).

But excessive groundwater extraction above its natural recharge has resulted in a sharp drop in water levels in most of the world, including shocking amounts in India's Indo-Gangetic plains, Punjab, and sections of Maharashtra (Pandey et al., 2022). In addition, groundwater quality is declining at an unprecedented level because of man-made activities like industrial effluent disposal, indiscriminate application of fertilizers and pesticides, improper sewage disposal, and significantly, toxic solid and hazardous waste landfilling (Ghosh et al., 2022). Landfills, particularly unlined or poorly managed landfills, are a major source of groundwater contamination by leachate percolation. This poisonous leachate usually has heavy metals, persistent organic pollutants, pathogens, and new contaminants such as PFAS (per- and polyfluoroalkyl substances) and pharmaceutical residues (Dagwar & Dutta, 2024). As this leachate percolates through layers of soil, it becomes able to seep into aquifers and lead to long-term, frequently permanent, contamination. Contrary to surface water, in which contaminants may be diluted and washed away over the years, contaminants in groundwater have a tendency to remain and spread slowly, greatly complicating and costing remediation (Gunarathne et al., 2024). In addition, the unseen character of groundwater contamination makes it usually unobserved until major harm is caused. Adding to this issue is the high variability of aquifer recharge and the dependency on factors like precipitation, land use, terrain, and permeability of geologic strata, which are further being debilitated by global warming and urbanization (Li et al., 2021). The strategic value of groundwater as a robust, decentralized water source emphasizes the imperative to protect it. As more threats are posed by hazardous landfilling techniques, groundwater pollution is not only an environmental problem but also an emerging public health, agricultural, and socio-economic issue that must be addressed through combined scientific, policy, and community-level actions to provide secure water for future generations (Obaideen et al., 2022).

1.2 Importance of Groundwater in Ecosystems and Human Consumption

Groundwater is essential for maintaining human societies and natural ecosystems alike, positioning it as one of the most important yet least respected aspects of the world's water cycle. As a source of safe drinking water for humans, groundwater is an integral primary supply, particularly where surface water is lacking, seasonally unavailable, or tainted (Scanlon et al., 2023). In most rural areas, especially in developing nations such as India, groundwater is the sole perennial and assured source of drinking water because of its relative availability and accessibility through wells, boreholes, and tube wells. It is also less prone to direct contamination than surface water bodies, thanks to the natural filtration process during its passage through strata of soil and rock. This makes it inherently shielded from microbial contamination and suspended solids, but chemical pollutants from anthropogenic activities can still enter and remain. Groundwater in urban areas is widely utilized for municipal water supply, sanitation, and industrial applications, accounting for much of the economic growth and improvement in standard of living. It is the major user of groundwater in agriculture across the world and supports irrigation, especially during periods of drought and in areas where rainfall is low or unreliable, ensuring food security. Groundwater-based irrigation systems were instrumental in the Green Revolution in converting water-deficient areas to agriculturally productive ones (Saito et al., 2021; Shaikh & Birajdar, 2024).

In addition to human needs, groundwater is naturally linked with the health of many ecosystems. It sustains wetlands, riparian ecosystems, and terrestrial vegetation cover by base flow in streams and lakes, especially during dry periods when the surface water supply is low. Aquatic and terrestrial organisms rely on groundwater-dependent ecosystems (GDEs) for survival, growth, and migration. Water is normally scarce in arid and semi-arid lands and the ground water is the sole source of water saving of biodiversity and ecosystem (Borowiecka et al., 2024). Changes or variation in the water level of the ground water may lead to destruction of eco system, lack of biodiversity and violation of ecosystem activity such as nutrient recycling, flood regulation, and carbon sank. Ground water also plays an important role in the thermo and chemical stability of the ecosystem by smoothing out temperature extremes and neutralizing pollution with a gradual speed of travel and spread capability (Land & Peters, 2023). This delicate balance is however, being disturbed by increasing pressure on the ground water through excessive pumping and contamination particularly through landfilling of risky hazardous and toxic wastes. The leachates that spread out of the improperly operated landfills can also change the chemical content of the ground water so as to render it unsafe to be consumed by humans and toxic to aquatic life. Such toxic agents do not only deteriorate the quality of water but these also affect the microbial activity in the soil and also decrease the vegetation cover, bio-accumulate in the food chain and threaten the health of human beings and wildlife as well on a chronic level (Parvin & Tareq, 2021). The utility of ground water thus extends much beyond the practical consequence of utilitarianism- it is a sustenance system of life to both people and the earth and requires dire attention, sound and unimpeded management and a safeguard against any form of pollution including but not limited to the landfilling process which imparts long-term and usually irreparable effects (Abiriga et al., 2020).

1.3 Definition and Nature of Toxic Wastes

Toxic Wastes: Any discarded substance, solid, liquid, or gaseous, which represents a high risk to human health or the environment by virtue of its chemical, biological or radionuclide characteristics. They are usually by-products of industrial operations as well as chemical production, mining operations, pharmaceuticals and agriculture and even household items (Sivaram et al., 2019).

According to the United States Environmental Protection Agency (USEPA), a waste is classified as hazardous if it exhibits one or more of the following characteristics: toxicity, reactivity, ignitability, or corrosivity (Singh & Yadav, 2022).

The chemistry of toxic wastes is characterized by their multifaceted origin, composition, and interaction with media of the environment. For instance, waste from electroplating facilities will have cyanides and chromium, whereas pesticide production facilities will release organophosphates and chlorinated hydrocarbons. Biomedical waste disposed of improperly can release pathogenic microorganisms, antibiotics, and radiological contaminants, whereas electronic waste has a blend of heavy metals and flame retardants (Varshney et al., 2022). When these toxic compounds are disposed of in landfills particularly in non-engineered or unlined facilities they produce leachate upon interaction with percolating rainwater or surface runoff. The leachate serves as a carrier, moving the toxic components into the subsurface environment, where they can move vertically and horizontally, ultimately polluting underlying aquifers (Clemente et al., 2024; Makhadi et al., 2020). The mobility, solubility, and persistence of these toxicants establish the severity and extent of groundwater contamination. Additionally, the additive toxicities of several pollutants may cause increased toxicity, and the pollution is thus more dangerous than would be anticipated from single substances individually. In most developing nations, however, the problem is enhanced by the lack of rigorous regulatory control, source segregation not being practiced, and poor landfill design (Liu et al., 2024). Therefore, hazardous wastes still remain a serious threat to environmental protection and public health, especially when landfilling is employed as the major disposal technique without appropriate precaution measures. The knowledge of definition, origin, chemical properties, and environmental consequences of hazardous wastes is important in the development of effective landfill management techniques and groundwater conservation policies (Li et al., 2021; Priyadarshanee et al., 2022).

1.4 Objectives and Scope of the Chapter

The major objective of this chapter is to critically examine the issue of groundwater contamination as a result of the landfilling of toxic wastes, a new trend in the context of industrialization, urbanization, and increased production of wastes. Based on the integration of scientific evidence, case studies, monitoring technologies, and regulatory information, this chapter tries to provide an integrated analysis of how landfilling particularly if inadequately designed or inadequately managed serves as a route for the migration of toxic substances into aquifers. The chapter tries to examine different routes through which toxic leachates migrate into groundwater systems and the parameters that control their mobility and persistence. The chapter further elucidates the chemical and biological properties of contaminants such as heavy metals, persistent organic pollutants (POPs), pathogens, and emerging chemicals such as PFAS and pharmaceutical residues that are responsible for long-term environmental and health risks. Another significant objective is to discuss actual-world contamination incidents occurring in India and elsewhere to identify the scale of the issue and its socio-environmental setting. Besides problematizing the issue, this chapter examines state-of-the-art monitoring and assessment techniques, such as groundwater sampling, leachate modeling, and the application of Geographic Information Systems (GIS) and remote sensing for contamination monitoring. The chapter also critically examines prevailing remediation and containment techniques ranging from pump-and-treat systems to bioremediation and reactive barriers based on sustainability, cost, and engineering practicability.

2 Landfilling: An Overview

Landfilling is one of the oldest and most common practices in the disposal of solid waste worldwide. Landfilling is controlled waste disposal on or in the land, typically in preferred locations or engineered pits (Cossu & Stegmann, 2018). In ideal circumstances, contemporary landfills are lined with protective systems such as impermeable liners, leachate collection and treatment systems, gas venting systems, and environmental monitoring equipment. These are intended to minimize the environmental impact and safeguard groundwater and surrounding ecosystems (Ma et al., 2022). However, in most of the world particularly in developing nations landfilling persists in open, unlined locations without minimal engineering requirements (Vaverková, 2019). These poorly managed landfills are principal sources of pollution, especially to groundwater, because toxic leachate formed by rainwater infiltration and waste decomposition percolates through the ground into aquifers. Environmental impact severity depends on landfill type, waste composition, climate, hydrogeology, and the presence or absence of protective barriers. Therefore, an explicit clarification of landfill types, operational practices, and leachate dynamics is necessary in assessing and mitigating their contribution to groundwater pollution (Siddiqua et al., 2022a, 2022b).

2.1 Types of Landfills

Landfills are specific sites for burying waste, and they are the most prevalent practice of solid waste management worldwide. The environmental consequences of landfilling, especially in terms of groundwater pollution, mostly depend on the nature of the landfill as well as the waste that is being buried (Abdel-Shafy et al., 2024).

Municipal Solid Waste Landfills

Municipal Solid Waste (MSW) landfills are facilities designed for the treatment of non-hazardous waste produced from residential, office, marketplace, school, and other similar activities. Typical constituents are biodegradable organic materials, plastics, paper, textiles, metals, and inert materials (Mor & Ravindra, 2023). Ideally, MSW landfills are designed with liner layers (most commonly clay or synthetic materials like HDPE), daily cover soils, leachate collection and treatment systems, and gas venting systems to manage methane and carbon dioxide emissions. In most of the world, especially in low- and middle-income nations, however, landfills are simply open dumps with no engineering controls. The absence of barriers permits rainwater to pass through the wastes, generating a highly contaminated leachate, which directly enters the soil and underlying aquifers (Ram & Kumar, 2021). Furthermore, the breakdown of organic waste in MSW landfills causes the generation of acids and gases, which further increase the mobility of heavy metals and other toxic materials. Experiments have proven that in areas where MSW landfills are not well maintained, groundwater surrounding landfill facilities usually has high concentrations of nitrate, ammonium, iron, manganese, and coliform bacteria, which imply serious leachate infiltration (Nanda & Berruti, 2021).

Industrial and Hazardous Waste Landfills

Industrial and toxic waste landfills are used to manage wastes that are of major concern to human health and the environment. These could be wastes from chemical production, metal processing, petroleum refining, electronic equipment, pharmaceutical companies, and other industries producing harmful by-products. Such wastes may comprise persistent pollutants like arsenic, chromium, cyanides, phenols, solvents, PCBs, and radioactive materials (Amirfazli et al., 2019). Because of its hazardous character, landfills for hazardous waste are regulated stringently and by requirements of engineering. These are required to be double-lined, have leachate detection and recovery systems, impermeable caps, and long-term monitoring programs. In spite of these measures, unauthorized dumping of hazardous waste into MSW landfills or unlined locations is a serious issue in most developing countries, including India. Insufficient segregation and disposal of biomedical waste, e-waste, and industrial sludge tend to generate highly toxic leachates, which find

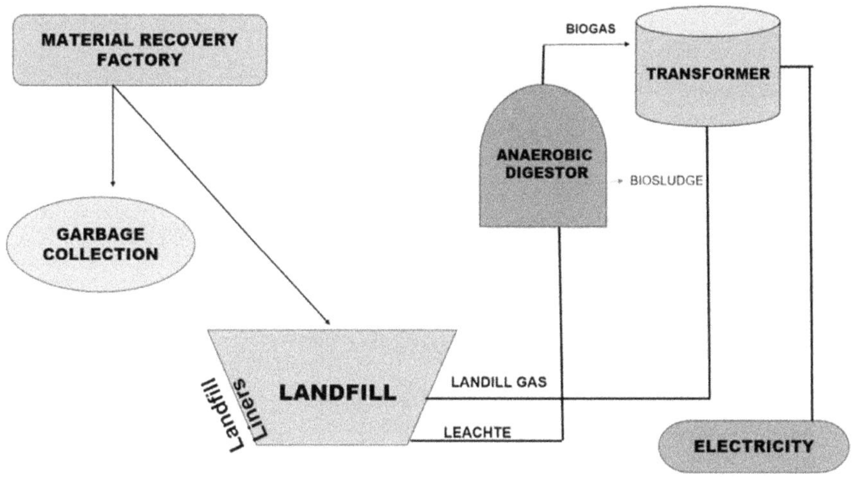

Fig. 1 A diagrammatic representation of industrial and hazardous waste landfills

their way into groundwater and get bioaccumulated in the food chain. Groundwater in such locations, once contaminated, loses all its useful purposes and creates long-term public health hazards (Gautam & Kumar, 2021; Ogunwumi & Salami, 2023). A diagrammatic representation of Industrial and Hazardous Waste Landfills mentioned in Fig. 1.

2.2 *Landfilling Practices Around the World*

Landfilling procedures range extensively throughout the world based on local legislation, economic means, environmental sensitivities, and technological facilities (Butti et al., 2018). Within developed countries like the United States, landfilling is regulated under the Resource Conservation and Recovery Act (RCRA), which ensures an elaborate framework for landfill design, operation, and closure. Such landfills are required to incorporate engineered liners, leachate treatment systems, and post-closure maintenance systems to track environmental effects (Ozymy & Ozymy, 2023). The European Union has implemented the Landfill Directive (1999/31/EC) with the objective of diminishing landfill reliance through diversion of biodegradable waste, recycling, and pre-treatment of waste prior to landfilling (Bugarski et al., 2021). Germany, Sweden, and Netherlands have already made significant progress towards minimal landfilling, with less than 1% of municipal waste being landfilled (Scarlat et al., 2019).

Conversely, a majority of developing nations continue to use open dumping or inadequately managed landfilling. In India alone, more than 70% of municipal solid

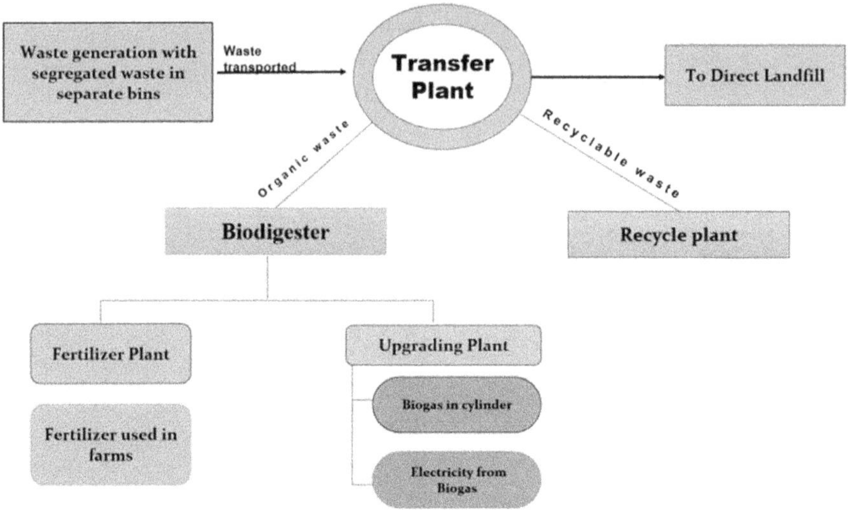

Fig. 2 Common landfilling practices around the world

waste is openly dumped or landfilled in legacy sites that do not have minimum environmental controls. Despite laws like the Solid Waste Management Rules (2016) and Hazardous Waste Management Rules (2016), enforcement is lacking because of the shortage of funds, the absence of trained staff, and the inefficiencies of institutions (Mor & Ravindra, 2023). Urbanization has compounded the problem by generating more waste and diminishing land availability for scientific landfilling. In China, there has been accelerated industrialization, leading to more generation of hazardous waste, and while design and management of landfills have improved since 2010, there are still residual problems (Kanwal et al., 2022). African countries also experience major limitations in sanitary landfilling due to limited financing and infrastructure. The variation in waste management practice between countries mirrors not just differences in economic level but also differences in public awareness, policy compliance, and dedication to sustainable development (Idowu et al., 2019). A common landfilling practices around the world is summarized in Fig. 2.

2.3 Waste Composition and Leachate Formation

The waste composition deposited in landfills is a key factor for determining the quality and quantity of leachate. When rain percolates through landfill waste, it comes into contact with decomposing substances and exists in intricate chemical and biological processes, leading to a leachate a very concentrated and poisonous fluid. This leachate generally has a broad range of contaminants such as dissolved organic matter (quantitated by chemical oxygen demand [COD] and biochemical oxygen demand [BOD]), inorganic ions (chloride, sulfate, nitrate), heavy metals (lead, chromium, cadmium, zinc), nutrients (ammonium, phosphate), and pathogens. In waste landfills containing hazardous wastes, leachate might include volatile organic compounds (VOCs), persistent organic pollutants (POPs), cyanides, and even radioactive materials (Roy et al., 2018; Vaverková et al., 2020b).

The waste composition is determined by parameters like the type of waste (municipal or industrial), level of segregation, landfill age, water content, temperature, and microbial activity. Leachate is more acidic and contains higher levels of volatile fatty acids and ammonia during the initial stages of landfill operation (Mishra et al., 2019). As stabilization follows decomposition, the leachate will become alkaline with higher heavy metal and refractory organic content. In highly rainy and hot tropical countries, such as portions of India, the rate of leachate generation is much higher because of intensive microbial action and more frequent percolation of water. In addition, the recent high demand for consumer products, e-waste, and pharmaceutical waste disposal has added new emerging pollutants in the form of PFAS, antibiotics, and microplastics to landfill leachates that pose new threats to groundwater that are not yet addressed by traditional treatment practices. When leachate is not treated and collected, it seeps downwards and pollutes subsurface soil and aquifers. The transport and mobility of such pollutants are governed by permeability of soil, hydraulic conductivity, and occurrence of natural or man-made barriers. Subsequently, this leads to widespread and frequently permanent groundwater pollution endangering drinking water security, agriculture, and ecosystems (Masoner et al., 2020).

3 Mechanisms of Groundwater Contamination

Landfilling is the process of contamination of groundwater by a variety of physical, chemical, biological, and geological factors. The prime force behind pollution is leachate, a poisonous liquid generated when water comes into contact with waste in the landfill. This generated leachate later moves vertically and horizontally in the subsurface and may advance to aquifers. The movement, dispersal, and subsequent contamination of groundwater are functions of the leachate characteristics, soil and rock conditions, and climatic factors such as rainfall and temperature. Knowledge of the processes is crucial in the design of prevention measures and remedial strategies (Ali et al., 2019).

3.1 Leachate Generation and Its Composition

Leachate is a very polluted liquid that occurs mostly because of rainwater percolation through the waste layers of the landfill. As water flows downwards, it dissolves and mobilizes many soluble and particulate materials, forming a complicated mixture of contaminants. Leachate composition varies depending on many factors such as waste type (municipal, industrial, hazardous), age of landfill, water content, and microbial processes (Zhang et al., 2023). In juvenile landfills, leachate is usually dominated by high levels of volatile fatty acids, ammonia, BOD (Biochemical Oxygen Demand), and COD (Chemical Oxygen Demand) (Abdel-Shafy et al., 2024). With increasing age of the landfill, leachate becomes alkaline and is laden with heavy metals (lead, cadmium, mercury, arsenic), inorganic ions (chloride, nitrate, sulfate), persistent organic pollutants (PCBs, PAHs), and novel substances like PFAS, microplastics, and pharmaceutical residues (Dwanga et al.). Leachate also contains microbial pathogens, especially in landfills accepting biomedical or municipal waste. The sheer toxicity and persistence of these compounds render leachate a serious hazard to underlying groundwater if not contained or treated correctly (Anouzla & Souabi, 2024).

3.2 *Leachate Migration Pathways*

After formation, leachate takes a number of migration routes through the subsurface. In properly designed landfills, geomembrane liners and leachate collection systems are used to keep this movement from happening or reduce it. In damaged or unlined landfills, leachate percolates through the vadose (unsaturated) zone and eventually finds its way to the saturated zone where aquifers exist (Jiang et al., 2019). It is gravity-driven vertical migration, and lateral migration may happen through groundwater movement or the movement of subsurface pressure gradients. Sandy soils, fractured rocks, or permeable sediments become conduits, hastening leachate flow. Clay soils or densely compacted geological materials can behave as natural barriers, retarding or preventing contaminant dispersal. Nevertheless, even these barriers tend to deteriorate with time by cracking, erosion, or chemical weakening. Also, preferential flow paths such as root channels, burrows, or fractures in the rock can provide leachate transport short-cuts through the natural filtration zones, and thus, contamination is hard to predict and track (Amiaz et al., 2011).

3.3 *Soil Permeability and Geologic Factors*

Soil and geological strata permeability is a critical factor to determine the speed and the distance leachate is capable of migrating. Highly permeable soils like sand and gravel provide easy infiltration and promote deeper aquifer contamination. Low-permeability soils like clay tend to retard the movement of leachate because of the dense composition of these soils and restrictive pore space (Alao, 2024). Yet, clay is also susceptible to swelling and cracking, particularly with repeated wet-dry cycles, that can develop new pore paths for contaminant travel. Karstic geology, consisting primarily of limestone and dolomite, is a major concern because it contains solution cavities and sinkholes that permit instantaneous leachate flow with little natural attenuation. Depth to the water table, permeability of underlying rock, and hydraulic conductivity of the soil layers also affect the vulnerability of the region to groundwater contamination. Where water tables are shallow and permeable soils prevail, small leaks of leachate can lead to extensive and rapid pollution of groundwater resources (Bera et al., 2021).

3.4 *Role of Climate and Rainfall in Contaminant Transport*

Climatic factors, specifically precipitation and temperature, greatly influence the rate of leachate formation and subsequent movement. In tropical and high-rainfall areas, the greater amount of rainwater percolating into the landfill results in more leachate generation, increasing the risk of groundwater pollution. Recurring or intensive rains have the potential to overcome the absorptive capacity of the landfill, compelling the pollutants to move downward or on the surface (Noyes et al., 2009). Conversely, dry areas might yield lower leachate content, but denser contaminants will still be present as a result of reduced degradation. Temperature further has an important function through its role in affecting microbial action and the rate of organic material breakdown in landfills. Increased temperatures stimulate biochemical breakdown, accelerating biochemical degradation of the products which contributes to the enhancement in the amount of leachate generated as well as the mobility of specific contaminants like ammonia and nitrate. Various external conditions such as seasonal changes like monsoons or snowmelts can cause sudden spikes in leachate releases and further landfill containment structures. These forces are additionally being aggravated by the occurrence of climate change that alters precipitation regimes, increases extreme events, and antibiotics disruption of natural recharge mechanism that makes groundwater more prone to pollution via landfill leachates (Anawar, 2013).

4 Types of Contaminants from Landfills

Landfill leachate comprises an extensive array of contaminants that vary in terms of origin, chemical nature, persistence, mobility, and toxicity. These contaminants can drastically impair groundwater quality once they move through soil profiles. They can be generally categorized into heavy metals, organic pollutants, microbial contaminants, and emerging contaminants as illustrated in Table 1 (Iravanian & Ravari, 2020).

4.1 *Heavy Metals (e.g., Lead, Arsenic, Mercury, Cadmium)*

Some of the most contaminating and toxic substances found in the landfill leachate are heavy metals. These materials contain the elements like lead (Pb), arsenic (As),

Table 1 Overview of major contaminants in landfill leachate and their environmental and health impacts

Contaminant type	Common examples	Primary sources in landfills	Environmental behavior	Health and ecological impacts	References
Heavy metals	Lead (Pb), arsenic (As), mercury (Hg), cadmium (Cd), chromium (Cr), nickel (Ni), zinc (Zn)	Batteries, paints, electronic waste, metal plating industries, industrial sludges, pesticides	Non-biodegradable; persist in soil and water; can adsorb onto soil particles but may leach depending on pH	Neurotoxicity, kidney damage, cancer, developmental disorders; bioaccumulation in food chains	Angon et al., (2024)
Organic pollutants	PCBs (polychlorinated biphenyls), PAHs (polycyclic aromatic hydrocarbons), VOCs (benzene, toluene, xylene)	Industrial chemicals, solvents, lubricants, paints, fuels, incinerator ash, plastics	Volatile, persistent, hydrophobic; can migrate long distances in soil vapor or dissolve in leachate	Carcinogenic, mutagenic, endocrine disruption; toxic to aquatic life	Santos et al., (2025)
Microbial contaminants	*E. coli*, *Salmonella*, *Shigella*, *Clostridium*, viruses, fungi, helminths	Household waste, biomedical waste, septic sludge, food residues	Easily transported in water; survival varies with temperature and soil moisture	Gastrointestinal diseases, hepatitis, infections; threat to immunocompromised individuals	Avery et al., (2012)
Emerging contaminants	PFAS (per- and polyfluoroalkyl substances), pharmaceuticals (antibiotics, hormones, NSAIDs), nanoparticles, microplastics	Personal care products, drug residues, packaging materials, textiles, firefighting foam, medical waste	Highly persistent, mobile; not removed by conventional treatments; can undergo long-range transport	Hormonal imbalance, antibiotic resistance, developmental toxicity, ecological imbalance	Nishmitha et al., (2025a, 2025b)

mercury (Hg), cadmium (Cd) and commonly are extracted as a result of battery disposal and electronic waste, paints, pigment and industrial wastes and by-products (Aucott, 2006; Iravanian & Ravari, 2020). Unlike the organic pollutants, heavy metals do not degrade with time and they tend to accumulate in earth and underground water, posing long-term risks to the ecosystem and the health of the human kind. The release of these metals may cause neurological damages, kidney damages, carcinogenesis, and developmental retardation, especially those in vulnerable groups such as children and pregnant women. Since heavy metals are mobile in certain soil, they may transfer to leachate into the aquifers and thus monitoring and confinement of heavy metals should be considered as landfill management concern (Dwanga et al.; Rahman & Singh, 2019).

4.2 Organic Pollutants (e.g., PCBs, PAHs, VOCs)

The organic pollutants in landfills are mostly synthetic chemicals which can also leach to the groundwater due to runoffs in the landfill. They are polychlorinated biphenyls (PCBs) in electrical devices, polycyclic aromatic hydrocarbons (PAHs) in burned items, and volatile organic compounds (VOCs) such as benzene, toluene, and trichloroethylene in solvents, plastics, and cleansers. The majority of these chemicals are toxic, carcinogenic, as well as mutagenic, and their lipophilicity property predisposes them to bioaccumulation in living organisms. Due to their mobility and volatility VOCs can easily contaminate the soil and water media and lead to widespread pollution. Such compounds are toxic and persistent and sophisticated detection and clean-up techniques are needed to restrain their impact on drinking water resources and environmental systems (Indelicato et al., 2017; Ochs et al., 2024).

4.3 Microbial Contaminants

Microbial contamination is caused by landfills receiving organic waste and sewage sludges as a major one, especially those that are poorly engineered or inspected. Pathogenic bacteria (e.g. *E. coli*, *Salmonella*), viruses (e.g. rotavirus, enterovirus) and protozoa (e.g. Giardia, Cryptosporidium) may be present in leachate of such landfills. Their microorganisms can penetrate through the pores of the soil and infiltrate the water of the ground and end up being waterbornes diseases such as gastroenteritis, cholera and hepatitis amongst others. Danger increases when there are unlined dumps or open dump areas in areas close to water, or a shallow water table. Microbial contamination does not only affect human health but upsets water balances in the aquatic system by altering the whole equilibrium of the microbial community in the groundwater as well as the oxygen demand in the ground water (Hui et al., 2023; Odonkor & Mahami, 2020).

4.4 Emerging Contaminants (e.g., PFAS, Pharmaceuticals)

Indeed, in the last decades, the landfills leachate has generated a number of emerging contaminants of interest on the grounds of their longevity, bioaccumulation potentials, or unsubstantiated long-term health effects. They include per- and polyfluoroalkyl substances (PFAS), colloquially known as forever chemicals, another product of non-stick cookware, fire retardant foam, and waterproof apparel (Barroso et al., 2019; Ruan et al., 2022). Other pollutants in landfill leachate are pharmaceutical remnants, personal care substances, endocrine disruptors (EDCs), and nanomaterials. These wastes have the tendency of passing through the usual waste treatment facilities and be active enough years later in the environment carrying dangers of hormonal interference, antibiotic resistance and long-term toxicity. Their appearance gives the picture of the enhanced complexity of the pollution by landfills and the high importance of advanced, complex detection, estimation, and advancing of the risks technologies (Rahman et al., 2023).

5 Case Studies of Groundwater Contamination

5.1 Notable Incidents in India

Due to poor municipal solid waste (MSW) landfilling activities, there are instances in India that are very alarming as far as the ground water contamination is concerned. These occurrences raise the issue of the impending need to have improved landfill engineering, leachate treatment, and government personnel oversight. A comparative overview of contamination levels across different landfill types is presented in Table 2 (Ratna et al., 2021).

Jawaharnagar Landfill, Hyderabad

Groundwater in the vicinity of the Jawaharnagar landfill operated by GHMC was found to be heavily polluted by leachate infiltration. Analysis of 15 groundwater samples revealed that TDS (1200–1800 mg/L), calcium (180–260 mg/L), nitrate (60–110 mg/L), and chloride (250–380 mg/L) exceeded the WHO permissible limits (TDS: 500 mg/L, calcium: 75 mg/L, nitrate: 50 mg/L, chloride: 250 mg/L). This indicates severe leachate percolation, making groundwater unfit for drinking and irrigation (Markus et al., 2022).

Table 2 Groundwater contamination due to different landfills compared with WHO/BIS permissible limits

Landfill type and location	Major contaminants detected	Reported groundwater levels	WHO/BIS permissible limits	References
Jawaharnagar MSW Landfill, Hyderabad (India)	TDS, calcium, nitrate, chloride	TDS: 1200–1800 mg/L; Ca: 180–260 mg/L; NO_3^-: 60–110 mg/L; Cl^-: 250–380 mg/L	TDS: 500 mg/L; Ca: 75 mg/L; NO_3^-: 50 mg/L; Cl^-: 250 mg/L	Choudhury et al. (2021)
Ranchi MSW Dumpsite, Jharkhand (India)	Lead, iron, manganese, coliforms	Pb: 0.097 mg/L; Fe: 0.97 mg/L; Mn: 0.36 mg/L; coliforms: $> 2 \times 10^4$ CFU/mL	Pb: 0.01 mg/L; Fe: 0.3 mg/L; Mn: 0.1 mg/L; coliforms: 0 CFU/100 mL	Dube and Kanido (2025)
Lianyungang MSW Landfill, China	Nitrate, lead, cadmium	NO_3^-: 85–130 mg/L; Pb: 0.05–0.12 mg/L; Cd: 0.01–0.03 mg/L	NO_3^-: 50 mg/L; Pb: 0.01 mg/L; Cd: 0.003 mg/L	Mathews et al. (2024)
Minnesota (USA) and UK Landfills	PFAS, pharmaceuticals, EDCs	PFAS: 0.05–0.25 μg/L; pharma residues: 0.1–0.6 μg/L; EDCs: detected (trace levels)	PFAS/EDCs—not specified; pharmaceuticals—not specified	Chima et al., (2025)

Ranchi Municipal Solid Waste (MSW) Dumpsite, Jharkhand

Groundwater analysis near Ranchi MSW dumpsite showed alarming contamination. Heavy metals such as lead (0.097 mg/L), iron (0.97 mg/L), and manganese (0.36 mg/L) were detected, far above the BIS permissible limits (lead: 0.01 mg/L, iron: 0.3 mg/L, manganese: 0.1 mg/L). In addition, coliform bacterial counts exceeded 2×10^4 CFU/mL, surpassing WHO's standard of zero coliforms per 100 mL. The overall Water Quality Index (WQI) classified the groundwater as "poor," confirming unsuitability for human consumption (Chakraborty & Kumar, 2016).

5.2 International Case Studies (USA, China, Europe)

Lianyungang, China—MSW Landfill Impact

Groundwater around the Lianyungang landfill showed nitrate levels (85–130 mg/L), lead (0.05–0.12 mg/L), and cadmium (0.01–0.03 mg/L), exceeding WHO permissible limits (nitrate: 50 mg/L, lead: 0.01 mg/L, cadmium: 0.003 mg/L). These values highlight the vulnerability of shallow aquifers in areas with high soil permeability (Chen et al., 2019).

Minnesota (USA) and UK Landfills

Long-term monitoring in Minnesota (USA) and the UK detected persistent PFAS (0.05–0.25 μg/L), pharmaceuticals (0.1–0.6 μg/L), and endocrine-disrupting chemicals (EDCs) in groundwater. While these levels may appear low, PFAS and EDCs are toxic even at trace concentrations due to their bioaccumulation and persistence. The findings highlight the urgent need for global regulation and remediation strategies to tackle emerging contaminants (Nishmitha et al., 2025a, 2025b). A comparative overview of contamination levels across different landfill types is presented in Table 2.

6 Monitoring and Assessment Techniques

6.1 Groundwater Sampling and Analysis

Groundwater sampling is a routine procedure in determining subsurface contamination surrounding landfill facilities. Monitoring wells are placed upgradient and downgradient to monitor the movement and concentration of pollutants (Liu & Wang, 2022). Sampling involves in situ measurement of parameters like pH, temperature, conductivity, and dissolved oxygen, with laboratory analysis for TDS, BOD, COD, nitrates, heavy metals (Pb, As, Cd), VOCs, PFAS, and microbial indicators like total and fecal coliforms. Techniques like grab sampling, purge-and-sample methods, and multilevel samplers are utilized as per ISO 5667 and EPA 300.0. Analyzing equipment includes AAS, ICP-MS, GC–MS, and IC, with output compared with BIS 10500:2012, WHO, and US EPA standards to assess water quality and risk (Agboola, 2024). A Systematic overview is mentioned in Fig. 3 about the various sources for the identification of ground water potential zones.

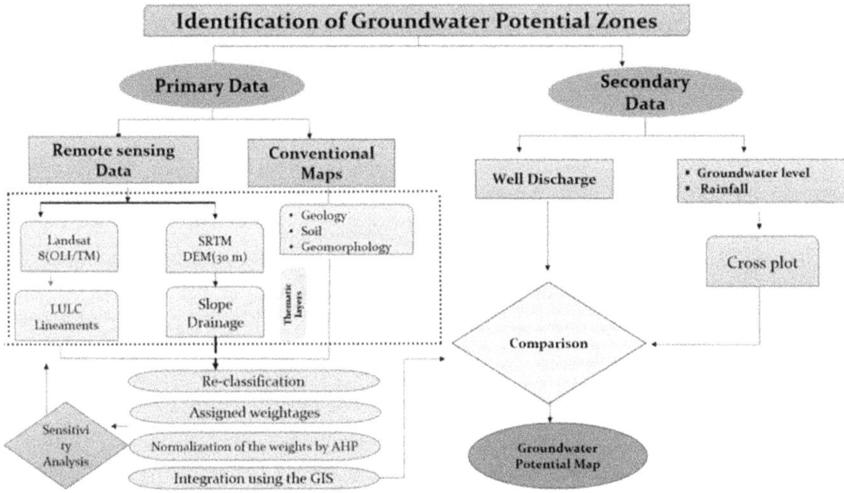

Fig. 3 Systematic overview about the various sources for the identification of ground water potential zones

6.2 *Leachate Monitoring and Modeling*

Monitoring of leachate is crucial for knowing the extent of the influence of the landfill on neighboring water bodies. Aids in collection systems or natural seepage locations cause data on the concentration of pollutants. Leachate models simulate generation rates and transport of pollutants in soil and aquifers based on climate, waste material, and design of the landfill. Common models used are HELP, MODFLOW, MT3DMS, LEACHM, and GMS, which assist in the prediction of contaminant plume movement, breakthrough times, and long-term effects on the environment. These models are important for designing successful remediation and landfill design planning (Ergene et al., 2022; Giakoumatos et al., 2024).

6.3 *Risk Assessment and Contamination Indexes*

Risk assessment converts chemical and hydrological information into possible human and environmental health effects. It involves hazard identification, exposure evaluation, toxicity analysis, and general risk characterization for carcinogenic and non-carcinogenic effects. Numerous indices like Water Quality Index (WQI), Contamination Index (Cd), Health Risk Index (HRI), Heavy Metal Pollution Index (HPI), and Leachate Pollution Index (LPI) are applied to estimate and convey risk levels. These indicators underpin environmental decision-making, rank polluted sites, and inform resource allocation for remediation (Biglari et al., 2025).

6.4 Remote Sensing and GIS Applications

Remote Sensing (RS) and Geographic Information Systems (GIS) are critical tools in spatial analysis and visualization of landfill-borne contamination. They assist in mapping groundwater vulnerability areas, tracing leachate plumes, monitoring land use changes, and establishing buffer zones around landfills. Satellite imagery (e.g., Landsat, Sentinel-2, MODIS), DEMs, and computer software such as ArcGIS, QGIS, ENVI, and GRASS GIS are frequently applied in hydrogeological modeling. By combining geospatial, chemical, and demographic information, RS and GIS improve monitoring precision and facilitate useful planning for mitigation and remediation activities (Erdanaev, 2024; Mohammadi-Behzad et al., 2019).

7 Health and Ecological Impacts

Landfill leachate-tainted groundwater releases a series of toxic substances into the environment, threatening human health, ecosystems, and socio-economic well-being with serious and frequently irreversible consequences (Dagwar & Dutta, 2024). The magnitude of such effects varies with the nature and concentration of the pollutants, the length of exposure, and the susceptibility of the affected people or ecological system. The main pathways of exposure are ingestion of contaminated drinking water, crop production using polluted irrigation sources, and bioaccumulation through the food web. This chapter describes the complex effects of groundwater contamination by landfills (Siddiqua et al., 2022a; Danthurebandara et al., 2013).

7.1 Human Health Hazards (Cancer, Neurotoxicity, Reproductive Effects)

Polluted groundwater employed for routine human use drinking, cooking, and agriculture exists as a direct pathway for toxic ingestion. Heavy metals such as arsenic, lead, mercury, and cadmium, as well as volatile organic compounds (VOCs) including benzene, toluene, and vinyl chloride, are commonly encountered in leachate. Arsenic, which is ranked a Group 1 human carcinogen by IARC, is implicated in skin, lung, and bladder cancers (Shakoor et al., 2017). Benzene also has a strong association with leukemia and bone marrow suppression, while vinyl chloride is a known cause of liver angiosarcoma. Neurotoxic effects, especially in children, are also notable because of lead and mercury, which damage neurodevelopment and cognitive function. VOCs like toluene also have an impact on memory and coordination. Nitrate contamination above 50 mg/L can induce methemoglobinemia or "blue baby syndrome," particularly hazardous to infants. Endocrine-disrupting chemicals (EDCs) such as PFAS, phthalates, and bisphenols disrupt hormonal equilibrium, reproductive function, fetal development, and thyroid function (Chaudhary et al., 2025). In rural and resource-constrained areas where there is only groundwater available as water and there is poor health infrastructure, these health threats are increased (Rajan et al., 2024).

7.2 Impact on Aquatic and Terrestrial Ecosystems

Leachate seeping into groundwater or finding its way into surface water through base-flow causes toxic stress on aquatic and terrestrial ecosystems. Aquatic ecosystems are benefited by the addition of nutrients (ammonia, nitrates, phosphates), which cause eutrophication in the form of algal growth and oxygen depletion, an agent of mass fish kills. Heavy metals such as cadmium, mercury, and lead accumulate in aquatic organisms, disrupting enzymic activity and reproductive well-being. Organic pollutants like PAHs and PCBs are noted to cause mutagenesis and lower fertility among aquatic life. On terrestrial ecosystems, leachate-exposed soil is acidified, has less microbial diversity, and lower agricultural yields. When contaminated groundwater is applied for irrigation, long-lasting pollutants like PFAS and microplastics find their way into the food chain and also impact flora and fauna. The additive impact might not be readily apparent but usually leads to long-term environmental degradation, loss of habitats, and possible collapse of ecosystems (Dutta et al., 2024; Häder & Barnes, 2019).

7.3 Socio-economic Consequences

The socio-economic consequences of groundwater contamination by landfill leachate are most acute in poor and peri-urban settings. Health conditions involve increased rates of chronic illnesses, including renal failure, gastrointestinal infections, and cancer, resulting in substantial out-of-pocket health costs and reduced productivity. In Indian cities such as Delhi and Hyderabad, people living close to large landfills such as Ghazipur and Jawahar Nagar experience repeated sickness, together with restricted healthcare access. Agriculture and fishery major livelihood in most developing areas migrate from the effects of polluted water and poor soil quality, curbing yield and marketability. Economic costs spill over to the real estate industry, where property close to polluted areas loses value, resulting in urban decay and marginalization. In most instances, impacted populations usually socially and economically vulnerable are resettled without rehabilitation. Protests, court cases, and policy logjams (e.g., Okhla, Deonar, Ghazipur) are routine, illustrating underlying concerns of environmental injustice and governance failure (Chandnani et al., 2022).

8 Remediation and Management Strategies

8.1 In Situ and Ex Situ Remediation Techniques

Landfill-leachate-contaminated groundwater remediation usually starts with choosing between in situ and ex situ methods, based on local conditions such as contaminant type, concentration, and geology. In situ remediation involves carrying out the remediation directly at the contaminated site without digging or excavation. One of the well-known in situ techniques is In Situ Chemical Oxidation (ISCO), where oxidizing chemicals like hydrogen peroxide, ozone, or permanganates are injected into the underground to chemically break down the contaminants such as VOCs and PAHs. On the other hand, ex situ remediation is done by physically removing contaminated media for treatment off-site. The most widespread ex situ approach is the pump-and-treat system, whereby contaminated groundwater is pumped to the ground surface and treated with technologies such as granular activated carbon filtration, air stripping, ion exchange, or reverse osmosis (Liao et al., 2022; Wei et al., 2022).

8.2 Natural Attenuation and Bioremediation

Natural attenuation is a passively executed remediation process by relying on the inherent occurrences of weather elements such as dilution, dispersion, sorption, volatilization, and bacterial biodegradation in order to reduce the number of contaminants in the environment in the long wave. It is cost-effective, non-invasive, but can only be used where there is low-risk since it requires a forceful monitoring to curb rebound or drift of pollutants. Bioremediation, in its turn, is an active remediation which encourages the activity of microorganisms to degrade harmful substances (Varjani & Upasani, 2019). Techniques provoke human-indigenous microorganisms or introduce nutrients or electron donors to particular microbial strains. As an example, petroleum hydrocarbons biodegrade under the action of aerobic bacteria whereas dechlorination occurs by anaerobic microorganisms of trichloroethylene (TCE) and perchloroethylene (PCE). The method as employed offers a green method and can be employed in situ as well as ex situ (Dutta et al., 2022; Joshi et al., 2025).

8.3 Use of Reactive Barriers and Pump-and-Treat Methods

The use of Permeable Reactive Barriers (PRBs) is a more lasting and environmental remediation alternative. These include systems installed underground that are stuffed with reactive media such as zero-valent iron (ZVI), activated carbon, compost or zeolite to intercept and treat the ground water moving systematically. PRBs can be used to treat nitrates, chlorinated solvents, and heavy metals and need little upkeep after installation. In a more complex scenario of contamination, pump and treat systems can also be used especially when it is necessary to have the contaminant removed almost immediately. They however have long lifecycle time and high cost of operation hence in most applications they are complemented by either PRBs or biological process to ensure efficiency and their lifecycle cost are reduced (Maitra, 2019).

8.4 Rehabilitation of Contaminated Sites

Once remediated the site, it is ensured that site reuse is carried out safely and returns the environment. The activities in the process include re-grading the ground, back-filling with clean soils, planting native vegetation and establishment of limits of land use where possible. In its turn, Monitored Natural Recovery (MNR) is generally transferred to the daily check of residual pollution and an assurance of the declining patterns of pollutants. To perform the ecological rehabilitation, techniques like the creation of wetlands, vegetative buffer strips or green belts at the periphery of the old landfill site can be used to enhance the presence of biodiversity and to stabilize the terrain. It is important to hold the awareness campaign in the community and involve the community at this stage to regain the user confidence in the community since mistrust was so central in the occupation of the land. A good rehabilitation plan does not only recover the quality of ground water but it also guarantees long term sustainability of the environment plus an economic boost to the society (Jorge, 2024).

9 Regulatory Framework and Policies

9.1 Global Guidelines (WHO, UNEP, USEPA)

Different organizations have developed elaborate environmental checklists and standards to prevent poor impact of landfills on the quality of ground water at international levels. The World Health Organization (WHO) has set universally accepted guidelines on the quality of drinking water, including the maximum allowable level of contaminants (including arsenic, lead, nitrates, pathogenic contaminants, and even emerging ones, like PFAS). These bulletins form a base of reference to the developed and developing nations in the formulation of the national drinking water policy. Some of international conventions developed by the United Nations Environment Programme (UNEP) include the Basel convention which governs trans boundary traffic of hazardous wastes and disposal of waste in order to prevent water and soil pollution. UNEP is also in the business of encouraging comprehensive waste management services and green landfill operations in its global environmental outlook documentation. Meanwhile, perhaps some of the most detailed land fill and groundwater preservation laws appear in the United States Environmental Protection Agency (USEPA) (Sanga et al., 2023). The Resource Conservation and Recovery Act (RCRA) also ensures that the USEPA implements the strict design and operation conditions of the municipal and hazardous waste landfills including the composite liner, leachate collection system, groundwater monitoring well, and closure/post-closure maintenance. The safe drinking water act (SDWA) also expects maximum contaminant level goals (MCLGs), drinking water sources at a particular level, necessary on establishing pollution level limits and pollution control planning. These global solutions do not only specify the technical standards of landfill safety and criteria of groundwater condition but also influence the policymakers of the surroundings in majority of the states, including India (Loveless & Skaggs, 2018).

9.2 Indian Regulations and CPCB Guidelines

The country of India has an extensive body of laws and policies in the Ministry of Environment, Forest and Climate Change (MoEFCC) and Central Pollution control Board (CPCB) to help contain waste and groundwater protection. Some of the salient laws include the Solid Waste Management Rules, 2016 that replaced its predecessor of 2000 and they cover segregation, transportation, processing, and safe disposal of solid waste (Das & Gargate, 2023). These guidelines need the scientific design of landfills to have protective lining, leachate treatment systems, and systems that monitor the groundwater. The generation, storage of the hazardous and industrial wastes and the disposal thereof are also regulated under Hazardous and Other Wastes (Management and Transboundary Movement) Rules, 2016, whereby the disposal facilities are not located near water bodies and the aquifer. CPCB has also come out with detailed "Guidelines for the Design and Construction of Landfill Facilities," stressing requirements for impermeable clay and artificial liners, gas venting systems, and ongoing environmental monitoring (Yang, 2020). The Environment (Protection) Act, 1986, is the parent legislation under which all pollution control laws function, including acceptable limits for pollutants in effluents and groundwater. For groundwater quality, India adopts IS 10500:2012 standards, which define acceptable and permissible values of different physicochemical and microbial parameters. Still, in spite of having these detailed laws and standards, there is a wide gap in the ground-level implementation, particularly in rural and peri-urban areas (Chand, 2018).

9.3 Role of Environmental Impact Assessment (EIA)

The Environmental Impact Assessment (EIA) is an important regulatory instrument for foretelling and mitigating the environmental impacts of landfill schemes prior to initiation. Under the EIA Notification, 2006, which is issued under the Environment (Protection) Act, any major waste treatment or disposal plant, including landfill facilities, is required to conduct an EIA study and secure prior environmental clearance. The EIA process involves baseline environmental surveys, impact forecasting, and development of mitigation strategies in areas ranging from groundwater, air, and soil to biodiversity and public health. An essential element in the process is the hydrogeological evaluation, which assesses groundwater flow direction, seasonal fluctuations, aquifer susceptibility, and the prevailing levels of pollution in the subject location for landfilling (Tanguilan & Peñaflor, 2024). The EIA report should also have a risk assessment, leachate management plan, and a Groundwater Monitoring Plan, all of which are scrutinized by the Expert Appraisal Committee (EAC) prior to clearance. As a part of the EIA process, there is obligatory public consultation which has to be done for the local people, giving them an opportunity to voice their problems considering water safety. However, India's system of EIA has been deeply criticized for such things as poor report quality, lack of transparency, limited public engagement courtesy of elitist capture, and pre-expedited approvals that go hand-in-hand with barely any examination factors which greatly harm its efficiency as a preventive mechanism (Sikdar, 2021).

9.4 Gaps and Challenges in Implementation

While there is a comprehensive legislative strategy to protect groundwater and landfills, this regulation is poorly implemented across India and other developing countries. One of the main difficulties stems from local enforcement within the region suffering from insufficient personnel, funding, or even technical skills. Most municipal authorities are unable to meet CPCB's technical requirements for landfill construction and maintenance which permits unscientific open dump yards to operate without leachate management systems or any form of groundwater monitoring and sampling. Another issue is that there is no defined regulatory framework in India that permits regular monitoring and sampling, especially concerning emerging contaminants such as pharmaceuticals, PFAS, and microplastics which pose great concern. Furthermore, corruption coupled with mismanagement tends to relocate legally defined landfill boundaries into environmentally sensitive zones due to political pressure that has no bounds or ceilings (Boro et al., 2025). Landfill institutional responsibility becomes fractured between municipal corporations, state pollution control boards, and urban development departments spiraling into layers of bureaucratic stagnation devoid of answerability. Active communities that do not possess adequate resources alongside lethargy precluding defiant action means public contribution and interest on environmental issues is minimal (Majid & Mir, 2021).

10 Sustainable Waste Management Approaches

10.1 Waste Minimization and Source Reduction

Source reduction and minimization of sources are integral in sustainable waste management which aims at confronting the problem at its origin. By minimizing the quantity of waste generated and its toxicity, the burden on landfills and thus on groundwater is significantly minimized. This can be achieved through measures such as eco-design, cleaner production technologies, minimalist packaging, and responsible consumption. Industries must adopt Extended Producer Responsibility (EPR), where the producer is held accountable for the entire life cycle of their product, including post-consumer disposal. Governments and municipalities also need to sponsor projects like green procurement policies and waste audits in public and private sectors. Proper adoption of these processes has been found to minimize the generation of hazardous waste by as much as 30–50% in some areas and demonstrates environmental as well as economic advantages (Mostaghimi & Behnamian, 2023; Paulraj et al., 2021).

10.2 Alternatives to Landfilling (Composting, Incineration, Recycling)

In order to prevent waste from ending up in landfills, a number of different treatment and disposal techniques are being implemented around the world. Composting is especially appropriate for organic municipal solid waste, turning food waste, garden clippings, and other biodegradable material into productive compost for soil amendment and agriculture (Karimi, 2023). Incineration with energy recovery, or waste-to-energy (WtE), can deal with non-recyclable and flammable waste, generating electricity or heat. Emission controls must, however, be stringent to counteract the risk of air pollution. Recycling is still the most commonly used way of processing plastics, paper, glass, and metals, cutting raw material extraction as well as landfill levels. Success with these alternatives depends on proper segregation at source and the development of an efficient collection and processing network. Scaled appropriately, these processes reduce landfill dependency and pollution and contribute to a circular economy (Vaverková et al., 2020a; Lima et al., 2018).

10.3 Circular Economy and Waste Valorization

The circular economy principle describes a systemic shift from the traditional linear "take-make-dispose" approach to a regenerative loop where the materials are maintained in constant reuse, repair, and transformation. Waste valorization, being one of the chief aspects of this principle, refers to converting waste into value-added products. Plastic waste, for instance, can be pyrolyzed to obtain fuel, whereas organic waste can be digested anaerobically to produce biogas and biofertilizers. These activities not only reduce landfill waste but also offer environmentally friendly alternatives for energy recovery, material conservation, and creation of green employment opportunities. Cities and industries, with sound policies and incentives for innovation, can adopt the principles of the circular economy to achieve optimal material flows and optimize waste environmental impacts (Ferronato et al., 2019; Kapoor et al., 2020).

10.4 Community Participation and Awareness Programs

Involvement of the community is a highly significant but often overlooked element in effective waste management (Kalra, 2019). It ensures effective use of policies and facilities, especially for activities like segregation of wastes, household composting, and monitoring of illegal dumping. School-level awareness campaigns, online campaigns, NGOs, and local bodies can collectively play a significant role towards behavioral changes towards more sustainable practices (Joshi et al., 2025) People's movements like zero-waste settlements, local composting dumps, and people's recycling centers have been quantifiably successful in urban areas like Pune, Bengaluru, and Ahmedabad. Not only does active participation enhance waste management outcomes but also develops a sense of ownership, transparency, and social justice in waste management. Engaging citizens through capacity building, incentives, and feedback has thus become imperative to develop more resilient, sustainable waste management systems (Debrah et al., 2021; Subri et al., 2025; Tarigan et al., 2020).

11 Future Perspectives and Research Needs

11.1 Advanced Monitoring and Early Warning Systems

The future of groundwater safeguard is in the deployment and establishment of real-time monitoring networks and early warning systems capable of detecting pollutants before they become critical. Breakthroughs like smart sensors, biosensors, and wireless sensor networks (WSNs) are transforming environmental monitoring through constant, automatic reporting of parameters such as pH, conductivity, nitrate, heavy metals, and microbial contaminants. Remote telemetry combined with GIS mapping and machine learning algorithms enables the quick detection of hotspots of pollution as well as the likely paths of leachate migration. These systems not only enhance the speed and accuracy of response but also increase the predictability of contamination events, enabling proactive management and mitigation (Pal et al., 2025).

11.2 Nanotechnology and Green Remediation Techniques

New nanotechnology-based solutions hold promising improvements in groundwater cleanup. Nanomaterials including nano zero-valent iron (nZVI), carbon nanotubes, and metal–organic frameworks (MOFs) have proved highly effective at adsorbing or degrading contaminants like chlorinated solvents, heavy metals, and dyes. These materials work at the molecular scale, enabling targeted, effective cleanup with negligible impact on the environment. In parallel with this, low-energy, low-impact green remediation techniques are becoming more popular. Techniques like phytoremediation, bioreactors, and natural wetlands are being enhanced to enable maximum contaminant breakdown while minimizing energy usage and preserving biodiversity. These strategies allow the onset of green-compatible cleanup methods that adhere to worldwide climatic and sustainability goals (Garg et al., 2024; Kuhn et al., 2022).

11.3 Policy and Technological Integration

One of the most critical near-term challenges in conserving groundwater is the mismatch between improving technology and policy practice. In the future, efforts must close the gap by creating evidence-based environmental policy that incorporates the latest scientific knowledge. Policies need to encourage the use of best available technologies (BATs) in landfill construction, monitoring, and clean-up, as well as the coordination across sectors of environmental, urban planning, and health agencies. Public–private partnerships (PPPs) can further be instrumental in funding and expanding emerging technologies. Harmonizing policy reforms with technological advancement will enhance compliance with regulations and propel the shift to sustainable waste and water management systems (Naworio, 2022).

11.4 Multidisciplinary Research Opportunities

Solving the multifaceted problem of groundwater contamination caused by landfills needs interdisciplinary research that includes environmental science, hydrology, toxicology, data science, public health, and policy. Future research needs to identify synergies between fields, including the use of AI and big data for prediction of contamination, genomics to understand microbial bioremediation pathways, and social science models to analyze community vulnerability and resilience. Longitudinal research on the long-term chronic health impacts of emerging pollutants such as PFAS and microplastics is also needed, particularly where little baseline data are available in specific areas. Cooperative national and international research programs will be key to developing coordinated, flexible solutions to protect groundwater assets for generations to come (Shuford, 2024).

12 Conclusion

Groundwater serves as a critical resource for drinking, agriculture, and ecosystem sustenance worldwide. However, improper disposal and landfilling of hazardous and toxic wastes, particularly in unlined or poorly engineered landfill sites, have become major contributors to groundwater contamination. Leachate infiltration introduces a broad spectrum of pollutants such as nitrates (130–150 mg/L), lead (0.097 mg/L), iron (~ 1 mg/L), and coliform bacteria (> 2×10^4 CFU/mL)—levels that far exceed WHO and BIS permissible limits for safe drinking water. This contamination is often chronic, cumulative, and difficult to detect, posing long-term risks to human and environmental health. Case studies across India and other countries consistently highlight this concern. For instance, groundwater near the Jawaharnagar landfill (Hyderabad) exhibits TDS levels between 1200 and 1800 mg/L, rendering the water unsuitable for both drinking and irrigation. Similarly, landfill sites in Ranchi report elevated manganese and lead concentrations exceeding acceptable thresholds. International investigations from China, the USA, and the UK corroborate these patterns, identifying persistent pollutants such as PFAS, pharmaceutical residues, and endocrine-disrupting chemicals (EDCs) that remain in aquifers for decades due to their bioaccumulative and non-biodegradable nature.

To address these escalating threats, policymakers must enforce stricter landfill regulations, promote scientifically engineered and climate-resilient landfill systems, and ensure periodic groundwater quality monitoring using GIS mapping, remote sensing, and leachate modeling tools. Simultaneously, regulatory frameworks should evolve to incorporate emerging contaminants and sustainability-driven waste management principles. On the scientific front, low-cost and eco-friendly remediation technologies—including bioremediation, phytoremediation, permeable reactive barriers, and nanotechnology-based adsorbents—show great potential in restoring contaminated groundwater. Furthermore, adopting circular economy approaches, such as waste minimization, source segregation, and valorization of waste into usable resources, can significantly reduce the burden on landfills.

Finally, achieving a clean groundwater is future demands an integrated and multidisciplinary effort combining scientific innovation, strong governance, and public participation. Empowering communities through education, transparency, and participatory monitoring will not only enhance groundwater protection but also foster trust and environmental resilience. Safeguarding groundwater is, therefore, not merely an environmental necessity—it is a social, economic, and ethical responsibility to ensure safe water security for current and future generations.

References

Abanyie, S. K., Apea, O. B., Abagale, S. A., Amuah, E. E. Y., & Sunkari, E. D. (2023). Sources and factors influencing groundwater quality and associated health implications: A review. *Emerging Contaminants, 9*, Article 100207.

Abdel-Shafy, H. I., Ibrahim, A. M., Al-Sulaiman, A. M., & Okasha, R. A. (2024). Landfill leachate: Sources, nature, organic composition, and treatment: An environmental overview. *Ain Shams Engineering Journal, 15*, Article 102293.

Abiriga, D., Vestgarden, L. S., & Klempe, H. (2020). Groundwater contamination from a municipal landfill: Effect of age, landfill closure, and season on groundwater chemistry. *Science of the Total Environment, 737*, Article 140307.

Agboola, O. E. (2024). *Assessment of physicochemical properties and heavy metals concentration of abattoir effluent on Oko Oba River in Agege*. Kwara State University (Nigeria).

Alao, J. (2024). *The factors influencing the landfill leachate plume contaminants in soils, surface and groundwater and associated health risks: A geophysical and geochemical view* (Vol. 1, pp. 20–43).

Al-Hashimi, O., Hashim, K., Loffill, E., Marolt Čebašek, T., Nakouti, I., Faisal, A. A., & Al-Ansari, N. (2021). A comprehensive review for groundwater contamination and remediation: Occurrence, migration and adsorption modelling. *Molecules, 26*, 5913.

Ali, W., Rasool, A., Junaid, M., & Zhang, H. (2019). A comprehensive review on current status, mechanism, and possible sources of arsenic contamination in groundwater: A global perspective with prominence of Pakistan scenario. *Environmental Geochemistry and Health, 41*, 737–760.

Amiaz, Y., Sorek, S., Enzel, Y., & Dahan, O. (2011). Solute-transport in the vadose zone and groundwater during flash floods. *Water Resources Research, 47*, W10513.

Amirfazli, M., Safarzadeh, S., & Samadi Khadem, R. (2019). Identification, classification and management of industrial hazardous waste in Ardabil province. *Anthropogenic Pollution, 3*, 29–36.

Anawar, H. M. (2013). Impact of climate change on acid mine drainage generation and contaminant transport in water ecosystems of semi-arid and arid mining areas. *Physics and Chemistry of the Earth, Parts a/b/c, 58*, 13–21.

Angon, P. B., Islam, M. S., Kc, S., Das, A., Anjum, N., Poudel, A., & Suchi, S. A. (2024). Sources, effects and present perspectives of heavy metals contamination: Soil, plants and human food chain. *Heliyon, 10*(7), Article e28357.

Anouzla, A., & Souabi, S. (2024). A review of landfill leachate: Characterization leachate environment impacts and sustainable treatment methods.

Aucott, M. (2006). *The fate of heavy metals in landfills: A review*. New Jersey Department of Environmental Protection Trenton.

Avery, L. M., Booth, P., Campbell, C., Tompkins, D., & Hough, R. L. (2012). Prevalence and survival of potential pathogens in source-segregated green waste compost. *The Science of the Total Environment, 431*, 128–138.

Banerjee, D., & Ganguly, S. (2023). A review on the research advances in groundwater–surface water interaction with an overview of the phenomenon. *Water, 15*, 1552.

Barroso, P. J., Santos, J. L., Martín, J., Aparicio, I., & Alonso, E. (2019). Emerging contaminants in the atmosphere: Analysis, occurrence and future challenges. *Critical Reviews in Environmental Science and Technology, 49*, 104–171.

Bera, A., Mukhopadhyay, B. P., Chowdhury, P., Ghosh, A., & Biswas, S. (2021). Groundwater vulnerability assessment using GIS-based DRASTIC model in Nangasai River Basin, India with special emphasis on agricultural contamination. *Ecotoxicology and Environmental Safety, 214*, Article 112085.

Biglari, H., Mousavi, S. M., Yaghmaeian, K., Dehghani, M. H., Nodehi, R. N., Nazmara, S., & Alimohammadi, M. (2025). Comprehensive health and environmental risk assessment of Zanjan lead-zinc hydrometallurgy tailings. *Results in Engineering, 26*, Article 105160.

Boro, D., Chirania, M., Verma, A. K., Chettri, D., & Verma, A. K. (2025). Comprehensive approaches to managing emerging contaminants in wastewater: Identification, sources, monitoring and remediation. *Environmental Monitoring and Assessment, 197*, 1–29.

Borowiecka, M., Alcaraz, M., & Manzano, M. (2024). Assessment of Ecosystem Services with numerical modelling to support groundwater dependent ecosystems and aquifer management: A demo study in the Medina del Campo Groundwater Body, Spain. *Case Studies in Chemical and Environmental Engineering, 10*, Article 100914.

Bugarski, T. D., Tubic, B. N., & Pisaric, M. M. (2021). Legal framework for municipal waste landfills in the European Union. *Zbornik Radova, 55*, 415.

Butti, L., Peres, F., & Lops, C. (2018). Legal framework of landfilling in different areas of the world. *Solid Waste Landfilling: Concepts, Processes, Technology, 41.*

Chakraborty, S., & Kumar, R. N. (2016). Assessment of groundwater quality at a MSW landfill site using standard and AHP based water quality index: A case study from Ranchi, Jharkhand, India. *Environmental Monitoring and Assessment, 188*, 335.

Chand, P. (2018). Environmental protection and regulations in India: Role of the central pollution control board. *Indian Journal of Public Administration, 64*, 645–663.

Chandnani, G., Gandhi, P., Kanpariya, D., Parikh, D., & Shah, M. (2022). A comprehensive analysis of contaminated groundwater: Special emphasis on nature-ecosystem and socio-economic impacts. *Groundwater for Sustainable Development, 19*, Article 100813.

Chaudhary, I. J., Chauhan, R., Kale, S. S., Gosavi, S., Rathore, D., Dwivedi, V., Singh, S., & Yadav, V. K. (2025). Groundwater nitrate contamination and its effect on human health: A review. *Water Conservation Science and Engineering, 10*, 33.

Chen, G., Sun, Y., Xu, Z., Shan, X., & Chen, Z. (2019). Assessment of shallow groundwater contamination resulting from a municipal solid waste landfill—A case study in Lianyungang, China. *Water, 11*, 2496.

Chima, E. N., Essandoh, H. M. K., Edziyie, R. E., Yaya, O. O., & Micheal, N. O. (2025). Removal of pharmaceutical contaminants from wastewater using novel ceramic nanomembrane filters. In *Advances in water security* (pp. 271–295).

Choudhury, A., Bolem, K. R., Duvva, L. K., Singh, N., Moota, P. K., Anagani, L., Boyina, L., Begum, A., & Umar, S. (2021). Techno-commercial assessment of concurrent municipal brown field reclamation procedures: A pivotal case study of Jawahar Nagar Dump Site. *Letters in Applied NanoBioScience, 11*(1), 3142–3154.

Clemente, E., Domingues, E., Quinta-Ferreira, R., Leitão, A., & Martins, R. (2024). European and African landfilling practices: An overview on MSW management, leachate characterization and treatment technologies. *Journal of Water Process Engineering, 66*, Article 105931.

Cossu, R., & Stegmann, R. (2018). *Solid waste landfilling: Concepts, processes, technology*. Elsevier.

Dagwar, P. P., & Dutta, D. (2024). Landfill leachate a potential challenge towards sustainable environmental management. *Science of the Total Environment, 926*, Article 171668.

Danthurebandara, M., Passel, S., Nelen, D., Tielemans, Y., & Van Acker, K. (2013). Environmental and socio-economic impacts of landfills.

Das, P., & Gargate, G. (2023). Legal jurisprudence of solid waste management in India: Development through the decades. *International Journal of Global Environmental Issues, 22*, 268–294.

Debrah, J. K., Vidal, D. G., & Dinis, M. A. P. (2021). Raising awareness on solid waste management through formal education for sustainability: A developing countries evidence review. *Recycling, 6*, 6.

Dube, A. D., & Kanido, C. K. (2025). Physicochemical qualities and health risk assessment of heavy metals in drinking water from selected areas of the GOFA zone, Ethiopia. *Journal of Chemistry, 2025*(1).

Dutta, N., Usman, M., Ashraf, M. A., Luo, G., & Zhang, S. (2022). A critical review of recent advances in the bio-remediation of chlorinated substances by microbial dechlorinators. *Chemical Engineering Journal Advances, 12*, Article 100359.

Dutta, S., Adhikary, S., Bhattacharya, S., Roy, D., Chatterjee, S., Chakraborty, A., Banerjee, D., Ganguly, A., Nanda, S., & Rajak, P. (2024). Contamination of textile dyes in aquatic environment: Adverse impacts on aquatic ecosystem and human health, and its management using bioremediation. *Journal of Environmental Management, 353*, Article 120103.

Dwanga, D. M., Abubakar, M. Y., Adam, A. B., Muhammad, S. A., Obasi, E. L. Heavy metal leaching from landfills: A review of groundwater contamination and long term environmental impacts.

Erdanaev, E. (2024). Classification of main irrigated crop types towards sustainable development using Landsat and Sentinel data by GIS and remote sensing techniques in a semi-arid area in Tashkent Province, Uzbekistan.

Ergene, D., Aksoy, A., & Sanin, F. D. (2022). Comprehensive analysis and modeling of landfill leachate. *Waste Management, 145*, 48–59.

Ferronato, N., Rada, E. C., Portillo, M. A. G., Cioca, L. I., Ragazzi, M., & Torretta, V. (2019). Introduction of the circular economy within developing regions: A comparative analysis of advantages and opportunities for waste valorization. *Journal of Environmental Management, 230*, 366–378.

Garg, M. C., Kumari, S., & Malik, N. (2024). Role of nanomaterials in advanced membrane technologies for groundwater purification. *Environmental Science: Water Research & Technology.*

Gautam, P., & Kumar, S. (2021). Characterisation of hazardous waste landfill leachate and its reliance on landfill age and seasonal variation: A statistical approach. *Journal of Environmental Chemical Engineering, 9*, Article 105496.

Ghosh, A., Manna, M. C., Jha, S., Singh, A. K., Misra, S., Srivastava, R. C., Srivastava, P. P., Laik, R., Bhattacharyya, R., Prasad, S. S., Singh, S. P., Singh, S. K., Kumar, V., Tiwari, S., & Singh, A. K. (2022). Chapter Six—Impact of soil-water contaminants on tropical agriculture, animal and societal environment. In: D. L. Sparks (Ed.), *Advances in agronomy*. Academic.

Giakoumatos, S. D., Siontorou, C., & Sidiras, D. (2024). An extensive review of leaching models for the forecasting and integrated management of surface and groundwater quality. *Water, 16*, 3348.

Gunarathne, V., Phillips, A. J., Zanoletti, A., Rajapaksha, A. U., Vithanage, M., Di Maria, F., Pivato, A., Korzeniewska, E., & Bontempi, E. (2024). Environmental pitfalls and associated human health risks and ecological impacts from landfill leachate contaminants: Current evidence, recommended interventions and future directions. *Science of the Total Environment, 912*, Article 169026.

Häder, D.-P., & Barnes, P. W. (2019). Comparing the impacts of climate change on the responses and linkages between terrestrial and aquatic ecosystems. *Science of the Total Environment, 682*, 239–246.

Hui, C., Yu, Q., Liu, B., Zhu, M., Long, Y., & Shen, D. (2023). Microbial contamination risk of landfilled waste with different ages. *Waste Management, 170*, 297–307.

Idowu, I. A., Atherton, W., Hashim, K., Kot, P., Alkhaddar, R., Alo, B. I., & Shaw, A. (2019). An analyses of the status of landfill classification systems in developing countries: Sub Saharan Africa landfill experiences. *Waste Management, 87*, 761–771.

Indelicato, S., Orecchio, S., Avellone, G., Bellomo, S., Ceraulo, L., Di Leonardo, R., Di Stefano, V., Favara, R., Candela, E. G., & La Pica, L. (2017). Effect of solid waste landfill organic pollutants on groundwater in three areas of Sicily (Italy) characterized by different vulnerability. *Environmental Science and Pollution Research, 24*, 16869–16882.

Iravanian, A., & Ravari, S. O. Types of contamination in landfills and effects on the environment: A review study. *IOP conference series: Earth and environmental science* (p. 012083). IOP Publishing.

Jiang, Y., Li, R., Yang, Y., Yu, M., Xi, B., Li, M., Xu, Z., Gao, S., & Yang, C. (2019). Migration and evolution of dissolved organic matter in landfill leachate-contaminated groundwater plume. *Resources, Conservation and Recycling, 151*, Article 104463.

Jorge, C. (2024). *Investigation, management and rehabilitation of contaminated sites.* Thomas Telford Ltd.

Joshi, D. C., Prasad, S., Sharma, A., & Joshi, N. (2025). Pathological contaminants in wastewater from hospital waste: Implications and mitigation strategies. In *Environmental science and engineering* (pp. 43–85). https://doi.org/10.1007/978-3-031-82579-8_3

Kalra, N. (2019). Community participation and waste management. In *Sustainable Waste Management: Policies and Case Studies: 7th IconSWM—ISWMAW 2017* (Vol. 1). Springer.

Kanwal, Q., Zeng, X., & Li, J. (2022). Drivers-pressures-state-impact-response framework of hazardous waste management in China. *Critical Reviews in Environmental Science and Technology, 52*, 2930–2961.

Kapoor, R., Ghosh, P., Kumar, M., Sengupta, S., Gupta, A., Kumar, S. S., Vijay, V., Kumar, V., Vijay, V. K., & Pant, D. (2020). Valorization of agricultural waste for biogas based circular economy in India: A research outlook. *Bioresource Technology, 304*, Article 123036.

Karimi, N. (2023). Assessing global waste management: Alternatives to landfilling in different waste streams—A scoping review. *Sustainability, 15*, 13290.

Khondoker, M., Mandal, S., Gurav, R., & Hwang, S. (2023). Freshwater shortage, salinity increase, and global food production: A need for sustainable irrigation water desalination—A scoping review. *Earth, 4*, 223–240.

Kuhn, R., Bryant, I. M., Jensch, R., & Böllmann, J. (2022). Applications of environmental nanotechnologies in remediation, wastewater treatment, drinking water treatment, and agriculture. *Applied Nano, 3*, 54–90.

Land, E., & Peters, C. N. (2023). Groundwater impacts on stream biodiversity and communities: A review. *Journal of Freshwater Ecology, 38*, 2260801.

Li, P., Karunanidhi, D., Subramani, T., & Srinivasamoorthy, K. (2021). Sources and consequences of groundwater contamination. *Archives of Environmental Contamination and Toxicology, 80*, 1–10.

Liao, X., Li, Y., Miranda-Avilés, R., Zha, X., Anguiano, J. H. H., Sánchez, C. D. M., Puy-Alquiza, M. J., González, V. P., & Garzon, L. F. R. (2022). In situ remediation and ex situ treatment practices of arsenic-contaminated soil: An overview on recent advances. *Journal of Hazardous Materials Advances, 8*, Article 100157.

Lima, P. D. M., Colvero, D. A., Gomes, A. P., Wenzel, H., Schalch, V., & Cimpan, C. (2018). Environmental assessment of existing and alternative options for management of municipal solid waste in Brazil. *Waste Management, 78*, 857–870.

Liu, X., Sathishkumar, K., Zhang, H., Saxena, K. K., Zhang, F., Naraginti, S., Anbarasu, K., Ramya, R., Rajasekar, A., & Guo, X. (2024). Frontiers in environmental cleanup: Recent advances in remediation of emerging pollutants from soil and water. *Journal of Hazardous Materials Advances, 16*, Article 100461.

Liu, X., & Wang, Y. (2022). Identification and assessment of groundwater and soil contamination from an informal landfill site. *Sustainability, 14*, 16948.

Loveless, S., & Skaggs, B. (2018). Resource Conservation and Recovery Act General Contingency Plan for Hazardous Waste Treatment, Storage, and Disposal Units at the Y-12 National Security Complex. Revision 12. Oak Ridge Y-12 Plant (Y-12), Oak Ridge, TN (United States). Y-12 Environment….

Ma, S., Zhou, C., Pan, J., Yang, G., Sun, C., Liu, Y., Chen, X., & Zhao, Z. (2022). Leachate from municipal solid waste landfills in a global perspective: Characteristics, influential factors and environmental risks. *Journal of Cleaner Production, 333*, Article 130234.

Maitra, S. (2019). Permeable reactive barrier: A technology for groundwater remediation—A mini review. *Biodegradation, 80*, 203.
Majid, M., & Mir, B. A. (2021). Landfill site selection using GIS based multi criteria evaluation technique. A case study of Srinagar city, India. *Environmental Challenges, 3*, 100031.
Makhadi, R., Oke, S. A., & Ololade, O. O. (2020). The influence of non-engineered municipal landfills on groundwater chemistry and quality in Bloemfontein, South Africa. *Molecules, 25*, 5599.
Markus, U. I., Ilori, O. P., Wada, I. M., Musa, S. T., & Peter, J. E. (2022). Hydrogeochemical evaluation and geospatial distribution modeling of the major ion chemistry of groundwater and their suitability for drinking and irrigation in Lagelu, Southwestern Nigeria. *Applied Water Science, 12*, 160.
Masoner, J. R., Kolpin, D. W., Cozzarelli, I. M., Smalling, K. L., Bolyard, S. C., Field, J. A., Furlong, E. T., Gray, J. L., Lozinski, D., & Reinhart, D. (2020). Landfill leachate contributes per-/poly-fluoroalkyl substances (PFAS) and pharmaceuticals to municipal wastewater. *Environmental Science: Water Research & Technology, 6*, 1300–1311.
Mishra, S., Tiwary, D., Ohri, A., & Agnihotri, A. K. (2019). Impact of Municipal Solid Waste Landfill leachate on groundwater quality in Varanasi, India. *Groundwater for Sustainable Development, 9*, Article 100230.
Mohammadi-Behzad, H., Charchi, A., Kalantari, N., Nejad, A. M., & Vardanjani, H. K. (2019). Delineation of groundwater potential zones using remote sensing (RS), geographical information system (GIS) and analytic hierarchy process (AHP) techniques: A case study in the Leylia-Keynow watershed, southwest of Iran. *Carbonates and Evaporites, 34*, 1307–1319.
Mor, S., & Ravindra, K. (2023). Municipal solid waste landfills in lower-and middle-income countries: Environmental impacts, challenges and sustainable management practices. *Process Safety and Environmental Protection, 174*, 510–530.
Mostaghimi, K., & Behnamian, J. (2023). Waste minimization towards waste management and cleaner production strategies: A literature review. *Environment, Development and Sustainability, 25*, 12119–12166.
Nanda, S., & Berruti, F. (2021). Municipal solid waste management and landfilling technologies: A review. *Environmental Chemistry Letters, 19*, 1433–1456.
Naworio, N. B. (2022). Waste-to-energy project—An integrated sustainable waste management in Yenagoa Metropolis–Numondogiyo.
Nguyen, G. T., & Huynh, N. T. H. (2023). Seasonal variations in groundwater quality under different impacts using statistical approaches. *Civil Engineering Journal, 9*, 497–511.
Nishmitha, P. S., Akhilghosh, K. A., Aiswriya, V. P., Ramesh, A., Muthuchamy, M., & Muthukumar, A. (2025a). Understanding emerging contaminants in water and wastewater: A comprehensive review on detection, impacts, and solutions. *Journal of Hazardous Materials Advances*, 100755.
Nishmitha, P. S., Akhilghosh, K. A., Aiswriya, V. P., Ramesh, A., Muthuchamy, M., & Muthukumar, A. (2025b). Understanding emerging contaminants in water and wastewater: A comprehensive review on detection, impacts, and solutions. *Journal of Hazardous Materials Advances, 18*, Article 100755.
Noyes, P. D., Mcelwee, M. K., Miller, H. D., Clark, B. W., Van Tiem, L. A., Walcott, K. C., Erwin, K. N., & Levin, E. D. (2009). The toxicology of climate change: Environmental contaminants in a warming world. *Environment International, 35*, 971–986.
Obaideen, K., Shehata, N., Sayed, E. T., Abdelkareem, M. A., Mahmoud, M. S., & Olabi, A. G. (2022). The role of wastewater treatment in achieving sustainable development goals (SDGs) and sustainability guideline. *Energy Nexus, 7*, Article 100112.
Ochs, C., Garrison, K., Saxena, P., Romme, K., & Sarkar, A. (2024). Contamination of aquatic ecosystems by persistent organic pollutants (POPs) originating from landfills in Canada and the United States: A rapid scoping review. *Science of the Total Environment*, 171490.
Odonkor, S. T., & Mahami, T. (2020). Microbial air quality in neighborhoods near landfill sites: Implications for public health. *Journal of Environmental and Public Health, 2020*, 4609164.

Ogunwumi, O. T., & Salami, L. (2023). Perspective chapter: Industrial waste landfills. *Solid waste and landfills management-recent advances*. IntechOpen.

Ozymy, J., & Ozymy, M. J. (2023). Beyond midnight dumping: Exploring the prosecution of companies for environmental crimes under the US Resource Conservation and Recovery Act (RCRA). *Journal of Law, Business & Ethics, 29*, 57.

Pal, P., Pandey, B., Prakash, A., Sarangi, P. K., Rathore, S. S., Agarwal, V., Gupta, A. K., & Singh, A. K. (2025). *Exploration of biosensors in innovative applications and sustainable solutions for environmental monitoring and management*. Springer.

Pandey, V., Singh, K., Ahmad, N., & Srivastava, S. (2022). Challenges and issues of groundwater management in India. *Current Science, 123*, 856–864.

Parvin, F., & Tareq, S. M. (2021). Impact of landfill leachate contamination on surface and groundwater of Bangladesh: A systematic review and possible public health risks assessment. *Applied Water Science, 11*, 100.

Paulraj, M. S., Nuzhat, S., & Hussain, C. M. (2021). *Source reduction and waste minimization*. Elsevier.

Priyadarshanee, M., Mahto, U., & Das, S. (2022). *Mechanism of toxicity and adverse health effects of environmental pollutants*. Elsevier.

Rahman, M. M., Sultan, M. B., & Alam, M. (2023). Microplastics and adsorbed micropollutants as emerging contaminants in landfill: A mini review. *Current Opinion in Environmental Science & Health, 31*, Article 100420.

Rahman, Z., & Singh, V. P. (2019). The relative impact of toxic heavy metals (THMs) (arsenic (As), cadmium (Cd), chromium (Cr)(VI), mercury (Hg), and lead (Pb)) on the total environment: An overview. *Environmental Monitoring and Assessment, 191*, 1–21.

Rajan, M., Karunanidhi, D., Jaya, J., Preethi, B., Subramani, T., & Aravinthasamy, P. (2024). A comprehensive review of human health hazards exposure due to groundwater contamination: A global perspective. *Physics and Chemistry of the Earth, Parts A/B/C*, 103637.

Ram, C., & Kumar, A. (2021). 2 Municipal solid waste. *Nanobiotechnology for Green Environment, 37*.

Ratna, M. V., Kumar, G. V., & Dileep, G. (2021). The effects of leachate from municipal solid waste landfill dump sites on ground water contamination. *Levant Journal, 20*, 194–207.

Roy, D., Azaïs, A., Benkaraache, S., Drogui, P., & Tyagi, R. D. (2018). Composting leachate: Characterization, treatment, and future perspectives. *Reviews in Environmental Science and Bio/technology, 17*, 323–349.

Ruan, T., Field, J., Cousins, I., Lohmann, R., & Jiang, G. (2022). *Emerging contaminants: Fluorinated alternatives to existing PFAS*. ACS Publications.

Saito, L., Christian, B., Diffley, J., Richter, H., Rohde, M. M., & Morrison, S. A. (2021). Managing groundwater to ensure ecosystem function. *Groundwater, 59*, 322–333.

Sanga, V. F., Fabian, C., & Kimbokota, F. (2023). Heavy metal pollution in leachates and its impacts on the quality of groundwater resources around Iringa municipal solid waste dumpsite. *Environmental Science and Pollution Research, 30*, 8110–8122.

Santos, M., Rebola, S., & Evtuguin, D. V. (2025). Soil remediation: Current approaches and emerging bio-based trends. *Soil Systems, 9*(2), 35.

Scanlon, B., Fakhreddine, S., Rateb, A., De Graaf, I., Famiglietti, J., Gleeson, T., Grafton, R., Jobbágy, E., Kebede, S., Kolusu, S., Konikow, L., Long, D., Mekonnen, M., Müller Schmied, H., Mukherjee, A., Macdonald, A., Reedy, R., Shamsudduha, M., Simmons, C., & Zheng, C. (2023). Global water resources and the role of groundwater in a resilient water future. *Nature Reviews Earth & Environment, 4*.

Scarlat, N., Fahl, F., & Dallemand, J.-F. (2019). Status and opportunities for energy recovery from municipal solid waste in Europe. *Waste and Biomass Valorization, 10*, 2425–2444.

Shaikh, M., & Birajdar, F. (2024). Groundwater and ecosystems: Understanding the critical interplay for sustainability and conservation. *EPRA International Journal of Multidisciplinary Research, 10*, 181–186.

Shakoor, M. B., Nawaz, R., Hussain, F., Raza, M., Ali, S., Rizwan, M., Oh, S.-E., & Ahmad, S. (2017). Human health implications, risk assessment and remediation of As-contaminated water: A critical review. *Science of the Total Environment, 601*, 756–769.

Sharma, P. K., Mayank, M., Ojha, C., & Shukla, S. (2020). A review on groundwater contaminant transport and remediation. *ISH Journal of Hydraulic Engineering, 26*, 112–121.

Shuford, J. (2024). Interdisciplinary perspectives: Fusing artificial intelligence with environmental science for sustainable solutions. *Journal of Artificial Intelligence General science (JAIGS), 1*, 106–123. ISSN: 3006-4023.

Siddiqua, A., Hahladakis, J. N., & Al-Attiya, W. (2022a). An overview of the environmental pollution and health effects associated with waste landfilling and open dumping. *Environmental Science and Pollution Research International, 29*, 58514–58536.

Siddiqua, A., Hahladakis, J. N., & Al-Attiya, W. A. K. (2022b). An overview of the environmental pollution and health effects associated with waste landfilling and open dumping. *Environmental Science and Pollution Research, 29*, 58514–58536.

Sikdar, P. K. (2021). Environmental impact assessment. *Environmental management: issues and concerns in developing countries* (pp. 253–283).

Singh, G., & Yadav, P. (2022). *Hazardous waste characteristics and standard management approaches*. Elsevier.

Sivaram, N., Gopal, P., & Barik, D. (2019). *Toxic waste from textile industries*. Elsevier.

Subri, U. S., Ghani, N. M., Rus, R. C., & Affandi, H. M. (2025). Waste no more: Empowering communities through education and participation in sustainable waste management. *Multidisciplinary Reviews, 8*, 2025204–2025204.

Tanguilan, A. T., & Peñaflor, B. L. (2024). Implementation of the Philippine environmental impact statement system for sanitary landfill facilities in Cagayan Valley, Philippines. *Journal of Environment, Climate, and Ecology, 1*, 46–56.

Tarigan, L. B., Rogaleli, Y. C., & Waangsir, F. W. (2020). Community participation in waste management. *International Journal of Public Health Science (IJPHS), 9*, 115–120.

Tom, C. N., & Adama, J. (2023). Surface water and groundwater in a changing environment. *International Journal of Technology & Engineering Studies, 9.*

Varjani, S., & Upasani, V. N. (2019). Influence of abiotic factors, natural attenuation, bioaugmentation and nutrient supplementation on bioremediation of petroleum crude contaminated agricultural soil. *Journal of Environmental Management, 245*, 358–366.

Varshney, R., Singh, P., & Yadav, D. (2022). *Hazardous wastes treatment, storage, and disposal facilities*. Elsevier.

Vaverková, M. D., Adamcová, D., Winkler, J., Koda, E., petrželová, L., & Maxianová, A. (2020a). Alternative method of composting on a reclaimed municipal waste landfill in accordance with the circular economy: Benefits and risks. *Science of the Total Environment, 723*, 137971.

Vaverková, M. D. (2019). Landfill impacts on the environment. *Geosciences, 9*, 431.

Vaverková, M. D., Elbl, J., Koda, E., Adamcová, D., Bilgin, A., Lukas, V., Podlasek, A., Kintl, A., Wdowska, M., & Brtnický, M. (2020b). Chemical composition and hazardous effects of leachate from the active municipal solid waste landfill surrounded by farmlands. *Sustainability, 12*, 4531.

Wei, K.-H., Ma, J., Xi, B.-D., Yu, M.-D., Cui, J., Chen, B.-L., Li, Y., Gu, Q.-B., & He, X.-S. (2022). Recent progress on in-situ chemical oxidation for the remediation of petroleum contaminated soil and groundwater. *Journal of Hazardous Materials, 432*, Article 128738.

Yang, S. (2020). Trade for the environment: Transboundary hazardous waste movements after the Basel Convention. *Review of Policy Research, 37*, 713–738.

Zhang, P., Chai, J., Cao, J., Qin, Y., Dang, M., Geng, K., & Wei, Y. (2023). Landfill leachate generation mechanism study: A review. *International Journal of Environmental Science and Technology, 20*, 9271–9290.

Nanomaterials in Solid Waste: Preparation and Environmental Applications

Aniket Walia, Samruddhi Mane, Naman Dosi, Mitali Mahajan, and Nidhi Gupta

1 Introduction

Solid waste is one of the most serious global problems that cost the most significant negative impact sustainable development goals. Construction and demolition waste generates for over a half of all solid waste collected globally each year (Daoud et al., 2020). The large amount of waste generated necessitates the strict implementation for waste collection, transportation, and disposal system. Understanding the garbage composition is crucial and aids the quick waste collection and disposal. With the rising population, the disposal of solid waste like of papers, plastic waste, and metals has increased, thereby making their disposal highly challenging. The local authorities have been powerless to collect and dispose of the massive amounts of solid waste discarded. The waste management authorities dump the large amounts of collected solid waste either in the landfilling or composting site or they perform open burning. However, these approaches are inefficient and harmful to the environment. In the present scenario, the waste management approach should shift from waste disposal to waste reuse, recycle, and recovery (Narayana, 2009). The various types of solid waste generated including their respective source are shown in Fig. 1.

Resource recovery is one of the methods to overcome from waste management problem. The components in municipal solid waste (MSW) includes papers, cardboards, glass, plastic, aluminum, ferrous, and non-ferrous metal components. Traditionally these materials are separated from MSW before they recycled. Traditional methods of recycling include material recycling, manual and mechanical separation, bioconversion of organic matter into industrial products like ethanol, sugar, and protein compost (Drobny et al., 1971; Tchobanoglous et al., 1993; Vesilind & Rimer, 1981; Wilson, 2023), incineration to reduce waste volume and energy recovery (Singh

A. Walia · S. Mane · N. Dosi · M. Mahajan · N. Gupta (✉)
TERI School of Advanced Studies, 10 Institutional Area, Vasant Kunj, New Delhi, India
e-mail: nidhi.gupta2@terisas.ac.in; nidhimic@gmail.com

H. Gupta et al. (eds.), *Advanced Technologies for Assessment, Recycling, and Management of Solid Waste*, Waste as a Resource,
https://doi.org/10.1007/978-3-032-24788-9_5

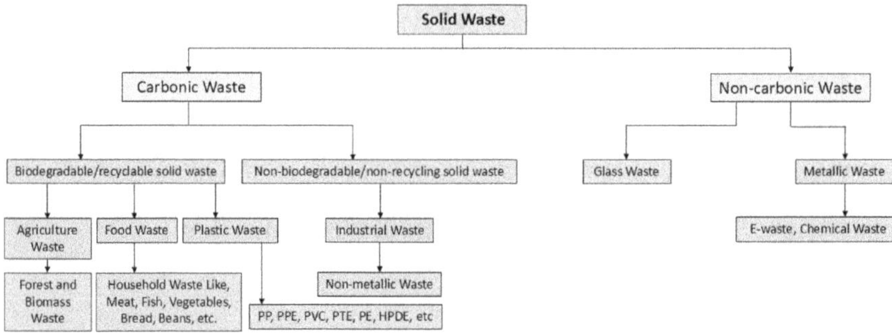

Fig. 1 Sources of various types of carbonic and non-carbonic solid waste generated by various human activities (Tatrari et al., 2021). The figure is adapted and re-created from original open access article by (Tatrari et al., 2021) under Creative Commons License CC BY https://creativecommons.org/licenses/by/3.0/

et al., 2014). Without proper treatment, industrial and municipal waste can be detrimental for both human and environment. It is of crucial importance that the 3R rules (Reduce, Reuse, and Recovery) must be adopted (Judith and Vasudevan, 2021).

Numerous research has been conducted to transform solid waste into important products like catalysts (Chen et al., 2017), pigments (Du et al., 2017; Gao et al., 2017; Liu et al., 2016), ceramics (Liao et al., 2017), etc. From the past few years, the synthesis of nanomaterials from solid waste and their applications in pollutant removal have generated immense attention because of their large surface area: volume ratio and highly active nature. The use of waste derived nanomaterials for commercial and industrial use resolves waste disposal problems and manage the resource exploitation.

This chapter emphasizes on introduction to nanomaterials, their types on the basis of their origin, dimensions, and composition. This is followed by comprehensive synthesis strategies and synthesis of nanomaterials from various sources of solid waste like plastics, batteries, e-waste, etc. Lastly, this chapter describes the recovery strategies of nanomaterials derived from solid waste and their potential applications for commercial use and environment remediation.

2 Nanomaterials and Their Classes

The systematic scientific study of nanostructures is relatively new, despite the fact that humans have incidentally used nanoscale effects, for example, the colors of medieval stained glass resulting from embedded metal nanomaterials. In 1959, Richard Feynman imagined atom-by-atom manipulation in his lecture There's Plenty of Room at the Bottom, which effectively founded modern nanotechnology (Feynman, 2018). According to the IUPAC convention, a nanomaterial has structural characteristics that range from 1 to 100 nm in length, such as exterior dimensions, interior pores, or surface structures (Alagarasi, 2013). At these scales, the proportion of atoms on the surface to those inside rises sharply, resulting in high surface energies that change the way chemicals react and interact with their surroundings. Targeted functionality is made possible by the controlled synthesis of engineered nanomaterials (ENMs), which have exact regulation over shape, composition, size, and surface chemistry. The discovery of C_{60} fullerenes/bucky balls (Kroto et al., 1985), the characterization of carbon nanotubes (CNTs) (Iijima, 1991), and the successful synthesis of graphene nanosheets (Novoselov et al., 2004) revealed a completely new allotropes of carbon at the nanoscale. These turning points sparked a rapid cross-disciplinary expansion that produced a wide range of nanomaterial classes and applications.

Nanomaterials are categorized in three major categories which is based on (i) its origin, (ii) the number of dimensions, and (iii) structural composition (Fig. 2). In case of origin, they are sub-categorized between the (i) natural nanomaterials and (ii) artificial nanomaterials. Natural nanomaterials are commonly found in nature in the form of viruses, protein molecules, ash, corals, bones, spider silk, and colloids (milk and blood), while artificial nanomaterials are materials synthesized by innovative mechanical and manufacturing processes like quantum dots, dendrimers, and nanocomposites (Cho et al., 2019; Khan & Hossain, 2022). On the other hand, the nanomaterials are often classified according to the dimensions confined to the nanoscale which are sub-categorized between (a) 0D nanomaterials, (b) 1D nanomaterials, (c) and 2D nanomaterials. 0D nanomaterials are category of nanomaterials which are restricted to all three dimensions and has zero degree of freedom for the free electron movement. The size is confined to < 100 nm in all the three dimensions of the Cartesian coordinates (Tang et al., 2020). This category includes quantum dots, carbon dots, nanospheres, etc. (Erfaninia et al., 2018; Veisi et al., 2021). 1D nanomaterials are characterized by confinement of 2 dimensions of a material < 100 nm range and one degree of freedom for the free electron movement (Seth & Rathinasabapathi, 2022). This category includes nanotubes, nanowires and, nanorods (Maleki et al., 2020). 2D nanomaterials have the restriction of one dimension < 100 nm and two degree of freedoms for the free electrons. Nanosheets (Ahmad et al., 2017) and nanoplatelets (Sharma et al., 2020) are the common examples of 2D nanomaterials. Lastly, the category of structural composition includes (i) carbon-based nanomaterials like fullerenes (Kroto et al., 1985), graphene (Luo et al., 2012; Nair et al., 2008; Novoselov et al., 2012; Rao et al., 2009), (ii) inorganic nanomaterials like copper (Chen et al., 2008), gold (Abd El-Aziz et al., 2007; Ahmad et al., 2003), silver (Anil Kumar et al., 2007; Durán et al., 2005), titanium dioxide (Bansal et al., 2005), silica dioxide (Bansal et al., 2005), molybdenum disulfide (Ahmad et al., 2017), etc., and (iii) various nanocomposites like inorganic nanocomposites of carbon nanotubes, gold, platinum, zinc oxide, zirconium oxide, cerium oxide, organic nanocomposites of polyaniline, polypyrrole, chitosan, and organic-inorganic hybrid nanocomposites like self-assembled monolayer of 8-amino polythiophene (Malhotra & Ali, 2018).

The above-mentioned categories of nanomaterials are highly dependent upon their size, structure, crystallinity, surface chemistry, etc. All of these properties together governed by their manufacturing and processing methods. Hence, it becomes very

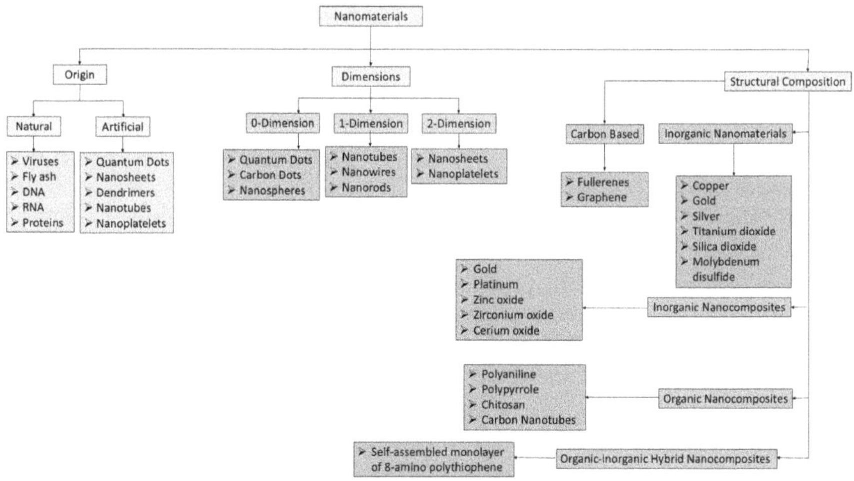

Fig. 2 Categorization of nanomaterials on the basis of their origin, number of dimensions and structural composition

important to understand the synthesis process so that properties of interest can be controlled and modified according to our application (Harish et al., 2022a).

3 Traditional Methods of Nanomaterial Synthesis

The synthesis tactics for nanomaterials are largely classified into two different categories; (a) top-down approach and (b) bottom-up approach. In the top-down approach, bulk material disintegrates into smaller sizes till it achieves the nanoscale (Rath et al., 2024). In the bottom-up approach, small atoms and molecules are assembled together to create nanosized materials utilizing chemical and physical forces (Jiang et al., 2022; Palagati & Reddy, 2024; Tripathy et al., 2023).

3.1 Top-Down Approach

The different techniques in top-down approach (Rath et al., 2024) of nanomaterial synthesis are described below.

Mechanical Milling

Mechanical milling is a technique where a bulk material breaks up till a nanoscale product is achieved. Techniques like ball milling or attrition milling cause repeated deformation, fracturing, and cold welding, hence making the product of interest (Suryanarayana, 2001). Although it is a very simple, effective and scalable method, it is very tough to have any control over the size of the product. Another potential drawback of this method is that there may be contamination in the product due to wear and tear of the mill.

Lithography

Lithography is used to do patterning on the nanomaterials. It is most commonly used in semiconductor fabrication. Methods like photolithography and electron beam lithography are used extensively. Photon beam allows patterning on the scale of 100 nm, whereas electron beams have much higher precision and resolution and can draw patterns in the range of less than 10 nm (Altissimo, 2010). This is a very expensive, complex, and low-yielding technique and is less frequently used despite its high precision and control.

Laser Ablation

In laser ablation, a high energy laser beam is used to vaporize the substrate into plasma. This plasma is then controlled and condensed into nanomaterials. This method permits the synthesis of nanomaterials in vacuum, air, or liquids. This provides high control over composition, providing high control over size and quality (de Anda Villa et al., 2019). Although very accurate, it is very expensive and energy-intensive.

3.2 Bottom-Up Approach

The different techniques in bottom-up approach (Rath et al., 2024) of nanomaterial synthesis are described below.

Sol-Gel Process

The sol-gel process utilizes the polycondensation and hydrolysis of metal oxides or alkaloids, or any other inorganic material, which form a colloid. This colloid will further make a gel-like network (Brinker & Scherer, 1990). Further, these networks are dried and heat-treated to get oxidized nanomaterials. This method is often used to make oxides like silica dioxide (SiO_2), titanium dioxide (TiO_2), and zinc oxide (ZnO) with controlled porosity and morphology (Keswani, 2020).

Coprecipitation

Coprecipitation is a simple and profitable technique used to synthesize oxides and hydroxide nanomaterials. It involves simultaneous precipitation of multiple atoms in an alkaline condition, followed by washing, drying and calcification of the material (Gupta & Gupta, 2005). Although widely used, this particular method lacks control of growth that may be required.

Chemical Vapor Diffusion

In chemical vapor deposition (CVD) technique, a volatile precursor molecule is deposited over a core-nanomaterial and forms a solid nanomaterial film (shell) (George, 2010). This method is often used to make carbon nanotubes, graphene, and films of metal oxides and nitrides. Although such methods give us very high-quality products and uniformity among the batches, it is very complex and difficult to expertise method.

Molecular Self Assembly

This method utilizes nature of molecules to organize in a particular ordered manner, forming nanostructures using non-covalent bonding like hydrogen bond, Vander Waals interaction, etc. (Lehn, 2009). This method is particularly useful when it comes to creating soft nanomaterials, biomimetic systems, drug delivery options, etc. Beside the bottom-up and top-down approaches, there is another category of nanoparticle synthesis which is based on the origin of the precursor material. Chemical synthesis of nanomaterials involved the utilization of chemical substrates and precursors while green synthesis of nanomaterials involved either a substrate or precursor derived from plants, microbes and algae parts. Chemical synthesis often includes the application of chemical species or molecules and due to their complexity and unpredictability of their composition, it is often detrimental to the environment. The emergence of sustainable green synthesis methods is not only safe for environment, but also energy-intensive, cost-effective and have high efficacy of resource recovery (Banjara et al., 2024). The utilization of green-synthesized nanomaterials can be used for wastewater and soil treatment. Therefore, there is a need of synthesis and blending of nanomaterials sourced from discarded solid waste material in order to address the challenges like climate change, waste management and sustainable development (Lyly et al., 2024).

4 Synthesis Methods of Nanomaterials Derived from Solid Waste

Household waste and industrial waste are the major components of solid waste (Deivasigamani et al., 2025) from which nanomaterials can be synthesized using techniques like the carbothermal (Zou et al., 2025), microwave-aided (Jabbar Khan et al., 2025), coprecipitation, solvothermal and hydrothermal methods (Kefeni et al., 2017). Synthesis of nanoparticles from solid waste can be a complex process due to the composition and surface properties of solid waste. Surface alteration of nanoparticles is one of the most important elements in pollutant removal applications, where the selectivity and stability of nanoparticles vary according to the application (Ahmad et al., 2018). This section provides a summary of common nanomaterial synthesis from solid waste along with their benefits and drawbacks.

4.1 Calcination

Calcination is process which converts solid waste into metal oxides and oxides. A simple metal oxide is generated when temperature stays under 400 °C and a spinel oxide when exceeds over 600 °C (Hou et al., 2021). Optimizing the calcination temperature can result in novel functional nanomaterials (Judith & Vasudevan, 2021). Calcination of precursor at lower temperature yields double layer of hydroxide and introducing this material in water or moist gas results in reconstruction which sometimes known as memory effect (Somanathan et al., 2024). A high calcination temperature may cause the spinel oxide unrecoverable (Hou et al., 2021).

4.2 Pyrolysis

Pyrolysis is a highly productive method for synthesizing nanomaterials having high stability, ion exchanging capability and functional groups like carboxylic acid (COOH), hydroxyl (OH), carbonyl (CO) and ketones (C=O) (Chen et al., 2018). The pyrolysis reaction time and temperature have a considerable impact on the surface chemistry of nanoparticles (Tiwari et al., 2024). A study by Bandosz and Block (2006a) demonstrate the rise in surface pH levels of nanomaterials by rising the temperature from 923 K to 1223 K. Adsorption process relies heavily on surface affinity (Kaur et al., 2024) and altering the conditions of pyrolysis can modify the surface properties, selectivity and catalytic activity (Somanathan et al., 2024). The catalytic activity on novel synthesized nanomaterials is attributed to their modified surface chemistry produced during pyrolysis. The drawbacks include the release of toxic gases and a energy extensive process (Wang et al., 2020).

4.3 Coprecipitation Method

Coprecipitation is a common wet chemical approach for synthesis of solid waste-derived nanoparticles. Its advantageous over the other ways because of its high yielding capability and simplicity of nanomaterials (Wang et al., 2020). Solid waste contains leachates which are composed of metal content with acids like HNO_3, H_2SO_4 and HCl. Alkali compounds NaOH and MgO are employed to increase the pH level of acid-extracted metal solution from pH 7 to 12 (Abdel-Aal et al., 2016). Various parameters influence the size and structure of nanoparticles including extractant type, metal molar ratio, pH, stirring rate, reaction temperature and speed of reaction (Chen et al., 2016; Huang et al., 2014; Zhang et al., 2017). Liang et al. (2020) synthesized silica nanoparticles (SiNP) by precipitating silica (Si), followed by extracting it with acid solution, treating it with alkaline solution to raise its pH to 7 and eventually drying it at 373 K for 12 h. The purity of synthesized silica nanoparticles (SiNP) from biomass fly ash ranges between 44.42 and 93.63%. The precipitation method has the disadvantage of yielding impure nanomaterials as compared to other synthesis methods (Judith & Vasudevan, 2021).

4.4 Microwave-Aided Synthesis

Microwave-aided synthesis has been used to produce various inorganic nanomaterials, like metallic nanomaterials, amorphous and nonporous nanomaterials, semiconducting nanomaterials, bioceramics, pure and mixed metal oxides (Mirzaei & Neri, 2016). Microwave energy influences the ionic interactions between reducing agents and precursors during synthesis process of nanomaterials (Mondal et al., 2020). Microwave radiation can remove slit and immobilized metal ions (Wu et al., 2009). This method is advantageous because of rapid heating, quick reaction process and an excellent thermal stability of synthesized nanomaterials (Zhang et al., 2012). Duan et al. (2017) developed magnetic biochar nanoparticles from cotton stalk and iron sludge to remove chromium (Cr(IV)) using a microwave-based chemical reactor. Zhang et al. (2012) has effectively synthesized hydroxyapatite nanoparticle from phospho-gypsum waste generated by phosphorus-based fertilizer industries and assessed its efficacy for the removal of fluoride ions (F^-) in aqueous solution. This approach produces crystalline nanomaterials quicky at a low temperature, outperforming prior synthesis techniques (Khan & Khan, 2020).

4.5 Hydrothermal Method

Solid waste is converted into value-added products through hydrothermal processing which involves heating the raw solid waste-derived waste materials at high temperature range of 393 K and 823 K and pressure between 20 bar and 150 bar, in an inert environment (Munir et al., 2018). This method is comparable to solvothermal method (Adul-Rasool et al., 2024); however, various solvents are utilized instead of water (Khan & Khan, 2020). This method yields nanomaterials with functional groups like C=O, C–O and OH groups. A ketone group (C=O) can produce oxygen $^1O_2$ by removing the organic pollutants while simultaneously resolving the environmental issues of carbon-rich waste (Zang et al., 2020). Ferrite catalyst derived from electroplating sludge (Chen et al., 2017), sludge carbon/TiO_2 nanocomposite from anaerobic sludge (Athalathil et al., 2015), tungsten oxide-FA (fly ash) nanocomposite (Visa, 2016; Visa et al., 2016), and magnetic biochar (Zhang et al., 2018) have been developed. This technique is advantageous over pyrolysis due to high efficiency. The low energy consumption allows the reaction conditions to stay mild and moisture-free conditions of the raw materials (Zang et al., 2020). This technique allows controlling the size, surface chemistry, particle shape and crystalline phase of nanomaterial synthesis by adjusting the reaction temperature, time, pressure and solvent content. However, primary disadvantage is the necessity of high pressure and for reaction (Khan & Khan, 2020).

4.6 Sol–Gel Process

The sol-gel method is used for the synthesis of those nanomaterials which are used for catalysis, chemical sensors, fibers, membranes, and many other scientific and technological applications. It involves hydrolysis, polycondensation, aging, drying, and thermal degradation of substrates to produce a colloidal suspension. Under comparable conditions, this method can produce highly crystalline materials with smaller crystallite size than hydrothermal method (Akpan & Hameed, 2010; Parashar et al., 2020). This method mostly employs either organometallic compounds or inorganic metal salts as precursors present in the solid waste (Saleh, 2022). The typical precursors used here are inorganic or mineral metal salts as well as organometallic compounds. The gel formation and quality influenced from various factors like water content, solvent type, precursor content, acid, or base content (Khan & Khan, 2020). This approach was used for the synthesis of single-phase chromium doped α-alumina nanoparticles. This involved recycling of chromium containing tannery waste having an initial pH of 4 and calcining it at 1373 K (da Costa Cunha et al., 2016). Similarly, a coral-like hierarchal magnesium oxide-incorporated fly ash nanocomposite was synthesized using this method and utilized as nanoadsorbent for removal of Reactive Black 5 azo dye from aqueous solution (Dogar et al., 2020).

5 Solid Waste as a Precursor for the Nanoparticle Synthesis

Nanoparticle synthesis from solid waste is the most efficient recycling method that results in minimal waste generation. Utilizing waste from diverse source as raw material can lower the production cost and promote sustainable development. Solid waste-derived nanomaterials have numerous applications like pollution control, fluoride removal, carbon nanotube synthesis, and wastewater treatment. Further research is required to enhance the recovery process, regulate the harmful gases release and toxic byproduct generation during the nanoparticle synthesis (Judith & Vasudevan, 2021).

The solid waste collected by municipalities and industries are pre-treated before they proceed for the synthesis of nanomaterials. Pre-treatment methods include chemical, physical, and combined processes. The treatment and chemical composition of waste-derived nanomaterials can impact the physical and chemical state of target nanomaterials (Hou et al., 2021). The synthesis is affected by the factor like chemical activation, temperature, time, and acid treatment (Huang et al., 2014). Initially, the solid waste was employed as a catalyst without prior treatment, but later on, the physical and chemical activation enhanced the reactivity of nanomaterial synthesis reactions (Huang et al., 2016; Ioana et al., 2015).

On the basis of source of origin, highly hazardous solid waste is categorized between industrial waste and consumer waste. Both types of wastes can be converted into valued products utilizing numerous procedures. Carbon (C), copper (Cu), lead (Pb), zinc (Zn), and palladium (Pd) are common heavy metals commonly found in solid waste like rubber tires, wastewater sludge, biosolids, and batteries (Rai & Saraswat, 2023; Yadav and Rana, 2024). The end-of-life consumer commodities create significant amounts of electronic waste (e-waste) and plastic waste. E-waste like batteries, integrated circuits (ICs), electric wires, etc., can be the most important source of recoverable precious metals and non-metals. Recycled plastic can be used as packaging material or as raw material in various commercial applications like paper making, construction material, novel polymers, fiber composites, and carbon nanomaterials (Abdelbasir et al., 2020). In this section, we discussed the applications and recovery techniques for the synthesis and modification of nanomaterials derived from discarded batteries, rubber tires, e-waste, and plastic waste.

5.1 Batteries

The lithium-ion batteries are regarded as the most promising commodity of the future requirements due to its high theoretical capacity. Its anode is made up of various silicone (Si) nanostructures. The study of He et al. (2018) demonstrates an innovative technique which involves the mixing of HCl gas with metallurgical-grade silica powder over a copper-based (Cu) catalyst at 573 K in a fixed-bed reactor, followed by its etching with HNO_3 that results in synthesis of silica nanoparticles (SiNP).

The recovered SinNPs from solid waste has a high reversible capacity and stability as an anode electrode material for lithium-ion batteries as compared to bulk silica. Their procedure can be simply optimized and scaled-up to produce large amounts of silica nanoparticles as starting material for anode production of lithium-ion based next generation batteries (He et al., 2018).

Pastes of lead (Pb) are composed of metal oxides, sulfides, and dioxides, which are very hard to recover and reuse due to its high degree of insolubility (Yadav & Rana, 2024). To overcome this problem, pyrometallurgical technique can be utilized to isolate lead from used lead-acid batteries. Zhou et al. (2017a) demonstrated a pyrolysis-based synthesis of lead nano-powder (PbNPD) from lead paste. Submicron size lead oxide particles were synthesized thermally using *N*-vinyl-2-pyrrolidone, which assist the reduction of particle size (Thakur & Kumar, 2023). Pyrometallurgical method can overcome lead paste's insolubility problem, but they may have detrimental effects due to high energy requirement and emission of contaminants like lead dust and sulfur dioxide (Zhou et al., 2017b).

Disposable batteries are increasingly being used to power everyday gadgets like clocks, remotes, toys, etc., adding to the piling up of e-waste. To address this problem, a long-term recycling strategy for discarded batteries has gained importance. The study from Farzana et al. (2018) demonstrated the detailed description and synthesis of zinc oxide (ZnONP) nanoparticles derived from zinc-carbon (Zn–C) batteries. ZnONP has a wide application including optoelectronics, gas sensors, photocatalysis, and energy storage (Parida & Dutta, 2022). ZnONP were synthesized from zinc-carbon (Zn–C) batteries utilizing a horizontal quartz tube furnace where Zn–C was decomposed thermally at 1173 K in an inert argon environment (Gopinath et al., 2021). UV excitation of ZnONP results in photoluminescence in the visible range of electromagnetic spectra. Their study paved the path for applications of ZnONP derived from used Zn–C-based batteries while minimizing the environmental issues by lowering the load of electronic waste in landfill sites (Farzana et al., 2018).

5.2 *Rubber Tires*

There are 1.6 billion new tires produced, and nearly 1 billion tire scrap waste are generated per year globally ("Global Tire Recycling Market Report—Edition 2020," 2020). A large proportion of discarded tires is burned openly or thrown away in the open landfill sites which lead to environment pollution (Moghaddasi et al., 2013) despite it has been completely prohibited in many countries (Turgut & Yesilata, 2008). Recently, there is a raising attention in the research and development of sustainable ways to control and manage rubber tire waste issue. A typical tire is composed of 45–47% natural rubber latex (NRL), 21–22% soot, 12–25% metals, 5–10% textile matter, 6–7% fixatives, 1–2% zinc and, ~ 1% sulfur (Evans & Evans, 2006; Samaddar et al., 2018). The studies by Al-Bassam et al. (2025) developed a two-step sonochemical technique for manufacturing of high-quality activated carbon nanoparticles derived from natural rubber latex (NRL) scraps loaded with iron(III) oxide (Fe_3O_4) or Fe_3O_4@AC. Discarded NRL was first used to manufacture carbon nanomaterials and then transformed into Fe_3O_4@AC via sonochemical process. This method enables a considerable reduction of processing time and energy consumption. The degradation of azo dye 4-((2-hydroxy naphthalenyl)diazenyl)antipyrine (HNDA) from Fe_3O_4@AC nanopartiles was examined and shows positive results in terms of deactivation of azo dye from water, enabling water treatment (Al-Bassam et al., 2025).

The study by Moghaddasi et al. (2013) demonstrated synthesis technique of nanoparticles from discarded rubber tires using mechanical ball milling method. Fly ash (FA) was synthesized after 5 h of ball milling whose particle size was under 500 nm. The addition of silicon (Si) waste further reduced the particle size under 50 nm. Additionally, their study also revealed that rubber FA and powdered tire rubber as an effective zinc source for plants as compared to conventional zinc sulfate ($ZnSO_4$) salt in fertilizers (Moghaddasi et al., 2013).

5.3 *Electronic Waste (E-Waste)*

Electronic waste or e-waste is a major problem in today's global scenario. It is any unwanted or outdated model of electronic equipment. Aging technology is inevitably phased out, resulting in an increasing in generation of Waste Electrical and Electronic Equipment (WEEE) waste. Large appliances like refrigerators, air conditioners, television sets, computers, dishwashers, mobile phones, etc., are all discarded in landfill sites. These discarded items have detrimental impact on environment and human health (Dhivya & Premalatha, 2024). Despite of practical challenges, the high profitable value of nobel metals like gold (Au), silver (Ag), and palladium (Pd) motivates recycling of e-waste. The percentage composition and concentration (ppm) of economically important noble metals present in e-waste items are shown in Fig. 3.

Waste Printed Circuit Boards (WPCB) consist an abundance of noble metals, making them an important secondary precursor for nanomaterial recycling from e-waste. Abdelbasir et al. (2020) demonstrated a combination of hydrometallurgy and ammonium chloride (NH_4Cl)/ammonia leaching and its reduction using reducing agents like ascorbic acid and copper oxide as nanoparticles. The leaching results demonstrates that 96.7% of copper (Cu) could be recovered when exposed under 308 K for 2 h. The concentration of ammonium chloride (NH_4Cl) was 2 M, and ammonia density was 1:10 g/cm^3. The shape of synthesized nanoparticles was spherical where the size of copper nanoparticles (CuNP) was 460 nm and copper oxide nanoparticles was (CuONP) 50 nm. Under the presence of visible light, the Cu/CuO blend demonstrated significant photocatalysis against rhodamine B dye. The demonstrated technique of recovering CuONP from WPCDs generates a wide variety of applications in order to minimize environmental problems (Abdelbasir et al., 2020).

The present techniques of resource recovery are insufficient. Industrial recycling is costly, inefficient, and left a large number of unwanted residues. Additionally, informal methods of e-waste processing in economically backward countries has

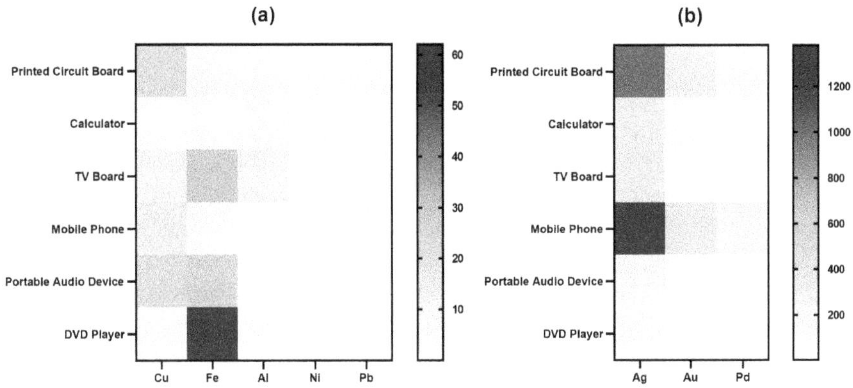

Fig. 3 Heatmaps showing **a** percentage composition of precious metals and **b** concentration of noble metals in the discarded home appliances and devices in electronic waste

the potential to contaminate the world's solid waste stockpile which can probably recycled for various applications (Judith & Vasudevan, 2021). Investigating thermal transformation mechanism shows that tin-lead (Sn–Pb) alloy present in abundance in e-waste can melt at 1173 K. Due to its high hydrophilicity with copper (Cu), it coated the surface of copper nanoparticles (CuNP) (Hossain et al., 2018). The diffusion of Sn into Cu was then activated because of good solubility of tin (Sn) in copper (Cu) than lead (Pb) and finally separated from the alloy (Shokri et al., 2017). The plastics present in the solid waste creates reducing conditions, while carbon reduces the agglomeration. These findings presented here are critically examined and detail for an innovative approach for safely converting problematic e-waste into precious metals by 'thermal micronizing' process (Shokri et al., 2017).

Microprocessors obtained during e-waste collection can possibly use as precursor for the synthesis of gold(III) or Au(III) nanoparticles (Panchatcharam, 2024). This method consists of physiochemical separation techniques that have been carefully designed to into account the complicated nature of the selected material and the application. The first step involves the mechanical separation, allowing Au(III) to be concentrated in metallic fraction followed by a two-step hydrometallurgical process that results in synthesis of Au(III)-enriched solution (Oestreicher et al., 2020). This enriched solution was utilized as secondary precursor to produce Au nanoparticles (AuNP). The addition of reducing agents like ascorbic acid, sodium citrate, and a polymeric stabilizer polyvinylpyrrolidine (PVP) result in generation of highly purified AuNP with well-defined triangular or spherical shape. These AuNP were finally formed on glass substrate and alter its refractive index, which is useful in designing optical sensors. The upcycling study shows that e-waste can substitute ore materials to generate a high valued nanoparticle. These findings emphasize urban mining as a sustainable method to technology development (Oestreicher et al., 2020).

5.4 Plastic Waste

Non-biodegradable plastics have become a major source of garbage globally. Plastic polymers have been extensively employed for more than a century to manufacture a wide range of items. Synthetic plastic production has expanded dramatically in recent decade (Gu & Ozbakkaloglu, 2016). Traditional plastics stays in the environment due to their durability and ability to sustain environmental changes (Pol & Thiyagarajan, 2010). Now, they account for the major composition of solid waste worldwide (Ritchie et al., 2023) and generated over 150 million metric tons annually (Aboulkas & El Bouadili, 2010). Synthetic polymers like polypropylene (PP), polystyrene (PS), high-density polyethene (HDPE), polyethylene terephthalate (PET), low-density polyethene (LDPE), linear low-density polyethene (LLDPE), nylon, and polyvinyl chloride (PVC) are some of the commonly found elements in household plasticwares solid waste (Abdelbasir et al., 2020).

The increasing growth of disposed plastic waste in the surroundings demands for development of novel strategies for plastic waste management and their recycling techniques into precious products. Nowadays, the alteration of plastic waste into recyclable items by chemical recycling has received a lot of consideration (Chan & Zinchenko, 2023). Plastic waste like polyethylene terephthalate (PET) can be utilized to synthesize a hydrogel-based catalyst by crosslinking PET-derived oligo terephthalamides followed by metallization. The synthesized hydrogel has various functions like (i) adsorption mediua for noble metal ions like gold (Au^{3+}), platinum (Pt^{2+}), and silver (Ag^{+}), (ii) reducing agent in conversion of gold (Au^{3+}), platinum (Pt^{2+}), and silver (Ag^{+}) ions to Au^{0}, $Pt^{0}$and, Ag^{0}, and (iii) matrix substratum for the controlled proliferation of gold (AuNP) and platinum (Pt) and silver (AgNP) nanoparticles (Chan & Zinchenko, 2023). Their catalytic efficacy was investigated in an aqueous model system which concludes them to be highly active for the reduction of *p*-nitrophenol according to first order kinetics. Additionally, AuNP and AgNP hybrid hydrogel matrix retain its original catalytic activity even after several cycles of *p*-nitrophenol reduction, inferring them to be a reusable hydrogel. The proposed technique broadens the range of plastic waste conversion routes into valuable resources while also developing novel type of polymeric matrices as a support material for growing metal nanoparticles (Chan & Zinchenko, 2023).

$Pt/SrTiO_3$ or Pt/STO is a nanomaterial which was produced in small scales using atomic layer deposition (ALD) technique. It is a suitable heterogenous nanocatalyst for hydrogenolysis of polyolefins into hydrocarbon oils, offering a sustainable strategy for recycling of plastic waste (Peczak, 2023). ALD is very costly and a resource exhaustive practice for producing novel Pt/STO for commercial use (McCullough et al., 2023). A scalable deposition approach was adopted for synthesizing Pt/STO nanoparticles where surface organometallic chemistry (SOMC) technique was employed containing Pt(II) acetylacetone or trimethyl(methylcyclopentadienyl)platinum as a substrate. The STO substratum was calcined at 823 K and then treatment and steaming by ozone (O_3) at 473 K. The precursors platinum (Pt) was mixed in toluene followed by deposition on the STO support and eventually reduced at 573 K. The extent and thickness of deposited layers in these nanoparticles were modified using a multi-cycle deposition method in redox conditions which is comparable to ALD process. At 573 K and hydrogen (H_2) gas pressure of 180 psi, the Pt/STO catalyst hydrolyze isotactic polypropylene into liquid products whose purity levels were above 95% with molar weight ranged between 200 Da and 300 Da for approximately 25 carbon atoms and limited size distributions (McCullough et al., 2023).

Polyethylene terephthalate (PET) has numerous applications which results in a large amount of waste generation worldwide; therefore, recycling of PET is important as it can be a precursor material for several industrial and economic importance (Judith & Vasudevan, 2021). Kigozi et al. (2023) has demonstrated an economical, emission-free course for conversion of PET polymers into carbon nanomaterials (CNMs) which is known as PET-nano powder (Manjunathan et al., 2024). The hydrothermal technique has been utilized to synthesize PET-nano powder. Once the temperature of PET reaches above its glass transition temperature (T_m), the PET polymers get successfully converted into PET-nano powder with crystal purity of 86.6% and average size distribution of 6.5 nm under ideal conditions. The PET-nano powder was tested for supercapacitance and electrochemical characterization which estimated the capacitance, energy density, current density, and power density of 250.8 F/g, 34.83 Wh/kg, 0.5 A/g, and 999.9 W/g, respectively. After 10,000 cycles, the device proved high cyclic stability and capacitance retention of 96.8% along with current density of 1.5 A/g (Kigozi et al., 2023).

6 Applications of Solid Waste-Derived Nanomaterials

Nanomaterials are used in electronics (Nazir et al., 2024; Nguyen and Zhao, 2014), sensors (Gloag et al., 2019), transistors (Feng et al., 2019; Vidor et al., 2016; Zhao et al., 2021), supercapacitors in batteries (Dhas et al., 2020; Meena et al., 2024; Yan et al., 2024), solar cells (Ahmad et al., 2017; Alkhalayfeh et al., 2021; Waketola et al., 2025), targeted drug delivery (Afzal et al., 2022; De Jong and Borm, 2008; Yusuf et al., 2023), imaging (Hang et al., 2024; Harish et al., 2022b), purification (Khdair et al., 2025; Le et al., 2023; Peng et al., 2022), pollutant removal (Wu et al., 2021; Yu et al., 2024; Zhang et al., 2023), etc. Metal oxide and semiconductor nanoparticles are employed for wastewater treatment in water treatment plants. Photocatalysts, Fenton-based catalysts and electrocatalysts are mixed along with metal oxide and semiconductor nanoparticles in order to enhance the efficacy in the form of enhanced antimicrobial properties and oxidation power in alleviation of organic contaminants from solid waste (Anawar et al., 2019). Recent research suggests that nanomaterials derived from solid waste can effectively remove organic and inorganic pollutants (Judith & Vasudevan, 2021). In this section, we highlight the mechanism of removal of harmful pollutants from solid waste-derived nanomaterials.

6.1 Catalytic Degradation of Pollutants Utilizing Solid Waste-Derived Nanomaterials

The utilization of solid waste as precursor material for nanomaterial synthesis with catalytic properties (nanocatalysts) has gained an vast popularity because of its high demand as catalyst in many industries (Judith & Vasudevan, 2021). The sulfate and hydroxyl radicles are utilized as oxidizing agents in the breakdown of organic compounds. Sulfate ion (SO_4^{2-}) is often liberated by the initiation of peroxymonosulfate (PMS). Zang et al. (2020) demonstrated the application of hydrothermal carbonization method to synthesize sludge-based magnetic biochar nanomaterial doped with iron(III) oxide (Fe_3O_4) and reacted with PMS, which degrades 50 ppm of rhodamine B in 10 min. Another study conducted by Sushil et al. (2011) demonstrated the role of catalytic oxidation of volatile organic compounds using propane gas as control. Solid waste from aluminum, tannery, and electroplating industries were successfully tested as catalyst. The proportion of ferrite in red mud and chromite in tannery shavings influences the catalytic reduction of propane. The mixing of chromium-containing waste with red mud increases the catalytic activity, and thereby, the propane conversion reached by 50% at temperature between 593 K and 653 K. Finally, it resulted in catalytic conversion of *p*-nitrophenol into aminophenols. Chromium (Cr(IV)) waste from tannery industries is mixed with alumina for synthesis of nanocatalyst that can convert *p*-nitrophenol into aminophenols. It is used in production of industrial coloring pigments, pharmaceuticals, and biologically active compounds (da Costa Cunha et al., 2016). The catalytic reduction of *p*-nitrophenol occurred rapidly ranging between 45 s to 45 min and was heavily affected from the amount of chromium ions (Cr^{3+}) present in the α-alumina matrix. Its efficacy is directly proportional to its purity, with the conversion rates ranging between 55% and 98%. Solid waste-derived nanocatalysts have catalytic power comparable to a highly purified bulk catalyst (da Costa Cunha et al., 2016). Nanocatalyst like TiO_2 and ZnO can degrade organic contaminants into innocuous products like CO_2 and water when exposed under visible light (Fujishima et al., 2000). Doping of TiO_2 with non-metal components or other semiconductors enhances photcatalytic activity in visible light, making it suitable for solar-based water treatment systems (Daniel et al., 2024). Very limited research has been conducted which are focused on catalytic degradation and conversion of harmful contaminants using solid waste-derived nanomaterials as a catalyst present in the literature till date.

6.2 Antimicrobial Nanocoating Derived from Solid Waste to Control Biofouling

Biofouling happens when biological organisms and molecules non-specifically adhere to membranes or surfaces like tissue scaffolds, biosensors, food storage containers, implanted devices, surgical equipment, and marine equipment

(Arabaghian et al., 2022). Biofouling of slurry-handling equipment, leachate-evaporation units, and landfill-gas collection membrane results in increased maintenance and decreased efficiency. Antimicrobial nanocoatings that consist of nanoparticles like copper (CuNP), silver (AgNP), or zinc oxide (ZnONP) embedded in a polymeric matrix which can prevent microbial colonization. Silver-based nanocoatings that are dispersed with chitosan or polyurethane films and AgNP of mean diameter between 10 and 50 nm tends to release modest amounts of Ag^+ ions which result in breakdown of bacterial cell membranes, achieving 99% of growth inhibition of microbes like *Staphylococcus aureus* and *Escherichia coli* up to 30 days. Titanium dioxide nanoparticles (TiO_2NP) integrate into membrane surface and generates reactive oxygen species (ROS) when exposed under UV light, eliminating adherent microorganisms, causing their membrane surface to break down and control microbial growth (Varghese et al., 2019). Studies by Hong and He (2014) synthesized a nanocomposite membrane composed of 0.1% wt ZnO/polyethersulfonate with a tensile strength of 2.1 MPa. Zinc oxide (ZnONP) nanoparticles were employed in the designing of polymer nanocomposites membranes with an improved hydrophilicity, porosity, high permeability, and improved anti-biofouling characteristics (Zahid et al., 2018). Netting materials employed in cage aquaculture are always prone to biofouling because of clogging of mesh lumen. The studies by Ashraf and Edwin (2016) presented a microwave-aided production of copper oxide nanoparticles (CuONP) combined with polyethyleneglycomethylacrylate (PMMA) hydrogel on a nylon-based fishing net cage and assessed its antifouling resistance. The synthesized hydrogel was reinforced with 0.004% (w/v) CuONP, and treated netting material was subjected to the estuarine conditions for 90 days resulted in a remarkable fouling resistance with a minimal amount of biomass buildup. CuONP in the matrix worked synergistically with hybrid hydrogel to prevent biofilm formation and thereby fouling organism adhesion to the nylon material. The composite is highlighted in there study as a possible contender and efficient nanomaterial derived from solid waste for prevention of fouling in aquaculture cage nets (Ashraf & Edwin, 2016).

6.3 Disinfection Using Solid Waste-Derived Nanomaterials

Antimicrobial compounds are small molecular chemicals which kill or control microbial growth and are used in a variety of industries, including textiles, food packaging, water treatment, and medicine (Mahmoodi et al., 2018). The antibacterial activity of iron ore tailing waste derived Ag–Fe_3O_4 nanoparticles was studied using the spot inoculation method to assess the growth rate of *E. coli* over solid agar media. The findings were compared with Fe_3O_4 as a control, hydrophobic, and hydrophilic Ag–Fe_3O_4 nanoparticles as test which revealed that Fe_3O_4 nanoparticles do not have a zone of inhibition, but hydrophobic and hydrophilic Ag–Fe_3O_4 nanoparticles have large zones of inhibition at a time interval between 60 min and 120 min. This indicates that silver nanoparticles (AgNP) and its complexes are effective against aerobic bacteria by degrading the cellular proteins and inhibiting the electron transport chain reaction. Additionally, magnetite nanomaterials derived from iron ore tailings provide AgNP a support for efficient water disinfection (Kumar et al., 2015). Similarly, Boruah et al. (2020) investigated the inactivation of biological pollutants using a photocatalytic technique in an aqueous media exposed to natural sunlight. An industry derived magnetic metal oxide effectively killed 96% of *E. coli* in 120 min at pH 5, where *E. coli* concentration was 200 mg/L and catalysis rate of 1 g/L. Electrospun nanofiber mats and nanoparticle filters can enhance the trapping and inactivation properties against airborne pathogens and particulate matter (Judith & Vasudevan, 2021). Silver and copper nanomaterials are commonly integrated into filter systems which protect the indoor environments from microbes and pollutants (Rai et al., 2009). Filter surfaces can also be coated with nano-engineered materials, which can be used to adsorb airborne pollutants. For example, photocatalytic surfaces are coated with TiO_2 which can degrade nitrogen-based oxides when exposed to sunlight, thereby reducing the overall pollution and smog formation (Chen & Poon, 2009).

6.4 Solid Waste-Derived Fenton and Photo-Fenton Degradation

Fenton's reagent is a mixture of ferrous ions (Fe^{2+}) and hydrogen peroxide (H_2O_2) and generates ROS (reactive oxygen species) which is capable of removing organic pollutants by oxidization (Hemond & Fechner, 2022). The photo-Fenton technique generates hydroxyl radicles by combining Fenton's reaction with UV-visible energy in order to convert ferric ions (Fe^{3+}) to ferrous ions (Fe^{2+}). This approach generates high amounts of hydroxy radicles as compared to typical Fenton reaction, which speeds up the oxidation process and is considered as favorable (Kahoush et al., 2018).

The synthesis of ferric ions (Fe^{3+}) by hydrothermal process and biological sludge from the pigment dye wastewater treatment produced magnetic biochar with an increased rate of degradation. The Fenton method catalyzes the decolorization of methylene blue dye. Magnetic biochar was employed to remove dye effluents that results in removal of methylene blue by 98% while its catalytic power stays stable throughout for up to four cycles. Zhang et al. (2018) reported COD removal rate of 47% and TOC removal rate of 49%. In recent study, Zhou et al. (2015) demonstrated that magnetic biochar, derived from paper mill sludge, can decompose methylene blue with an efficiency of 93%. Zhou et al. (2017b) investigated the catalytic role of ferric sludge in removal of bisphenol A (BPA) mixed with calcium peroxide and, the capability of chelating agents like citric acid, oxalic acid, ethylenediaminetertraacetic acid (EDTA), and tartaric acid to accelerate bisphenol A (BPA) degradation while reducing Fe^{3+} dosage. Meanwhile, Kong et al. (2016) described the importance of carbothermal process in transformation of iron-rich sludge to Fe–SC hybrid Fenton-like catalyst for AOII removal. The complete degradation of AOII was achieved in 20 min at 30 °C and pH 7. Magnetically separable nanoparticles $MnFe_2O_4$ and $CuFe_2O_4$ were synthesized via coprecipitation of pickling waste liquor (PWL) with electroplating wastewater. Rhodamine B is a carcinogenic dye which went decomposition in a Fenton-like manner, with up to 99% of remaining as it is for up to 8 cycles. Huang et al. (2014) described that Fe_3O_4 nanoparticles eliminate bisphenol A from steel PWL at a Fenton-like catalysis rate of almost 100% and 45% TOC.

The effective catalysis of methylene blue was carried out by a zinc-ferrite nanocatalyst which was derived and synthesized from calcining electroplating sludge. In the presence of UV light and H_2O_2, the photo-Fenton catalyst degrades methylene blue by 83%. The sludge contains trace amounts of aluminum (Al) and magnesium (Mg), which may help in catalysis in decolorization of methylene blue dye from waste water (Cao et al., 2017). The ferrite-based Fenton catalysts derived from solid waste is good at removing dyes from contaminated water and remains stable even after several cycles of recycling (Judith & Vasudevan, 2021).

6.5 Electrocatalysis Using Solid Waste-Derived Nanomaterials

Electrocatalysis features charge transfer, which aids to promote sustainable development and technological advancement (Masaud et al., 2023). The primary process includes conversion of carbon dioxide (CO_2) electrochemically into synthetic liquid fuels, hydrogen production from water splitting, and energy recovery in solar and fuel cells which makes solid waste-derived nanomaterials helpful in development of electrocatalysts (Gawande et al., 2016). Nano-based electrocatalysts are frequently derived from solid waste that consist of high organic content and low metal content, as metal dopants help improving the electrocatalytic activity. Waste sludge containing complex organic matter can be used as a low-cost precursor to synthesize nano-electrocatalysts that liberate H_2O_2. A porous carbon electrode with a high organic matter content was developed by converting solid waste components like nitrogen (N), sulfur (S), and phosphorous (P) into heteroatoms on the electrode's surface (Hou et al., 2021). The performance and efficiency of nano-electrocatalyst were controlled by its structure, components, and surface properties, which can vary depending upon the synthesis methodology adopted. Nano-electrocatalyst synthesized from thermally treated electroplating sludge exhibits maximum potential of $-$ 0.3 V inferring high electrocatalytic activity which is estimated by a linear sweep voltammogram for carbon dioxide (CO_2) reduction (Yuan et al., 2016).

6.6 Adsorption of Pollutants and Heavy Metals by Nanomaterials Derived from Solid Waste

Pollution is a major concern to humanity, and many studies are concentrating on utilizing solid waste to eliminate it (Duta & Visa, 2015). To prevent environmental contamination, numerous ways for eliminating toxins were developed. Adsorption is a significant physicochemical process that relies on recovering and recycling consumed adsorbents (MiarAlipour et al., 2018). Non-traditional adsorbents, such as fly ash, zeolites, kaolinite clay, agriculture waste, leaf dust, neem bark, wood dust, rice bran, are becoming increasingly popular (MiarAlipour et al., 2018).

Magnetic nanoparticles, carbon nanotubes, and graphene oxides have a strong degree of adsorption of heavy metals like cadmium (Cd), lead (Pb), and mercury (Hg) (Judith & Vasudevan, 2021). They are also used in removal of organic pollutants like pesticides, dyes in contaminated water (Wang & Chen, 2009). For example, magnetite/iron(III) oxide (Fe_3O_4) nanoparticles are used for their high sorption efficiency and easy recovery (Laurent et al., 2008). Functionalized silica and titania are also used as they have the excellent affinity toward arsenic and fluoride ions present as pollutants in the environment.

Several studies have demonstrated the solid waste-derived nanoadsorbent for heavy metal adsorption and dye removal. Previous studies demonstrate that tea-waste/Fe_3O_4 nanocomposite adsorbent (Fan et al., 2017), magnetic chitosan nanocomposite (Liu et al., 2017), Fe_2O_3 in red mud and Cr_2O_3 in tannery shavings (Chen et al., 2018), alum-based nanoadsorbent (Jeon et al., 2018), calcium hydroxyapatite nanoparticles (Mousa et al., 2016), and Fe-loaded magnetic biochar (Duan et al., 2017) have the potential to adsorb chromium (Cr(IV)), arsenic (As(V)), lead (Pb(II)) and, copper (Cu(II)) ions. Iron oxide nanoparticles (Fe_3O_4) have been used to decolorize congo red and methylene blue dyes (Giri et al., 2011), sludge-based nanocomposite have been employed for H_2S removal (Bandosz & Block, 2006a, 2006b) and nanosized hydroxyapatite (nHAp) has been reported to remove floride from the contaminated water (Zhang et al., 2012). Nickel (Ni(II)) ions from aqueous solution gets adsorbed on steel slag because of high degree of magnetite particles. The rate of adsorption increases at the starting concentrations but it gets reduced with rising temperature because of competence from different other ions present in steel slag (Anawar et al., 2019).

Adsorption of harmful chemicals on nanocomposite material occurs via adsorption, electrostatic interactions, and surface interactions. Research studies of Tu et al. (2017) demonstrated the physical adsorption of arsenic (As(III)) and lead (Pb(II)) on copper ferrite nanocomposite derived from printed circuit board (PCB) present in the industrial sludge. The high levels of pH levels lowered copper ferrite adsorption capabilities due to high repulsive forces between arsenic (AS(III)) and the nanoadsorbent surface (Tu et al., 2017, 2013). Studies of Manna et al. (2019) assessed that introducing graphene oxide nanosheets on pre-treated carbonaceous solid waste enhanced the performance of phenol separation in contrast with bulk graphene oxide. Mesopore and macropore adsorbents, synthesized from a blend of chemical and biological sludge, outperformed the surface adsorption capabilities of commercially available activated carbon because of well-accomplished mesospore and macropore structure, as well as the presence of acidic functional groups on the surface of nanomaterial (Pan et al., 2011). Cation adsorption is based on electrostatic interactions; hence, the surface energy of substrate material has a substantial impact on the adsorption process (Visa et al., 2012). The pH of adsorbent influences its zeta potential, ionization level, and functional group dissociation, all of which affect its adsorption efficacy (Giri et al., 2011). At pH 4.2, arsenic (As(III)) has a far greater adsorption capability (42.2 mg/g) than other nanoadsorbents. Copper ferrite nanoparticles were employed at pH 4.5 and remove 90% of of lead (Pb(II)) particles (Tu et al., 2017, 2013). Research studies of Visa et al. (2016) evaluated the usage of zeolites as an effective nanoadsorbent to remove heavy metal ions like Cd^{2+}, Cu^{2+}, Ni^{2+}, Pb^{2+}, and Zn^{2+}. Increasing the amount of substrate creates more active centers, which aids in heavy metal adsorption (Visa, 2016). The new active centers like silica oxide (SiO) and aluminum oxide (AlO) on the surface of zeolites developed complex structures with metal cations, which replace exchangeable cations like sodium ions (Na^+) and potassium ions (K^+). Fan et al. (2017) discovered that eliminating chromium (Cr(IV)) from tea-waste/Fe_3O_4 nanocomposite improved its ion exchange capability.

6.7 Photocatalytic Degradation Using Solid Waste-Derived Nanomaterials

Heterogenous photocatalysis has been identified as a viable method for clearing organic contaminants from wastewater (Diez et al., 2018). It occurs on the surface of catalyst; therefore, impurities must be pre-adsorbed before photocatalysis commences. Nanoparticle like titanium dioxide (TiO_2NP) has a high adsorbing strength which is high enough to facilitate diffusion and eventually, photocatalysis to take place. Hybrid photocatalytic nanoparticles were synthesized using a variety of support materials like zeolites, activated carbon, stainless steel, silica and, clay (Szczepanik, 2017). There has been a renewed interest in employing waste material as a surface support for TiO_2NP based photocatalysts (Judith & Vasudevan, 2021).

Photocatalytic Efficiency of Solid Waste-Derived Nanocomposite Materials

The surface of semiconductor nanomaterials creates redox agents when exposed to light resulting in creation of electron-hole pairs. Under influence of light photons, these redox agents assist the breakdown of organic contaminants. Various semiconductor, core-shell, metal oxide nanomaterials like Cu_2O/Ti_2O, CdO/SnO_2, TiO_2/CuS, $NaNBO_3/CdS$, SnO_2/TiO_2, ZnO-Cu_2O, SnO_2-graphene, and SnS_2-SnO_2, have been synthesized in order to boost the photocatalytic efficiency by extending the absorption range to the visible spectra and suppressing electron-hole pair recombination (Devi et al., 2019). Magnetic iron oxide nanoparticles (Fe_3O_4NP) synthesized from solid waste exhibit a positive zeta potential (ζ-potential) at pH level between 2 and 3 and a negative zeta potential (ζ-potential) between pH 5 and 10. The catalytic surface adsorbed hydronium ions (H^+) in acidic and basic media due to the presence of hydroxyl ions (OH^-), resulting in the deactivation of congo red, methylene blue (anionic dyes), and rhodamine B (cationic dye). The photocatalytic degradation of magnetic iron oxide nano-powder resulted in recycling of 78% nano-powder back after 10 cycles of photocatalysis (Boruah et al., 2020).

When exposed to light, tungsten trioxide nanoparticles (WO_3NP), which are semiconductor photocatalysts that become photoactive and have attracted considerable interests as an alternative to titanium dioxide (TiO_2NP) nanoparticles. The pairing concept could be broadened by combining WO_3NP with various semiconductors which have similar band gaps (Judith & Vasudevan, 2021). Fly ash (FA) containing TiO_2NP was used as a photocatalyst, together with Fenton precursors like Fe_2O_3 and MnO, to produce hydroxyls. Visa and colleagues employed a hydrothermally synthesized TiO_2 and FA nanocomposite as a photocatalytic material to degrade bemacid balu-BB and bemacid rot-BR colors along with Cu^{2+} heavy metal ions (Visa et al., 2015b). Zeolite FA bead/TiO_2 nanocomposite activated with alkali and combined with TiO_2 is used to breakdown rhodamine B dye (Yang et al., 2017). Furthermore, a carbon-based nanomaterial derived from anaerobic sludge and mixed with TiO_2 functions as a photocatalyst in order to degrade bisphenol A dye. The addition of TiO_2 to sludge carbon surfaces increases the charge mobility, affinities, and electronegativity resulting in improved level of photocatalysis (Athalathil et al., 2015).

Photocatalytic Efficiency of Solid Waste-Derived Fly ash/TiO2 Nanocomposite Materials

Polyaromatic hydrocarbons, chlorinated solvents, and pharmaceutical residues are examples of persistent organic pollutants (POPs) that are unaffected to both traditional oxidation and biodegradation. Utilizing solar or ultraviolet light, semiconductor nanophotocatalysts generates electron-hole pairs that yield reactive oxygen species (OH, O_2^-), which mineralize organic materials into CO_2, H_2O, and inorganic ions. Titanium oxide (TiO_2) is the most frequently used nanophotocatalyst due to its high level of redox potential (Lee and Park, 2013; Rajaambal et al., 2015), chemical stability, high temperature resilient, cost-effectiveness, and environmental friendliness (Nakata and Fujishima, 2012; Schneider et al., 2014). TiO_2 and doped oxides: Anatase-phase TiO_2 nanoparticles between 10 nm and 20 nm decolorize methylene blue by > 90% in 1 hour under the influence of UV light. When exposed to sunlight, N- or C-doping results in 70% elimination of light absorption, extending it into the visible spectrum. Quantum dot Titanium oxide (QD-TiO_2) heterojunctions show improved charge separation, boosting photocatalytic kinetics by 2–3 times as compared to bulk TiO_2. Cadmium sulfate (CdS) and cadmium selenide (CdSe) quantum dots (QDs) linked to TiO_2 increase visible-light harvesting (Varghese et al., 2019). Gold (Au) and silver (Ag)-coated TiO_2 nanoparticles tend to produce Localized Surface Plasmon Resonance (LSPR) that produce "hot" electrons when presented under visible light. Au/TiO_2 nanocomposites have quantum efficiencies of up to 0.1% in the degradation of chlorophenols, which is five times more than that of nano TiO_2 (Ikhmayies, 2014). By incorporating photocatalytic nanomaterials into floating membrane modules or packed-bed reactors, concentrated waste streams, like pharmaceutical effluents or textile industry sludges, can be treated decentralized manner. The reuse of solid waste incineration FA by converting it to NaP1 zeolites degrades methylene blue by photocatalysis (Chen et al., 2023). Zinc oxide and Zinc oxide-based nanomaterials like Ag/ZnO, Au/ZnO, Mg doped ZnO-Al_2O_3 are utilized in alleviation of pharmaceutical drug compounds mixed in waste slurry (Majumder et al., 2020).

Using fly ash as an adsorption substrate in conjunction with TiO_2 photocatalyst is an effective and environmentally beneficial approach. Nanocomposites effectively treat wastewater containing heavy metals, dyes, and surfactants by removing pollutants with distinct chemical structures (Anawar et al., 2019; Visa et al., 2015c). Fly ash (FA) contains unburned carbon and metal oxides like Fe_2O_3 and MnO, which can be used as in situ Fenton systems along with TiO_2NP for photocatalysis. Waste FA is an appropriate support material because of its appropriate size, charge, and surface chemistry (Visa et al., 2015a). In various studies by Visa and colleagues, they thoroughly investigated the photodegradation and adsorption of bemacid rot, methylene blue, bemacid blau, surfactant sodium dodecylbenzene sulfonate (SDBS), cadmium (Cd), and copper (Cu) in FA/TiO_2 nanocomposites (Visa et al., 2016, 2015a, 2011; Visa & Chelaru, 2014). Additionally, alkali activation generates new active sites for SiO and AlO on FA surfaces, which allows metals to form complexes. Surface affinity for Cd^{2+} and Cu^{2+} varies before and after treatment with 2 N NaOH due to structural and chemical (Visa et al., 2009). Yang et al. (2017) investigated decolorization of rhodamine B dye by utilizing zeolite fly ash beads (ZFAB) nanocomposites and TiO_2NP, along with benzene with Ag-TiO_2/ZFAB. The nanocomposite gets synthesized by activating it with NaOH and then introducing TiO_2 nanoparticles. Rhodamine B dye degraded by 83% after 5 cycles. This shows that the synthesized nanomaterials are highly recyclable in terms of rhodamine B degradation. After 3 cycles, the Ag-TiO_2/ZFAB photocatalytic concrete material eliminated 96.3% of benzene gas (Yang et al., 2017).

7 Conclusion and Future Prospects

Nanomaterials provide a harmless, cost-effective, and environment friendly method of synthesis. Nanotechnology introduces a new and unique resourcefulness to create and test various pristine inventions based on metallic nanomaterials, which has potential in a variety of applications. This chapter discussed a variety of metal and non-metallic nanomaterials that are derived from solid waste and their potential applications. Nanomaterials possessed a growing hazard to public health and the environment, prompting research into their toxicity and development techniques. The usage of nanomaterials with added benefits has significantly improved the synthesis efficiency.

The increased dumping of industrial and municipal solid waste has compelled scientists to emphasis on solid waste mining and recycling in order to accomplish the notion of waste treatment and wealth creation. Nanotechnology is a rapidly emerging science with several applications, and nanomaterial synthesis from solid waste is regarded as a successful method that provides valuable element recovery as well as economic benefits. Research into improving and developing the recovery process is required to address the issue of resource exploitation and waste generation. Various solid wastes, including, vegetable waste, meat waste, agriculture waste, plastic waste, e-waste, tannery waste, steel industry waste, fly ash, electroplating waste, jute mill waste, papermill waste, and silicon waste can be reused to synthesize and modify solid waste-derived nanomaterials. More particular, waste pretreatment changes the characteristics of nanomaterials by altering their surface chemistry. Physicochemical parameters during nanomaterial synthesis have a significant influence on nanomaterial development. To produce waste-derived products on large scale, synthesis procedures like photocatalysis, Fenton degradation, electrocatalysis, etc., must be feasible, practical, and optimized. Additional research is needed to address the issue of subsidiary product generation like hazardous gases during thermal synthesis. The solid waste-derived nanomaterials synthesized using physicochemical processes offers a variety of applications in environmental remediation. Solid waste can become precursor material for nanomaterials like carbon nanotubes (CNT), graphene nanosheets, Fe_3O_4, $MgFe_2O_4$ (Mn, Cu, Zn, Ag), magnetic biochar, $MgCr_2O_4$, Cr_2O_3, CuO, and Copper ferrite. Iron-rich waste has been shown to effectively remove pollutants through Fenton degradation, photocatalytic degradation, and adsorption methods. While they are effective for pollutant cleanup, it is necessary to study and assess the potential risks from by-products and leakage of original contaminants. To expand the application, LCA should consider pollution reduction, energy consumption, and cost efficiency. The reuse and recovery of economically and commercially important metals and non-metals offers tremendous potential for reducing time and costs along while facilitating the restoration of contaminated environments. The usage of heavy metal and polymer nanomaterials should be strictly controlled to minimize the environmental risks. The long-term impact of persistence application of nanomaterials must be extensively assessed and proper research assessment required before its large-scale execution in order to minimize the future possible environment risks.

Acknowledgement Prof. Chaithanya Madhurantakam, Principal Investigator-SMBL, Department of Biotechnology, TERI SAS, Vasant Kunj New Delhi is acknowledged for his scientific insights and laboratory facility.

References

Abd El-Aziz, M., Badr, Y., Mahmoud, M.A. (2007). Biosynthesis of gold nanoparticles using *Pseudomonas aeruginosa.* In *AIP Conference Proceedings. Presented at the Modern Trends in Physics Research: Second International Conference on Modern Trends in Physics Research MTPR-06*, AIP, Cairo (Egypt), pp. 177–181. doi:https://doi.org/10.1063/1.2711108.

Abdel-Aal, E., Farghaly, F., Tahawy, R., & El-Shahat, M. (2016). A comparative study on recovery of chromium from tannery wastewater as nano magnesium chromite. *Physicochemical Problems of Mineral Processing, 52*, 821–834.

Abdelbasir, S. M., McCourt, K. M., Lee, C. M., & Vanegas, D. C. (2020). Waste-derived nanoparticles: Synthesis approaches, environmental applications, and sustainability considerations. *Frontiers in Chemistry, 8*, 782. https://doi.org/10.3389/fchem.2020.00782

Aboulkas, A., & El Bouadili, A. (2010). Thermal degradation behaviors of polyethylene and polypropylene. Part I: Pyrolysis kinetics and mechanisms. *Energy Conversion and Management, 51*, 1363–1369.

Adul-Rasool, A. A., Athair, D. M., Zaidan, H. K., Rheima, A. M., Al-Sharify, Z. T., & Mohammed, S. H. (2024). 0, 1, 2, 3D nanostructures, types of bulk nanostructured materials, and drug nanocrystals: An overview. *Cancer Treatment and Research Communications, 40*, 100834.

Afzal, O., Altamimi, A. S. A., Nadeem, M. S., Alzarea, S. I., Almalki, W. H., Tariq, A., Mubeen, B., Murtaza, B. N., Iftikhar, S., Riaz, N., & Kazmi, I. (2022). Nanoparticles in drug delivery: From history to therapeutic applications. *Nanomaterials, 12*, 4494. https://doi.org/10.3390/nano12244494

Ahmad, A., Senapati, S., Khan, M. I., Kumar, R., & Sastry, M. (2003). Extracellular biosynthesis of monodisperse gold nanoparticles by a novel Extremophilic Actinomycete, *Thermomonospora* sp. *Langmuir, 19*, 3550–3553. https://doi.org/10.1021/la0267721

Ahmad, I., Siddiqui, W. A., Qadir, S., & Ahmad, T. (2018). Synthesis and characterization of molecular imprinted nanomaterials for the removal of heavy metals from water. *Journal of Materials Research and Technology, 7*, 270–282.

Ahmad, R., Srivastava, R., Yadav, S., Chand, S., & Sapra, S. (2017). Functionalized 2D-MoS2-incorporated polymer ternary solar cells: Role of Nanosheet-induced Long-range ordering of polymer chains on charge transport. *ACS Applied Materials & Interfaces, 9*, 34111–34121. https://doi.org/10.1021/acsami.7b08725

Akpan, U., & Hameed, B. (2010). The advancements in sol–gel method of doped-TiO_2 photocatalysts. *Applied Catalysis A, General, 375*, 1–11.

Alagarasi, A. (2013). Chapter-introduction to nanomaterials. Indian Institute of Technology Madras 1–24.

Al-Bassam, A. Z., Rabeea, M. A., Sarhan, A. A., Toohi, H. T. A.-S., & Yingnakorn, T. (2025). Rapid sonochemical synthesis of Fe_3O_4@ AC from waste rubber tires to use for azo dye removal. *Inorganic Chemistry Communications, 174*, 113968.

Alkhalayfeh, M. A., Aziz, A. A., & Pakhuruddin, M. Z. (2021). An overview of enhanced polymer solar cells with embedded plasmonic nanoparticles. *Renewable and Sustainable Energy Reviews, 141*, 110726. https://doi.org/10.1016/j.rser.2021.110726

Altissimo, M. (2010). E-beam lithography for micro−/nanofabrication. *Biomicrofluidics, 4.*

Anawar, H. M., Strezov, V., & Hossain, M. Z. (2019). Comparison of different nanoprocesses and industrial waste-based adsorbents such as red mud, steel slag, and fly ashes for treating wastewater nanomaterial contaminants. In *Emerging and nanomaterial contaminants in wastewater* (pp. 107–136). Elsevier.

Anil Kumar, S., Abyaneh, M. K., Gosavi, S. W., Kulkarni, S. K., Pasricha, R., Ahmad, A., & Khan, M. I. (2007). Nitrate reductase-mediated synthesis of silver nanoparticles from $AgNO_3$. *Biotechnology Letters, 29*, 439–445. https://doi.org/10.1007/s10529-006-9256-7

Arabaghian, H., Wang, M., Ordonez, J., & Rodrigues, D. F. (2022). Zwitterionic polymers in biofouling and inorganic fouling mechanisms. In *60 years of the Loeb-Sourirajan membrane* (pp. 33–70). Elsevier.

Ashraf, P. M., & Edwin, L. (2016). Nano copper oxide incorporated polyethylene glycol hydrogel: An efficient antifouling coating for cage fishing net. *International Biodeterioration & Biodegradation, 115*, 39–48.

Athalathil, S., Erjavec, B., Kaplan, R., Stüber, F., Bengoa, C., Font, J., Fortuny, A., Pintar, A., & Fabregat, A. (2015). TiO_2-sludge carbon enhanced catalytic oxidative reaction in environmental wastewaters applications. *Journal of Hazardous Materials, 300*, 406–414.

Bandosz, T. J., & Block, K. (2006a). Effect of pyrolysis temperature and time on catalytic performance of sewage sludge/industrial sludge-based composite adsorbents. *Applied Catalysis B: Environmental, 67*, 77–85.

Bandosz, T. J., & Block, K. (2006b). Municipal sludge– industrial sludge composite desulfurization adsorbents: Synergy enhancing the catalytic properties. *Environmental Science & Technology, 40*, 3378–3383.

Banjara, R. A., Kumar, A., Aneshwari, R. K., Satnami, M. L., & Sinha, S. (2024). A comparative analysis of chemical vs green synthesis of nanoparticles and their various applications. *Environmental Nanotechnology, Monitoring & Management, 22*, 100988.

Bansal, V., Rautaray, D., Bharde, A., Ahire, K., Sanyal, A., Ahmad, A., & Sastry, M. (2005). Fungus-mediated biosynthesis of silica and titania particles. *Journal of Materials Chemistry, 15*, 2583. https://doi.org/10.1039/b503008k

Boruah, P. K., Yadav, A., & Das, M. R. (2020). Magnetic mixed metal oxide nanomaterials derived from industrial waste and its photocatalytic applications in environmental remediation. *Journal of Environmental Chemical Engineering, 8*, 104297.

Brinker, C. J., & Scherer, G. W. (1990). *Sol-gel science: The physics and chemistry of sol-gel processing*. Gulf Professional Publishing.

Cao, Z., Zhang, J., Zhou, J., Ruan, X., Chen, D., Liu, J., Liu, Q., & Qian, G. (2017). Electroplating sludge derived zinc-ferrite catalyst for the efficient photo-Fenton degradation of dye. *Journal of Environmental Management, 193*, 146–153.

Chan, K., & Zinchenko, A. (2023). Templating of catalytic gold and silver nanoparticles by waste plastic PET-derived hydrogel playing a dual role of a reductant and a matrix. *Waste Management, 164*, 20–28.

Chen, D., Li, Q., Shao, L., Zhang, F., & Qian, G. (2016). Recovery and application of heavy metals from pickling waste liquor (PWL) and electroplating wastewater (EPW) by the combination process of ferrite nanoparticles. *Desalination and Water Treatment, 57*, 29264–29273.

Chen, J., & Poon, C. (2009). Photocatalytic construction and building materials: From fundamentals to applications. *Building and Environment, 44*, 1899–1906.

Chen, M., Zhou, J., Zhang, J., Zhang, J., Chen, Z., Ding, J., Kong, F., Qian, G., & Chen, J. (2017). Ferrite catalysts derived from electroplating sludge for high-calorie synthetic natural gas production. *Applied Catalysis A: General, 534*, 94–100.

Chen, Q., Zhao, Y., Qiu, Q., Long, L., Liu, X., Lin, S., & Jiang, X. (2023). Zeolite NaP1 synthesized from municipal solid waste incineration fly ash for photocatalytic degradation of methylene blue. *Environmental Research, 218*, 114873.

Chen, W.-J., Hsiao, L.-C., & Chen, K. K.-Y. (2008). Metal desorption from copper(II)/nickel(II)-spiked kaolin as a soil component using plant-derived saponin biosurfactant. *Process Biochemistry, 43*, 488–498. https://doi.org/10.1016/j.procbio.2007.11.017

Chen, Y., Ho, S.-H., Wang, D., Wei, Z., Chang, J.-S., & Ren, N. (2018). Lead removal by a magnetic biochar derived from persulfate-ZVI treated sludge together with one-pot pyrolysis. *Bioresource Technology, 247*, 463–470.

Cho, G., Park, Y., Hong, Y.-K., & Ha, D.-H. (2019). Ion exchange: An advanced synthetic method for complex nanoparticles. *Nano Convergence, 6*, 17. https://doi.org/10.1186/s40580-019-0187-0

da Costa Cunha, G., Peixoto, J. A., de Souza, D. R., Romão, L. P. C., & Macedo, Z. S. (2016). Recycling of chromium wastes from the tanning industry to produce ceramic nanopigments. *Green Chemistry, 18*, 5342–5356.

Daniel, L. S., Uahengo, V., Rahman, A. (2024). Factors influencing TiO_2-based composites for water Decolorization: A systematic review.

Daoud, A. O., Othman, A. A. E., Robinson, H., & Bayyati, A. (2020). An investigation into solid waste problem in the Egyptian construction industry: A mini-review. *Waste Management & Research, 38*, 371–382. https://doi.org/10.1177/0734242X20901568

de Anda Villa, M., Gaudin, J., Amans, D., Boudjada, F., Bozek, J., Evaristo Grisenti, R., Lamour, E., Laurens, G., Macé, S., & Nicolas, C. (2019). Assessing the surface oxidation state of free-standing gold nanoparticles produced by laser ablation. *Langmuir, 35*, 11859–11871.

De Jong, W. H., & Borm, P. J. (2008). Drug delivery and nanoparticles: Applications and hazards. *International Journal of Nanomedicine, 3*, 133–149. https://doi.org/10.2147/ijn.s596

Deivasigamani, P., Gajendiran, V., Chitra, B., Kumar, P. S., Balasubramanian, N., Sundararaman, S., Dhananjeyan, V., Subramani, L., Jagadeesan, A. K., Sivamani, S., 2025. Magnetic nanoparticles: Synthesis, characterization and application based on environmental perspective. Results in chemistry 102023.

Devi, M. M., Singh, H., Kaur, K., Gupta, A., Nishanthi, S., Bera, C., Ganguli, A. K., & Jha, M. (2019). New approach for the transformation of metallic waste into nanostructured Fe_3O_4 and SnO_2-Fe_3O_4 heterostructure and their application in treatment of organic pollutant. *Waste Management, 87*, 719–730.

Dhas, S. D., Maldar, P. S., Patil, M. D., Nagare, A. B., Waikar, M. R., Sonkawade, R. G., & Moholkar, A. V. (2020). Synthesis of NiO nanoparticles for supercapacitor application as an efficient electrode material. *Vacuum, 181*, 109646. https://doi.org/10.1016/j.vacuum.2020.109646

Dhivya, K., & Premalatha, G. (2024). E-waste challenges & solutions. In *Sustainable Management of Electronic Waste* (pp. 255–275). Wiley. https://doi.org/10.1002/9781394166923.ch13

Diez, A. M., Moreira, F. C., Marinho, B. A., Espindola, J. C., Paulista, L. O., Sanromán, M., Pazos, M., Boaventura, R. A., & Vilar, V. J. (2018). A step forward in heterogeneous photocatalysis: Process intensification by using a static mixer as catalyst support. *Chemical Engineering Journal, 343*, 597–606.

Dogar, S., Nayab, S., Farooq, M. Q., Said, A., Kamran, R., Duran, H., & Yameen, B. (2020). Utilization of biomass fly ash for improving quality of organic dye-contaminated water. *ACS Omega, 5*, 15850–15864.

Drobny, N. L., Hull, H., & Testin, R. F. (1971). *Recovery and utilization of municipal solid waste.* Public Health Service Publication.

Du, M., Du, Y., Chen, Z., Li, Z., Yang, K., Lv, X., & Feng, Y. (2017). Synthesis and characterization of black ceramic pigments by recycling of two hazardous wastes. *Applied Physics A, Materials Science & Processing, 123*, 575.

Duan, S., Ma, W., Pan, Y., Meng, F., Yu, S., & Wu, L. (2017). Synthesis of magnetic biochar from iron sludge for the enhancement of Cr (VI) removal from solution. *Journal of the Taiwan Institute of Chemical Engineers, 80*, 835–841.

Durán, N., Marcato, P. D., Alves, O. L., De Souza, G. I., & Esposito, E. (2005). Mechanistic aspects of biosynthesis of silver nanoparticles by several fusarium oxysporum strains. *Journal of Nanobiotechnology, 3*, 8. https://doi.org/10.1186/1477-3155-3-8

Duta, A., & Visa, M. (2015). Simultaneous removal of two industrial dyes by adsorption and photocatalysis on a fly-ash–TiO_2 composite. *Journal of Photochemistry and Photobiology A: Chemistry, 306*, 21–30.

Erfaninia, N., Tayebee, R., Foletto, E. l., Amini, M. m., Dusek, M., & Zonoz, F. m. (2018). Preparation of magnetically recyclable ZnFe2O4 nanoparticles by easy single-step co-precipitation method and their catalytic performance in the synthesis of 2-aminothiophenes. *Applied Organometallic Chemistry, 32*, e4047. https://doi.org/10.1002/aoc.4047

Evans, A., & Evans, R. (2006). The composition of a tyre: typical components. *The Waste & Resources Action Programme, 5.*

Fan, S., Wang, Y., Li, Y., Tang, J., Wang, Z., Tang, J., Li, X., & Hu, K. (2017). Facile synthesis of tea waste/Fe_3O_4 nanoparticle composite for hexavalent chromium removal from aqueous solution. *RSC Advances, 7*, 7576–7590.

Farzana, R., Rajarao, R., Behera, P. R., Hassan, K., & Sahajwalla, V. (2018). Zinc oxide nanoparticles from waste Zn–C battery via thermal route: Characterization and properties. *Nanomaterials, 8*, 717. https://doi.org/10.3390/nano8090717

Feng, Q., Wilhelm, J., & Gao, J. (2019). Transistor-like ultra-pH-sensitive polymeric nanoparticles. *Accounts of Chemical Research, 52*, 1485–1495.

Feynman, R. (2018). There's plenty of room at the bottom. In *Feynman and computation* (pp. 63–76). CRC Press.

Fujishima, A., Rao, T. N., & Tryk, D. A. (2000). Titanium dioxide photocatalysis. *Journal of Photochemistry and Photobiology C: Photochemistry Reviews, 1*, 1–21.

Gao, Q., Wu, X., Xia, Z., & Fan, Y. (2017). Coating mechanism and near-infrared reflectance property of hollow fly ash bead/TiO_2 composite pigment. *Powder Technology, 305*, 433–439.

Gawande, M. B., Goswami, A., Felpin, F.-X., Asefa, T., Huang, X., Silva, R., Zou, X., Zboril, R., & Varma, R. S. (2016). Cu and cu-based nanoparticles: Synthesis and applications in catalysis. *Chemical Reviews, 116*, 3722–3811.

George, S. M. (2010). Atomic layer deposition: An overview. *Chemical Reviews, 110*, 111–131.

Giri, S., Das, N., & Pradhan, G. (2011). Synthesis and characterization of magnetite nanoparticles using waste iron ore tailings for adsorptive removal of dyes from aqueous solution. *Colloids and Surfaces A: Physicochemical and Engineering Aspects, 389*, 43–49.

Gloag, L., Mehdipour, M., Chen, D., Tilley, R. D., & Gooding, J. J. (2019). Advances in the application of magnetic nanoparticles for sensing. *Advanced Materials, 31*, 1904385.

Global Tire Recycling Market Report—Edition 2020. (2020). WWW Document. https://www.goldsteinresearch.com/report/global-tire-recycling-industry-market-trends-analysis. Accessed 10 Sep 2025.

Gopinath, K. P., Vo, D.-V. N., Gnana Prakash, D., Adithya Joseph, A., Viswanathan, S., & Arun, J. (2021). Environmental applications of carbon-based materials: A review. *Environmental Chemistry Letters, 19*, 557–582.

Gu, L., & Ozbakkaloglu, T. (2016). Use of recycled plastics in concrete: A critical review. *Waste Management, 51*, 19–42.

Gupta, A. K., & Gupta, M. (2005). Synthesis and surface engineering of iron oxide nanoparticles for biomedical applications. *Biomaterials, 26*, 3995–4021.

Hang, Y., Wang, A., & Wu, N. (2024). Plasmonic silver and gold nanoparticles: Shape- and structure-modulated plasmonic functionality for point-of-caring sensing, bio-imaging and medical therapy. *Chemical Society Reviews, 53*, 2932–2971. https://doi.org/10.1039/D3CS00793F

Harish, V., Ansari, M. M., Tewari, D., Gaur, M., Yadav, A. B., García-Betancourt, M.-L., Abdel-Haleem, F. M., Bechelany, M., & Barhoum, A. (2022a). Nanoparticle and nanostructure synthesis and controlled growth methods. *Nanomaterials, 12*, 3226. https://doi.org/10.3390/nano12183226

Harish, V., Tewari, D., Gaur, M., Yadav, A. B., Swaroop, S., Bechelany, M., & Barhoum, A. (2022b). Review on nanoparticles and nanostructured materials: Bioimaging, biosensing, drug delivery, tissue engineering, antimicrobial, and agro-food applications. *Nanomaterials, 12*, 457. https://doi.org/10.3390/nano12030457

He, Q., Yu, J., Wang, Y., Zhong, Z., Jiang, J., & Su, F. (2018). Silicon nanoparticles prepared from industrial wastes as high-performing anode materials for lithium ion batteries. *Solid State Ionics, 325*, 141–147. https://doi.org/10.1016/j.ssi.2018.08.001

Hemond, H. F., & Fechner, E. J. (2022). *Chemical fate and transport in the environment*. Academic press.

Hong, J., & He, Y. (2014). Polyvinylidene fluoride ultrafiltration membrane blended with nano-ZnO particle for photo-catalysis self-cleaning. *Desalination, 332*, 67–75.

Hossain, R., Nekouei, R. K., Mansuri, I., & Sahajwalla, V. (2018). Sustainable recovery of cu and Sn from problematic global waste: Exploring value from waste printed circuit boards. *ACS Sustainable Chemistry & Engineering, 7*, 1006–1017.

Hou, H., Liu, Z., Zhang, J., Zhou, J., & Qian, G. (2021). A review on fabricating functional materials by heavy metal–containing sludges. *Environmental Science and Pollution Research, 28*, 133–155.

Huang, R., Fang, Z., Fang, X., & Tsang, E. P. (2014). Ultrasonic Fenton-like catalytic degradation of bisphenol a by ferroferric oxide (Fe_3O_4) nanoparticles prepared from steel pickling waste liquor. *Journal of Colloid and Interface Science, 436*, 258–266.

Huang, X., Wang, Z., Liu, Y., Hu, W., & Ni, W. (2016). On the use of blast furnace slag and steel slag in the preparation of green artificial reef concrete. *Construction and Building Materials, 112*, 241–246.

Iijima, S. (1991). Helical microtubules of graphitic carbon. *Nature, 354*, 56–58.

Ikhmayies, S. J. (2014). Characterization of nanomaterials. *JOM, 66*, 28–29.

Ioana, A., Constantin, N., & Moldovan, P. (2015). About EAF and environment. In *Presented at the IOP conference series: Materials science and engineering* (p. 012015). IOP Publishing.

Jabbar Khan, A., Gao, L., Numan, A., Khan, S., Hussain, I., Sajjad, M., Shah, S. S., Mateen, A., & Zhao, G. (2025). Recent advancements in the tailoring of nanomaterials via microwave-assisted synthesis: A comprehensive review. *Critical Reviews in Solid State and Materials Sciences, 50*, 515–538.

Jeon, E.-K., Ryu, S., Park, S.-W., Wang, L., Tsang, D. C., & Baek, K. (2018). Enhanced adsorption of arsenic onto alum sludge modified by calcination. *Journal of Cleaner Production, 176*, 54–62.

Jiang, Z., Li, L., Huang, H., He, W., & Ming, W. (2022). Progress in laser ablation and biological synthesis processes:"top-down" and "bottom-up" approaches for the green synthesis of au/ag nanoparticles. *International Journal of Molecular Sciences, 23*, 14658.

Judith, J. V., & Vasudevan, N. (2021). Synthesis of nanomaterial from industrial waste and its application in environmental pollutant remediation. *Environmental Engineering Research, 27*, 200672–200670. https://doi.org/10.4491/eer.2020.672

Kahoush, M., Behary, N., Cayla, A., & Nierstrasz, V. (2018). Bio-fenton and bio-electro-Fenton as sustainable methods for degrading organic pollutants in wastewater. *Process Biochemistry, 64*, 237–247.

Kaur, H., Phanden, R. K., Singh, R. K., & Sikarwar, B. S. (2024). *Discovery, disruption, and future implications of nanomaterials*. IGI Global.

Kefeni, K. K., Mamba, B. B., & Msagati, T. A. (2017). Application of spinel ferrite nanoparticles in water and wastewater treatment: A review. *Separation and Purification Technology, 188*, 399–422.

Keswani, C. (2020). *Intellectual property issues in nanotechnology*. CRC Press.

Khan, F. A., & Khan, F. (2020). *Applications of nanomaterials in human health*. Springer.

Khan, S., & Hossain, M. (2022). *Nanoparticle-based polymer composites* (p. 15). Woodhead.

Khdair, A. I., Aburumman, G. A., Gholipour, S., & Afrand, M. (2025). Nanoparticles in water purification: Multifunctional roles, challenges, and sustainable applications. *Environmental Science: Nano, 12*, 3871–3895. https://doi.org/10.1039/D5EN00268K

Kigozi, M., Kasozi, G. N., Balaso Mohite, S., Zamisa, S., Karpoormath, R., Kirabira, J. B., & Tebandeke, E. (2023). Non-emission hydrothermal low-temperature synthesis of carbon nanomaterials from poly (ethylene terephthalate) plastic waste for excellent supercapacitor applications. *Green Chemistry Letters and Reviews, 16*, 2173025. https://doi.org/10.1080/17518253.2023.2173025

Kong, L., Zhu, Y., Liu, M., Chang, X., Xiong, Y., & Chen, D. (2016). Conversion of Fe-rich waste sludge into nano-flake Fe-SC hybrid Fenton-like catalyst for degradation of AOII. *Environmental Pollution, 216*, 568–574.

Kroto, H. W., Heath, J. R., O'Brien, S. C., Curl, R. F., & Smalley, R. E. (1985). C60: Buckminsterfullerene. *Nature, 318*, 162–163.

Kumar, R., Sakthivel, R., Behura, R., Mishra, B., & Das, D. (2015). Synthesis of magnetite nanoparticles from mineral waste. *Journal of Alloys and Compounds, 645*, 398–404.

Laurent, S., Forge, D., Port, M., Roch, A., Robic, C., Vander Elst, L., & Muller, R. N. (2008). Magnetic iron oxide nanoparticles: Synthesis, stabilization, vectorization, physicochemical characterizations, and biological applications. *Chemical Reviews, 108*, 2064–2110.

Le, T.-D., Suttikhana, I., & Ashaolu, T. J. (2023). State of the art on the separation and purification of proteins by magnetic nanoparticles. *Journal of Nanobiotechnology, 21*, 363. https://doi.org/10.1186/s12951-023-02123-7

Lee, S.-Y., & Park, S.-J. (2013). TiO_2 photocatalyst for water treatment applications. *Journal of Industrial and Engineering Chemistry, 19*, 1761–1769.

Lehn, J.-M. (2009). Supramolecular chemistry and self-organisation. In *Presented at the supramolecular chemistry and self-organisation.*

Liang, G., Li, Y., Yang, C., Zi, C., Zhang, Y., Hu, X., & Zhao, W. (2020). Production of biosilica nanoparticles from biomass power plant fly ash. *Waste Management, 105*, 8–17.

Liao, C.-Z., Tang, Y., Lee, P.-H., Liu, C., Shih, K., & Li, F. (2017). Detoxification and immobilization of chromite ore processing residue in spinel-based glass-ceramic. *Journal of Hazardous Materials, 321*, 449–455.

Liu, B., Zhang, S., Pan, D., & Chang, C. (2016). Synthesis and characterization of micaceous iron oxide pigment from oily cold rolling mill sludge. *Procedia Environmental Sciences, 31*, 653–661.

Liu, T., Han, X., Wang, Y., Yan, L., Du, B., Wei, Q., & Wei, D. (2017). Magnetic chitosan/anaerobic granular sludge composite: Synthesis, characterization and application in heavy metal ions removal. *Journal of Colloid and Interface Science, 508*, 405–414.

Luo, B., Liu, S., & Zhi, L. (2012). Chemical approaches toward graphene-based nanomaterials and their applications in energy-related areas. *Small, 8*, 630–646.

Lyly, L. H. T., Chang, Z. H., & Teow, Y. H. (2024). Nanomaterials synthesis from the industrial solid wastes. In *Green and sustainable approaches using wastes for the production of multifunctional nanomaterials* (pp. 177–190). Elsevier.

Mahmoodi, S., Elmi, A., & Hallaj-Nezhadi, S. (2018). Copper nanoparticles as antibacterial agents. *Journal of Molecular Pharmaceutics & Organic Process Research, 6*, 1–7.

Majumder, S., Chatterjee, S., Basnet, P., & Mukherjee, J. (2020). ZnO based nanomaterials for photocatalytic degradation of aqueous pharmaceutical waste solutions—A contemporary review. *Environmental Nanotechnology, Monitoring & Management, 14*, 100386. https://doi.org/10.1016/j.enmm.2020.100386

Maleki, B., Natheghi, H., Tayebee, R., Alinezhad, H., Amiri, A., Hossieni, S. A., & Nouri, S. M. M. (2020). Synthesis and characterization of Nanorod magnetic Co–Fe mixed oxides and its catalytic behavior towards one-pot synthesis of Polysubstituted pyridine derivatives. *Polycyclic Aromatic Compounds, 40*, 633–643. https://doi.org/10.1080/10406638.2018.1469519

Malhotra, B. D., & Ali, M. A. (2018). Nanocomposite Materials. In *Nanomaterials for Biosensors* (pp. 145–159). Elsevier. https://doi.org/10.1016/B978-0-323-44923-6.00005-4

Manjunathan, J., Pavithra, K., Nangan, S., Prakash, S., Saxena, K. K., Sharma, K., Muzammil, K., Verma, D., Gnanapragasam, J. R., & Ramasubburayan, R. (2024). Polyethylene terephthalate waste derived nanomaterials (WDNMs) and its utilization in electrochemical devices. *Chemosphere, 353*, 141541.

Manna, S., Prakash, S., & Das, P. (2019). Synthesis of graphene oxide nano-materials coated bio-char using carbonaceous industrial waste for phenol separation from water. *Colloids and Surfaces A: Physicochemical and Engineering Aspects, 581*, 123818.

Masaud, Z., Liu, G., Roseng, L. E., & Wang, K. (2023). Progress on pulsed electrocatalysis for sustainable energy and environmental applications. *Chemical Engineering Journal, 475*, 145882.

McCullough, K. E., Peczak, I. L., Kennedy, R. M., Wang, Y.-Y., Lin, J., Wu, X., Paterson, A. L., Perras, F. A., Hall, J., & Kropf, A. J. (2023). Synthesis of platinum nanoparticles on strontium titanate nanocuboids via surface organometallic grafting for the catalytic hydrogenolysis of plastic waste. *Journal of Materials Chemistry A, 11*, 1216–1231.

Meena, J., shankari Sivasubramaniam, S., David, E., & K, S. (2024). Green supercapacitors: Review and perspectives on sustainable template-free synthesis of metal and metal oxide nanoparticles. *RSC Sustainability, 2*, 1224–1245. https://doi.org/10.1039/D4SU00009A

MiarAlipour, S., Friedmann, D., Scott, J., & Amal, R. (2018). TiO_2/porous adsorbents: Recent advances and novel applications. *Journal of Hazardous Materials, 341*, 404–423.
Mirzaei, A., & Neri, G. (2016). Microwave-assisted synthesis of metal oxide nanostructures for gas sensing application: A review. *Sensors and Actuators B: Chemical, 237*, 749–775.
Moghaddasi, S., Khoshgoftarmanesh, A. H., Karimzadeh, F., & Chaney, R. L. (2013). Preparation of nano-particles from waste tire rubber and evaluation of their effectiveness as zinc source for cucumber in nutrient solution culture. *Scientia Horticulturae, 160*, 398–403.
Mondal, P., Anweshan, A., & Purkait, M. K. (2020). Green synthesis and environmental application of iron-based nanomaterials and nanocomposite: A review. *Chemosphere, 259*, 127509.
Mousa, S., Ammar, N., & Ibrahim, H. (2016). Removal of lead ions using hydroxyapatite nanomaterial prepared from phosphogypsum waste. *Journal of Saudi Chemical Society, 20*, 357–365.
Munir, M. T., Mansouri, S. S., Udugama, I. A., Baroutian, S., Gernaey, K. V., & Young, B. R. (2018). Resource recovery from organic solid waste using hydrothermal processing: Opportunities and challenges. *Renewable and Sustainable Energy Reviews, 96*, 64–75.
Nair, R. R., Blake, P., Grigorenko, A. N., Novoselov, K. S., Booth, T. J., Stauber, T., Peres, N. M., & Geim, A. K. (2008). Fine structure constant defines visual transparency of graphene. *Science, 320*, 1308–1308.
Nakata, K., & Fujishima, A. (2012). TiO_2 photocatalysis: Design and applications. *Journal of Photochemistry and Photobiology C: Photochemistry Reviews, 13*, 169–189.
Narayana, T. (2009). Municipal solid waste management in India: From waste disposal to recovery of resources? *Waste Management, 29*, 1163–1166. https://doi.org/10.1016/j.wasman.2008.06.038
Nazir, S., Zhang, J.-M., Junaid, M., Saleem, S., Ali, A., Ullah, A., & Khan, S. (2024). Metal-based nanoparticles: Basics, types, fabrications and their electronic applications. *Zeitschrift für Physikalische Chemie, 238*, 965–995.
Nguyen, K. T., & Zhao, Y. (2014). Integrated graphene/nanoparticle hybrids for biological and electronic applications. *Nanoscale, 6*, 6245–6266.
Novoselov, K. S., Fal′ ko, V. I., Colombo, L., Gellert, P. R., Schwab, M. G., & Kim, K. (2012). A roadmap for graphene. *Nature, 490*, 192–200.
Novoselov, K. S., Geim, A. K., Morozov, S. V., Jiang, D., Zhang, Y., Dubonos, S. V., Grigorieva, I. V., & Firsov, A. A. (2004). Electric field effect in atomically thin carbon films. *Science, 306*, 666–669.
Oestreicher, V., García, C. S., Pontiggia, R., Rossi, M. B., Angelomé, P. C., & Soler-Illia, G. J. (2020). E-waste upcycling for the synthesis of plasmonic responsive gold nanoparticles. *Waste Management, 117*, 9–17.
Palagati, S., & Reddy, J. (2024). Synthesis by top-down and bottom-up. In *Advanced materials: production, characterization and multidisciplinary applications* (Vol. 201).
Pan, Z., Tian, J., Xu, G., Li, J., & Li, G. (2011). Characteristics of adsorbents made from biological, chemical and hybrid sludges and their effect on organics removal in wastewater treatment. *Water Research, 45*, 819–827.
Panchatcharam, P. (2024). Recent advancements in the green synthesis of bioactive metallic nanoparticles from biological entities and their biomedical applications. In *Biogenic wastes-enabled nanomaterial synthesis: applications in environmental sustainability* (pp. 239–256).
Parashar, M., Shukla, V. K., & Singh, R. (2020). Metal oxides nanoparticles via sol–gel method: A review on synthesis, characterization and applications. *Journal of Materials Science: Materials in Electronics, 31*, 3729–3749.
Parida, S., & Dutta, D. P. (2022). Nanostructured materials from biobased precursors for renewable energy storage applications. In *Biorenewable nanocomposite materials, vol. 1: Electrocatalysts and Energy Storage* (pp. 307–366). ACS Publications.
Peczak, I.L., 2023. Syntheses of platinum catalysts on strontium Titanate Nanocuboids for selective polyolefin hydrogenolysis.
Peng, Y., Azeem, M., Li, R., Xing, L., Li, Y., Zhang, Y., Guo, Z., Wang, Q., Ngo, H. H., Qu, G., & Zhang, Z. (2022). Zirconium hydroxide nanoparticle encapsulated magnetic biochar composite

derived from rice residue: Application for as(III) and as(V) polluted water purification. *Journal of Hazardous Materials, 423*, 127081. https://doi.org/10.1016/j.jhazmat.2021.127081
Pol, V. G., & Thiyagarajan, P. (2010). Remediating plastic waste into carbon nanotubes. *Journal of Environmental Monitoring, 12*, 455–459.
Rai, J. P. N., & Saraswat, S. (2023). *Green technologies for waste management: A wealth from waste approach.* CRC Press.
Rai, M., Yadav, A., & Gade, A. (2009). Silver nanoparticles as a new generation of antimicrobials. *Biotechnology Advances, 27*, 76–83.
Rajaambal, S., Sivaranjani, K., & Gopinath, C. S. (2015). Recent developments in solar H2 generation from water splitting. *Journal of Chemical Sciences, 127*, 33–47.
Rao, M. M., Reddy, D. H. K. K., Venkateswarlu, P., & Seshaiah, K. (2009). Removal of mercury from aqueous solutions using activated carbon prepared from agricultural by-product/waste. *Journal of Environmental Management, 90*, 634–643. https://doi.org/10.1016/j.jenvman.2007.12.019
Rath, P., Bhardwaj, L. K., Yadav, P., & Bhardwaj, A. K. (2024). A synthesis of biogenic nanoparticles (NPs) for the treatment of wastewater and its application: A review. In *Biogenic wastes-enabled nanomaterial synthesis: Applications in environmental sustainability* (pp. 127–148).
Ritchie, H., Samborska, V., & Roser, M. (2023). *Plastic pollution.* Our World in Data.
Saleh, T. A. (2022). Development and synthesis of nanoparticles and nanoadsorbents. In *Interface science and technology* (pp. 127–165). Elsevier.
Samaddar, P., Ok, Y. S., Kim, K.-H., Kwon, E. E., & Tsang, D. C. (2018). Synthesis of nanomaterials from various wastes and their new age applications. *Journal of Cleaner Production, 197*, 1190–1209.
Schneider, J., Matsuoka, M., Takeuchi, M., Zhang, J., Horiuchi, Y., Anpo, M., & Bahnemann, D. W. (2014). Understanding TiO_2 photocatalysis: Mechanisms and materials. *Chemical Reviews, 114*, 9919–9986.
Seth, S., & Rathinasabapathi, P. (2022). A short review on detection of antibiotics in Milk using nanomaterial-based biosensor. *Food Analytical Methods, 15*, 2181–2192. https://doi.org/10.1007/s12161-022-02291-6
Sharma, B., Walia, A., Singh, A., Yadav, S., & Soni, U. (2020). Impact of water-soluble 2D CdSe/CdS nanoplatelets on seed germination and early growth in Vigna radiata. *FlatChem, 23*, 100182. https://doi.org/10.1016/j.flatc.2020.100182
Shokri, A., Pahlevani, F., Levick, K., Cole, I., & Sahajwalla, V. (2017). Synthesis of copper-tin nanoparticles from old computer printed circuit boards. *Journal of Cleaner Production, 142*, 2586–2592.
Singh, G. K., Gupta, K., & Chaudhary, S. (2014). Solid waste management: Its sources, collection, transportation and recycling. *IJESD, 5*, 347–351. https://doi.org/10.7763/IJESD.2014.V5.507
Somanathan, A., Mathew, N., & Arfin, T. (2024). Environmental impacts and developments in waste-derived nanoparticles for air pollution control. In *Waste-Derived Nanoparticles* (pp. 281–318).
Suryanarayana, C. (2001). Mechanical alloying and milling. *Progress in Materials Science, 46*, 1–184.
Sushil, S., Scholz, P., Pollok, K., Ondruschka, B., & Batra, V. (2011). Application of industrial waste based catalysts for total oxidation of propane. *Chemical Engineering Journal, 166*, 568–578.
Szczepanik, B. (2017). Photocatalytic degradation of organic contaminants over clay-TiO_2 nanocomposites: A review. *Applied Clay Science, 141*, 227–239.
Tang, C., He, Z., Liu, H., Xu, Y., Huang, H., Yang, G., Xiao, Z., Li, S., Liu, H., Deng, Y., Chen, Z., Chen, H., & He, N. (2020). Application of magnetic nanoparticles in nucleic acid detection. *Journal of Nanobiotechnology, 18*, 62. https://doi.org/10.1186/s12951-020-00613-6
Tatrari, G., Karakoti, M., Tewari, C., Pandey, S., Bohra, B. S., Dandapat, A., & Sahoo, N. G. (2021). Solid waste-derived carbon nanomaterials for supercapacitor applications: A recent overview. *Materials Advances, 2*, 1454–1484. https://doi.org/10.1039/D0MA00871K

Tchobanoglous, G., Theisen, H., Vigil, S. (1993). Integrated solid waste management: Engineering principles and management issues.

Thakur, A., & Kumar, A. (2023). *Ecotoxicity analysis and risk assessment of nanomaterials for the environmental remediation* (p. 2100438). Wiley Online Library.

Tiwari, A. K., Jha, S., Shukla, M., Shukla, R., Singh, R., Singh, A. K., Pathak, A. K., Ojha, R. P., & Dikshit, A. (2024). Scope to improve the synthesis of nanomaterial's using industrial waste. In *Green and sustainable approaches using wastes for the production of multifunctional nanomaterials* (pp. 117–129). Elsevier.

Tripathy, S., Rodrigues, J., & Shimpi, N. G. (2023). Top-down and bottom-up approaches for synthesis of nanoparticles. In *Nanobiomaterials—perspectives for medical applications in the diagnosis and treatment of diseases* (Vol. 145, pp. 92–130).

Tu, Y.-J., You, C.-F., Chang, C.-K., Wang, S.-L., & Chan, T.-S. (2013). Adsorption behavior of as (III) onto a copper ferrite generated from printed circuit board industry. *Chemical Engineering Journal, 225*, 433–439.

Tu, Y.-J., You, C.-F., Chen, M.-H., & Duan, Y.-P. (2017). Efficient removal/recovery of Pb onto environmentally friendly fabricated copper ferrite nanoparticles. *Journal of the Taiwan Institute of Chemical Engineers, 71*, 197–205.

Turgut, P., & Yesilata, B. (2008). Physico-mechanical and thermal performances of newly developed rubber-added bricks. *Energy and Buildings, 40*, 679–688.

Varghese, R. J., Parani, S., Thomas, S., Oluwafemi, O. S., & Wu, J. (2019). Introduction to nanomaterials: Synthesis and applications. In *Nanomaterials for solar cell applications* (pp. 75–95). Elsevier.

Veisi, H., Karmakar, B., Tamoradi, T., Tayebee, R., Sajjadifar, S., Lotfi, S., Maleki, B., & Hemmati, S. (2021). Bio-inspired synthesis of palladium nanoparticles fabricated magnetic Fe_3O_4 nanocomposite over Fritillaria imperialis flower extract as an efficient recyclable catalyst for the reduction of nitroarenes. *Scientific Reports, 11*, 4515. https://doi.org/10.1038/s41598-021-83854-1

Vesilind, P.A., Rimer, A.E. (1981). *Unit operations in resource recovery engineering.*

Vidor, F. F., Meyers, T., & Hilleringmann, U. (2016). Inverter circuits using ZnO nanoparticle based thin-film transistors for flexible electronic applications. *Nanomaterials, 6*, 154.

Visa, M. (2016). Synthesis and characterization of new zeolite materials obtained from fly ash for heavy metals removal in advanced wastewater treatment. *Powder Technology, 294*, 338–347.

Visa, M., Andronic, L., & Duta, A. (2015a). Fly ash-TiO_2 nanocomposite material for multi-pollutants wastewater treatment. *Journal of Environmental Management, 150*, 336–343.

Visa, M., Andronic, L., & Enesca, A. (2016). Behavior of the new composites obtained from fly ash and titanium dioxide in removing of the pollutants from wastewater. *Applied Surface Science, 388*, 359–369.

Visa, M., Andronic, L., Lucaci, D., & Duta, A. (2011). Concurrent dyes adsorption and photodegradation on fly ash based substrates. *Adsorption, 17*, 101–108.

Visa, M., Bogatu, C., & Duta, A. (2015b). Tungsten oxide–fly ash oxide composites in adsorption and photocatalysis. *Journal of Hazardous Materials, 289*, 244–256.

Visa, M., Carcel, R. A., Andronic, L., & Duta, A. (2009). Advanced treatment of wastewater with methyl orange and heavy metals on TiO_2, fly ash and their mixtures. *Catalysis Today, 144*, 137–142.

Visa, M., & Chelaru, A.-M. (2014). Hydrothermally modified fly ash for heavy metals and dyes removal in advanced wastewater treatment. *Applied Surface Science, 303*, 14–22.

Visa, M., Isac, L., & Duta, A. (2015c). New fly ash TiO_2 composite for the sustainable treatment of wastewater with complex pollutants load. *Applied Surface Science, 339*, 62–68.

Visa, M., Isac, L., & Duta, A. (2012). Fly ash adsorbents for multi-cation wastewater treatment. *Applied Surface Science, 258*, 6345–6352.

Waketola, A. G., Tegegne, N. A., & Hone, F. G. (2025). Recent progress in silver and gold nanoparticle-based plasmonic organic solar cells. *Plasmonics, 20*, 5521–5556. https://doi.org/10.1007/s11468-024-02430-1

Wang, J., & Chen, C. (2009). Biosorbents for heavy metals removal and their future. *Biotechnology Advances, 27*, 195–226.

Wang, S., Yan, W., & Zhao, F. (2020). Recovery of solid waste as functional heterogeneous catalysts for organic pollutant removal and biodiesel production. *Chemical Engineering Journal, 401*, 126104.

Wilson, D. C. (2023). Learning from the past to plan for the future: An historical review of the evolution of waste and resource management 1970–2020 and reflections on priorities 2020–2030–the perspective of an involved witness. *Waste Management & Research, 41*, 1754–1813.

Wu, A., Zhao, X., Wang, J., Tang, Z., Zhao, T., Niu, L., Yu, W., Yang, C., Fang, M., Lv, H., Liu, S., & Wu, F. (2021). Application of solid-phase extraction based on magnetic nanoparticle adsorbents for the analysis of selected persistent organic pollutants in environmental water: A review of recent advances. *Critical Reviews in Environmental Science and Technology, 51*, 44–112. https://doi.org/10.1080/10643389.2020.1720493

Wu, C.-H., Kuo, C.-Y., & Lo, S.-L. (2009). From waste to resource: A CASE study of heavy-metal sludge by microwave treatment. *Journal of Environmental Engineering and Management, 19*.

Yadav, K. K., & Rana, S. (2024). Application of waste-synthesized nanoparticles in energy. In *Waste-Derived Nanoparticles* (pp. 319–340). Elsevier.

Yan, A., Wang, X., & Cheng, J. (2024). Roles of metal nanoparticles for supercapacitors: A review. *Particle & Particle Systems Characterization, 41*, 2400002. https://doi.org/10.1002/ppsc.202400002

Yang, L., Wang, F., Hakki, A., Macphee, D. E., Liu, P., & Hu, S. (2017). The influence of zeolites fly ash bead/TiO_2 composite material surface morphologies on their adsorption and photocatalytic performance. *Applied Surface Science, 392*, 687–696.

Yu, T., Chen, H., Hu, T., Feng, J., Xing, W., Tang, L., & Tang, W. (2024). Recent advances in the applications of encapsulated transition-metal nanoparticles in advanced oxidation processes for degradation of organic pollutants: A critical review. *Applied Catalysis B: Environmental, 342*, 123401. https://doi.org/10.1016/j.apcatb.2023.123401

Yuan, H., Deng, L., Cai, X., Zheng, T., Zhou, S., Chen, Y., & Yuan, Y. (2016). Recycling electroplating sludge to produce sustainable electrocatalysts for the efficient conversion of carbon dioxide in a microbial electrolysis cell. *Electrochimica Acta, 222*, 177–184.

Yusuf, A., Almotairy, A. R. Z., Henidi, H., Alshehri, O. Y., & Aldughaim, M. S. (2023). Nanoparticles as drug delivery systems: A review of the implication of nanoparticles' physicochemical properties on responses in biological systems. *Polymers, 15*, 1596. https://doi.org/10.3390/polym15071596

Zahid, M., Rashid, A., Akram, S., Rehan, Z. A., & Razzaq, W. (2018). A comprehensive review on polymeric nano-composite membranes for water treatment. *Journal of Membrane Science & Technology, 8*, 1–20.

Zang, T., Wang, H., Liu, Y., Dai, L., Zhou, S., & Ai, S. (2020). Fe-doped biochar derived from waste sludge for degradation of rhodamine B via enhancing activation of peroxymonosulfate. *Chemosphere, 261*, 127616.

Zhang, D., Luo, H., Zheng, L., Wang, K., Li, H., Wang, Y., & Feng, H. (2012). Utilization of waste phosphogypsum to prepare hydroxyapatite nanoparticles and its application towards removal of fluoride from aqueous solution. *Journal of Hazardous Materials, 241*, 418–426.

Zhang, H., Liu, J., Ou, C., Shen, J., Yu, H., Jiao, Z., Han, W., Sun, X., Li, J., & Wang, L. (2017). Reuse of Fenton sludge as an iron source for $NiFe_2O_4$ synthesis and its application in the Fenton-based process. *Journal of Environmental Sciences, 53*, 1–8.

Zhang, H., Xue, G., Chen, H., & Li, X. (2018). Magnetic biochar catalyst derived from biological sludge and ferric sludge using hydrothermal carbonization: Preparation, characterization and its circulation in Fenton process for dyeing wastewater treatment. *Chemosphere, 191*, 64–71.

Zhang, K., Song, X., Liu, M., Chen, M., Li, J., & Han, J. (2023). Review on the use of magnetic nanoparticles in the detection of environmental pollutants. *Water, 15*, 3077. https://doi.org/10.3390/w15173077

Zhao, X., Yang, L., Guo, J., Xiao, T., Zhou, Y., Zhang, Y., Tu, B., Li, T., Grzybowski, B. A., & Yan, Y. (2021). Transistors and logic circuits based on metal nanoparticles and ionic gradients. *Nature Electronics, 4*, 109–115.

Zhou, G., Fang, F., Chen, Z., He, Y., Sun, H., & Shi, H. (2015). Facile synthesis of paper mill sludge-derived heterogeneous catalyst for the Fenton-like degradation of methylene blue. *Catalysis Communications, 62*, 71–74.

Zhou, H., Su, M., Lee, P.-H., & Shih, K. (2017b). Synthesis of submicron lead oxide particles from the simulated spent lead paste for battery anodes. *Journal of Alloys and Compounds, 690*, 101–107.

Zhou, Y., Fang, X., Wang, T., Hu, Y., & Lu, J. (2017a). Chelating agents enhanced CaO2 oxidation of bisphenol a catalyzed by Fe^{3+} and reuse of ferric sludge as a source of catalyst. *Chemical Engineering Journal, 313*, 638–645.

Zou, J., Tang, L., & Kang, L. (2025). Innovative heating for the Nano age: Exploring the potentials of Carbothermal shock. *ACS Nano, 19*, 152–186.

Applications of Solid Wastes in Environmental Friendly Processes

Shalini Tiwari, Shreya Kotnala, Richa Saxena, Rashmi Verma, and Mohd Amin Mir

1 Introduction

Due to ever-increasing population, urbanization, industrialization, and economic development, the production of solid waste has significantly increased (Voukkali et al., 2023). It has been estimated by the World Bank that about 2.01 billion tons of solid waste are generated by various industrial, agricultural as well as local activities, which has resulted in the contamination of the environment (Varjani et al., 2021). Waste generated globally comprises 44% food waste, 2% waste from wood, 2% from leather and rubber, 12% waste from plastics, 2% from paper waste, 5% from waste glass, and about 4% metal waste (Ferdous et al., 2021). The majority of the waste is solid waste and it is obtained from various industrial, agricultural, municipal sectors as well as from animal waste (Abdel-Shafy & Mansour, 2018; Gupta, 2018; Gupta & Gupta, 2016, 2018). Since solid waste is a rich source of sugars, proteins, and minerals, it may be used for the production of green energy sources as well as other biologically important by-products.

S. Tiwari · S. Kotnala (✉)
Department of Chemistry, School of Basic and Applied Sciences, Shri Guru Ram Rai University, Dehradun, Uttarakhand, India
e-mail: shreya.kotnala001@gmail.com

R. Saxena
Department of Physics, School of Basic and Applied Sciences, Shri Guru Ram Rai University, Dehradun, Uttarakhand, India

R. Verma
Department of Biotechnology, School of Basic and Applied Sciences, Shri Guru Ram Rai University, Dehradun, Uttarakhand, India

M. A. Mir
Department of Mechanical Engineering, Prince Mohammad Bin Fahd University, Al-Khobar, Saudi Arabia

H. Gupta et al. (eds.), *Advanced Technologies for Assessment, Recycling, and Management of Solid Waste*, Waste as a Resource,
https://doi.org/10.1007/978-3-032-24788-9_6

Various conventional as well as modern technologies can be applied for the transformation of the solid waste into clean fuel. This not only leads to the production of eco-friendly fuel but at the same time helps in the disposal and management of waste (Un, 2023). There are several categories of solid waste based on their origin as well as on the basis of their physical, chemical, and biological characteristics (Abdel-Shafy & Mansour, 2018). These may be biodegradable, non-biodegradable, combustible, non-combustible, commercial, industrial, agricultural, hazardous waste, etc. There are several methods that can be employed for the management of waste such as landfilling, composting, incineration, and anaerobic degradation that can be utilized for the production of bio-based products (Yaashikaa et al., 2020). Several bio-based products are biosurfactants, biopolymers, bioethanol, and biohydrogen that can serve as an alternative to conventional fuel. There are several reasons why the waste generated is not managed properly; a few of them are a lack of awareness, limited knowledge about technology, and financial constraints (Atnkut et al., 2025). The waste has a detrimental effect not only on the plants but also on the environment, so there is a pressing need for the generation of technologies for effective management of waste (Agboola et al., 2025).

This chapter provides a comprehensive overview of various types of waste that are generated via different sources. Various waste management methods, such as landfilling, incineration, composting, pyrolysis, and biomethanation, that can be employed for the effective management of waste are discussed in detail. Additionally, various useful products derived from solid waste are explained in the chapter. The chapter also provides detailed insight into various environmental as well as industrial applications of solid waste. The chapter also highlights important obstacles such as a lack of infrastructure, financial constraints, and regulatory gaps and opportunities for improving waste utilization through research and innovation. In order to reduce negative environmental effects, the chapter also highlights how crucial it is to incorporate the idea of waste valorization into strategies for sustainable development.

2 Types of Solid Wastes

As a result of rapid growth in production and consumption processes, societies regularly produce and reject solid materials from a variety of sectors, including agricultural, commercial, residential, industrial, and institutional (Peng et al., 2023). The most significant amount of waste generated is solid waste. Solid waste is generally the solid by-products that result from human and animal activity that are discarded as undesirable material (Abdel-Shafy & Mansour, 2018). Due to the presence of solid waste, there exists enormous strain on the natural resources as a result of which effective and sustainable development is substantially jeopardized (Shah et al., 2021). Solid wastes include both organic and inorganic waste materials that are produced in a society and typically have no value to the initial users. These may include materials such as product packaging, lawn clippings, furniture, clothing, bottles, kitchen trash, paper, appliances, paint cans, and batteries (El-Khateeb, 2022). Therefore,

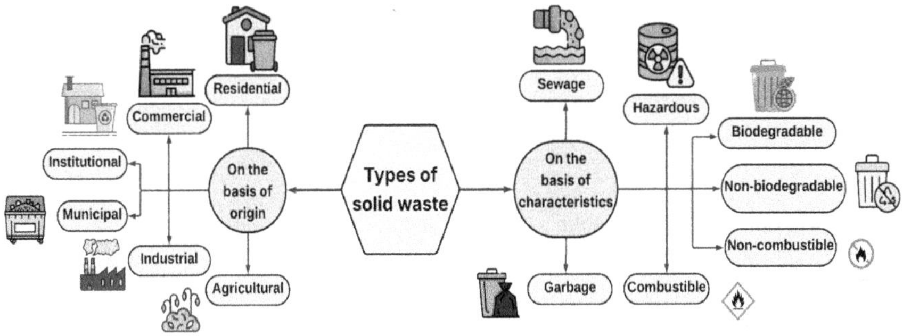

Fig. 1 Various types of solid waste

solid wastes include both a more uniform buildup of industrial, mineral, and agricultural wastes and a diverse mass of trash from the urban community (Noor et al., 2020). Solid waste is generally of several types based on the source of their origin as well as on the basis of physical, chemical, and biological characteristics of waste and is depicted in Fig. 1 (Dehghani et al., 2021). The detailed classification of solid waste is presented in Table 1.

On the basis of source of their origin, solid wastes are of various types such as residential, commercial, institutional, municipal, industrial, and agricultural. Residential solid waste is generated from residential areas and may include leftover food, vegetable peels, plastic, clothing, ashes, e-waste, and other garbage from homes, flats, etc. (Gupta, 2015; Gupta & Shiju, 2025; Kumar & Gupta, 2020; Singh & Arora, 2018). Another type of waste is commercial waste which originates from commercial sources such as restaurants, hotels, motels, shops, medical stores, and other business sites and may include garbage such as leftover food, glassware, metals, and ashes (Vambol et al., 2024). Waste generated from offices, colleges, schools, other administrative, and educational sectors are considered under institutional solid waste and these may include paper, plastic, glassware, and other materials (Onyilokwu et al., 2024). In addition to this, municipal solid waste comprises dust, leafy matter, building debris, residual sludge from treatment plants, and other materials produced by a variety of municipal operations, such as street cleaning, landscaping, construction, and demolition (Arenibafo, 2023). The hazardous waste generated from industrial operations, ashes, demolition and construction debris, and process waste are known as industrial waste (Ramli et al., 2021). Agricultural waste comprises agricultural waste from fields, orchards, vineyards, farms, etc., as well as rotten food grains and vegetables, litter, etc. (Khan et al., 2024a).

Based on the physical, chemical, and biological characteristics, solid waste is classified into various forms. Combustible wastes are materials such as paper, cloth, rubber, cardboard, material obtained from garden, etc., that can easily burn, whereas non-combustible waste includes things like glasses, crockery, dust, dirt, and aluminum cans that do not burn easily (Jabeen et al., 2022). This type of waste is usually generated from homes, businesses, institutions, etc. The waste that can easily decompose by the action of microorganisms is termed as biodegradable waste, and it

Table 1 Classification of solid waste

Category	Type	Description	Examples	References
Based on source of origin	Residential solid waste	Waste from residential areas	Leftover food, plastic, clothing, ashes	Singh and Arora (2018)
	Commercial solid waste	Waste from commercial sources	Food waste, glassware, metals, ashes	Vambol et al. (2024)
	Institutional solid waste	Waste from schools, offices, and other institutions	Paper, plastic, glassware	Onyilokwu et al. (2024)
	Municipal solid waste	Waste from municipal activities	Dust, leaves, building debris, sludge	Arenibafo (2023)
	Industrial solid waste	Waste from industrial operations	Hazardous materials, process waste	Ramli et al. (2021)
	Agricultural solid waste	Waste from farming and related activities	Rotten food, crop residue, litter	Khan et al. (2024a)
Based on physical, chemical, and biological characteristics				
Physical property	Combustible waste	Easily burnable materials	Paper, cloth, rubber, garden waste	Jabeen et al. (2022)
	Non-combustible waste	Materials that don't burn easily	Glass, crockery, aluminum cans	Jabeen et al. (2022)
Biological degradability	Biodegradable Waste	Easily decomposed by microbes	Food waste, vegetable peels, paper	Ghai et al. (2022)
	Non-biodegradable Waste	Not easily decomposed	Plastics, metals, glass	Ghai et al. (2022)
Chemical and toxic properties	Hazardous waste	Risk to health and environment	Paint cans, pesticides, radioactive waste	Sanito et al. (2022)
Treatment residue	Sewage waste	Solid by-products from sewage treatment	Sludge, grit, organic debris	Cheong et al. (2022)
Organic waste	Garbage	Organic matter from food handling	Vegetable peels, animal waste	Adetunji et al. (2023)

includes materials made of organic matter from a variety of domestic and commercial processes, such as leftover food, fruit and vegetable peels, paper, textiles, wood, etc. On the other hand, non-biodegradable waste cannot be easily decomposed by the action of microorganisms, and it may include materials such as plastics, glass, cans, metals, and other recyclable and inorganic elements (Ghai et al., 2022). The waste obtained from various sectors such as industrial, institutional, or consumer origin, that poses a risk to human health and the environment, either immediately or over time, is classified as hazardous wastes. This is because of their toxicity, reactivity, corrosivity, and other physical, chemical, and biological or radioactive properties (Ma et al., 2024). Although these wastes might be harmful gases or liquids, they are classified as solid waste because they are stored in solid containers. Empty solvent, paint, and pesticide containers are common examples of hazardous wastes; these materials are often combined with municipal trash and end up in the urban waste stream (Faiz et al., 2024). Certain hazardous wastes have the potential to ignite landfill fires and explode in incinerators. Pathological waste generated from hospitals and radioactive waste are also hazardous and harmful and require proper management strategies to combat their harmful effect (Sanito et al., 2022). Sewage wastes are the solid by-products of sewage treatment. They are produced by treating organic sludge that has been removed from both raw and treated sewage, and they are primarily organic. At the initial stage of treatment, the inorganic portion of raw sewage, such as grit and eggshells, is separated because it may entrain pathogens with putrescible organic debris and needs to be buried right away. Although the majority of treated, dewatered sludge is beneficial as a soil conditioner, it is not always cost-effective. Therefore, unless particular arrangements are made for its disposal, solid sludge ends up in the stream of municipal waste (Cheong et al., 2022). Garbage includes vegetable and animal wastes that come from handling, selling, storing, preparing, cooking, and serving food. The putrescible organic matter in the garbage that makes up these wastes gives off an unpleasant smell and draws rats and other rodents. As a result, it needs particular care when being handled, stored, and disposed of (Adetunji et al., 2023).

3 Solid Waste Management Methods

Solid waste is generated from various sources. Their proper collection and disposal is essential in order to avoid their negative effect on the environment. Various methods involved in the management of solid waste (depicted in Fig. 2) are described below:

3.1 Landfilling

The process of systematically disposing of biodegradable and non-biodegradable waste in a landfill or terrestrial burial site that is situated outside the municipal

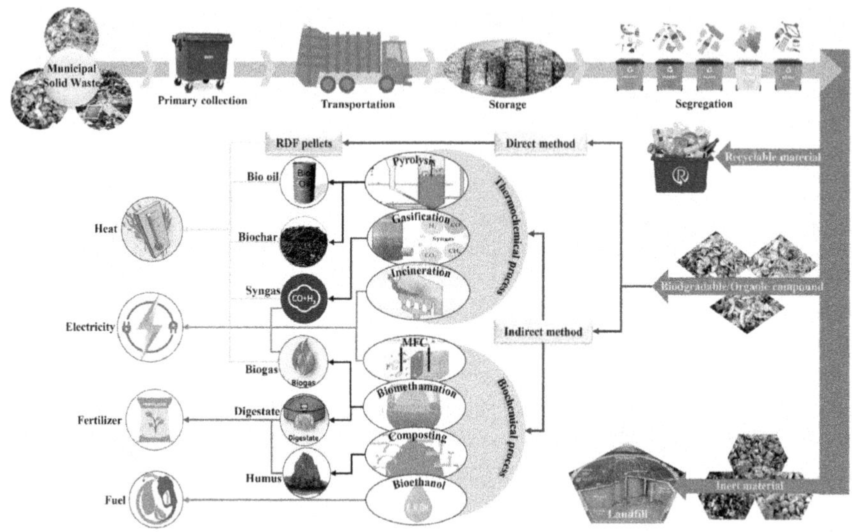

Fig. 2 Various methods involved in management of solid waste and the important value-added products obtained from solid waste (Adapted from original article by Hoang et al., 2022 after taking permission from Elsevier)

suburban regions is known as landfilling (Dagwar & Dutta, 2024; Nanda & Berruti, 2021). It is the most common and profitable method of disposing of waste in many countries. This technology is preferred over other solid waste management methods as it offers advantages such as low cost and fewer labor-intensive processes (Singh et al., 2024). Furthermore, by using leachate and landfill gas for energy production, a consolidated landfill can also generate revenue (Khan et al., 2024b). The location of the landfill and the remediation strategies depend upon the type of waste (such as biomedical waste, radioactive waste, household waste, waste from construction or demolition sites, and chemical waste) (Nalladiyil & Babu, 2024). Landfills are categorized into six classes based on the waste refuse: Class 1 landfills are generally used for the soil disposal, Class 2 are utilized for disposal of minerals and the waste obtained from construction and demolition sites, Class 3 on the other hand is used for disposal of municipal solid waste, Class 4 landfills are used for industrial and chemical waste, Class 5 are utilized for hazardous waste disposal, and lastly class 6 is used for underground disposal of hazardous waste (Ehrig et al., 2011). The landfills are also classified on the basis of their functionality type into secure, monofil, reusable, and bioreactor landfills (Mor & Ravindra, 2023; Nanda & Berruti, 2021). The secure landfill involves the burial of waste so as to delay the harmful impact on the environment. The monofil landfill holds the waste that cannot be handled by incineration, resource recovery, or composting. Reusable landfills enable garbage stability for an extended amount of time, followed by excavation to recover compost, metals, glass, and plastics, and the bioreactor landfill is specially designed landfill with liners, leachate collection, and recirculation systems in order to obtain maximum degradation of waste by microbial activity, thereby reducing the environmental impact.

3.2 *Incineration*

It is a thermal process that involves the heating of waste in the presence of air at around 850°C, which allows the raw waste material to be recovered as feedstock (Meena et al., 2023). It produces solid residues, carbon dioxide, water and non-combustible products (Khan et al., 2022a). Usually in India, the incineration process is followed for the disposal of hospital and biological waste due to low CV content (800–1100 kcal/kg), a high organic percentage (40–60%), and a high moisture contact (40–60%) (Parmar & Rai, 2017).

3.3 *Composting*

The process of breaking down of organic waste by the action of microorganisms is known as composting (Khan et al., 2022b; Potysz et al., 2018; Zomorodian et al., 2018). The first step of degradation involves the breaking down of organic waste into smaller components via earthworms and soil insects (such as ants, mites, and beetles). The second steps involve the breaking down of organic material by the action of soil microbes (Kumar et al., 2018). There are several composting methods such as vermicomposting, windrow composting, co-composting, aerobic and anaerobic composting that are widely used in the rural regions of developing nations (Kumar et al., 2018; Yuvaraj et al., 2021). Before choosing a particular composting method, several factors such as moisture content, temperature, C/N ratio, size of compost material, and aeration must be taken into consideration as all these factors are known to affect the process of composting (Gruchlik et al., 2018; Khan et al., 2022b).

Vermicomposting is the method that employs earthworms to break down the organic waste into granular and nitrogen-rich material (Patwa et al., 2020). The temperature requirement in vermicomposting is usually lower as compared to composting as in the former process, less microbial activity is involved (Kumar & Agrawal, 2020). It is one of the sustainable practices involved in organic waste processing. Another type of composting is aerobic composting that involves the breaking down of organic waste, thereby releasing CO_2, NO_2, and NO_3 by using the oxygen-dependent microorganisms. On the other hand, when the breakdown occurs in the absence of oxygen by the microorganisms, the composting is termed as anaerobic composting and the by-products are hydrogen sulfide (H_2S) and methane (CH_4). The process of aerobic composting is faster when compared to anaerobic composting (Schnell et al., 2020). The presence of harmful substances such as batteries and heavy metals is known to interrupt the process of composting (Wuana & Okieimen, 2011). The other form of composting is windrow composting which involves the breaking down of organic waste by separating it and placing it in a long and narrow pile that is either trapezoidal or triangular in shape. The two types of wind composting methods are traditional wind composting and forced-ventilated wind composting (Das et al., 2021). The turning of composting material results in increased aerobic conditions of the pile and the ventilation (Khan et al., 2022b).

3.4 Reduce, Reuse, and Recycle (3Rs)

The very first step is to reduce the use of solid waste products that cannot be degraded easily. The amount of waste generated can be decreased by using green materials as raw materials, extending the waste lifecycle, lowering energy consumption, optimizing design, and minimizing heat losses (Meena et al., 2023). In the solid waste management, this step is very crucial as it emphasizes on reducing the cost and environmental impact that are generally associated with the handling of waste. The second step involved is reusing of the material again and again. For example, a plastic bag which is used to carry fruits and vegetables can be used again and again for the same purpose as it can lead to less waste generation. The last step involves the recycling of the material which involves the collection and separation of material of a similar type that can be recycled again to make other useful products (Meena et al., 2023). A diagrammatic illustration of recycling and proper management of plastic waste is depicted in Fig. 3.

3.5 Biomethanation

It is a process that involves the treatment of solid waste in the absence of air to convert the organic matter present in the waste into methane and manure (Kumar et al., 2025a, 2025b). This method is commonly utilized for the degradation of waste obtained from agricultural sources as it produces biogas as well as manure rich in nutrients, and the agricultural waste is known to have high organic content (Phillip et al., 2024). The major byproduct of this process is biogas, which is mainly composed of methane and carbon dioxide, which can be extensively utilized as a source of fuel for cooking and power generation purposes (Alengebawy et al., 2024). The biomethanation process is also implemented in various regions of India for the disposal of fruit and vegetable waste (Meena et al., 2023).

3.6 Pyrolysis and Gasification

Gasification is a thermochemical method which is utilized to convert the biomass and other types of solid waste into gaseous products such as carbon monoxide, hydrogen,

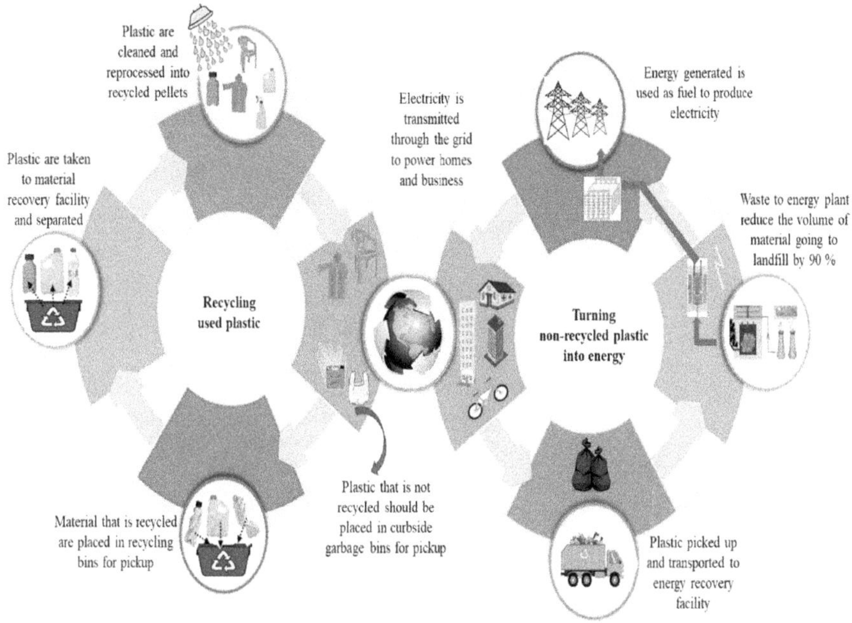

Fig. 3 Recycling and management of plastic waste (Adapted from original article by Kumar et al., 2024 after taking permission from Elsevier)

carbon dioxide, and methane by using gasifying agents such as air, steam, oxygen or supercritical water inside a gasifier in the temperature range of 500–1200 °C (Nguyen et al., 2024). This method offers several advantages over the traditional methods such as landfill and incineration as it could use a wide variety of feedstock and convert those into valuable products (Kong et al., 2022). Gasification involves several steps, including drying of feedstock initially, then pyrolyzing, and finally gasification to obtain the value-added products (Wu et al., 2022) (Table 2).

4 Useful Products from Solid Wastes

Various products that are useful and are obtained from solid waste are depicted in Fig. 4 and are described in detail below.

4.1 Bioethanol

Bioethanol is one of the biofuels that can be obtained from various sources such as sugars, starch, and lignocellulosic waste (Du et al., 2018; Varjani et al., 2021). The

Table 2 Overview of various solid management methods with their key features

Method	Description	Key features	References
Landfilling	Systematic disposal of biodegradable and non-biodegradable waste in designated landfill sites outside municipal regions	• Most common and cost-effective method • Generates revenue via leachate and gas recovery • Six landfill classes (Class 1–6) • Types: Secure, monofil, reusable, Bioreactor	Dagwar and Dutta (2024), Khan et al. (2024b) and Singh et al. (2024)
Incineration	Thermal treatment of waste at ~ 850 °C in the presence of air	• Produces CO_2, water, residues • Suitable for hospital and biological waste • Common in India for low calorific value wastes	Khan et al. (2022b) and Meena et al. (2023)
Composting	Microbial breakdown of organic waste	• Types: Aerobic, anaerobic, vermicomposting, windrow, co-composting • Influenced by C/N ratio, moisture, temperature, etc • Vermicomposting uses earthworms; windrow uses long piles	Gruchlik et al. (2018), Zomorodian et al. (2018), Patwa et al. (2020) and Yuvaraj et al. (2021)
Reduce, reuse, recycle (3Rs)	Waste minimization strategy	• Reduce: Use of eco-friendly raw materials • Reuse: Reuse items like plastic bags • Recycle: Collect, separate, and remake new items	Meena et al. (2023)

(continued)

Table 2 (continued)

Method	Description	Key features	References
Biomethanation	Anaerobic digestion of organic waste to produce methane and manure	• Ideal for agricultural waste • Produces biogas ($CH_4 + CO_2$) • Used for cooking, electricity • Implemented for fruit/vegetable waste disposal	Meena et al. (2023) and Kumar et al., (2025a, 2025b)
Pyrolysis and gasification	Thermochemical conversion of solid waste into gas using high temperatures and gasifying agents	• Products: CO, H_2, CH_4, CO_2 • Feedstocks: biomass, municipal waste • Steps: drying, pyrolysis and gasification • Efficient and versatile vs. landfilling/incineration	Kong et al. (2022) and Wu et al. (2022)

lignocellulosic waste is easily available and is present as peels from barley, corns, grains, branches, leaves, stems, grasses, husks, shells, straws, sawdust, etc. (Gupta et al., 2023). The production of bioethanol is highly advantageous for supplying energy in countries where there is a large production of lignocellulosic and forestry waste. This fuel offers several advantages such as being less toxic and biodegradable when compared to fossil fuels. Bioethanol can be mixed with gasoline and can be used in diesel machines as a source of fuel as it has high octane as well as a low cetane number (Tan et al., 2017). Gasohol is the mixture of bioethanol (24%) and gasoline (76%). Bioethanol can also be derived from food waste as the food waste is an easily available, cheap, and recyclable material (Singh et al., 2022). Obtaining bioethanol from various waste sources has resulted in energy conservation, thereby increasing the growth of the country. Also, using bioethanol has resulted in the minimization of carbon emissions (Li & Cheng, 2020).

The potential of various types of solid waste for the production of bioethanol via the process of microbial fermentation is highlighted in Table 3. From the data, it is evident that bread waste produces the highest yield of ethanol (106.9 g/L) via using *Saccharomyces cerevisiae* KL17, demonstrating its potential as a low-cost substrate for the generation of biofuel. On the other hand, pretreated lignocellulosic waste, by-products of rice milling, dairy waste, and household food waste generated bioethanol yield of 55.2, 51.88, 30.6, and 6 g/L. From this data, it is seen that the type of waste, microorganism used, as well as conditions play a vital role in the generation of biofuel from various organic substrates that can be used as a feedstock for sustainable biofuel production.

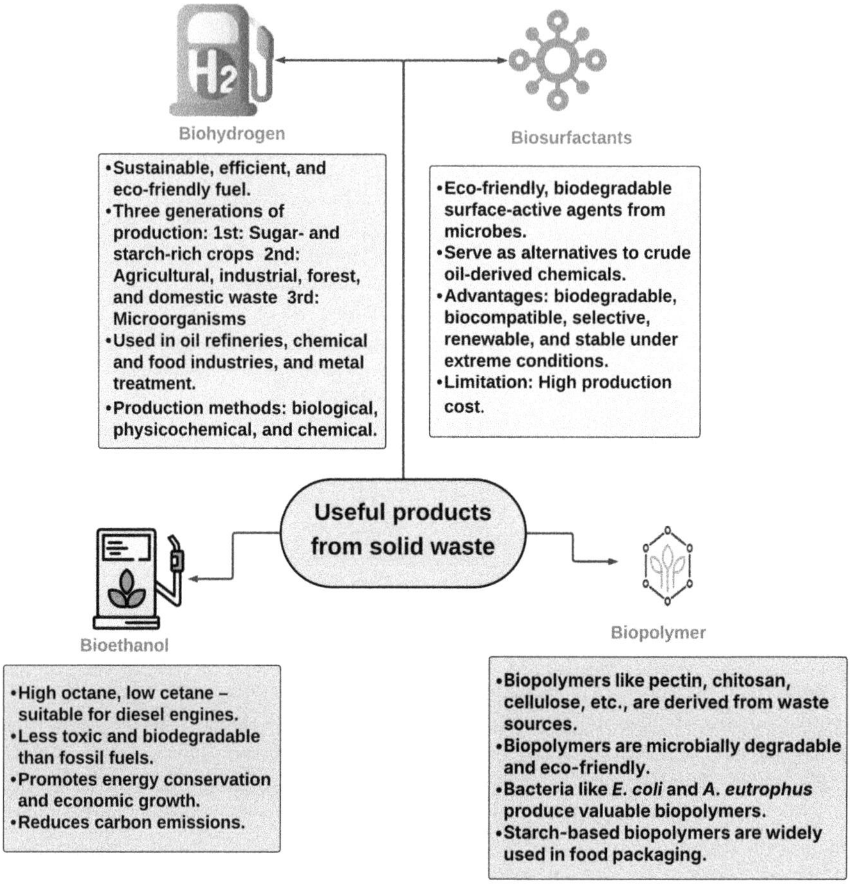

Fig. 4 Various useful products obtained from solid waste

4.2 Biopolymer

The generation of biopolymers such as pectin, chitosan, cellulose, starch, and collagen from various waste sources has gained significant attention in recent times (Jegal, 2018). Collagen, a proteinaceous biopolymer is utilized in several industries such as pharmaceuticals, cosmetics, food, and leather, and is usually obtained from mammals (cattle, bones, fish waste, chicken processing waste, and pig skin) (Varjani et al., 2021). The production rate of biopolymers is quite high in the global plastic market. The waste materials obtained from various sources are used for the generation of biogas as well as biohydrogen (Kabir et al., 2020). Several bacterial species such as *Alcaligenes eutrophus, Azotobacter beijerincki*, and *Escherichia coli* can be utilized to produce biopolymers for waste conversion into bioproducts like exopolysaccharides and polyhydroxyalkanoates (Pagliano et al., 2017). The most

Table 3 Various types of solid waste utilized for the production of bioethanol (Adapted from original article by Adetunji et al., 2023 after taking permission from Elsevier)

Solid waste type	Microbe used	Fermentation conditions	Bioethanol yield (g/L)	References
Pretreated lignocellulosic sawdust	*Saccharomyces cerevisiae*	Agitation 330 rpm; substrate: 45.14 mL/L; incubation 102 h; 36.5 °C	55.2	Saravanakumar and Kathiresan (2014)
Palm kernel cake hydrolysate	Recombinant *Geobacillus thermoglucosidasius*	Inoculum 10% (v/w); pH 7.0; agitation 250 rpm; 30 °C	9.9	Raita et al. (2016)
Dairy waste	Recombinant *Lactococcus lactis* subsp. *cremoris* MG1363	30 °C	30.6	Liu et al. (2016)
Rice milling by-products	Recombinant *Saccharomyces cerevisiae* M2	30 °C; pH 5.5	51.88	Favaro et al. (2017)
Organic municipal solid waste	*Mucor indicus* CCUG 22,424	Inoculum 0.02% (w/v); 37 °C; agitation 150 rpm; pH 5.5	27.4	Mahmoodi et al. (2018)
Potato peel waste	*Wickerhamia* sp. SD1	Inoculum 2% (v/v); 30 °C; pH 7.0; agitation 300 rpm	21.7	Hossain et al. (2018)
Household food waste	*Saccharomyces cerevisiae*	30 °C	6.0	Matsakas et al. (2015)
Bread waste	*Saccharomyces cerevisiae* KL17	Inoculum 2% (v/v); pH 6.0; agitation 200 rpm; 30 °C	106.9	Narisetty et al. (2022)

commonly used biopolymers are starch-based and they are used for the packaging of food items. Biopolymers can easily be degraded by microbes. These microorganisms convert the insoluble biopolymer into a soluble one and utilize their components for their metabolic development (Siracusa, 2019).

Various types of waste specially dairy waste (cheese whey and whey permeates) have demonstrated high potential for the production of biopolymers utilizing the microbial strains and are presented in Table 4. Xanthan gum was obtained in high yield from *Xanthomonas campestris* pv. *Mangiferaeindicae* (46.8 g L^{-1}) and *X. campestris* pv. *campestris* (21.91 g L^{-1}), whereas PHA was obtained from *Haloferax mediterranei* (9.6 g L^{-1}) and *Pseudomonas hydrogenovora* (1.27 g L^{-1}). From the results depicted in table, it is seen that some strains such as *Ralstonia eutropha* and

Lactococcus lactis produce relatively low yields, thereby highlighting the importance of microbial selection as well as substrate composition as two prominent factors that influence the production of biopolymer. Overall, the valorization of dairy wastes into value-added biopolymers offers an environmentally sustainable strategy for waste management and bioproduct generation.

4.3 Biohydrogen

Biohydrogen is gaining lots of attention as it is produced via biomass and biomass serves as a sustainable and inexhaustible resource with minimal energy consumption and greater efficiency for the preservation of food items (Malode et al., 2025). Production of biohydrogen via biomass is categorized into three generations, namely first (includes crops with high sugar and starch content), second (includes waste from agricultural fields, organic, industrial, domestic and forest waste), and third (includes microorganisms that catalyze the production of biohydrogen) (Bolatkhan et al., 2019; Ren et al., 2019; Stamenković et al., 2020). The production of biohydrogen via biomass is a cost-effective and energy-efficient process and has application in several sectors such as oil refineries, chemical production units, metal treatment industries and food production industries (Santiago et al., 2020). Various factors such as volatile solid content, COD, BOD, and inhibiting compounds play an important role in the generation of biohydrogen. The production of biohydrogen can be done by biological, physicochemical, and chemical methods (Boodhun et al., 2017; Gopalakrishnan et al., 2019; Sampath et al., 2020). The biological process is divided into three parts: dark fermentation, which involves the utilization of complex waste, photofermentation that utilizes the photosynthetic bacteria for the production at higher rates, and electrobio hydrogenation, which involves the release of protons and electrons via exo-electrogenic bacteria in specific reactors. Physicochemical methods involve the use of liquid hot water and steam explosion for the generation of biohydrogen, and the chemical technique involves acid, alkaline, and oxidative treatments for the enhanced yield of biohydrogen.

The amount of biohydrogen produced depends upon the type of organic waste, the type of reactor system as well as the temperature. Various types of solid waste utilized for the generation of biohydrogen are depicted in Table 5. Among all the substrates depicted in table, it is seen that maximum yield of biohydrogen (754 mL H_2/L d) is obtained when pineapple peel solid waste was used in ASBR reactor under mesophilic conditions, whereas dry fermentation of municipal solid waste produced about 7.2 L H_2/L d of biohydrogen under thermophilic conditions. The moderate yield ranging from 12.9 mL H_2/g COD to 100 mL H_2/g VS was obtained with food waste under mesophilic conditions. These results highlight that both the choice of feedstock and reactor design are crucial in optimizing biohydrogen production efficiency.

Table 4 Various microbial strains used for the production of biopolymers from dairy wastes (Adapted from original article by Ranganathan et al., 2020 after taking permission from Elsevier)

Source	Microbial strain	Biopolymer	Yield	References
Cheese whey lactose	*Xanthomonas campestris* and *Xanthomonas pelargonii*	Xanthan gum	0.42 g of xanthan g^{-1} of lactose; 0.27 g of xanthan g^{-1} of lactose	Niknezhad et al. (2015)
Cheese whey	*Haloferax mediterranei*	P(3HB-co-3HV)	9.6 g L^{-1}	Pais et al. (2016)
Whey	Dairy wastewater as innoculum	PHA	0.284 g L^{-1}	Bosco and Chiampo (2010)
Whey supernatant	*Thermus thermophilus* HB8	PHA	0.57 g L^{-1}	Pantazaki et al. (2009)
Cheese whey	*Xanthomonas campestris* pv. Mangiferaeindicae IBSF 1230	Xanthan gum	46.8 g L^{-1} of gum	Mesomo et al. (2009)
Whey permeate	*Pseudomonas hydrogenovora*	PHA	1.27 g L^{-1}	Koller et al. (2017)
Milk whey	*X. campestris* pv. campestris—2149	Xanthan gum	21.91 g L^{-1}	Nery et al. (2008)
Whey lactose	*Haloferax mediterranei* *Pseudomonas hydrogenovora* *Hydrogenophaga pseudoflava*	PHA	5.5 g L^{-1} 1.3 g L^{-1} 2.7 g L^{-1}	Koller et al. (2017)
Lactose and whey permeate	*Sinorhizobium* 41 *meliloti*, *Hydrogenophage pseudoflava* DSM 1034	PHA	*Sinorhizobium* 41 *meliloti*: 0.483 g L^{-1} *Hydrogenophage pseudoflava*: 0.375 g L^{-1}	Povolo and Casella (2003)
Whey	*Ralstonia eutropha*	PHB	0.17 g L^{-1}	Marangoni et al. (2002)

Table 5 Various types of waste used in the production of biohydrogen (Adapted from original article by Keskin et al., 2019 after taking permission from Elsevier)

Waste Type	Reactor Configuration	Temperature	Yield/rate	References
Municipal solid waste	ABR	Mesophilic	250 mL H_2/g VS	Tawfik et al. (2011)
Food waste	ABR	Mesophilic	12.9 mL H_2/g COD	Tawfik and El-Qelish (2012)
Wheat starch	ABR	Mesophilic	1.2 L H_2/L d	Jürgensen et al. (2015)
Soybean	ABR	Mesophilic	3 L H_2/L d	Zhu et al. (2012)
Deoiled jatropha	ASBR	Mesophilic	1.48 L H_2/L d	Kumar and Lin (2014)
Food waste	ASBR	Mesophilic	62.6 mL H_2/g VS	Kim et al. (2010)
Food waste	ASBR	Mesophilic	100 mL H_2/g VS	Kim et al. (2010)
Swine manure + glucose	ASBR	Thermophilic	5 L H_2/L d	Wu et al. (2009)
Pineapple waste	ASBR	Mesophilic	754 mL H_2/L d	Wang et al. (2006)
Municipal solid waste + paper mill waste	Dry fermenter	Thermophilic	4.5–7.2 L H_2/L d	Elsamadony and Tawfik (2015)
Food waste + cardboard waste	Dry fermentation	Mesophilic	89 mL H_2/g VS	Capson-Tojo et al. (2018)

4.4 Biosurfactant

Surfactants derived from microbes are termed as biosurfactants. They are gaining attention due to their biodegradable nature and organic characteristics (Das & Kumar, 2018). Several microbial species utilized for the production of biosurfactants are *Rhodococcus, Pseudomonas, Bacillus, and Candida spp (*Markande et al., 2021*).* Surface-active agents like trehalose lipids are produced from *Rhodococcus* species. These biosurfactants are more beneficial and competent than other surface-active agents (Markande et al., 2021). Biosurfactants serve as a substitute for various chemical products of crude oil. These are produced from various chemicals, milk products, hydrocarbons as well as oil wastes. The production of biosurfactant is dependent on carbon sources. Various conditions like pH, incubation period, temperature, aeration, and agitation speed are known to affect the production of biosurfactants (López-Prieto et al., 2019). Several advantages of using surfactants from biological sources are they are biodegradable, biocompatible, harmless, highly selective and have the ability to get renewed from the feedstock and also demonstrate activity under unfavorable conditions of temperature, salinity, and pH. The major drawback associated with biosurfactants is that they are not cost-effective as their production requires ample money (López-Prieto et al., 2019). The potential of various organic waste for the generation of biosurfactant by microorganisms is highlighted in Table 6.

Table 6 Biosurfactant production from various feedstocks and microorganisms used [Adapted from Mohanty et al., 2021 under the terms of the Creative Commons Attribution License (CC BY 4.0)]

Feedstock	Microorganism used	Type of biosurfactant	Applications	References
Fruit and vegetable waste	*Halobacteriaceae archaeon*	Lipopeptides	Bioremediation of oil-contaminated sites, cosmetic industry, pharmaceutical industries	Kumar et al. (2016) and Paraszkiewicz et al. (2018)
Starch rich waste	*Pseudomonas aeruginosa*	Rhamnolipids	Oil recovery, environmental protection, pharmaceutical industries	Varjani and Upasani (2021)
Frying oil waste	*Bacillus subtilis*	Surfactin	Environmental applications, bioremediation of oil and hydrocarbon-contaminated sites	Andrade et al. (2017)
Oil processing waste and by-products	*Bacillus subtilis*	Surfactin	Remediation of oil and hydrocarbon-contaminated sites	Sharma et al. (2018)
Lignocellulosic wastes	*Pseudomonas aeruginosa*	Rhamnolipids	Environmental applications	Hasanizadeh et al. (2017) and Ozdal et al. (2017)
Kitchen organic waste	*Candida bombicola, Pseudomonas aeruginosa, Bacillus subtilis, Starmerella bombicola*	Sophorolipids, Rhamnolipids, Lipopeptides	Development of waste-based bioprocesses, environmental applications, remediation of oil sites	Li et al. (2017), Zhao et al. (2017), Chen et al. (2018), Lourenço et al. (2018), Jadhav et al. (2019), Kaur et al. (2019), Liu et al. (2019), Marcelino et al. (2019), Santos et al. (2019) and Varjani et al. (2020)

5 Environmental Applications

There are several environmental applications of solid waste that are discussed in detail below.

5.1 Water and Air Purification

The use of agricultural residues and industrial by-products such as biochar and fly ash in environmental clean-up has been growing because of their high surface area and porosity. Water pollution by heavy metals (like Pb^{2+} and Cd^{2+}), organic dyes, and residue from pharmaceuticals can be handled using biochar which comes from biomass (Trivedi et al., 2025). Recent studies have shown that surface oxygenated functional groups of biochar (–COOH, –OH) and π-π interaction with aromatic pollutants are major contributors to adsorptive removal and that the pore structure of biochar can be modified to reach adsorption capacities of over 200 mg/g with such dyes as methylene blue (Liu et al., 2023a). Recent literature has pointed out that these fly ash composites with iron oxides can remove over 95% of arsenic and fluoride within the wastewater besides acting as a low-cost adsorbent, which can be recycled (Mir & Rather, 2024). Just like slag, fly ash, which is formed during coal combustion, contains silicate and alumina phases that render it a desirable site of ion exchange and sticking (Zhang et al., 2021). They are more cost-effective and sustainable compared to normal activated carbon or synthetic resins because these waste-based adsorbents are cheaper and more eco-friendly in areas that are still developing (Gupta et al., 2021). Solid waste materials have also been included in systems designed to clean air. Activated carbon from coconut shells and sawdust is used to catch hazardous volatile organic compounds (VOCs) and greenhouse gases such as CO_2 and CH_4 (Khedulkar et al. 2025). Adsorption of VOCs is primarily achieved by filling micropores and hydrophobic forces, and the efficiencies of the removal are commonly greater than 90% when the conditions are optimal. Moreover, activated carbons made of biochar and waste material have been investigated in terms of combined carbon capture with competitive CO_2 uptake rates of (23 mmol g^{-1} at ambient pressure) with dramatically lower costs of production and environmental impacts (Wang et al., 2023). These innovations do not only enhance the air quality, but also ensure that there is reduction in the greenhouse gases in the atmosphere, which also cause climate changes.

5.2 Soil Enhancement and Land Reclamation

Solid organic wastes, especially composted food waste, green waste, manure, and biosolids, have a lot of organic matter and nutrients like nitrogen, phosphorus, and potassium. If they are added to soil, it gets stronger, drier soil is kept wetter, and microbes are more active. By having these benefits, lands can be restored, farming can become better and less chemical fertilizer is needed (Manea et al., 2024). The scientific reasons indicate that organic amendments enhance soil aggregation, cation exchange capacity, which leads to higher nutrient regaining and lowering the leaching losses. Application of compost and vermicompost also enhances microbial biomass carbon and enzymatic activities that will improve the nutrient cycling and enhance long-term soil fertility (Liu et al., 2023b).

Using organic amendments from waste can make contaminated or emptied land more secure for growing crops and stabilizing the pH. As a result of these actions, ecosystems improve and the soil holds more carbon (Amoah-Antwi et al., 2020). Recent researches in biochar made out of agricultural residues emphasize on its double purpose in improving soil fertility and stabilizing heavy metals including Cd, Pb, and Zn, thereby ensuring food safety and soil well-being (Gao et al., 2024). Also, long-term field experiment shows that such amendments made of wastes can enhance soil organic carbon stocks by 10–20% in a decade, which will play a significant role in carbon sequestration and mitigation of climatic changes (Peng et al., 2024). The practices do not only restore the degraded or polluted lands, but also they meet the objective of sustainable agriculture through minimizing the use of synthetic inputs, enhancing soil biodiversity, and increasing the resilience to drought and erosion.

5.3 Reduction in Landfill Burden and Pollution

Using solid waste to produce biofuels, compost, reused construction materials, and fertilizers prevents a great deal of waste from going to landfills. This approach is friendly to the environment: it cuts greenhouse gases, prevents the pollution of groundwater, and makes land use more efficient (Abdel-Shafy & Mansour, 2018). Landfilling is a significant source of methane (CH_4) gases, which carry a global warming potential 28 times higher than that of CO_2; hence, by shifting biodegradable components of municipal solid waste (MSW) to composting and bioenergy generation, it can significantly reduce the amount of greenhouse gases. An example of these is that lifecycle assessment has revealed that composting and anaerobic digestion can cut the amount of methane emission of organic waste by nearly 90% relative to traditional landfilling (Gavrila et al., 2025).

In addition, composting, anaerobic digestion, and mechanical–biological treatment (MBT) are good options for handling a lot of municipal solid waste (MSW). With these technologies, the environment is protected and waste can be used to provide new energy and materials (Fei et al., 2018). Recent developments of MBT systems combine aerobic composting and material recovery facilities, which result in a diversion rate of 60–70% of entering MSW, and also produce refuse-derived fuels (RDF) to power industrial kilns (Bian et al., 2025). Furthermore, anaerobic digestion of food, agricultural waste results into biogas that has a yield of 350–500 L/kg volatile solids that can be upgraded to biomethane and introduced into gas grids (Marín et al., 2022). Not only can such valorization strategies decrease the landfill load and isolated cases of leachate contamination, but also meet resource loops, turning waste management into an important source of circular economy and climate mitigation activities.

5.4 *Carbon Sequestration and Climate Change Mitigation*

Carbon sequestration or long-term storage of carbon dioxide in the atmosphere is one of the most serious techniques of reducing climate change. The valorization of solid waste is becoming an increasingly significant part in this process as it allows the transformation of organic residues into a stable form of carbon or capturing greenhouse gases. One of the most common climate-smart practices that is thoroughly researched is the application of biochar in agriculture. In pyrolysis of organic solid wastes, biomass carbon is converted to biochar, one of the most recalcitrant materials in the form of biochar, which has mean residence times in the soil between hundreds and thousands of years, and therefore biomass carbon is locked away, and the CO_2 in the atmosphere is held in a stable form (Shoudho et al., 2024). The increase in the soil fertility, soil nutrient retention, and water-holding capacity is also enhanced by biochar, and this presents a two-fold advantage of improving the soil and reducing climate. Lifecycle assessments demonstrate that, on top of CO_2, biochar is able to reduce non-CO_2 gases including methane (CH_4) and nitrous oxide (N_2O), and research indicates that net decreases of 2–5 t CO_2-eq per ton of applied feedstock are achievable (Ashiq et al., 2020).

The use of waste-based solutions is facilitated by other resources besides biochar that lead to carbon sequestration and reduction of emissions. Recycling of organic waste also helps in the reduction of unregulated emission of methane into the atmosphere, and yet it increases the pools of carbon in the soil. Moreover, solid waste can be integrated with the carbon capture and storage (CCS) or carbon capture and utilization (CCU) technologies, providing some extra ways of mitigation. An example is that the construction and demolition waste deposits can contain alkaline residues that can serve as passive carbonation agents, trapping atmospheric CO_2 into stable carbonates (Alli et al., 2024). These approaches with them emphasize the fact that solid waste valorization cannot only remove landfill materials but also make them long-term carbon sinks, which solidifies the position of waste management on climate policy and circular economy systems.

5.5 *Circular Economy and Resource Conservation*

Circular economy approach to solid waste management aims at reducing waste and maximizing resource use by following the 4Rs of reduce, reuse, recycle, and recover. The idea is presented in contrast to the traditional linear approach to the so-called take-make-dispose path because it prolongs the life cycle of the materials to the greatest extent and reduces the amount of virgin resources being extracted. As an illustration, the natural build materials, such as the natural stone and sand, can be replaced by the construction and demolition waste, such as the recycled concrete, bricks and ceramics, which can save up to 60–70% of natural aggregates extraction, thus saving geological resources and decreasing the carbon footprint of a building (Zhang et al., 2024). In the same manner, industrial wastes such as fly ash and slag are also being introduced in geopolymer cements; not only they recycle wastes but also reduce greenhouse gases emitted due to Portland cement manufacturing.

Farm wastes on the agricultural front are being converted to bioplastics, bio-packaging, and biofuels which are renewable substitutes to products based on petroleum. One such thing is the manufacture of polyhydroxyalkanoates (PHAs) and polylactic acid (PLA) biodegradable polymers based on the use of lignocellulosic biomass and has a growing niche in the area of sustainable packaging (Moshood et al., 2022). Also, waste valorization within the city and industrial systems leads to the creation of eco-novation, job creation, and the sustainable economic development, which aligns with the UN Sustainable Development Goals (SDGs) (Sharma et al., 2021). Lifecycle assessments verify that circular practices can lower greenhouse gas emissions by 30–40% in comparison with linear techniques, which point to their contribution to the reduction of climate change and resource security in the long term. Collectively, these routes demonstrate that solid waste, through being managed under the components of the circular economy, is not only a by-product but also a beneficial secondary resource in terms of environmental sustainability and socio-economic resiliency (Yang et al., 2023).

5.6 *Pollution Control Through Waste-Based Materials*

The application of waste-based materials in environmentally sustainable processes has a great potential future in pollution control and resource efficiency. These by-products of industries and agricultural wastes are found undergoing more and more engineering into functional material that can address air, water, and soil pollution as well as help in the sustainable construction activities. Another example would be the use of biochar made of crop residues effectively absorbing up to 90% of the heavy metals like Pb^{2+} and Cd^{2+} in wastewater because of strong adsorption capacity of the biochar and high porosity (Liu et al., 2024). Likewise, pharmaceuticals and dyes in contaminated water streams have been eliminated using modified biochars and fly ash composites, which are small-scale and economical substitutes to activated carbon.

The industrial solid wastes like fly ash, blast furnace slag, red mud, and phosphogypsum are currently being reused to form geopolymer materials that are more mechanically and chemically stable. Not only do such geopolymers trap toxic elements such as arsenic, chromium, and lead, thus preventing pollution of groundwater, but these polymers are a sustainable and environmentally friendly alternative to Portland cement (Li et al., 2024b). Interestingly, geopolymeric binders may be used as an alternative to Portland cement, which will reduce the amount of CO_2 emitted into the environment during construction by 40–80%, and therefore, they are the focus of green buildings (Barbhuiya et al., 2024). Recent studies also indicate the potential of such geopolymer composites to absorb volatile organic compounds (VOCs) in indoor air, and therefore, its usage into air purification system (Soon et al., 2023). Therefore, waste-based materials can not only suppress environmental degradation by using waste materials but also fit the circular economy by transforming waste liability to valuable resources toward a sustainable infrastructure and cleaner ecosystems.

6 Industrial Applications

Effective management of solid waste means it can be a useful input for eco-friendly industrial growth. Many industries now use new ideas to transform solid waste into useful raw materials or energy, which supports environmental protection, resource conservation, and the idea of circular economy. There are important industrial fields that make good use of solid waste for environmental and economic reasons that are depicted in Fig. 4 and are discussed in detail below:

6.1 Construction Industry

The construction sector has an immense potential of cutting its environmental slack with the use of environmentally friendly procedures which involve the use of solid waste in construction. Among the most notable industrial uses of solid waste one may take a construction industry where construction and demolition (C&D) waste, including concrete debris, bricks, gypsum, wood, and glass, is being reused and transformed into new building materials (Abdel-Shafy & Mansour, 2018). Aggregates are commonly recycled which are used in road bases, in the sub-basis layers, as well as in the fresh concrete production. Recent studies show that recycled concrete aggregates (RCA) have compressive strengths similar to natural ones, 10–25% less, but with appropriate treatment methods, such as leaching adhered mortar or surface finishes, compressive strengths and durability can be greatly increased, and thus these materials can be used in construction (Peiris et al., 2025). Moreover, industrial wastes, fly ash (coal combustion), and ground granulated blast furnace slag (GGBFS) in the iron industry are used as supplementary cementitious materials in blended and geopolymer cements (Yang et al., 2024). The options minimize the use of such natural resources as limestone and sand, as well as minimize the carbon footprint associated with cement manufacture. Besides, the ashes of the municipal solid waste are also being considered to be used to manufacture ecologically sound building materials such as bricks and cement. It should be noted that geopolymers produced from these wastes are highly durable and chemically resistant, and they can be immobilized against toxic components such as heavy metals, thus highly qualified as components of sustainable and resilient infrastructure (Barbhuiya et al., 2024).

6.2 Energy Production and Biofuels

Waste-to-energy and bio-fuels can be good alternatives to managing solid waste and a sustainable solution given the potential to convert this type of waste to energy which would reduce the production of clean energy (Lee et al., 2019). Food processing wastes, municipal solid wastes, and agricultural residues are most abundant sources of organic matter, thus are good feedstocks to bioenergy. Recent technologies have shown effectiveness in converting these wastes into biogas, bio-oil, pyrolysis, and gasification which can in turn be used in generation of electricity and as fuels by industries (Rajkumar & Kurinjimalar, 2022). Figure 5 represents the various types of organic solid waste for the production of biofuel and its applications. These renewable energy products can be utilized to produce electricity or as industrial fuel and this will minimize dependence on fossil fuel. As an example, sugarcane bagasse and rice husks are generally used as agricultural wastes in cogeneration plants in the sugar and rice milling sectors, which not only leads to self-signed energy operations but also reduces the cost of disposal. Burning of wastes to produce energy also compliments the principles of a circular economy through up-cycling of waste on top of helping to curb greenhouse gas emissions and ensuring energy security (Awasthi et al., 2024). Also, waste systems have been subjected to lifecycle assessment (LCAs), which show great potential in mitigation of net greenhouse gases over landfilling, especially on mitigation of methane released through anaerobic decomposition of the waste (Cho et al., 2024). Biofuel systems based on algae and microbial fuel cells, as emerging technologies, expand the range of energy derived through waste to combine biological and chemical pathways to produce energy more efficiently (Brennan & Owende, 2010).

Therefore, the idea of valorization of waste into energy and biofuels promotes the principles of the circular economy when converting waste streams into useful energy carriers to provide the long-term sustainability of waste management and energy industries.

6.3 Metal and Resource Recovery

Environmentally friendly processes metal and resource recovery of industrial solid waste is concerned with the extraction of valuable resources such as metals out of waste streams in a manner that has minimal impact on the environment. Some valuable metals such as copper, aluminum, gold, lithium, cobalt, germanium, and rare earth elements can be found in the electronic and industrial wastes (Tejaswini

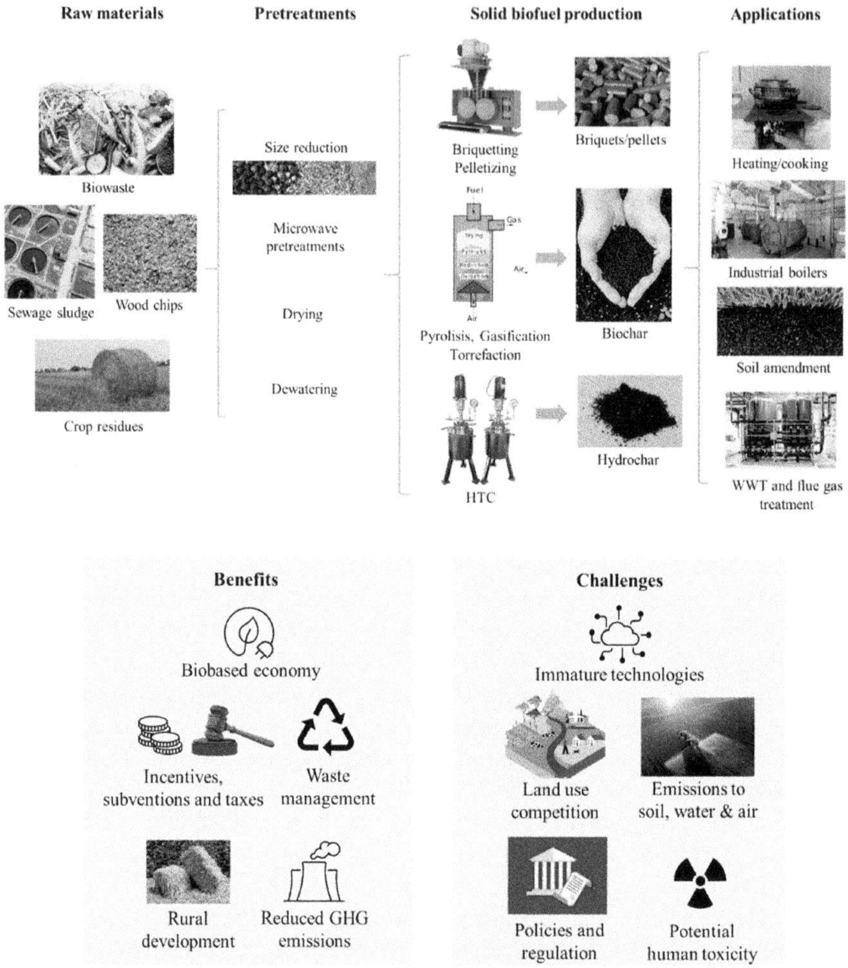

Fig. 5 Overview of solid biofuel production from organic solid waste and its applications (Adapted from original article by Angulo-Mosquera et al., 2021 after taking permission from Elsevier)

et al., 2022; Agarwal et al., 2024; Dhiman et al., 2024; Gupta et al., 2025, Kumar et al., 2025a, 2025b). Recycling of e-waste has become one of the important industrial processes in which the metals are obtained through mechanical separation, hydrometallurgical leaching, and biohydrometallurgical methods which are more efficient and less energy-demanding than the traditional mining (Wang et al., 2024).

Simultaneously, industrial waste, like the red mud (bauxite byproduct of the aluminum industry) and slag of steel production are also being considered as secondary sources of iron, titanium, and rare earths recovery by applying advanced pyrometallurgicals and electrochemical treatment (Li et al., 2024a, 2024b). Recent researches have also focused on the possibilities of microbial consortia in bioleaching processes, which allow to selectively extract metals by using less chemicals and less environmental impact (Zhang et al., 2023). The utilization of these wastes not only minimizes the impact on the environment and the risk of contamination but also leads to the green mining development as the industrial and electronic wastes are transformed into a resource base. This is in line with the principles of a circular economy assisting in conserving the limited mineral resources and contributing to the sustainable development of high-tech and energy-intensive industries.

6.4 Manufacturing of Ceramics, Glass, and Tiles

Innovative use of industrial solid waste products provides a viable solution toward the environmentally friendly production of ceramics, glass, and tiles in environmentally sustainable ways as an alternative to traditional resource-intensive techniques (Vidak Vasić et al., 2024). Other raw materials like fly ash, red mud, glass cullet, and ceramic shards may be integrated in production processes to substitute traditional raw materials. All these raw materials that are of secondary nature not only allow the diversion of waste that is found in the land filling sites but also maximizes the mechanical and thermal features of the end products i.e., in the case of glass–ceramics, ceramic tiles (Silva et al., 2017). They may also be used to greatly reduce the amount of energy required in processing, by using fly ash, or glass cullet may allow the sintering temperatures to be lowered, saving energy and reducing emissions of carbon dioxide. The strategy is resource-conserving and very limited toward the environmental loading of the ceramics and glass sector (Franus et al., 2024).

6.5 Agricultural Inputs and Organic Fertilizers

Organic and agricultural solid wastes are being considered important resources in the manufacturing of environmentally friendly inputs like organic fertilizers and soil conditioners using environmentally sustainable industrial operations (Gamage et al., 2023). Composting, vermicomposting, anaerobic digestion, and biochar production can be used to turn these waste materials, which contain organic matter, nitrogen, phosphorus, potassium, and micronutrients, into highly efficient soil amendments which have a significant effect in enhancing the soil structure, nutrient availability, and microbial diversity. In the example of compost and vermicompost produced by agriculture, municipal, and food discard, soil fertility is reinstated, and crop production is enhanced through their extensive application in organic agriculture and landscaping (Goswami et al., 2024). Also, the bone waste and plant residues that are rich in phosphorus are being converted into pelletized biofertilizers. The method decreases the demand in artificial fertilizers, enhances soil vitality, and assists industries to achieve sustainable farming and zero-waste targets (Majee et al., 2023). Figure 6 depicts the use of agricultural waste in the production of organic fertilizer.

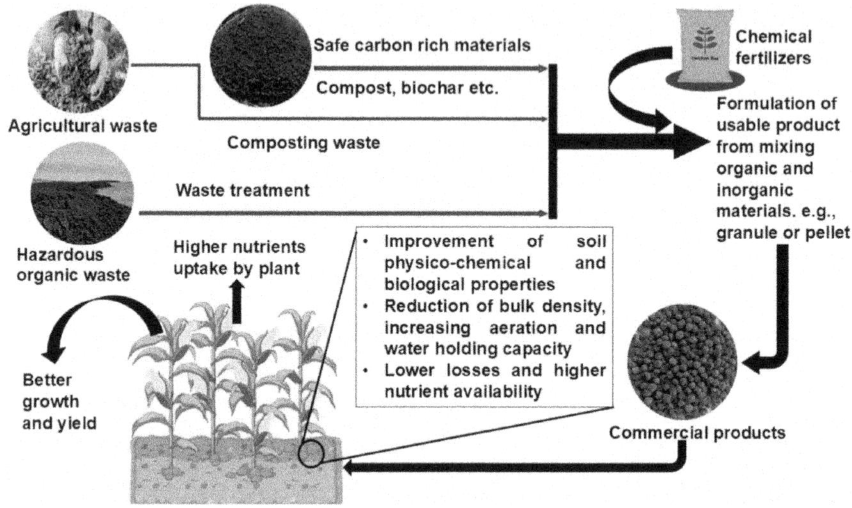

Fig. 6 Application of agricultural waste in production of organic fertilizers [Adapted from Uddin et al., 2025 under the terms of the Creative Commons Attribution License (CC BY 4.0)]

Biochar is a result of pyrolyzing the wastes of agriculture that not only enhances the retention and water storage, but also acts as a permanent store of carbon, thereby reducing climate change (Shoudho et al., 2024). It is important to note that these approaches lead to a reduction in the use of artificial fertilizers, reduce the environmental contamination of the factors of fertilizing the soil and greenhouse gases, and help in the transition to the circular bioeconomy and the zero-waste agriculture (Liu et al., 2025). Overall, the concept of reusing the farm wastes into organic fertilizers not only facilitates the reduction of the strains within the landfills, but also boosts the proper farming systems, which leads to the reinstatement of the soil health, food security, and environmental stability.

6.6 *Packaging and Bioplastics*

Bio-based packaging materials and bioplastics are becoming one of the most important ingredients of environmentally friendly industrial processes not least in terms of solid waste management (Zhao et al., 2023). More industries are turning agricultural and food wastes like starch, cellulose, lignin (e.g., stems of banana, peels of potato, husks of corn), into biodegradable plastic and other packaging materials (Vigneswari et al., 2024). These bio-derived polymers such as polylactic acid (PLA), polyhydroxyalkanoates (PHA), and starch-based composites have several benefits compared to the petroleum-based plastics: they are biodegradable, compostable, less dependent on fossil fuels, and linked with lower greenhouse gas emissions. Recent investigation has demonstrated that bioplastics would better end up in food packaging and consumer goods by incorporating the lignocellulosic residues to improve the mechanical strength and barrier properties of the product (Nizamuddin et al., 2024). Furthermore, biochar and nanocellulose additives based on solid waste streams have also been employed in enhancing thermal stability and lifespan of bio-based films to have a wider range of application than short-term packaging (Khedulkar et al. 2025).

These greener alternatives of plastics made of petroleum have numerous benefits such as it easily decomposes, it has less reliance on fossil fuel, and it produces fewer greenhouse emissions. Their increasing use in packaging and consumer goods industry solves the burning problem of plastic pollution and helps to reach the objectives of climate change mitigation and the circular economy. (Moshood et al., 2022).

6.7 Textile and Paper Industry

Solid waste is produced in large volume in the textile and paper industries and newer and greener processes are increasingly used to reuse, recycle, and up-cycle solid waste. The cellulosic waste products used in the paper mills or agricultural waste products (cotton stalks and jute waste lands) can be converted into paper pulp or could be used to generate blended textiles (Abolore et al., 2024). These activities enable it to minimize the use of virgin raw materials, which is a factor that saves forests, water, and energy that is normally used in making new paper or synthetic fibers. This sustainable and circular industrial ecology can be enhanced by these waste-to-resource methods that conserve landfill mass and are more resource efficient (Chen et al., 2024). The use of solid waste as a resource in the different industries reflects the idea that waste can be recycled to a resource factor, whereby industries work toward achieving the objectives of environmental sustainability (Quan et al., 2024). In construction, energy production, agriculture, and manufacturing, the adoption of waste-generated material and processes eliminates environmental pollution and contributes to the circular economy as part of the value chain of wastes. A cleaner and more resource-efficient industrial future requires further government policy support and industry collaboration to help solid waste to play a key role in ensuring a cleaner and more resource-efficient industrial future (Norouzi et al., 2021).

7 Benefits and Challenges

Utilization of the solid waste in environmentally sustainable operations has many advantages in the industrial, agricultural, and energy sectors. It promotes recovery of the resources and prevents the demand of virgin materials, as it uses waste streams of agricultural wastes, construction wastes, and industrial by-products (Peng et al., 2023). It encourages a closed-loop economy, reduces landfill, and reduces greenhouse gases. Such processes as composting, biochar production, healthier soils, and carbon sequestration, as well as such technologies as anaerobic digestion turn organic waste into renewable energy. Such methods do not only reduce the footprint on the environment but also develop cost and scalable ways of sustainable development (Afshar & Mofatteh, 2024).

Nonetheless, these practices are faced with a number of challenges to be widely established. The accessibility to technology, availability of infrastructure, and segregations of waste are also critical factors that limit efficiency with regard to processing of waste, which is more common in developing parts of the world (Abdel-Shafy & Mansour, 2018). Moreover, the economic competitiveness of waste products is also an issue because they tend to be more expensive than conventional products because they are expensive to process and transport. There are also gaps in the policy and the absence of regulatory incentives that make the industry slow. Besides this, quality issues and health threats that improperly handled hazardous waste and e-waste pose do not encourage acceptance by the population. To mitigate these barriers, a combined approach based on the development of waste processing technologies, favorable policies, awareness among the population, and effective monitoring structures is needed (Kevin van Langen et al., 2021). Only through such a holistic approach can the full potential of solid waste utilization be realized as a cornerstone of sustainable and climate-resilient societies.

Future Scope

The future of using solid wastes in the environmentally sustainable processes is bright with huge potential brought by the development of technology, enhanced environmental consciousness, and the current global tendency to adopt the principles of the circular economy. New waste-to-value technologies (advance bio-refineries, carbon capture-enhanced composting, smart sorting systems, etc.) are likely to make the conversion of solid waste into valuable resources (biofuels, green chemicals, construction materials, bioplastics, etc.) more efficient (Islam et al., 2024). Incorporation of AI, automation, and IoT in the waste management systems will enhance the waste segregation, tracking, and acting on the recycling processes in an economically viable and scalable way. Also, the rising trends in nature-based solutions and sustainable agriculture will bring an increase in demand on waste-derived soil amendment, fertilizers, and energy sources (Ahmed et al., 2024).

In future, there is a potentiality of solid wastes having a ground breaking role in a sustainable industrial ecosystem. Pilot projects have to be scaled up by policy support, research funds, and multi-lateral partnerships in order to make eco-friendly waste use become mainstream (Shahbaz et al., 2023). The emerging industries like green building, e-waste metallurgy, and biodegradable packaging will absorb the waste-based products, thus minimizing the use of the finite resources. Also, incorporation of solid waste valorization into the national climate action plans and sustainable development strategies will enhance the efforts to reduce the pollution, energy security, and climate resilience. Therefore, when properly invested into and guided by the appropriate policy, the scope of solid waste practices may even develop into an essential building block in the process of sustainable development in the global context (Quinto et al., 2025).

8 Conclusion

The sustainable advancement in the field of eco-friendly processes via the efficient valorization of solid wastes provides an avenue to address global issues of resource depletion, pollution, and climate change. The conversion of agricultural, industrial, municipal, and electronic wastes into value-added products such as biofuels, fertilizers, construction materials, and adsorbents, not only mitigate the environmental burden but also strengthen the principles of circular economy. The amalgamation of such waste-derived solutions in industries contributes to the conservation of natural resources, reduction of greenhouse emissions, and promotion of eco-friendly innovations. To explore the full potential of valorization of solid waste, combined efforts must be made to overcome the challenges like technological constraints, inadequate policy frameworks, and lack of public awareness. By continued investments in scientific research, technological innovations, as well as the establishment of sustainable waste management infrastructure, the solid waste can be transformed into a valuable resource. Implementation of eco-friendly strategies in the process of mitigation of solid waste can lead to an eco-friendly and sustainable future.

References

Abdel-Shafy, H. I., & Mansour, M. S. M. (2018). Solid waste issue: Sources, composition, disposal, recycling, and valorization. *Egyptian Journal of Petroleum, 27*(4), 1275–1290. https://doi.org/10.1016/j.ejpe.2018.07.003

Abolore, R. S., Jaiswal, S., & Jaiswal, A. K. (2024). Green and sustainable pretreatment methods for cellulose extraction from lignocellulosic biomass and its applications: A review. *Carbohydrate Polymer Technologies and Applications, 7*, 100396. https://doi.org/10.1016/j.carpta.2023.100396

Adetunji, A. I., Oberholster, P. J., & Erasmus, M. (2023). From garbage to treasure: A review on biorefinery of organic solid wastes into valuable biobased products. *Bioresource Technology Reports, 24*, 101610. https://doi.org/10.1016/j.biteb.2023.101610

Afshar, M., & Mofatteh, S. (2024). Biochar for a sustainable future: Environmentally friendly production and diverse applications. *Results in Engineering, 23*, 102433. https://doi.org/10.1016/j.rineng.2024.102433

Agarwal, S., Dhiman, S., & Gupta, H. (2024). Recovery of lithium and copper from anode electrode materials of spent LIBs by acidic leaching. *Environmental Science and Pollution Research, 31*, 34249–34257. https://doi.org/10.1007/s11356-024-33537-x

Agboola, S. O., Inetabor, G. M., Bello, O. O., & Bello, O. S. (2025). *Waste pollution and management: Current challenges and future perspectives* (pp. 3–20). https://doi.org/10.1007/978-981-97-8673-2_1

Ahmed, K., Kumar Dubey, M., Kumar, A., & Dubey, S. (2024). Artificial intelligence and IoT driven system architecture for municipality waste management in smart cities: A review. *Measurement: Sensors, 36*, 101395. https://doi.org/10.1016/j.measen.2024.101395

Alengebawy, A., Ran, Y., Osman, A. I., Jin, K., Samer, M., & Ai, P. (2024). Anaerobic digestion of agricultural waste for biogas production and sustainable bioenergy recovery: A review. *Environmental Chemistry Letters, 22*(6), 2641–2668. https://doi.org/10.1007/s10311-024-01789-1

Alli, Y. A., et al. (2024). Transforming waste to wealth: Harnessing carbon dioxide for sustainable solutions. *Results in Surfaces and Interfaces, 17*, 100321. https://doi.org/10.1016/j.rsurfi.2024.100321

Amoah-Antwi, C., Kwiatkowska-Malina, J., Thornton, S. F., Fenton, O., Malina, G., & Szara, E. (2020). Restoration of soil quality using biochar and brown coal waste: A review. *Science of the Total Environment, 722*, 137852. https://doi.org/10.1016/j.scitotenv.2020.137852

Angulo-Mosquera, L. S., Alvarado-Alvarado, A. A., Rivas-Arrieta, M. J., Cattaneo, C. R., Rene, E. R., & García-Depraect, O. (2021). Production of solid biofuels from organic waste in developing countries: A review from sustainability and economic feasibility perspectives. *Science of the Total Environment, 795*, 148816. https://doi.org/10.1016/j.scitotenv.2021.148816

Arenibafo, F. E. (2023). The 3Rs (reduce, reuse, recycle) of waste management—An effective and sustainable approach for managing municipal solid waste in developing countries. *Proceedings of the International Conference of Contemporary Affairs in Architecture and Urbanism-ICCAUA, 6*(1), 383–398. https://doi.org/10.38027/iccaua2023en0108

Ashiq, W., et al. (2020). Biochar amendment mitigates greenhouse gases emission and global warming potential in dairy manure based silage corn in boreal climate. *Environmental Pollution, 265*, 114869. https://doi.org/10.1016/j.envpol.2020.114869

Atnkut, B., et al. (2025). Factors influencing solid waste management practices and challenges in Awi administrative zone, Northwestern Ethiopia. *BioMed Research International, 2025*(1). https://doi.org/10.1155/bmri/1311674

Awasthi, S., Mishra, A., & Pal, D.B. (2024). *Energy production from sugarcane bagasse and rice husk* (pp. 157–181). https://doi.org/10.1007/978-981-97-0840-6_7

Barbhuiya, S., Kanavaris, F., Das, B. B., & Idrees, M. (2024). Decarbonising cement and concrete production: Strategies, challenges and pathways for sustainable development. *Journal of Building Engineering, 86*, 108861. https://doi.org/10.1016/j.jobe.2024.108861

Bian, Z., et al. (2025). Environmental benefits of airport waste classification and recycling strategies: An analysis of behavior in multi-agent systems. *Resources, Conservation & Recycling Advances, 25*, 200239. https://doi.org/10.1016/j.rcradv.2024.200239

Bolatkhan, K., Kossalbayev, B. D., Zayadan, B. K., Tomo, T., Veziroglu, T. N., & Allakhverdiev, S. I. (2019). Hydrogen production from phototrophic microorganisms: Reality and perspectives. *International Journal of Hydrogen Energy, 44*(12), 5799–5811. https://doi.org/10.1016/j.ijhydene.2019.01.092

Boodhun, B. S. F., Mudhoo, A., Kumar, G., Kim, S.-H., & Lin, C.-Y. (2017). Research perspectives on constraints, prospects and opportunities in biohydrogen production. *International Journal of Hydrogen Energy, 42*(45), 27471–27481. https://doi.org/10.1016/j.ijhydene.2017.04.077

Bosco, F., & Chiampo, F. (2010). Production of polyhydroxyalcanoates (PHAs) using milk whey and dairy wastewater activated sludge. *Journal of Bioscience and Bioengineering, 109*(4), 418–421. https://doi.org/10.1016/j.jbiosc.2009.10.012

Brennan, L., & Owende, P. (2010). Biofuels from microalgae—A review of technologies for production, processing, and extractions of biofuels and co-products. *Renewable and Sustainable Energy Reviews, 14*(2), 557–577. https://doi.org/10.1016/j.rser.2009.10.009

Capson-Tojo, G., et al. (2018). Cardboard proportions and total solids contents as driving factors in dry co-fermentation of food waste. *Bioresource Technology, 248*, 229–237. https://doi.org/10.1016/j.biortech.2017.06.040

Chen, C., Sun, N., Li, D., Long, S., Tang, X., Xiao, G., & Wang, L. (2018). Optimization and characterization of biosurfactant production from kitchen waste oil using *Pseudomonas aeruginosa. Environmental Science and Pollution Research, 25*(15), 14934–14943. https://doi.org/10.1007/s11356-018-1691-1

Chen, L., et al. (2024). Conversion of waste into sustainable construction materials: A review of recent developments and prospects. *Materials Today Sustainability, 27*, 100930. https://doi.org/10.1016/j.mtsust.2024.100930

Cheong, W. L., et al. (2022). Anaerobic co-digestion of food waste with sewage sludge: simulation and optimization for maximum biogas production. *Water, 14*(7), 1075. https://doi.org/10.3390/w14071075

Cho, H. H., Strezov, V., & Evans, T. J. (2024). Life cycle assessment of power-to-methane and renewable methane production technologies. *Renewable and Sustainable Energy Reviews, 206*, 114856. https://doi.org/10.1016/j.rser.2024.114856

Dagwar, P. P., & Dutta, D. (2024). Landfill leachate a potential challenge towards sustainable environmental management. *Science of the Total Environment, 926*, 171668. https://doi.org/10.1016/j.scitotenv.2024.171668

Das, A. K., Islam, Md. N., Billah, Md. M., & Sarker, A. (2021). COVID-19 pandemic and healthcare solid waste management strategy—A mini-review. *Science of the Total Environment, 778*, 146220. https://doi.org/10.1016/j.scitotenv.2021.146220

Das, A. J., & Kumar, R. (2018). Utilization of agro-industrial waste for biosurfactant production under submerged fermentation and its application in oil recovery from sand matrix. *Bioresource Technology, 260*, 233–240. https://doi.org/10.1016/j.biortech.2018.03.093

de Andrade, C. J., de Andrade, L. M., Rocco, S. A., Sforça, M. L., Pastore, G. M., & Jauregi, P. (2017). A novel approach for the production and purification of mannosylerythritol lipids (MEL) by Pseudozyma tsukubaensis using cassava wastewater as substrate. *Separation and Purification Technology, 180*, 157–167. https://doi.org/10.1016/j.seppur.2017.02.045

Dehghani, M.H., Omrani, G.A., & Karri, R.R. (2021). Solid waste—Sources, toxicity, and their consequences to human health. In: *Soft computing techniques in solid waste and wastewater management*. Elsevier, pp. 205–213. https://doi.org/10.1016/B978-0-12-824463-0.00013-6.

Dhiman, S., Agarwal, S., & Gupta, H. (2024). Application of phosphonium ionic liquids to separate Ga, Ge and In utilizing solvent extraction: A review. *Journal of Ionic Liquids, 4*(1), 100080.

Du, C., et al. (2018). Valorization of food waste into biofertiliser and its field application. *Journal of Cleaner Production, 187*, 273–284. https://doi.org/10.1016/j.jclepro.2018.03.211

Ehrig, H., Schneider, H., & Gossow, V. (2011). Waste, 7. Deposition. In: *Ullmann's encyclopedia of industrial chemistry*. Wiley. https://doi.org/10.1002/14356007.o28_o07

El-Khateeb, M. (2022). Household hazardous waste: handling, precaution and hazard reduction. *Egyptian Journal of Chemistry*. https://doi.org/10.21608/ejchem.2022.110445.5024

Elsamadony, M., & Tawfik, A. (2015). Dry anaerobic co-digestion of organic fraction of municipal waste with paperboard mill sludge and gelatin solid waste for enhancement of hydrogen production. *Bioresource Technology, 191*, 157–165. https://doi.org/10.1016/j.biortech.2015.05.017

Faiz, I., et al. (2024). Hazardous waste management (Buxus papillosa) investment for the prosperity of environment and circular economy: Response surface methodology-based simulation. *Journal of Environmental Management, 350*, 119567. https://doi.org/10.1016/j.jenvman.2023.119567

Favaro, L., Cagnin, L., Basaglia, M., Pizzocchero, V., van Zyl, W. H., & Casella, S. (2017). Production of bioethanol from multiple waste streams of rice milling. *Bioresource Technology, 244*, 151–159. https://doi.org/10.1016/j.biortech.2017.07.108

Fei, F., Wen, Z., Huang, S., & De Clercq, D. (2018). Mechanical biological treatment of municipal solid waste: Energy efficiency, environmental impact and economic feasibility analysis. *Journal of Cleaner Production, 178*, 731–739. https://doi.org/10.1016/j.jclepro.2018.01.060

Ferdous, W., et al. (2021). Recycling of landfill wastes (tyres, plastics and glass) in construction—A review on global waste generation, performance, application and future opportunities. *Resources, Conservation and Recycling, 173*, 105745. https://doi.org/10.1016/j.resconrec.2021.105745

Franus, M., Madej, J., Panek, R., & Grabias-Blicharz, E. (2024). Effect of microwave radiation temperature and content of different solid waste on the microstructure and physicomechanical properties of lightweight aggregates. *Ceramics International, 50*(2), 2871–2886. https://doi.org/10.1016/j.ceramint.2023.11.021

Gamage, A., Gangahagedara, R., Gamage, J., Jayasinghe, N., Kodikara, N., Suraweera, P., & Merah, O. (2023). Role of organic farming for achieving sustainability in agriculture. *Farming System, 1*(1), 100005. https://doi.org/10.1016/j.farsys.2023.100005

Gao, L., et al. (2024). Physiological responses of low- and high-cadmium accumulating *Robinia pseudoacacia*-rhizobium symbioses to cadmium stress. *Environmental Pollution, 345*, 123456. https://doi.org/10.1016/j.envpol.2024.123456

Gavrila, G. O., et al. (2025). Assessment of CH_4 and CO_2 emissions from a municipal waste landfill: Trends, dispersion, and environmental implications. *Atmosphere, 16*(7), 752. https://doi.org/10.3390/atmos16070752

Ghai, H., et al. (2022). An overview on co-pyrolysis of biodegradable and non-biodegradable wastes. *Energies, 15*(11), 4168. https://doi.org/10.3390/en15114168

Gopalakrishnan, B., Khanna, N., & Das, D. (2019.) Dark-fermentative biohydrogen production. In *Biohydrogen* (pp. 79–122). Elsevier. https://doi.org/10.1016/B978-0-444-64203-5.00004-6

Goswami, L., Ekblad, A., Choudhury, R., & Bhattacharya, S. S. (2024). Vermi-converted tea industry coal ash efficiently substitutes chemical fertilization for growth and yield of cabbage (Brassica oleracea var. capitata) in an alluvial soil: A field-based study on soil quality, nutrient translocation, and metal-risk remediation. *Science of The Total Environment, 907*, 168088. https://doi.org/10.1016/j.scitotenv.2023.168088

Gruchlik, Y., Linge, K., & Joll, C. (2018). Removal of organic micropollutants in waste stabilisation ponds: A review. *Journal of Environmental Management, 206*, 202–214. https://doi.org/10.1016/j.jenvman.2017.10.020

Gupta, A. P., Upadhyay, P., Sen, T., & Dutta, J. (2023). Agricultural waste as a resource. In *Agriculture waste management and bioresource* (pp. 1–20). Wiley. https://doi.org/10.1002/9781119808428.ch1

Gupta, H., Dhiman, S., Stam, P.G., Chaudhary, S., Murugavel, R., & Shiju, N. R. (2025) Sustainable approach to cobalt recycling: Levulinic acid leaching of waste lithium-ion battery cathode materials. *Environmental Technology & Innovation,* 104675. https://doi.org/10.1016/j.eti.2025.104675

Gupta, H. (2015). Removal of phenanthrene from water using activated carbon developed from orange rind. *International Journal of Scientific Research in Environmental Sciences, 3*(7), 28–255.

Gupta, H. (2018). PAH determination in effluent and sludge samples of paper industry. *Environmental Technology and Innovation, 9*, 115–121.

Gupta, H., & Gupta, B. (2016). Adsorption of polycyclic aromatic hydrocarbons on banana peel activated carbon. *Desalination and Water Treatment, 57*(20), 9498–9509.

Gupta, H., & Gupta, B. (2018). Vehicular tire as potential adsorbent for the removal of polycyclic aromatic hydrocarbons. *Polycyclic Aromatic Compounds, 38*(4), 354–368.

Gupta, H., & Shiju, N. R. (2025). Recycling and reuse of olefinic separators in lithium-ion batteries: A mini review. *Current Opinion in Green and Sustainable Chemistry, 56*, Article 101054. https://doi.org/10.1016/j.cogsc.2025.101054

Gupta, P., Nagpal, G., & Gupta, N. (2021). Fly ash-based geopolymers: An emerging sustainable solution for heavy metal remediation from aqueous medium. *Beni-Suef University Journal of Basic and Applied Sciences, 10*(1), 89. https://doi.org/10.1186/s43088-021-00179-8

Hasanizadeh, P., Moghimi, H., & Hamedi, J. (2017). Biosurfactant production by *Mucor circinelloides* on waste frying oil and possible uses in crude oil remediation. *Water Science and Technology, 76*(7), 1706–1714. https://doi.org/10.2166/wst.2017.338

Hoang, A. T., et al. (2022). Perspective review on Municipal Solid Waste-to-energy route: Characteristics, management strategy, and role in circular economy. *Journal of Cleaner Production, 359*, 131897. https://doi.org/10.1016/j.jclepro.2022.131897

Hossain, T., Miah, A. B., Mahmud, S. A., & Mahin, A.-A.-. (2018). Enhanced bioethanol production from potato peel waste via consolidated bioprocessing with statistically optimized medium. *Applied Biochemistry and Biotechnology, 186*(2), 425–442. https://doi.org/10.1007/s12010-018-2747-x

Islam, N. F., Gogoi, B., Saikia, R., Yousaf, B., Narayan, M., & Sarma, H. (2024). Encouraging circular economy and sustainable environmental practices by addressing waste management and biomass energy production. *Regional Sustainability, 5*(4), 100174. https://doi.org/10.1016/j.regsus.2024.100174

Jabeen, F., et al. (2022). Trash to energy: A measure for the energy potential of combustible content of domestic solid waste generated from an industrialized city of Pakistan. *Journal of the Taiwan Institute of Chemical Engineers, 137*, 104223. https://doi.org/10.1016/j.jtice.2022.104223

Jadhav, J. V., Pratap, A. P., & Kale, S. B. (2019). Evaluation of sunflower oil refinery waste as feedstock for production of sophorolipid. *Process Biochemistry, 78*, 15–24. https://doi.org/10.1016/j.procbio.2019.01.015

Jegal, J. (2018). Biopolymers based on raw materials from biomass. In *Emerging areas in bioengineering* (pp. 399–427). Wiley. https://doi.org/10.1002/9783527803293.ch23

Jürgensen, L., Ehimen, E. A., Born, J., & Holm-Nielsen, J. B. (2015). Hydrogen production using an anaerobic baffled reactor—Mass balances for pathway analysis and gas composition profiles. *International Journal of Hydrogen Energy, 40*(36), 12154–12161. https://doi.org/10.1016/j.ijhydene.2015.07.068

Kabir, E., Kaur, R., Lee, J., Kim, K.-H., & Kwon, E. E. (2020). Prospects of biopolymer technology as an alternative option for non-degradable plastics and sustainable management of plastic wastes. *Journal of Cleaner Production, 258*, 120536. https://doi.org/10.1016/j.jclepro.2020.120536

Kaur, G., Wang, H., To, M. H., Roelants, S. L. K. W., Soetaert, W., & Lin, C. S. K. (2019). Efficient sophorolipids production using food waste. *Journal of Cleaner Production, 232*, 1–11. https://doi.org/10.1016/j.jclepro.2019.05.326

Keskin, T., Nalakath Abubackar, H., Arslan, K., & Azbar, N. (2019). Biohydrogen production from solid wastes. In *Biohydrogen* (pp. 321–346). Elsevier. https://doi.org/10.1016/B978-0-444-64203-5.00012-5

Kevin van Langen, S., Vassillo, C., Ghisellini, P., Restaino, D., Passaro, R., & Ulgiati, S. (2021). Promoting circular economy transition: A study about perceptions and awareness by different stakeholders groups. *Journal of Cleaner Production, 316*, 128166. https://doi.org/10.1016/j.jclepro.2021.128166

Khan, M. N., Sial, T. A., Ali, A., & Wahid, F. (2024a). *Impact of agricultural wastes on environment and possible management strategies* (pp. 79–108). https://doi.org/10.1007/978-3-031-50503-4_4

Khan, A. H., et al. (2022a). Municipal solid waste generation and the current state of waste-to-energy potential: State of art review. *Energy Conversion and Management, 267*, 115905. https://doi.org/10.1016/j.enconman.2022.115905

Khan, A. H., López-Maldonado, E. A., Khan, N. A., Villarreal-Gómez, L. J., Munshi, F. M., Alsabhan, A. H., & Perveen, K. (2022b). Current solid waste management strategies and energy recovery in developing countries—State of art review. *Chemosphere, 291*, 133088. https://doi.org/10.1016/j.chemosphere.2021.133088

Khan, O., Parvez, M., Yahya, Z., Alhodaib, A., Yadav, A. K., Hoang, A. T., & Ağbulut, Ü. (2024b). Waste-to-energy power plants: Multi-objective analysis and optimization of landfill heat and methane gas by recirculation of leachate. *Process Safety and Environmental Protection, 186*, 957–968. https://doi.org/10.1016/j.psep.2024.04.022

Khedulkar, A. P., et al. (2025). Bio-based nanomaterials as effective, friendly solutions and their applications for protecting water, soil, and air. *Materials Today Chemistry, 46*, 102688. https://doi.org/10.1016/j.mtchem.2025.102688

Kim, D.-H., Kim, S.-H., Kim, K.-Y., & Shin, H.-S. (2010). Experience of a pilot-scale hydrogen-producing anaerobic sequencing batch reactor (ASBR) treating food waste. *International Journal of Hydrogen Energy, 35*(4), 1590–1594. https://doi.org/10.1016/j.ijhydene.2009.12.041

Koller, M., Maršálek, L., de Sousa Dias, M. M., & Braunegg, G. (2017). Producing microbial polyhydroxyalkanoate (PHA) biopolyesters in a sustainable manner. *New Biotechnology, 37*, 24–38. https://doi.org/10.1016/j.nbt.2016.05.001

Kong, G., Wang, K., Zhang, X., Li, J., Han, L., & Zhang, X. (2022). Torrefaction/carbonization-enhanced gasification-steam reforming of biomass for promoting hydrogen-enriched syngas production and tar elimination over gasification biochars. *Bioresource Technology, 363*, 127960. https://doi.org/10.1016/j.biortech.2022.127960

Kumar, A. P., Janardhan, A., Viswanath, B., Monika, K., Jung, J.-Y., & Narasimha, G. (2016). Evaluation of orange peel for biosurfactant production by *Bacillus licheniformis* and their ability to degrade naphthalene and crude oil. *3 Biotechnology, 6*(1), 43. https://doi.org/10.1007/s13205-015-0362-x

Kumar, A., & Agrawal, A. (2020). Recent trends in solid waste management status, challenges, and potential for the future Indian cities—A review. *Current Research in Environmental Sustainability, 2*, 100011. https://doi.org/10.1016/j.crsust.2020.100011

Kumar, A., Dhiman, S., Kumar, R., & Gupta, H. (2025a). Recovery of germanium from waste optical fibres using phosphonium ionic liquid. *Journal of Cleaner Production, 505*, 145466. https://doi.org/10.1016/j.jclepro.2025.145466

Kumar, A., & Gupta, H. (2020). Activated carbon from sawdust for naphthalene removal from contaminated water. *Environmental Technology and Innovation, 20*, 101080.

Kumar, G., & Lin, C.-Y. (2014). Biogenic hydrogen conversion of de-oiled jatropha waste via anaerobic sequencing batch reactor operation: process performance, microbial insights, and CO_2 reduction efficiency. *The Scientific World Journal, 2014*, 1–9. https://doi.org/10.1155/2014/946503

Kumar, K. K., Cema, G., Ziembińska-Buczyńska, A., Hassan, G. K., Hellal, M. S., & Surmacz-Górska, J. (2025b). Sustainable biomethane production from waste biomass: Challenges associated with process optimization in improving the yield. *Environmental Science and Pollution Research*. https://doi.org/10.1007/s11356-024-35864-5

Kumar, M., et al. (2024). A review on value-addition to plastic waste towards achieving a circular economy. *Science of the Total Environment, 921*, 171106. https://doi.org/10.1016/j.scitotenv.2024.171106

Kumar, S., Negi, S., Mandpe, A., Singh, R. V., & Hussain, A. (2018). Rapid composting techniques in Indian context and utilization of black soldier fly for enhanced decomposition of biodegradable wastes—A comprehensive review. *Journal of Environmental Management, 227*, 189–199. https://doi.org/10.1016/j.jenvman.2018.08.096

Lee, S. Y., Sankaran, R., Chew, K. W., Tan, C. H., Krishnamoorthy, R., Chu, D.-T., & Show, P.-L. (2019). Waste to bioenergy: A review on the recent conversion technologies. *BMC Energy, 1*(1), 4. https://doi.org/10.1186/s42500-019-0004-7

Li, G., Liu, J., Yi, L., Luo, J., & Jiang, T. (2024a). Bauxite residue (red mud) treatment: Current situation and promising solution. *Science of the Total Environment, 948*, 174757. https://doi.org/10.1016/j.scitotenv.2024.174757

Li, J., & Cheng, W. (2020). Comparison of life-cycle energy consumption, carbon emissions and economic costs of coal to ethanol and bioethanol. *Applied Energy, 277*, 115574. https://doi.org/10.1016/j.apenergy.2020.115574

Li, P., et al. (2017). Effect of pretreatment on the enzymatic hydrolysis of kitchen waste for xanthan production. *Bioresource Technology, 223*, 84–90. https://doi.org/10.1016/j.biortech.2016.10.035

Li, Z., et al. (2024b). Designing low-carbon fly ash based geopolymer with red mud and blast furnace slag wastes: Performance, microstructure and mechanism. *Journal of Environmental Management, 354*, 120362. https://doi.org/10.1016/j.jenvman.2024.120362

Liu, H., Long, J., Zhang, K., Li, M., Zhao, D., Song, D., & Zhang, W. (2024). Agricultural biomass/waste-based materials could be a potential adsorption-type remediation contributor to environmental pollution induced by pesticides—A critical review. *Science of the Total Environment, 946*, 174180. https://doi.org/10.1016/j.scitotenv.2024.174180

Liu, J., Kan, T., Lou, J., Xiang, L., Zhu, X., & Tang, Y. (2016). Localized damage response of carbon fiber reinforced polymer composite sandwich panel after thermal exposure. *Polymer Testing, 50*, 33–40. https://doi.org/10.1016/j.polymertesting.2015.12.006

Liu, W., Dong, Z., Sun, D., Chen, Y., Wang, S., Zhu, J., & Liu, C. (2019). Bioconversion of kitchen wastes into bioflocculant and its pilot-scale application in treating iron mineral processing wastewater. *Bioresource Technology, 288*, 121505. https://doi.org/10.1016/j.biortech.2019.121505

Liu, X., Wang, X., Xue, Q., Tian, Y., & Feng, Y. (2023a). Inhalation bioaccessibility and risk assessment for PM-bound organic components: Co-effects of component physicochemical properties, PM properties, and sources. *Journal of Hazardous Materials, 459*, 132291. https://doi.org/10.1016/j.jhazmat.2023.132291

Liu, Y., et al. (2023b). Geochemical factors controlling the mobilization of geogenic cadmium in soils developed on carbonate bedrocks in Southwest China. *Geoderma, 437*, 116606. https://doi.org/10.1016/j.geoderma.2023.116606

Liu, Y., Liang, F., Yan, S., Tu, T., Shi, Z., Pezzuolo, A., & Feng, Q. (2025). Optimizing fertilization strategies for low-carbon agriculture: Balancing greenhouse gas mitigation, soil health, and productivity. *Results in Engineering, 27*, 107094. https://doi.org/10.1016/j.rineng.2025.107094

López-Prieto, A., Martínez-Padrón, H., Rodríguez-López, L., Moldes, A. B., & Cruz, J. M. (2019). Isolation and characterization of a microorganism that produces biosurfactants in corn steep water. *CyTA—Journal of Food, 17*(1), 509–516. https://doi.org/10.1080/19476337.2019.1607909

Lourenço, L. A., Alberton Magina, M. D., Tavares, L. B. B., Ulson, G., de Souza, S. M. A., García Román, M., & Altmajer Vaz, D. (2018). Biosurfactant production by *Trametes versicolor* grown on two-phase olive mill waste in solid-state fermentation. *Environmental Technology, 39*(23), 3066–3076. https://doi.org/10.1080/09593330.2017.1374471

Ma, S., et al. (2024). Exploiting bifunctional 3D-Printed geopolymers for efficient cesium removal and immobilization: An approach for hazardous waste management. *Journal of Cleaner Production, 437*, 140599. https://doi.org/10.1016/j.jclepro.2024.140599

Mahmoodi, P., Karimi, K., & Taherzadeh, M. J. (2018). Efficient conversion of municipal solid waste to biofuel by simultaneous dilute-acid hydrolysis of starch and pretreatment of lignocelluloses. *Energy Conversion and Management, 166*, 569–578. https://doi.org/10.1016/j.enconman.2018.04.067

Majee, S., Sarkar, K. K., Sarkhel, R., Halder, G., Mandal, D. D., Rathinam, N. K., & Mandal, T. (2023). Bio-organic fertilizer production from industrial waste and insightful analysis on release kinetics. *Journal of Environmental Management, 325*, 116378. https://doi.org/10.1016/j.jenvman.2022.116378

Malode, S. J., Pandiaraj, S., Alodhayb, A., & Shetti, N. P. (2025). Review on sustainable biohydrogen production from biomass: Recent developments and prospects toward a promising source. *Energy & Fuels, 39*(17), 7967–7989. https://doi.org/10.1021/acs.energyfuels.5c00315

Manea, E. E., Bumbac, C., Dinu, L. R., Bumbac, M., & Nicolescu, C. M. (2024). Composting as a sustainable solution for organic solid waste management: Current practices and potential improvements. *Sustainability, 16*(15), 6329. https://doi.org/10.3390/su16156329

Marangoni, C., Furigo, A., & de Aragão, G. M. F. (2002). Production of poly(3-hydroxybutyrate-co-3-hydroxyvalerate) by Ralstonia eutropha in whey and inverted sugar with propionic acid feeding. *Process Biochemistry, 38*(2), 137–141. https://doi.org/10.1016/S0032-9592(01)00313-2

Marcelino, P. R. F., et al. (2019). Biosurfactants production by yeasts using sugarcane bagasse hemicellulosic hydrolysate as new sustainable alternative for lignocellulosic biorefineries. *Industrial Crops and Products, 129*, 212–223. https://doi.org/10.1016/j.indcrop.2018.12.001

Marín, D., Méndez, L., Suero, I., Díaz, I., Blanco, S., Fdz-Polanco, M., & Muñoz, R. (2022). Anaerobic digestion of food waste coupled with biogas upgrading in an outdoors algal-bacterial photobioreactor at pilot scale. *Fuel, 324*, 124554. https://doi.org/10.1016/j.fuel.2022.124554

Markande, A. R., Patel, D., & Varjani, S. (2021). A review on biosurfactants: Properties, applications and current developments. *Bioresource Technology, 330*, 124963. https://doi.org/10.1016/j.biortech.2021.124963

Matsakas, L., Rova, U., & Christakopoulos, P. (2015). Sequential parametric optimization of methane production from different sources of forest raw material. *Frontiers in Microbiology, 6*. https://doi.org/10.3389/fmicb.2015.01163

Meena, M. D., et al. (2023). Municipal solid waste: Opportunities, challenges and management policies in India: A review. *Waste Management Bulletin, 1*(1), 4–18. https://doi.org/10.1016/j.wmb.2023.04.001

Mesomo, M., et al. (2009). Xanthan gum produced by *Xanthomonas campestris* from cheese whey: Production optimisation and rheological characterisation. *Journal of the Science of Food and Agriculture, 89*(14), 2440–2445. https://doi.org/10.1002/jsfa.3743

Mir, D. H., & Rather, M. A. (2024). Kinetic and thermodynamic investigations of copper (II) biosorption by green algae *Chara vulgaris* obtained from the waters of Dal Lake in Srinagar (India). *Journal of Water Process Engineering, 58*, 104850. https://doi.org/10.1016/j.jwpe.2024.104850

Mohanty, S. S., et al. (2021). A critical review on various feedstocks as sustainable substrates for biosurfactants production: A way towards cleaner production. *Microbial Cell Factories, 20*(1), 120. https://doi.org/10.1186/s12934-021-01613-3

Mor, S., & Ravindra, K. (2023). Municipal solid waste landfills in lower- and middle-income countries: Environmental impacts, challenges and sustainable management practices. *Process Safety and Environmental Protection, 174*, 510–530. https://doi.org/10.1016/j.psep.2023.04.014

Moshood, T. D., Nawanir, G., Mahmud, F., Mohamad, F., Ahmad, M. H., & AbdulGhani, A. (2022). Sustainability of biodegradable plastics: New problem or solution to solve the global plastic pollution? *Current Research in Green and Sustainable Chemistry, 5*, 100273. https://doi.org/10.1016/j.crgsc.2022.100273

Nalladiyil, A., & Babu, G. L. S. (2024). Dumpsite remediation through landfill mining and rehabilitation: A circular economy perspective. *Journal of the Indian Institute of Science, 104*(2), 489–518. https://doi.org/10.1007/s41745-024-00445-5

Nanda, S., & Berruti, F. (2021). Municipal solid waste management and landfilling technologies: A review. *Environmental Chemistry Letters, 19*(2), 1433–1456. https://doi.org/10.1007/s10311-020-01100-y

Narisetty, V., et al. (2022). Process optimization for recycling of bread waste into bioethanol and biomethane: A circular economy approach. *Energy Conversion and Management, 266*, 115784. https://doi.org/10.1016/j.enconman.2022.115784

Nery, T. B. R., Brandão, L. V., Esperidião, M. C. A., & Druzian, J. I. (2008). Biossíntese de goma xantana a partir da fermentação de soro de leite: Rendimento e viscosidade. *Química Nova, 31*(8), 1937–1941. https://doi.org/10.1590/S0100-40422008000800004

Nguyen, V. G., et al. (2024). Recent advances in hydrogen production from biomass waste with a focus on pyrolysis and gasification. *International Journal of Hydrogen Energy, 54*, 127–160. https://doi.org/10.1016/j.ijhydene.2023.05.049

Niknezhad, S. V., Asadollahi, M. A., Zamani, A., Biria, D., & Doostmohammadi, M. (2015). Optimization of xanthan gum production using cheese whey and response surface methodology. *Food Science and Biotechnology, 24*(2), 453–460. https://doi.org/10.1007/s10068-015-0060-9

Nizamuddin, S., Baloch, A. J., Chen, C., Arif, M., & Mubarak, N. M. (2024). Bio-based plastics, biodegradable plastics, and compostable plastics: Biodegradation mechanism, biodegradability standards and environmental stratagem. *International Biodeterioration & Biodegradation, 195*, 105887. https://doi.org/10.1016/j.ibiod.2024.105887

Noor, T., Javid, A., Hussain, A., Bukhari, S. M., Ali, W., Akmal, M., & Hussain, S. M. (2020). Types, sources and management of urban wastes. In *Urban ecology* (pp. 239–263). Elsevier. https://doi.org/10.1016/B978-0-12-820730-7.00014-8

Norouzi, M., Chàfer, M., Cabeza, L. F., Jiménez, L., & Boer, D. (2021). Circular economy in the building and construction sector: A scientific evolution analysis. *Journal of Building Engineering, 44*, 102704. https://doi.org/10.1016/j.jobe.2021.102704

Onyilokwu, T. E., Ogwuche, J. A., Shabu, T., Ali, P., & Onah, M. A. (2024). Assessment of solid waste management practices in Higher Educational Institutions in Makurdi Metropolis Benue State, Nigeria. *International Journal of Research and Innovation in Social Science, VIII*(III), 1848–1867. https://doi.org/10.47772/IJRISS.2024.803131

Ozdal, M., Gurkok, S., & Ozdal, O. G. (2017). Optimization of rhamnolipid production by *Pseudomonas aeruginosa* OG1 using waste frying oil and chicken feather peptone. *3 Biotech, 7*(2), 117. https://doi.org/10.1007/s13205-017-0774-x

Pagliano, G., Ventorino, V., Panico, A., & Pepe, O. (2017). Integrated systems for biopolymers and bioenergy production from organic waste and by-products: A review of microbial processes. *Biotechnology for Biofuels, 10*(1), 113. https://doi.org/10.1186/s13068-017-0802-4

Pais, J., Serafim, L. S., Freitas, F., & Reis, M. A. M. (2016). Conversion of cheese whey into poly(3-hydroxybutyrate-co-3-hydroxyvalerate) by Haloferax mediterranei. *New Biotechnology, 33*(1), 224–230. https://doi.org/10.1016/j.nbt.2015.06.001

Pantazaki, A. A., Papaneophytou, C. P., Pritsa, A. G., Liakopoulou-Kyriakides, M., & Kyriakidis, D. A. (2009). Production of polyhydroxyalkanoates from whey by *Thermus thermophilus* HB8. *Process Biochemistry, 44*(8), 847–853. https://doi.org/10.1016/j.procbio.2009.04.002

Paraszkiewicz, K., Bernat, P., Kuśmierska, A., Chojniak, J., & Płaza, G. (2018). Structural identification of lipopeptide biosurfactants produced by *Bacillus subtilis* strains grown on the media obtained from renewable natural resources. *Journal of Environmental Management, 209*, 65–70. https://doi.org/10.1016/j.jenvman.2017.12.033

Parmar, D., & Rai, M. (2017). Performance evaluation of combined biomedical waste treatment plants with specific focus on incinerators using data envelope analysis in Uttar Pradesh, India. *Journal of Hazardous, Toxic, and Radioactive Waste, 21*(1). https://doi.org/10.1061/(ASCE)HZ.2153-5515.0000321

Patwa, A., Parde, D., Dohare, D., Vijay, R., & Kumar, R. (2020). Solid waste characterization and treatment technologies in rural areas: An Indian and international review. *Environmental Technology & Innovation, 20*, 101066. https://doi.org/10.1016/j.eti.2020.101066

Peiris, D., Gunasekara, C., W. Law, D., Patrisia, D. Y., Tam, W. Y. V., & Setunge, S. (2025). Impact of treatment methods on recycled concrete aggregate performance: A comprehensive review. *Environmental Science and Pollution Research, 32*, 14405–14438.

Peng, X., Jiang, Y., Chen, Z., Osman, A. I., Farghali, M., Rooney, D. W., & Yap, P.-S. (2023). Recycling municipal, agricultural and industrial waste into energy, fertilizers, food and construction materials, and economic feasibility: A review. *Environmental Chemistry Letters, 21*(2), 765–801. https://doi.org/10.1007/s10311-022-01551-5

Peng, Y., Xu, H., Shi, J., Wang, Z., Lv, J., Li, L., & Wang, X. (2024). Soil microbial composition, diversity, and network stability in intercropping versus monoculture responded differently to drought. *Agriculture, Ecosystems & Environment, 365*, 108915. https://doi.org/10.1016/j.agee.2024.108915

Phillip, A., Bhatt, S. M., Sharma, N., & Poudel, B. (2024). Enhanced biogas production using anaerobic co-digestion of animal waste and food waste: A review. *Journal of Scientific Research and Reports, 30*(8), 761–781. https://doi.org/10.9734/jsrr/2024/v30i82297

Potysz, A., van Hullebusch, E. D., & Kierczak, J. (2018). Perspectives regarding the use of metallurgical slags as secondary metal resources—A review of bioleaching approaches. *Journal of Environmental Management, 219*, 138–152. https://doi.org/10.1016/j.jenvman.2018.04.083

Povolo, S., & Casella, S. (2003). Bacterial production of PHA from lactose and cheese whey permeate. *Macromolecular Symposia, 197*(1), 1–10. https://doi.org/10.1002/masy.200350701

Quan, Z., Xu, X., Wang, W., Jiang, J., & Gao, S. (2024). Do industrial solid waste recycling and technological innovation promote low-carbon development in China? New insights from NARDL approach. *Science of the Total Environment, 916*, 170446. https://doi.org/10.1016/j.scitotenv.2024.170446

Quinto, S., Law, N., Fletcher, C., Le, J., Antony Jose, S., & Menezes, P. L. (2025). Exploring the e-waste crisis: Strategies for sustainable recycling and circular economy integration. *Recycling, 10*(2), 72. https://doi.org/10.3390/recycling10020072

Raita, M., Ibenegbu, C., Champreda, V., & Leak, D. J. (2016). Production of ethanol by thermophilic oligosaccharide utilising Geobacillus thermoglucosidasius TM242 using palm kernel cake as a renewable feedstock. *Biomass and Bioenergy, 95*, 45–54. https://doi.org/10.1016/j.biombioe.2016.08.015

Rajkumar, R., & Kurinjimalar, C. (2022). Food wastes/residues: Valuable source of energy in circular economy. In *Handbook of biofuels* (pp. 147–163). Elsevier. https://doi.org/10.1016/B978-0-12-822810-4.00007-5.
Ramli, H., Aziz, H. A., & Hung, Y.-T. (2021). *Practices of solid waste processing and disposal* (pp. 625–673). https://doi.org/10.1007/978-3-030-84180-5_10
Ranganathan, S., Dutta, S., Moses, J. A., & Anandharamakrishnan, C. (2020). Utilization of food waste streams for the production of biopolymers. *Heliyon, 6*(9), e04891. https://doi.org/10.1016/j.heliyon.2020.e04891
Ren, H.-Y., Kong, F., Zhao, L., Ren, N.-Q., Ma, J., Nan, J., & Liu, B.-F. (2019). Enhanced co-production of biohydrogen and algal lipids from agricultural biomass residues in long-term operation. *Bioresource Technology, 289*, 121774. https://doi.org/10.1016/j.biortech.2019.121774
Sampath, P., Brijesh, Reddy, K. R., Reddy, C. V., Shetti, N. P., Kulkarni, R. V., & Raghu, A. V. (2020). Biohydrogen production from organic waste—A review. *Chemical Engineering & Technology, 43*(7), 1240–1248. https://doi.org/10.1002/ceat.201900400
Sanito, R. C., Bernuy-Zumaeta, M., You, S.-J., & Wang, Y.-F. (2022). A review on vitrification technologies of hazardous waste. *Journal of Environmental Management, 316*, 115243. https://doi.org/10.1016/j.jenvman.2022.115243
Santiago, S. G., Morgan-Sagastume, J. M., Monroy, O., & Moreno-Andrade, I. (2020). Biohydrogen production from organic solid waste in a sequencing batch reactor: An optimization of the hydraulic and solids retention time. *International Journal of Hydrogen Energy, 45*(47), 25681–25688. https://doi.org/10.1016/j.ijhydene.2019.11.224
Santos, E. F., Teixeira, M. F. S., Converti, A., Porto, A. L. F., & Sarubbo, L. A. (2019). Production of a new lipoprotein biosurfactant by Streptomyces sp. DPUA1566 isolated from lichens collected in the Brazilian Amazon using agroindustry wastes. *Biocatalysis and Agricultural Biotechnology, 17*, 142–150. https://doi.org/10.1016/j.bcab.2018.10.014
Saravanakumar, K., & Kathiresan, K. (2014). Bioconversion of lignocellulosic waste to bioethanol by Trichoderma and yeast fermentation. *3 Biotech, 4*(5), 493–499. https://doi.org/10.1007/s13205-013-0179-4
Schnell, M., Horst, T., & Quicker, P. (2020). Thermal treatment of sewage sludge in Germany: A review. *Journal of Environmental Management, 263*, 110367. https://doi.org/10.1016/j.jenvman.2020.110367
Shah, A. V., Srivastava, V. K., Mohanty, S. S., & Varjani, S. (2021). Municipal solid waste as a sustainable resource for energy production: State-of-the-art review. *Journal of Environmental Chemical Engineering, 9*(4), 105717. https://doi.org/10.1016/j.jece.2021.105717
Shahbaz, M., Rashid, N., Saleem, J., Mackey, H., McKay, G., & Al-Ansari, T. (2023). A review of waste management approaches to maximise sustainable value of waste from the oil and gas industry and potential for the State of Qatar. *Fuel, 332*, 126220. https://doi.org/10.1016/j.fuel.2022.126220
Sharma, H. B., Vanapalli, K. R., Samal, B., Cheela, V. R. S., Dubey, B. K., & Bhattacharya, J. (2021). Circular economy approach in solid waste management system to achieve UN-SDGs: Solutions for post-COVID recovery. *Science of the Total Environment, 800*, 149605. https://doi.org/10.1016/j.scitotenv.2021.149605
Sharma, R., Singh, J., & Verma, N. (2018). Production, characterization and environmental applications of biosurfactants from *Bacillus amyloliquefaciens* and *Bacillus subtilis*. *Biocatalysis and Agricultural Biotechnology, 16*, 132–139. https://doi.org/10.1016/j.bcab.2018.07.028
Shoudho, K. N., Khan, T. H., Ara, U. R., Khan, M. R., Shawon, Z. B. Z., & Hoque, M. E. (2024). Biochar in global carbon cycle: Towards sustainable development goals. *Current Research in Green and Sustainable Chemistry, 8*, 100409. https://doi.org/10.1016/j.crgsc.2024.100409
Silva, R. V., de Brito, J., Lye, C. Q., & Dhir, R. K. (2017). The role of glass waste in the production of ceramic-based products and other applications: A review. *Journal of Cleaner Production, 167*, 346–364. https://doi.org/10.1016/j.jclepro.2017.08.185

Singh, A., & Arora, K. (2018). Solid waste management in flats of a residential campus in Panchli Khurd, Meerut. *Bulletin of Pure & Applied Sciences- Zoology, 37*a(1), 33. https://doi.org/10.5958/2320-3188.2018.00007.4

Singh, A., et al. (2022). Production of bioethanol from food waste: Status and perspectives. *Bioresource Technology, 360*, 127651. https://doi.org/10.1016/j.biortech.2022.127651

Singh, D., Dikshit, A. K., & Kumar, S. (2024). Smart technological options in collection and transportation of municipal solid waste in urban areas: A mini review. *Waste Management & Research: THe Journal for a Sustainable Circular Economy, 42*(1), 3–15. https://doi.org/10.1177/0734242X231175816

Siracusa, V. (2019). Microbial degradation of synthetic biopolymers waste. *Polymers, 11*(6), 1066. https://doi.org/10.3390/polym11061066

Soon, Z. Y., Kim, T., Jung, J.-H., & Kim, M. (2023). Metals and suspended solids in the effluents from in-water hull cleaning by remotely operated vehicle (ROV): Concentrations and release rates into the marine environment. *Journal of Hazardous Materials, 460*, 132456. https://doi.org/10.1016/j.jhazmat.2023.132456

Stamenković, O. S., et al. (2020). Production of biofuels from sorghum. *Renewable and Sustainable Energy Reviews, 124*, 109769. https://doi.org/10.1016/j.rser.2020.109769

Tan, Y. H., Abdullah, M. O., Nolasco-Hipolito, C., Zauzi, N. S. A., & Abdullah, G. W. (2017). Engine performance and emissions characteristics of a diesel engine fueled with diesel-biodiesel-bioethanol emulsions. *Energy Conversion and Management, 132*, 54–64. https://doi.org/10.1016/j.enconman.2016.11.013

Tawfik, A., & El-Qelish, M. (2012). Continuous hydrogen production from co-digestion of municipal food waste and kitchen wastewater in mesophilic anaerobic baffled reactor. *Bioresource Technology, 114*, 270–274. https://doi.org/10.1016/j.biortech.2012.02.016

Tawfik, A., Salem, A., & El-Qelish, M. (2011). Two stage anaerobic baffled reactors for bio-hydrogen production from municipal food waste. *Bioresource Technology, 102*(18), 8723–8726. https://doi.org/10.1016/j.biortech.2011.03.028

Tejaswini, M. S. S. R., Pathak, P., & Gupta, D. K. (2022). Sustainable approach for valorization of solid wastes as a secondary resource through urban mining. *Journal of Environmental Management, 319*, 115727. https://doi.org/10.1016/j.jenvman.2022.115727

Trivedi, Y., et al. (2025). Biochar potential for pollutant removal during wastewater treatment: A comprehensive review of separation mechanisms, technological integration, and process analysis. *Desalination, 600*, 118509. https://doi.org/10.1016/j.desal.2024.118509

Uddin, Md. K., Saha, B. K., Wong, V. N. L., & Patti, A. F. (2025). Organo-mineral fertilizer to sustain soil health and crop yield for reducing environmental impact: A comprehensive review. *European Journal of Agronomy, 162*, 127433. https://doi.org/10.1016/j.eja.2024.127433

Un, C. (2023). A Sustainable approach to the conversion of waste into energy: Landfill gas-to-fuel technology. *Sustainability, 15*(20), 14782. https://doi.org/10.3390/su152014782

Vambol, S., et al. (2024). Type of industries, waste options, and their potential. In: *Advances in energy from waste* (pp. 175–221). Elsevier. https://doi.org/10.1016/B978-0-443-13847-8.00006-3

Varjani, S., Lee, D.-J., & Zhang, Q. (2020). Valorizing agricultural biomass for sustainable development: Biological engineering aspects. *Bioengineered, 11*(1), 522–523. https://doi.org/10.1080/21655979.2020.1759185

Varjani, S., Shah, A. V., Vyas, S., & Srivastava, V. K. (2021). Processes and prospects on valorizing solid waste for the production of valuable products employing bio-routes: A systematic review. *Chemosphere, 282*, 130954. https://doi.org/10.1016/j.chemosphere.2021.130954

Varjani, S., & Upasani, V. N. (2021). Bioaugmentation of *Pseudomonas aeruginosa* NCIM 5514—A novel oily waste degrader for treatment of petroleum hydrocarbons. *Bioresource Technology, 319*, 124240. https://doi.org/10.1016/j.biortech.2020.124240

Vidak Vasić, M., et al. (2024). State and perspectives of sustainable production of traditional silicate ceramics. *Open Ceramics, 17*, 100537. https://doi.org/10.1016/j.oceram.2024.100537

Vigneswari, S., et al. (2024). Turning agricultural waste streams into biodegradable plastic: A step forward into adopting sustainable carbon neutrality. *Journal of Environmental Chemical Engineering, 12*(2), 112135. https://doi.org/10.1016/j.jece.2024.112135

Voukkali, I., Papamichael, I., Loizia, P., & Zorpas, A. A. (2023). Urbanization and solid waste production: Prospects and challenges. *Environmental Science and Pollution Research, 31*(12), 17678–17689. https://doi.org/10.1007/s11356-023-27670-2

Wang, C.-H., Lin, P.-J., & Chang, J.-S. (2006). Fermentative conversion of sucrose and pineapple waste into hydrogen gas in phosphate-buffered culture seeded with municipal sewage sludge. *Process Biochemistry, 41*(6), 1353–1358. https://doi.org/10.1016/j.procbio.2006.01.016

Wang, X., Chen, M., & Zhao, L. (2023). Development of a colorimetric sensing assay for ascorbic acid and sarcosine utilizing the dual-class enzyme activity of Fe_3O_4@SiO_2@$NiCo_2S_4$. *Chemical Engineering Journal, 468*, 143612. https://doi.org/10.1016/j.cej.2023.143612

Wang, X., Xu, W., & Cui, W.-R. (2024). Modulating conjugated microporous polymers via cyclization as a remarkable photo-enhanced uranium recovery platform. *Journal of Hazardous Materials, 463*, 132846. https://doi.org/10.1016/j.jhazmat.2023.132846

Wu, X., Zhu, J., Dong, C., Miller, C., Li, Y., Wang, L., & Yao, W. (2009). Continuous biohydrogen production from liquid swine manure supplemented with glucose using an anaerobic sequencing batch reactor. *International Journal of Hydrogen Energy, 34*(16), 6636–6645. https://doi.org/10.1016/j.ijhydene.2009.06.058

Wu, Y., et al. (2022). Applications of catalysts in thermochemical conversion of biomass (pyrolysis, hydrothermal liquefaction and gasification): A critical review. *Renewable Energy, 196*, 462–481. https://doi.org/10.1016/j.renene.2022.07.031

Wuana, R. A., & Okieimen, F. E. (2011). Heavy metals in contaminated soils: A review of sources, chemistry, risks and best available strategies for remediation. *ISRN Ecology, 2011*, 1–20. https://doi.org/10.5402/2011/402647

Yaashikaa, P. R., Kumar, P. S., Saravanan, A., Varjani, S., & Ramamurthy, R. (2020). Bioconversion of municipal solid waste into bio-based products: A review on valorisation and sustainable approach for circular bioeconomy. *Science of the Total Environment, 748*, 141312. https://doi.org/10.1016/j.scitotenv.2020.141312

Yang, M., et al. (2023). Circular economy strategies for combating climate change and other environmental issues. *Environmental Chemistry Letters, 21*(1), 55–80. https://doi.org/10.1007/s10311-022-01499-6

Yang, Y., Lu, P., Shao, R., Zhao, Q., Yang, T., & Wu, C. (2024). A comprehensive review of multi-source solid wastes in sustainable concrete: From material properties to engineering application. *Construction and Building Materials, 435*, 136775. https://doi.org/10.1016/j.conbuildmat.2024.136775

Yuvaraj, A., Thangaraj, R., Ravindran, B., Chang, S. W., & Karmegam, N. (2021). Centrality of cattle solid wastes in vermicomposting technology—A cleaner resource recovery and biowaste recycling option for agricultural and environmental sustainability. *Environmental Pollution, 268*, 115688. https://doi.org/10.1016/j.envpol.2020.115688

Zhang, J., et al. (2023). Achieving advanced nitrogen removal and excess sludge treatment via single nitritation/anammox-fermentation combined system. *Bioresource Technology, 387*, 129550. https://doi.org/10.1016/j.biortech.2023.129550

Zhang, J., Bhuiyan, M., Zhang, G., Sandanayake, M., & Navaratnam, S. (2024). Circular economy life cycle cost for kerbside waste material looping process. *Waste Management, 186*, 307–317. https://doi.org/10.1016/j.wasman.2024.06.023

Zhang, X., Du, T., & Jia, H. (2021). Efficient activation of coal fly ash for silica and alumina leaches and the dependence of Pb(II) removal capacity on the crystallization conditions of Al-MCM-41. *International Journal of Molecular Sciences, 22*(12), 6540. https://doi.org/10.3390/ijms22126540

Zhao, F., Shi, R., Cui, Q., Han, S., Dong, H., & Zhang, Y. (2017). Biosurfactant production under diverse conditions by two kinds of biosurfactant-producing bacteria for microbial enhanced oil recovery. *Journal of Petroleum Science and Engineering, 157*, 124–130. https://doi.org/10.1016/j.petrol.2017.07.022

Zhao, X., et al. (2023). Sustainable bioplastics derived from renewable natural resources for food packaging. *Matter, 6*(1), 97–127. https://doi.org/10.1016/j.matt.2022.11.006

Zhu, G., Li, J., & Liu, C. (2012). Fermentative hydrogen production from soybean protein processing wastewater in an anaerobic baffled reactor (ABR) using anaerobic mixed consortia. *Applied Biochemistry and Biotechnology, 168*(1), 91–105. https://doi.org/10.1007/s12010-011-9357-1

Zomorodian, M., Lai, S. H., Homayounfar, M., Ibrahim, S., Fatemi, S. E., & El-Shafie, A. (2018). The state-of-the-art system dynamics application in integrated water resources modeling. *Journal of Environmental Management, 227*, 294–304. https://doi.org/10.1016/j.jenvman.2018.08.097

Advanced Technologies for the Detection of Contaminants in Solid Wastes

Shaily and Danish Khan

1 Introduction

In current ages, concerns about ecological contamination have strengthened owing to consumer demand and high alive standards. For instance, the use of fertilizers, pesticide spreading, oil spillages, the assessment of poisonous greenhouse gases, discarding of poisonous hefty metals, use of non-biodegradable substances, slush and unused of additional dangerous ingredients are posing grave intimidations to the atmosphere (Bilal et al., 2017; Gavrilescu, 2010; Rasheed et al., 2018a, 2018b, 2018c; Rasheed et al., 2019a, 2019b, 2019c). Micro-pollutants or emerging contaminants (ECs) are a huge and growing array of human-caused compounds that generally exist in aquatic and nonetheless have lone newly been recognized as substantial aquatic contaminants and comprise a diversity of widely used artificial synthetic composites, counting personal care products, pesticides, cosmetics, steroidal hormones and pharmaceuticals, etc., that are essential in contemporary culture around the world (Gomes et al., 2018; Luo et al., 2014). The worldwide production of these contaminants has been estimated, and it was observed that it augmented from one million to five hundred million tonnes per year (Thomaidis et al., 2012). Nevertheless, these synthetically or naturally occurring materials (i.e. ECs) have been related to exerting identified or assumed adverse consequences on the environment or human health (Fig. 1) (Rasheed et al., 2019a, 2019b, 2019c). For example, the pharmaceutical substances that are particularly synthesized to obtain a biological response in respect to the target

Shaily (✉)
Department of Chemistry, D. B. S. (P.G.) College, Dehradun 248001, Uttarakhand, India
e-mail: shailydbs@gmail.com

D. Khan
Department of Chemistry, R&D Division, Gene to Protein Pvt. Ltd, Hyderabad 500040, Telangana, India

H. Gupta et al. (eds.), *Advanced Technologies for Assessment, Recycling, and Management of Solid Waste*, Waste as a Resource,
https://doi.org/10.1007/978-3-032-24788-9_7

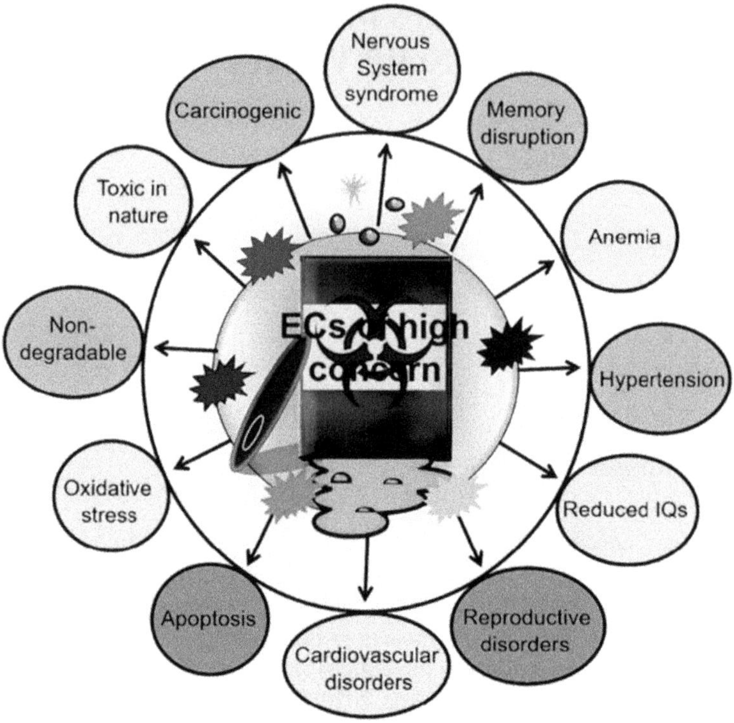

Fig. 1 Effect of ECs on human health (Rasheed et al., 2019a, 2019b, 2019c), 'reused after taking permission from Elsevier'

organism may too obtained identical retort in non-specific objects resulting chronic contact to even minute amounts of these materials (Wilkinson et al., 2016).

To the authorities of both huge and minor municipalities in evolving nations, the solid leftover organization problem is the biggest challenge. This issue is principally owing to the load stood on the civic budget, as the production of such solid leftover is growing very quickly. Moreover, along with the high costs, SWM is connected with the absence of understanding of the diverse influences that affect the complete management structure. In developing countries, the analysis of reported literature related to waste management revealed that only a few articles provided measurable data. The main aim of the said studies was to determine the stakeholders' behaviour/act that has part in the SWM and to analyse diverse aspects that disturb the arrangement. The studies remained directed on four landmasses, in twenty-two evolving republics, and in extra than thirty city zones. In the studied cities, to measure the issues affecting the recital of SWM, and to encourage the stakeholders, a blend of variable methods was revealed in detail (Guerrero et al., 2013). In developing countries, the rise in the living standard, rapid urbanization, population increase, and booming economy has significantly enhanced the quality, quantity, and rate of the public solid waste production (Minghua et al., 2009).

Table 1 Divers analytical methods used for the analysis of solid waste

Methods	ECs	References
LC-MS	Simvastatin acid, estriol, codeine, olanzapine, 2,4-dinitrotoluene, triclosan, caffeine, lincomycin, Ibuprofen	Aslam et al. (2023), Encarnaç˜ ao et al. (2021), He et al. (2019)
GC-MS	17β-estradiol, 17a-ethinylestradiol, Estrone, ibuprofen, estriol, naproxen, diclofenac, fenofibrate, clofibric acid	Albero et al. (2017), Bristow et al. (2023); Cincotta et al. (2018)
NMR spectroscopy	Heparin, Polyfluoroalkyl substances, olmesartan acid	Camdzic et al. (2021)
ELISA	Staphylococcus, 17 α-ethinylestradiol, Salmonella, 17β-estradiol, cetirizine, caffeine, pesticides, carbamazepine	Aslam et al. (2023), Silva et al. (2020); Wang et al. (2023)
FT-IR spectroscopy	Microplastics, Tetracycline, bisphenol, polycyclic aromatic hydrocarbons	Khoubi et al. (2024)
Biosensors	Microcystins, proteins, mycotoxins, RNAs	Jubeen et al. (2024)

For effective environmental assessment, monitoring and remediation strategies, analysing the major pollutants of soil and water is crucial. A variety of analytical techniques are discovered to sense and measure these impurities (Li et al., 2024; Rasheed et al., 2019a, 2019b, 2019c). The most often used analytical methods for identifying contaminants in water include analysis of Temperature, Phosphate, and Nitrate, Total Dissolved Solids (TDS), Conductivity, pH measurement, Liquid Chromatography-Mass Spectrometry (LC-MS), Voltammetry, Gas Chromatography-Mass Spectrometry (GC-MS), Microbial Analysis, FT-IR, NIR, High-Performance Liquid Chromatography (HPLC), NMR, Raman, and others. The most advanced techniques are discussed here, shown in Table 1.

2 Sources, Composition, and Characterization of the Solid Waste

Municipal solid waste (MSW) is ace of the most significant tasks for environmental protection. Municipalities are mostly accountable for waste management. An efficient and effective system for the residents has to be provided by them. However, they often face numerous difficulties that exceed the capacity of the municipal authority to handle MSW, primarily due to a lack of organization, economic resources, and intricacy (Sujauddin et al., 2008; Burntley et al., 2007). From country to country and from one municipality to another, the arrangement of MSW contrasts meaningfully. Waste organization guidelines, economic situation, manufacturing structure, and lifestyle all mainly govern these variations. For the proper management and appropriate handling of these wastes, the composition and quantity of the MSW are critical. For setting up a solid waste to energy conversion facility in a particular municipality, such information is useful and essential. Based on the elemental composition and the calorific value of MSW, the scientists and engineers can agree upon its efficacy as a fuel, and such information will help to forecast the composition of gaseous emissions. Subsequently, this MSW is subjected to the energy conversion technologies, including incineration, gasification, etc. Although the ash may contain potentially hazardous materials, it should be handled with appropriate caution (ASME, 2014). Thus, for the effectiveness of the waste materials for also biogas manufacture as fuel via living organism change or for composting, their composition will provide valuable information (Kumar et al., 2010). In the interim, the composition of MSW has a greatly influenced by the time. Over time, the biodegradation of such MSW is a significant influence that controls the quantity of ecological substance, especially organic material. In 2013, MSW generation in the United States was estimated by the EPA at 254 million tonnes (U.S. Environmental Protection Agency, 2013). The classification of such MSW according to its composition and content is given in Fig. 2. Municipal or domestic waste is typically produced after various resources, anywhere various living organism actions take place. According to the literature, public solid waste produced in emerging states comes primarily from families (55–80%), trailed by marketplaces or profitable parts (10–30%). The concluding comprises mutable measures produced from numerous sources, including institutions, streets, industries, and many others, and is highly heterogeneous in nature (Nabegu et al., 2010). Hence, depending on their sources, they have variable chemical and physical characteristics. It would be a tedious process to classify them because of their varied composition, such as construction and demolition materials, inert materials, batteries, paint containers, plastics, yard waste, rubbers, metals, leather, papers, textiles, wood, food waste, and many others. Thus, sorting and utilization of such produced dense as material, heterogeneity is the chief delay. Therefore, earlier any expressive action procedure of this waste, proper sorting and fractionation are needed. To offer information on the eminence of the segregated portions for any possible operation, separating and sorting of such wastelands are one of the greatest significant and outdated approaches as vital stages in SWM. However, the success of any plan for solid waste segregation depends primarily on the active participation and public awareness of such leftover producers in the diverse societies (i.e. how they track the principles and fundamentals of waste separation and sorting (Nabegu et al., 2010). All over the world, especially in all urban centres, solid waste production (SWG) is a problem and a matter of anxiety. Maximum emerging nations that agonize from serious conservational effluence complications, like SWG are measured one of the most stimulating matters due to their large quantities (Al-Khatib et al., 2010). The basic services, including waste management, sanitation facilities, transport infrastructure, water supply, and sanitary-related problems in urban cities, are affected intensely due to the increased solid waste generation (Liyala et al., 2011). Numerous studies revealed that transportation, storage, collection, and disposal of solid wastes are the main crises in town areas and metropolises (Okot-Okumu et al., 2011). Greatest emerging republics, as well as cities in North and East Africa, are also fronting the identical severe difficulties associated with SWG. The poor economy of these areas is accredited as the main reason for these problems because it accounts for the small success in SWM (Okot-Okumuet al., 2011). Owing to the inadequate obtainable sources and the rival urgencies for the resources, most of these developing countries fail in SWM. Hence, the SWG is certainly one of the major and serious issues handled by many municipalities in the world. Additionally, other socioeconomic factors, like the employ position, regular revenue, number of room(s), average family size, etc., influenced the SWG and composition (Sankoh et al., 2012). Moreover, it was also observed that there is a straight connection between the communal actions in the public and the solid waste composition (Gidarakos et al., 2006). Furthermore, factors such as consumption of goods and alteration in the source-sorting behaviour are among other reasons affecting the quantity in households and the composition of the solid waste (Dahlén et al., 2008). In all countries, available resources, socio-cultural, political, lawful, economic, and atmospheric factors are the core subjects that impact the MSW administration (Al-Khatib et al., 2010). Therefore, implementation of innovative technology for SWG and MSW management should take into consideration the influence and effect on the economy and the socio-cultural aspects of the community. According to a study conducted by the European Environmental Agency, to investigation the per capita annual MSW formed by 32 European nations throughout the years 2001–2010, they observed that this leftover decreased in 11 nations and increased in 21 nations (EEA, 2013). Between the years 2001 and 2008, the quantity of waste in 26 countries was also studied, and in 6 countries, decreased amounts were observed (EEA, 2013). Therefore, the characteristics and amounts of the wastes diverse from one republic to additional and from area to area, uniform within the similar town, rendering to the stated features, counting the practice traditions of human beings (Khan et al., 2016; Shaily & Khan 2026) Fig. 2 (U.S. Environmental Protection Agency, 2013).

2.1 Food Solid Waste

In the world, large amounts of the generated waste can be an important and maintainable foundation for certain industrial substances. Surplus, within kitchen refuse, mash, garbage, and food residues, is generally described as the solid wastes and by-products of edible sustenance, and such wastes are generated from the food preparations, consumption, manufacture, distribution, and processing of edible things (Zhang et al., 2007; Gupta & Gupta, 2016; Singh et al., 2018; 2020; 2022). From municipalities and republics to other towns and nations, the edible wastes and their definition significantly varied. In the European Union, food wastes are summarized as 'uncooked or cooked of any edible substances that are rejected, or envisioned or necessary to be castoff'. While, U.S. Environmental Protection Agency (EPA) defines the food wastes as ''Un-eaten foods and food preparation wastes from residences and commercial establishments including restaurants, grocery provisions, and harvest stances, institutional mass and edible material stale, as well as industrial resources such as public food rooms'. Moreover, in the United Nations, 'Food loss'

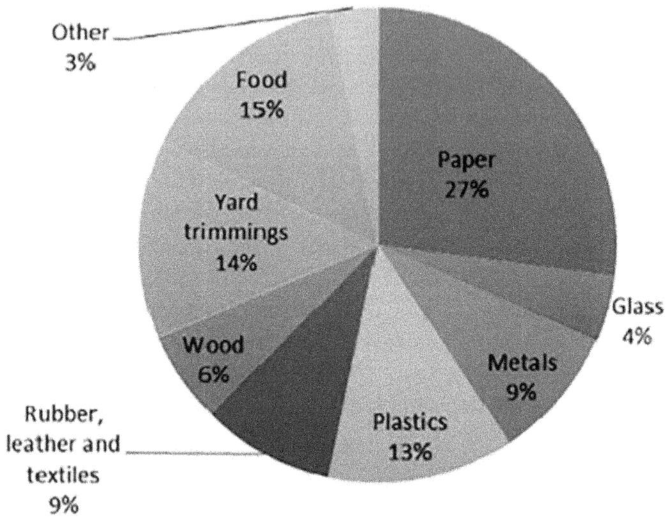

Fig. 2 Categories and types of solid waste, generated by the United States in 2013

and 'Food waste' are recognized differently. The term 'food waste' refers to the food losses due to retailers' and/or consumers' behaviour. In contrast, 'food losses' are defined as reductions in food quantity and/or quality that occur during production, post-harvest handling, processing, storage, and distribution (Russ et al., 2004). However, food wastes comprise the wasted foodstuffs, uncooked raw materials, and also the edible materials from the wet market or groceries.

2.2 *Wastes as a Source of Income*

The characterization of compact waste has been studied extensively (Banar et al., 2008; Nas et al., 2008; Rigamonti et al., 2015; Thanh et al., 2010; Ferreira et al., 2014; Gu et al. 2015). Moreover, the socioeconomic utilization of solid leftover was studied to explore potential salary from these wastes (Cheng et al., 2010; Geng et al., 2010). In this regard, waste management issues in the Sakarya area, Turkey were investigated and addressed (Yay et al., 2015). In this study, over a year, the author investigates one-tonne examples of solid waste and proposed the most effective waste management strategies. Moreover, throughout the year, during four different seasons, research also focused on characterizing the collected solid waste. Founded on the socioeconomic lifestyles in Lahore, Pakistan, these wastes were categorized into trio distinct collections. Related to socioeconomic conditions and income, significant differences in waste composition were revealed (Jadoon et al., 2014). In the area of Eskisehir, Turkey, the description of solid leftover was also studied, and they divided their investigation into the diverse income classes (Banar et al., 2008). Their categories classes the collections into high-, medium-, and low-income category. Thus, they estimated the parts of solid waste and their proportion according to the income of each group. Created on the stages of income differences, singly low, middle, and high, a further characterization study was conducted. Their recommendations and suggestions were ended with esteem to leftover organization (Miezah et al., 2015). Alternatively, to categorize the features of the solid wastes, seasonal variations were considered by Gómez et al., and they absorbed on three dissimilar socioeconomic collections in their investigations (Gómez et al. 2009). In such waste management, the classification and study of composition play an important character in planning and selecting the utmost apposite scheme of disposal, stowage, and transportation of solid waste. Meanwhile, to regulate any conceivable ecological influences, with nature and civilization, characterization is important (Alamgir et al., 2007). In most MSW, the usual herbal nutrient component innards are between 0.5 and 0.7 for nitrogen, 0.5 and 0.8 for phosphorus, and 0.5 and 0.8% for potassium. The calorific worth varieties between 200 and 3000 Btu/lb (Alamgir et al., 2007).

2.3 Disposal of Solid Waste

It has been stated that inappropriate container group applies, group, transmission, and/or conveyance systems have a great effect on the characteristics of the solid wastes. Besides, the poor route of planning, lack of information concerning the collection schedule (Hazra et al., 2009), number of vehicles for solid waste collection, and poor roads (Henry et al., 2006), and insufficient infrastructure (Moghadam et al., 2009) can also affect the solid waste characteristics. A studied revealed the affordable and effective ways of waste collection services (Sharholy et al., 2008). For promoting microenterprises and organizing the informal sector, information on treatment by the system is the only significant features distressing the management of solid leftover (Chung et al., 2008). Tadesse et al. analysed the factors that influence the disposal of household waste, and results showed that the source of waste services meaningly impacts the selection of waste removal (Tadesse et al., 2008). They disclosed that the lengthier distance of conveying waste vessels and the inadequate supply of these flasks increase the chances of discarding such wastes on roadsides and in exposed parts sideways the excursion. The absence of legislation, inadequate financial resources, and engineered and well-equipped landfills is all reasons that restrict the safe disposal of solid waste (Pokhrel et al., 2005).

2.4 Plastics Waste Disposal

Removal of the material made from plastic waste is the most important ecological problem worldwide. Japan, the United States, and Europe generate a total of 50 million tonnes of post-consumer plastic waste yearly. As of the ecological opinion, it is considered non-sustainable to dispose of these plastic wastelands in a landfill. However, legislation is strict worldwide, and the capacity and sites of landfills are declining rapidly. Several European directives and the USA legislation are worried with the administration and discarding of plastic waste (Brems et al., 2012). Since plastics are fundamentally hydrocarbons, they have a calorific value between 30 and 40 MJ/kg. Hence, they can be incinerated or burned in the urban or other devoted waste with heat and power group.

3 Disposal of Municipal Solid Waste

In the city zones, disposal, collection, and managing of the MSW are one of the foremost environmental issues. Lack of MSW disposal and management results in significant environmental problems, including aesthetic, water, air, and soil pollution. Due to the growth of greenhouse gas emissions, these pollutants in the environment are linked with human health disorders (Weigand et al., 2003). The dangerous materials produced in household waste are different than waste line generated from commercial resources (Abdel-Shafy et al., 2015, 2016; Abdel-Shafy & Mansour 2017). Under the hazardous waste regulations, such as the U.S. Resource Conservation and Recovery Act of 1976 (RCRA) (US Code, 1976) and the European Hazardous Waste Directive 91/689/EEC, these wastes are not strictly controlled (US Code. 1976). Along with general household waste (HW), the household hazardous wastes (HHW) are disposed of in landfills. The significance, quality, and amounts of such disposal are inadequately understood. Generally, it is expected that the quantity of HHWs is small, and thus, the risks of disposal are insignificant. However, the importance of the poisonous and dangerous elements limited in such wastelands is elevated due to the distinct removal of MSW, trade, and other leftover (Slack et al., 2005). In landfills, around 71% of MSW's are globally inclined of (Zacarias-Farah et al., 2003). MSW mostly contains hazardous substances such as mercury-containing waste, paints, some batteries, automobile upkeep products, pharmaceuticals, and many additional inventions (Slack et al., 2004). However, more than 53% of the landfilled waste comprises papers, yard waste, hardboard paper, and edible materials that are decomposable by the anaerobic bacteria (Barlaz et al., 2010). In the United States and Europe, due to this, landfills have become the main technique of removing of waste. A characteristic cross-section showing a landfill design is given in Fig. 3.

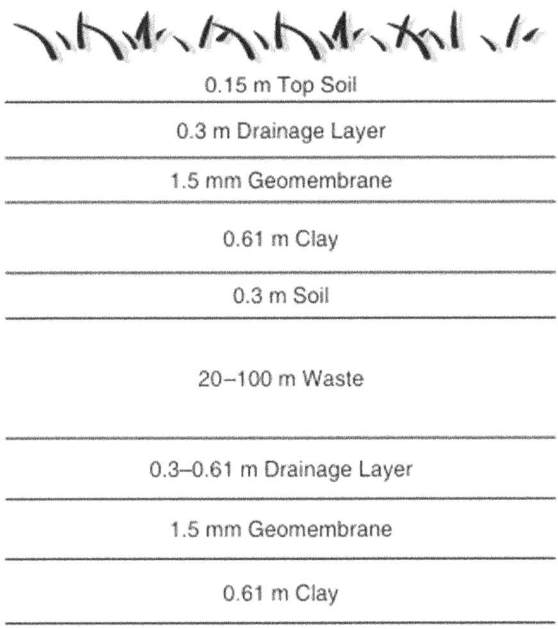

Fig. 3 Cross-section of a typical landfill collecting MSW (Barlaz et al., 2010), reused after taking permission from Wiley Blackwell

4 Issues Arising for Disposal for Disposal of Solid Waste in Rural Areas

In several growing nations, trash removal as solid leftover is a tough and extensive problem in together rural and urban areas. Numerous drains and canals are extensively utilized as exposed places to junkyard a variety of debris, including national organic and inorganic waste. Owing to the lack of nonstop debris gathering schemes, suitable landfills, open waterways, and pipes are being congested by discarding enormous quantities of solid waste and trash. Thus, they are no lengthier in purpose. These trash wastes are typically papers, plastic, and a slight poisonous material. Owing to the collapse of degradable ingredients of such noxious substance, a danger influence to the atmosphere is posed, a material that increases substantial tonnes of the BOD to the resident ecosystem (Abdel-Shafy & Mansour 2018).

5 Management and Recycling of Solid Waste

According to the reports that in terms of solid waste disposal pricing, warning signs of high recycling recovery rates at disposal sites are related to tipping fees (Scheinberg et al., 2010). When recovering the produced solid waste, high disposal prices indicate a positive impact. A investigation revealed some important reasons, including regulatory factors, influences, and altruistic factors, among others, explaining why certain societies develop strong recycling habits (González-Torre et al., 2005). Although waste management with suitable methods can definitely save or reduce GHG emissions through various ways, such as storing carbon in landfills, applying compost to soils as fertilizers, producing energy, and avoiding chief substance through substance retrieval from waste, the UNEP stated that a forthcoming waste arrangement approach based on the 3 Rs (Reduce, Reprocess, and Reutilize) is recommended by internationally recognized institutions. These 3Rs promote waste reduction, spherical economy formation, cleaner manufacturing, and the valorization of leftover by converting it into sources of materials and energy (UNEP, 2010). Improper waste management can lead to environmental degradation, including soil, aquatic, and air contamination, thus posing a actual threat to people's well-being. Additionally, the impacts of incineration and dumping of MSW on community healthiness have not been completely investigated. In a report, it was revealed that few investigations have provided sign that residents near MSW services experience congenital anomalies, little birth mass, and certain types of cancer (Rushton et al., 2003). Though, the effects on resident populations appear to differ based on the investigation's community. Regarding epidemiological research, clarification of this method deserves more focus, especially concerning the potential links to human diseases (Giusti et al., 2009). In developing countries, the issue of inadequate waste treatment is a serious concern owing to partial economic sources, and many of these nations' junkyard MSW without appropriate regulator. As a result, water, soil, and air pollution are accelerated.

Waste management is a critical issue facing humanity today. Though, leftover must not solitary be seen as a basis for material retrieval (metals, glass, plastics, and fibres) and energy, but also as a means to save oil and protect the environment. If we deliberate the entire worldwide energy that could be generated solely from farming organic leftover, with crop deposits, it is projected to be about 50 billion tonnes of oil equal (UNEP, 2010). According to UNEP (UNEP, 2010), proper separation of organic and non-organic leftover is essential for actual energy production. In detail, organic deposits can compromise the efficiency of thermal methodology in producing energy and increase GHG emissions. The way this waste is managed is shown in Fig. 4 (U.S. Environmental Protection Agency, 2013).

6 Solid Waste Management Scenario in the Global Situation

SWM is of great concern in both developing and developed countries, and urbanization is increasing at a record rate. Such variations mark a encounter to municipalities, which are accountable for the arrangement of waste in an environmentally and socially accountable way. For the solutions of operative waste management, resident waste features, which fluctuate with climatological, socioeconomic, cultural

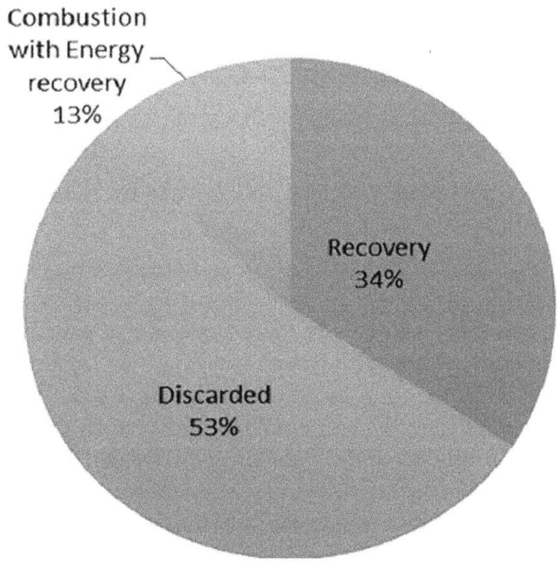

Fig. 4 MSW'S management, generated by the United States in 2013

rudiments, and institutional capabilities, are critical. On a worldwide ruler, leftover management is becoming more structured and regionalized. Surplus is characteristically controlled officially at the civic or local level in advanced countries, where populations generate meaningfully more leftover than additional residents. In less-commercialized nations, leftover is achieved by a combination of informal and formal groups, and the residents produce less waste, which is mostly biogenic (Vergara et al., 2012). Ten years ago, the 2.9 billion town inhabitants used to produce approximately 0.64 kg of MSW per day each person (0.68 billion tonnes per year). The volume has now enlarged to nearly 3 billion inhabitants, producing 1.2 kg per day each being (1.3 billion tonnes per year). By 2025, it will surely be 4.3 billion municipal inhabitants producing around 1.42 kg/capita/day of municipal solid waste (Hoornweg et al., 2012). Landfills continue as the cost-effective and most practical technique of garbage disposal despite incidents of environmental disasters that have impacted Americans. A few of the developed nations, including Canada and Japan, are inescapably changing their strategies of waste disposal away from landfills and moving to burning (Karim et al., 2020).

7 Case Study-1: Case Study in Tulsipur

More than 100 Tulsipur Municipality homes were researched using cluster sampling approaches to describe survey waste management practices and household solid waste. Based on the conducted composition study, 5% home solid waste is rubber and leather, 6% is paper and paper made materials, 7% is glass, 10% is plastics, 11%

is dust and house debris, and 46% is organic waste, and the remaining comprises of hazardous trash (1%), textiles (1%), and others (8%). Tulsipur, per person per day, makes 330.4 g of domestic solid garbage (Dangi et al., 2013).

8 Case Study-2: A Case Study in Kirtipur

Particularly in Kirtipur, the family magnitude varieties from 4 to 7 associates, with 35.57% of the total respondents having a maximum of 5 members. As 72.5% of respondents reported uncertain frequency of garbage collection, there is ambiguity in waste collection by private companies. In the Kirtipur municipality, the information collected from the survey is in contrast with the statistics obtained from the research. The results of a survey conducted by the Kirtipur Municipality show that the constituents of the waste are 77% organic, and this is almost double the result. This difference in statistics is owing to the timing of the garbage gathering review. This examination was completed throughout the winter months, which resulted in inferior dry mass, while the survey conducted by the Kirtipur municipality took place during the rainy season, causing fluctuations in the data (Awasthi et al., 2023).

9 Advanced Detection Methods of Emerging Contaminants

The methods used for the determination in the solid waste are describe below in Table 2. These methods used for the selective detection of the leftover in the different ways. The selectivity of the methods and the investigation of the compounds is based on the component available in the waste. The selectivity is given in the Table 2 and each and every method for the leftover investigation is described in details below.

9.1 Liquid Chromatography (LC)

For analysing numerous organic composites such as medicinal compounds, herbicides, and pesticides, in soil and water samples, Liquid Chromatography, principally High-Performance Liquid Chromatography (HPLC), is the expansively adopted method. The application of HPLC for the analysis of water contaminants was explored, and very low amounts of trace organic chemicals were detected in water (Kasiske et al., 1978; Gupta et al., 2025). The HPLC is largely explored for high-accuracy quality monitoring as it plays a substantial part in analysing contaminants, including organic compounds, pesticides, and heavy metals. In 2021, A HPLC method was discovered for environmental samples monitoring likes food controls and drug inspection (Chen et al., 2021).

Table 2 Detection limit of the contaminants examined by ICP-MS

Element	Quadrupole ICP-MS (ng/ml)	ICP-MS/MS (pg/ml)	Mh-ICP-MS (ng/ml)	HR-ICP-MS (pg/ml)	ICP-TOF-MS (ng/ml)
Li	12.25	–	0.14	1.031	–
B	13.12	20	0.27	1.055	5.00
V	12.69	3	0.021	0.099	0.16
Cr	6.13	4.2	0.0074	0.099	0.10
Mn	6.74	–	0.011	0.091	–
Co	12.25	–	0.016	0.140	0.17
Ni	12.37	30	0.021	0.002	11.74
Cu	6.13	10	0.044	0.012	1.20
Se	55.56	30	0.49	2.674	0.03
As	5.56	90	0.045	1.241	1.66
Pt	5.95	–	0.011	0.049	–
Pd	12.11	–	0.012	0.057	–
Ag	6.13	31	0.0067	0.507	–
Au	7.19	–	0.021	0.053	–
Cd	6.43	–	0.038	0.040	0.12
Hg	10.22	40	0.020	0.547	0.01
Te	4.90	–	0.108	0.049	0.29
Sb	6.13	–	0.013	0.021	0.16
La	12.74	–	0.0092	0.128	0.07
Ba	12.25	–	0.0032	2.233	1.56
Lu	0.02	–	0.0014	0.026	0.02
Gd	0.98	–	0.016	0.156	0.11
Bi	0.61	–	0.0069	0.008	0.02

A RP-HPLC approach was developed for the concurrent quantification of formation in the framework of microalgae bioremediation, with LOQ and LOD 0.09 and 0.03 ng $\times$ g^{-1} for ibuprofen, 0.10 and 0.03 ng $\times$ g^{-1} for paracetamol, 0.14 and 0.05 ng $\times$ g^{-1} for simvastatin acid, 0.83 and 0.27 ng $\times$ g^{-1} for simvastatin, 0.13 and 0.04 ng $\times$ g^{-1} for olanzapine, respectively (Encarnaç˜ao et al., 2021). Moreover, for high-throughput quantification of 11 ECs in three typical crops, wheat, tomato, and corn, a original fast freeze-centrifugation removal trailed by HPLC-MS/MS, with the LOD varied from 0.01 to 2.0 ng $\times$ g^{-1} (He et al., 2019).

9.2 Gas Chromatography-Mass Spectrometry (GC-MS)

An investigations on contaminant analysis in drinking and groundwater in Chandigarh, India was done by the using GC-MS, they analysed novel pollutants hexachlorohexane (HCH) and endosulfan (ES) in aquatic samples, and it was concluded that the concentration of HCH and ES in all aquatic aliquots was less than the detection limits. Moreover, it was also determined that some impurities in water, including total hardness, TDS, DO, and EC, exceeded the boundary established by the Bureau of Indian Standards (BIS). The post-summer season was chosen to monitor various parameters such as COD, BOD, pH, DO, EC, total, and faecal coliforms. The pH and EC values exhibited an elevated leaning from the dam source to the blending points. The bicarbonate, magnesium, and calcium ions were observed to be raised along with an increased count of pathogens and Enterobacteriaceae at the densely populated sites (Vashisht et al., 2020). A headspace-solid stage micro-extraction-gas chromatography-mass spectrometry (HS-SPME-GC-MS) method was developed for volatile ECs determination in fruit, which displayed the limits of quantification (LOQ) and limits of detection (LOD) values ranging between 0.037–0.098 ng $\times$ g^{-1} and 0.011 and 0.033 ng $\times$ g^{-1}, respectively (Cincotta et al., 2018). Further, a multi-residue approach was developed, to determine the 17 ECs in vegetables on the basis of ultrasound-assisted matrix solid-phase dispersion, for the different vegetables analysed the LOQ ranges from 0.1–0.4 ng $\times$ g^{-1} (Albero et al., 2017). To detect the geosmin and 2-methylisoborneol in drinking water, a further study was conducted, and established by using the methods a micro-solid phase extraction (SPE) coupled with GC-MS, with LODs is 2.0 ng $\times$ g^{-1} and 4.3 ng $\times$ g^{-1}, respectively (Bristow et al., 2023).

9.3 Liquid Chromatography-Mass Spectrometry (LC-MS)

Numerous organic substances, including pesticides, environmental, and pharmaceutical contaminants, can be analysed with LC-MS. The contaminants negatively affect instrument resolution, sensitivity, efficiency, reproducibility, selectivity, speed, and column lifetime. The study assessed a diverse range of microcontaminants in solid and aqueous environmental matrices through the integration of Liquid Chromatography (LC) with Mass Spectrometry (MS) (Tomšíková et al., 2012). Based on these performances, the authors deliberated on diverse features of procedures, including matrix effects and sample preparation (pre-concentration or extraction). Later, explored the utility of LC-MS for the measurement of chemical pollutants in food (Hird et al., 2014). The concentrations of chemical pollutants in chicken eggs were analyzed using Liquid Chromatography–Tandem Mass Spectrometry (LC–MS/MS) (Zhang et al., 2022). Thus, the current strategy can be effectively utilized for egg quality control. The technique identified approximately 244 contaminants, including some banned chemicals, in the egg samples. A research group executed a study on modern innovations in harmful contaminants examination in food by mass spectrometry (Sun et al., 2023). The author concluded that the technique is highly sensitive, precise, and quick for examining contaminants in food packaging materials and nutritional products.

9.4 Enzyme-Linked Immunosorbent Assay (Elisa)

A well-developed analytical method, ELISA is widely used for investigation of definite pollutants in numerous sample types. It uses antibody-antigen reactions to accurately quantify and similar embattled analytes, including heavy metals, hormones, and pesticides. Water quality determination to analyse possibly injurious elements, like endocrine-disrupting materials, herbicides, and pesticides, is a recent application of ELISA. To monitor caffeine, carbamazepine, 17β-estradiol and 17α-ethinylestradiol, and cetirizine in aquas matrices, applied ELISA (Silva et al., 2020). For accurately and conveniently measurement attention stages likes pesticide residues, hormone disruptors, and another chemicals of controlling concern, commercially available ELISA kits offer an efficient tool. The high specificity and sensitivity of ELISA make it compatible for robust impurity monitoring in food safety and environmental applications. Despite the high popularity of ELISA, including cost-effectiveness, effortlessness of use, and high specificity, it has numerous boundaries in the investigation of the environment. The possible for cross-reactivity is one of the primary challenges when the antibodies utilized in the evaluation may attach to molecules another than the target analyte, resulting in untrue optimistic outcomes. This is mainly challenging in complicated ecological media where many identical substances may exist. Moreover, ELISA is mostly fewer subtle associated to extra sophisticated analytical methods such as GC-MS or LC-MS/MS, which can recognize pollutants at much lesser concentrations (Wang et al., 2023). For each target contaminant, ELISA characteristically needs the obtainability of specific antibodies, thus not providing the ELISA with the flexibility to detect nonspecific or new ECs, except that these antibodies can be developed or are already available. Additionally, matrix effects and sample preparation methods affect the quantifiable consequences produced from ELISA, potentially leading to inconsistency in reproducibility and information correctness (Aslam et al., 2023).

9.5 Inductively Coupled Plasma Mass Spectrometry (ICP-MS)

ICP-MS is an extremely sensitive approach to measure trace elements and metalloids such as selenium, arsenic, and rare earth metals in soil and water samples. A plethora of ICP-MS analyses was documented in the literature. The study investigated the use of Multi-Collector High-Resolution Inductively Coupled Plasma Mass Spectrometry (MH-ICP-MS) for the analysis of saltwater and freshwater environmental resources on Upolu Island, Samoa. In each sample, 69 elements extending from Li (3) to U (92) were determined by the MH-ICP-MS technique. Numerous anthropogenic actions, including the usage of herbicides, pesticides, and certain pharmaceutical chemicals related to obesity, can also lead to the collapse of coral reefs. Additionally, examination of some saltwater samples revealed a high salt level, which may be due to climate modification, which further leads to harmful effects on the biodiversity and health of the coral reefs (Rabieh et al., 2020). Recently, a comprehensive report was published by on the pollution of water resources and application of ICP-MS techniques for monitoring and management (Balaram et al. 2023). The report explores the use of diverse ICP-MS techniques, including classical ICP-MS, ICP-TOF-MS, ICP-MS/MS, MC-ICP-MS, HR-ICP-MS, and MH-ICP-MS. The key areas covered in the study were quality determination and detection of metal species, metals, isotopes, and quality control of numerous contaminants in water supply, wastewater, underground water, groundwater, household water, drinking water, and numerous industries. The detection limit of the contaminants examined by ICP-MS is given in Table 2 (Satyanarayanan et al., 2018; Balaram et al., 2013).

9.6 Atomic Absorption Spectroscopy (AAS)

For the investigation of heavy metals, including cadmium, mercury, and lead in soil and water, ASS is most commonly employed technique. AAS works on the principle of absorption of a particular light wavelength by metal in a sample. The study revealed the documentation of various investigations conducted on water pollution. For instance, analyse the heavy metals such as iron, copper, chromium, and zinc in the groundwater of Thirukoviloor, Villuppuram, Tamil Nadu, India, was done using the atomic absorption spectrophotometer (Sirajudeen et al., 2021). After comparing the data with WHO guidelines, the author suggested that water quality is consistent with drinking water, suitable, and good for consumption. The evaluation of heavy metals in fish and water tasters of terminated mining ponds in Barkin Ladi, Plateau State, Nigeria was also done through the AAS (Gungsat et al., 2020). The metal contents were observed to be increased in the subsequent order: Cd<As<Pb<Cr<Ni<Hg. Moreover, in the catfish samples, the elemental concentrations were observed to be improved in the order for gills as Cd<As<Pb<Cr<Hg<Ni, for the flesh Cd<As<Hg<Pb<Cr<Ni, for liver, As<Hg< Pb <Cd <Cr<Ni, and for kidney Cd<As <Cr<Hg<Pb<Ni. Based on the above data, the authors deduced that the pond water is inappropriate for agricultural and drinking needs. From Nigeria's Daura gypsum mining area, water was collected and metal content, including Zn, As, Cd, Cu, Ni, and Pb, was analysed (Idris et al., 2021). The study concluded that determined metal concentration increases in the order Zn<Pd<Ni<As<Cd, indicating

high contamination of Ni, As, and Cd in drinking water, leading to kidney-related risk in people through prolonged exposure.

9.7 UV-Visible Spectroscopy

Organic compounds, including some pesticides and aromatic hydrocarbons, can be analysed by UV-Vis spectroscopy through their UV or visible light absorption measurement. Applications and working principles of UV-visible spectroscopy for water quality analysis is based on the excitation or the organic hydrocarbons. Water quality, affected by various contaminants, including heavy metal ions, dissolved carbon monoxide, nitrate nitrogen, and toxic oxygen demand, can be measured with the help of numerous UV-visible technologies. It is crucial to develop more advances in UV-visible spectroscopic technology in the future to detect water quality parameters (Guo et al., 2020).

9.8 Nuclear Magnetic Resonance (NMR) Spectroscopy

It is a progressive and significant analytical technique used to investigate the stereochemical, dynamical, functional, and structural features of molecules across diverse scientific fields. Almost every branch of science, such as environmental science, has explored the use of NMR spectroscopy. The significant information in liquids and liquid mixtures, such as the nature of molecular interactions, can be obtained by analysing numerous parameters of NMR, including relaxation times and chemical shifts (Burger et al., 2022; Marega et al., 2009; Mani et al., 2023; Marega et al., 2009). Moreover, analysis of molecules' translational motion, which provides details about their shape, size, and interactions, can be done with the help of diffusion NMR. It is a significant tool explored in numerous scientific areas to study the spatial distribution and dynamics of molecules. The various pollutants present in environmental samples, and pharmaceuticals can be identified with the use of highly sensitive diffusion NMR without the need of prior sample preparation (Simpson et al., 2012; Zheng et al., 2013). By creating new pulse sequences, the potential of this technique should be additionally studied. Exploration and advancement in the characterization of wastewater via NMR spectroscopy is very common nowadays (Filho et al., 2015). The conducted study presents the NMR spectroscopy application in the research of wastewater treatment, offers a comprehensive literature overview, and guides interested readers. Additionally, a diverse range of wastewater, including sludge and organic matter (humic and fulvic acids), can be analysed by the most widely used NMR spectroscopy. In 2021, a research group performed a study on the implementation of NMR spectroscopy for wastewater (Anaraki et al., 2021). They examine the facts on the various metals such as ^{113}Cd, ^{51}V, ^{27}Al, and ^{207}Pb, heteronuclear nuclei, including ^{29}Si, ^{31}P, ^{15}N, ^{19}F, and ^{13}C, and other nuclei. Furthermore, a blend of NMR approaches, including NMR microscopy, multiphase NMR, and hyphenated methods (e.g., LC-SPE-NMR-MS) was explored to subsidize the existing wastewater consideration.

9.9 Fourier Transform Infrared Spectroscopy (FT-IR) Spectroscopy

Among the advanced techniques, FT-IR is a powerful systematic technique explored to quantify and identify the chemical structure based on infrared radiation absorption (Veerasingam et al., 2020). This technology can be utilized to quantify and identify the chemicals, including nitrogen oxides (NOx), sulphur dioxide (SO_2), volatile organic compounds (VOCs), carbon monoxide (CO), and other gases. A plethora of research on the chemical investigation of inorganic and organic substances in water samples has been documented in the literature (Louati et al., 2020; Pejcic et al., 2009; Binish et al., 2022). Moreover, the procedure can be explored to analyse contaminants and pollutants in soil samples and thus contribute to understanding the effect of financial action on soil quality and healthcare services. In a conducted study on the simultaneous investigation of heavy metals by exploring FT-IR technology in soil samples was investigated (Śliwińska et al., 2019). The design of the PLS model used in FT-IR spectroscopy enables FT-IR to be explored as a supplementary approach for monitoring and predicting soil contamination. The FT-IR spectroscopy is a non-destructive and quick approach and can be utilized to recognize several chemical substances. Despite identifying a wide range of compounds, it is important to bring in attention that the FT-IR may be less precise than other methods, such as GC-MS, in certain applications. Thus, to overcome this problem and for a more comprehensive detection, it is often used in combination with other analytical approaches.

9.10 Biosensors

For rapid on-site recognition of pollutants, novel biosensor approaches are being established. For precise quantification, next-generation biosensors are utilizing biological reactions, presenting a transformative means of monitoring numerous atmospheric and health contaminants, including RNAs, Proteins, Mycotoxins, and Microcystins (Magee et al., 2023). In detecting chemical food contaminants, graphene-based biosensors have established potential, display powerfully for trans-portable utility (Bobrinetskiy & Knezevic, 2018). Furthermore, developments in electrochemical biosensing have improved the ability to detect pollutants in water and food supplies (Gong et al., 2024). These developing platforms take advantage of the more selectivity of biological detection for adulteration recognition. Their establishment aims to enable on-demand examination in varied situations such as water quality management, food processing, and agricultural production (Shaily & Khan 2026; Umapathi et al., 2022). To quantify and identify the ECs in numerous surroundings, biosensors are advanced analytical strategies that associate with a biological element with a physicochemical sensor. For ECs detection, despite the rapid response times, potential, and portability of biosensors for real-time examination, they face numerous boundaries. A substantial task is their sensitivity to atmospheric circumstances such as pH and temperature, which can disturb the functionality and steadiness of the biological component, foremost to varying outcomes. Furthermore, biosensors repeatedly have a restricted range of target specificity; they characteristically need careful design and calibration for each definite contaminant, which can be resource-intensive. Interpreting results becomes difficult because false positives or negatives can occur due to cross-reactivity with non-target materials found in multifaceted ecological samples.

10 Conclusion and Future Perspective

According to the global west management, 2.2 billion tonnes of MSW were generated in 2024. Owing to the extensive effects and indefinite threat, ECs pose substantial public health and environmental science concerns. The cross-border character of ECs offers barriers for their management and regulation, as these materials can transcend national frontiers and impact provinces beyond their birthplace. Moreover, the deficiency of standardized recognition approaches and reference constituents for various ECs hinders their precise quantification and identification. To resolve these concerns, such as microplastics, industrial chemicals and by-products, nanomaterials, pharmaceuticals and personal care products (PPCPs), endocrine-disrupting chemicals (EDCs), and polyfluoroalkyl substances (PFAS), cooperative research, innovation, and policy expansion efforts are needed. Refining recognition approaches and implementing preventive guidelines are critical for handling these fears successfully. Moreover, worldwide collaboration is vital for a comprehensive comeback. The present work demonstrated the advanced technologies for the detection of contaminants in solid wastes. The recent advances in detection technology, including LC, GC-MS, LC-MS, ICP-MS, AAS, HPLC, UV-Visible, NMR, and FT-IR, were reviewed. There is an urgent need to replace the conventional sample preparation methods with advanced detection techniques for significant removal of contaminants from various samples. Directing on improved recognition, sensible guidelines, and cooperative authority is vital in protecting the public health and the environment. Additionally, it suggests approvals for upcoming research, emphasizing areas demanding further examination to understand the consequences of these pollutants. Thus, by tackling these critical concerns and implementing potential solutions to overcome the contaminants threat, the public health and environmental risk can be reduced by enhancing the ability to manage, monitor, and detect contaminants.

References

Abdel-Shafy, H. I. (2015). Chemical treatment for removal of heavy metals from industrial wastewater. *Egyptian Journal of Chemistry, 58*, 1.

Abdel-Shafy, H. I., El-Khateeb, M. A., & Mansour, M. S. M. (2016). Treatment of leather industrial wastewater via combined advanced oxidation and membrane filtration. *Water Science and Technology, 74*, 586.

Abdel-Shafy, H. I., & Mansour, M. S. M. (2017). Treatment of pharmaceutical industrial wastewater via anaerobic/aerobic system for unrestricted reuse. *Journal of Scientific Industrial Research, 76*, 119.

Abdel-Shafy, H. I., & Mansour, M. S. M. (2018). Solid waste issue: Sources, composition, disposal, recycling, and valorization. *Egyptian Journal of Petroleum, 27*, 1275.

Alamgir, M., & Ahsan, A. (2007). Characterization of MSW and nutrient contents of organic components in Bangladesh. *Electron. J. Environ. Agric. Food Chem., 6*, 1945.

Albero, B., Sánchez-Brunete, C., Miguel, E., & Tadeo, J. L. (2017). Application of matrix solid-phase dispersion followed by GC–MS/MS to the analysis of emerging contaminants in vegetables. *Food Chemistry, 217*, 660.

Al-Khatib, I. A., Monou, M., Abdul, S. F., Hafez, Q. S., & Despo, K. (2010). Solid, waste characterization, quantification and management practices in developing countries. A case study Nablus district-Palestine. *Journal of Environmental Management, 91*, 1131.

American Society of Mechanical Engineers (ASME). (2014). Waste to Energy, Summary Report. Accessed July 5, 2014 http://www.asme.org

Anaraki, M. T., Lysak, D. H., Downey, K., Kock, F. V. C., You, X., Majumdar, R. D. A. B., Lião, L. M., Ferreira, A. G., Decker, V., Goerling, B., Spraul, M., Godejohann, M., Helm, P. A., Kleywegt, S., Jobst, K., Soong, R., Simpson, M. J., & Simpson, A. J. (2021). NMR spectroscopy of wastewater: A review, case study, and future potential. *Progress in Nuclear Magnetic Resonance Spectroscopy, 126*, 121.

Aslam, S. (2023). A review on validation of enzyme linked immunosorbent assay (Elisa) techniques for detection and quantification of different contaminant in aquatic environment. *Journal of Bioresource Management, 10*, 6.

Awasthi, P., Chataut, G., & Khatri, R. (2023). Solid waste composition and its management: A case study of Kirtipur Municipality-10. *Heliyon, 9*, e21360.

Balaram, V., Copia, L., Kumar, U. S., Miller, J., & Chidambaram, S. (2023). Pollution of water resources and application of ICP-MS techniques for monitoring and management-A comprehensive review. *Geosystems and Geoenvironment, 2*, 100210.

Balaram, V., Satyanarananan, M., Murthy, P. K., Mohapatra, C., & Prasad, K. L. (2013). Quantitative multi-element analysis of cobalt crust from Afanasy-Nikitin seamount in the North Central Indian ocean by inductively coupled plasma time-of-flight mass spectrometry. *Journal of Metrology Society of India, 28*, 63.

Banar, M., & Ozkan, A. (2008). Characterization of the Municipal Solid Waste in Eskisehir City, Turkey. *Environmental Engineering Science, 25*, 1213.

Barlaz, M. A., Staley, B. F., & de los Reyes, F.L. (2010). Anaerobic biodegradation of solid waste. In R. Mitchell & J. Gu (Eds.), *Environmental Microbiology* (pp. 281–299). Wiley-Blackwell.

Bilal, M., Asgher, M., Parra-Saldivar, R., Hu, H., Wang, W., Zhang, X., & Iqbal, H. M. N. (2017). Immobilized ligninolytic enzymes: an innovative and environmental responsive technology to tackle dye-based industrial pollutants-a review. *Science of the Total Environment, 576*, 646.

Binish, M. B., Binu, P., Gopikrishna, V. G., & Mohan, M. (2022). Potential of anaerobic bacteria in bioremediation of metal-contaminated marine and estuarine environment. In S. Das & H. R. Dash (Eds.), *Microbial Biodegradation and Bioremediation* (2nd ed., p. 305). Elsevier.

Bobrinetskiy, I. I., & Knezevic, N. Z. (2018). Graphene-based biosensors for on-site detection of contaminants in food. *Analytical Methods, 10*, 5061.

Brems, A., Baeyens, J., & Dewil, R. (2012). Recycling and recovery of post-consumer plastic solid waste in a European context. *Thermal Science, 16*, 669.

Bristow, R. L., Haworth-Duff, A., Young, I. S., Myers, P., Hampson, M. R., Williams, J., & Maher, S. (2023). An automated micro solid phase extraction gas chromatography–mass spectrometry (μ SPE-GC–MS) detection method for geosmin and 2-methylisoborneol in drinking water. *Scientific Reports, 13*, 1768.

Burger, R., Lindner, S., Rumpf, J., Do, X. T., Diehl, B. W. K., Rehahn, M., Monakhova, Y. B., & Schulze, M. (2022). Benchtop versus high field NMR: Comparable performance found for the molecular weight determination of lignin. *Journal of Pharmaceutical and Biomedical Analysis, 212*, 114649.

Burntley, S. J. (2007). A review of municipal solid waste composition in the United Kingdom. *Journal of Waste Management, 27*, 1274.

Camdzic, D., Dickman, R. A., & Aga, D. S. (2021). Total and class-specific analysis of per- and polyfluoroalkyl substances in environmental samples using nuclear magnetic resonance spectroscopy. *Journal of Hazardous Materials Letters, 2*, 100023.

Cao, X., Williams, P. N., Zhan, Y., Coughlin, S. A., McGrath, J. W., Chin, J. P., & Xu, Y. (2003). Municipal solid waste compost: Global trends and biogeochemical cycling. *Soil & Environmental Health, 1*, 100038.

Chen, X., & Wang, L. (2021). Analysis of the application of high performance liquid chromatography. *Institute of Physics Conference Series: Earth and Environmental Science, 769*, 032020.

Cheng, H., & Hu, Y. (2010). Municipal solid waste (MSW) as a renewable source of energy: Current and future practices in China. *Bioresource Technology, 101*, 3816.

Chung, S., & Lo, C. (2008). Local waste management constraints and waste administrators in China. *Waste Management, 28*, 272.

Cincotta, F., Verzera, A., Tripodi, G., & Condurso, C. (2018). Volatile emerging contaminants in melon fruits, analysed by HS-SPME-GC-MS. *Food Additives & Contaminants: Part A, 35*, 512.

Colombo, E., Mauri, L., Marinozzi, M., Rudd, T. R., Yates, E. A., Ballabio, D., & Guerrini, M. (2022). NMR spectroscopy and chemometric models to detect a specific non-porcine ruminant contaminant in pharmaceutical heparin. *Journal of Pharmaceutical and Biomedical Analysis, 214*, 114724.

Dahlén, L. (2008). Household Waste Collection Factors and Variations, Department of Civil, Mining and Environmental Engineering Division of Waste Science and, Technology Luleå University of Technology.

Dangi, M. B., Urynowicz, M. A., & Belbase, S. (2013). Characterization, generation, and management of household solid waste in Tulsipur, Nepal. *Habitat international, 40*, 65.

Encarnação, T., Santos, D., Ferreira, S., Valente, A. J. M., Pereira, J. C., Campos, M. G., Burrows, H. D., & Pais, A. A. C. C. (2021). Removal of imidacloprid from water by microalgae Nannochloropsis sp. and its determination by a validated RP-HPLC method. *Bulletin of Environmental Contamination and Toxicology, 107*, 131.

European Environmental Agency (EEA). (2013). *Managing Municipal Solid Waste-A Review of Achievements in 32 European Countries EEA Report No. 2/2013*. Publications Office of the European Union.

Ferreira, S., Cabral, M., Cruz, N. F., Simões, P., & Marques, R. C. (2014). Life cycle assessment of a packaging waste recycling system in Portugal. *Waste Management, 34*, 1725.

Filho, E. G. A., e Silva, L. M. A., & Ferreira, A. G. (2015). Advancements in waste water characterization through NMR spectroscopy: review. *Magnetic Resonance in Environmental Research, 53*, 648.

Frank, O., Blumberg, S., Kunert, C., Zehentbauer, G., & Hofmann, T. (2007). Structure determination and sensory analysis of Bitter-Tasting 4-Vinylcatechol oligomers and their identification in roasted coffee by means of LC-MS/MS. *Journal of Agricultural and Food Chemistry, 6*, 1945.

Gavrilescu, M. (2010). Environmental biotechnology: achievements, opportunities and challenges. *Dynamic Biochemistry, Process Biotechnology and Molecular Biology, 4*, 1.

Geng, Y., Tsuyoshi, F., & Chen, X. (2010). Evaluation of innovative municipal solid waste management through urban symbiosis: a case study of Kawasaki. *Journal of Cleaner Production, 18*, 993.

Gidarakos, E., Havas, G., & Ntzamilis, P. (2006). Municipal solid waste composition determination supporting the integrated solid waste management system in the island of Crete. *Waste Management, 26*, 668.

Giusti, L. (2009). A review of waste management practices and their impact on human health. *Waste Management, 29*, 2227.

Gomes, I. B., Simões, L. C., & Simões, M. (2018). The effects of emerging environmental contaminants on Stenotrophomonas maltophilia isolated from drinking water in planktonic and sessile states. *Science of the Total Environment, 643*, 1348.

Gómez, G., Meneses, M., Ballinas, L., & Castells, F. (2009). Seasonal characterization of municipal solid waste (MSW) in the city of Chihuahua, Mexico. *Waste Management, 29*, 2018.

Gong, H., Bao, C., Luo, X., Yu, Y., & Yang, W. (2024). Reusable electrochemical sensor for quinine detection via β-cyclodextrin-based indicator displacement assay. *Microchemical Journal, 198*, 110109.

González-Torre, P. L., & Adenso-Díaz, B. (2005). Influence of distance on the motivation and frequency of household recycling. *Waste Management, 25*, 15.

Gu, B., Wang, H., Chen, Z. Z., Jiang, S., Zhu, W., Liu, M., Chen, Y., Wu, Y., He, S., Cheng, R., Yang, J., & Bi, J. (2015). Characterization, quantification and management of household solid waste: A case study in China. *Resources, Conservation and Recycling, 98*, 67.

Guerrero, L. A., Maas, G., & Hogland, W. (2013). Solid waste management challenges for cities in developing countries. *Waste management, 33*(1), 220–232. https://doi.org/10.1016/j.wasman.2012.09.008

Gungsat, J. N., Naanogot, D. N., & Deban, L. T. (2020). Assessment of heavy metals in water and fish samples of abandoned mining ponds in Barkin Ladi, Plateau State, Nigeria. *MJBAS, 4*, 18.

Guo, Y., Liu, C., Ye, R., & Duan, Q. (2020). Advances on water quality detection by UV-Vis spectroscopy. *Applied Sciences, 10*, 6874.

Gupta, H., Dhiman, S., Madhav, S., & Mishra, R. (2025). Emerging contaminants in water and wastewater: An introduction. In H. Gupta, S. Madhav, S. Dhiman, & R. Mishra (Eds.), *Emerging contaminants in water and wastewater environmental science and engineering*. Springer. https://doi.org/10.1007/978-3-031-82579-8_1

Gupta, H., & Gupta, B. (2016). Adsorption of polycyclic aromatic hydrocarbons on banana peel activated carbon. *Desalination and Water Treatment, 57*(20), 9498–9509.

Hazra, T., & Goel, S. (2009). Solid waste management in Kolkata, India: Practices and challenges. *Journal of Waste Management, 29*, 470.

He, X., Zhang, H., Xue, R., Liu, W., Bagheri, M., Limmer, M. A., Burken, J. G., & Shi, H. (2019). Green analysis: High throughput analysis of emerging pollutants in plant sap by freeze-thaw-centrifugal membrane filtration sample preparation-HPLC-MS/MS analysis. *Journal of Agricultural and Food Chemistry, 67*, 12927.

Henry, R. K., Yongsheng, Z., & Jun, D. (2006). Municipal solid waste management challenges in developing countries-Kenyan case study. *Waste Management, 26*, 92.

Hird, S. J., Lau, B. P. Y., Schuhmacher, R., & Krska, R. (2014). Liquid chromatography-mass spectrometry for the determination of chemical contaminants in food. *TrAC Trends in Analytical Chemistry, 59*, 59.

Hoornweg, D., Perinaz, B.-T. (2012). What a waste: A global review of solid waste management. Urban development series; knowledge papers no. 15. http://hdl.handle.net/10986/17388.

Huang, Q., Parshotam, L., Wang, H., Bampfylde, C., & Lewis, M. A. (2013). A model for the impact of contaminants on fish population dynamics. *Journal of Theoretical Biology, 334*, 71.

Jadoon, A., Batool, S. A., & Chaudhry, M. N. (2014). Assessment of factors affecting household solid waste generation and its composition in Gulberg Town, Lahore, Pakistan. *Journal of Material Cycles and Waste Management, 16*, 73.

Jubeen, F., Batool, A., Naz, I., Sehar, S., Sadia, H., Hayat, A., & Kazi, M. (2024). Mycotoxins detection in food using advanced, sensitive and robust electrochemical platform of sensors: A review. *Sensors and Actuators A: Physical, 367*, 115045.

Karim, M. A., & Wetterhan, J. T. (2020). A comparative study of solid waste management in the United States, Europe and Asia. *Annals of civil and environmental Engineering, 4*(1), 003–011. https://doi.org/10.29328/journal.acee.1001019

Kasiske, D., Klinkmuller, K. D., & Sonneborn, M. (1978). Application of high-performance liquid chromatography to water pollution analysis. *Journal of Chromatography A, 149*, 703.

Khan, D., Kumar, A., & Samadder, S. R. (2016). Impact of socioeconomic status on municipal solid waste generation rate. *Waste Management, 49*, 15.

Khoubi, J., Ghiasvand, A., Bahrami, A., Shahna, F. G., & Farhadian, M. (2024). Needle trap devices packed with an imine-based 2D COF: An innovative tool for the sampling of polycyclic aromatic hydrocarbons in air. *Microchemical Journal, 199*, 110018.

Kumar, S. J., Subbaiah, K. V., & Rao, P. V. V. (2010). Waste to Energy-A case study of Elurucity, Andhra Pradesh. *International Journal of Environmental Sciences, 1*, 26.

Li, H., Sheng, W., Haruna, S. A., Hassan, M. M., & Chen, Q. (2023). Recent advances in rare earth ion-doped upconversion nanomaterials: from design to their applications in food safety analysis. *Comprehensive Reviews in Food Science and Food Safety, 22*, 3732.

Li, X., Shen, X., Jiang, W., Xi, Y., & Li, S. (2024). Comprehensive review of emerging contaminants: Detection technologies, environmental impact, and management strategies. *Ecotoxicology and Environmental Safety, 278*, 116420.

Liyala, C.M. (2011). Modernizing Solid Waste Management at Municipal Level: Institutional arrangements in urban centers of East Africa. PhD Thesis, Environmental Policy Series. Wageningen University.

Louati, H., Maria, S., Rocci, J. F., & Doumenq, P. (2020). Determination of total hydrocarbons in contaminated soil with thin layer sorptive extraction coupled with attenuated total reflectance–Fourier transform infrared spectroscopy. *Analytical Chemistry, 92*, 15344.

Luo, Y., Guo, W., Ngo, H. H., Nghiem, L. D., Hai, F. I., Zhang, J., Liang, S., & Wang, X. C. (2014). A review on the occurrence of micropollutants in the aquatic environment and their fate and removal during wastewater treatment. *Science of the Total Environment, 473*, 619.

Magee, H. Y., Maurer, M. M., Cobos, A., Adhikari, S., & Halden, R. U. (2023). Impact of LC-MS/MS quantification methods on analytical results by example of nationwide U.S. environmental samples. *Science of the Total Environment, 892*, 164425.

Mani, M., Pachaiappan, R., Gowda, R. V. M., & Aroulmoji, V. (2023). Environmental and health effects of emerging contaminants—A critical review. *International Journal of Advanced Science and Engineering, 10*, 3449.

Marega, R., Aroulmoji, V., Dinon, F., Vaccari, L., Giordani, S., Murano, E., & Prato, M. (2009). Diffusion-Ordered NMR Spectroscopy (DOSY) as a promising method to elucidate the structural modifications of functionalized carbon nanotubes. *Journal of the American Chemical Society, 131*, 9086.

Miezah, K., Obiri-Danso, K., Kádár, Z., Fei-Baffoe, B., & Mensah, M. Y. (2015). Municipal solid waste characterization and quantification as a measure towards effective waste management in Ghana. *Waste management, 46*, 15–27. https://doi.org/10.1016/j.wasman.2015.09.009

Minghua, Z., Xiumin, F., Rovetta, A., Qichang, H., Vicentini, F., Bingkai, L., Giusti, A., & Yi, L. (2009). Municipal solid waste management in Pudong New Area, China. *Journal Waste management, 29*, 1227.

Moghadam, M. R. A., Mokhtarani, N., & Mokhtarani, B. (2009). Municipal solid waste management in Rasht City. *Iranian Journal of Waste Management, 29*, 485.

Nabegu, A. B. (2010). An analysis of municipal solid waste in Kano metropolis, Nigeria. *Journal of Human Ecology, 31*, 111.

Nas, S. S., & Bayram, A. (2008). Municipal solid waste characteristics and management in Gümüşhane, Turkey. *Waste Management, 28*, 2435.

Okot-Okumu, J., & Nyenje, R. (2011). Municipal solid waste management under decentralisation in Uganda. *Habitat International, 35*, 537.
Pathak, D.R. (2019). Status and potential of resource recovery from municipal solid waste in Kathmandu valley, Nepal. https://doi.org/10.3126/joetp.v1i0.38239.
Pejcic, B., Myers, M., & Ross, A. (2009). Mid-infrared sensing of organic pollutants in aqueous environments. *Sensors, 9*, 6232.
Pokhrel, D., & Viraraghavan, T. (2005). Municipal solid waste management in Nepal: practices and challenges. *Waste Management, 25*, 555.
Rabieh, S., Bayaraa, O., Romeo, E., Amosa, P., Calnek, K., Idaghdour, Y., Ochsenkühn, M. A., Amin, S. A., Goldstein, G., & Bromage, T. G. (2020). MH-ICP-MS analysis of the freshwater and saltwater environmental resources of Upolu island, Samoa. *Molecules, 25*, 4871.
Rasheed, T., Bilal, M., Nabeel, F., Adeel, M., & Iqbal, H. M. N. (2019a). Environmentally-related contaminants of high concern: Potential sources and analytical modalities for detection, quantification, and treatment. *Environment International, 122*, 52.
Rasheed, T., Bilal, M., Nabeel, F., Iqbal, H. M., Li, C., & Zhou, Y. (2018a). Fluorescent sensor based models for the detection of environmentally-related toxic heavy metals. *Science of the Total Environment, 615*, 476.
Rasheed, T., Li, C., Bilal, M., Yu, C., & Iqbal, H. M. (2018c). Potentially toxic elements and environmentally-related pollutants recognition using colorimetric and ratiometric fluorescent probes. *Science of the Total Environment, 640*, 174.
Rasheed, T., Li, C., Fu, L., Nabeel, F., Yu, C., Gong, L., & Zhou, Y. (2018b). Development and characterization of newly engineered chemosensor with intracellular monitoring potentialities and lowest detection of toxic elements. *Journal of Molecular Liquids, 272*, 440.
Rasheed, T., Li, C., Nabeel, F., Huang, W., & Zhou, Y. (2019c). Self-assembly of alternating copolymer vesicles for the highly selective, sensitive and visual detection and quantification of aqueous Hg^{2+}. *Chemical Engineering Journal, 358*, 101.
Rasheed, T., Nabeel, F., Li, C., & Zhang, Y. (2019b). Rhodol assisted alternating copolymer based chromogenic vesicles for the aqueous detection and quantification of hydrazine via switch-on strategy. *Journal of Molecular Liquids, 274*, 461.
Rigamonti, L., Ferreira, S., Grosso, M., Cunha, R., & Marques, R. C. (2015). Economic-financial analysis of the Italian packaging waste management system from a local authority's perspective. *Journal of Cleaner Production, 87*, 533.
Rushton, L. (2003). Health hazards and waste management. *British Medical Bulletin, 68*, 183.
Russ, W., & Meyer-Pittroff, R. (2004). Utilizing waste products from the food production and processing industries. *Critical Reviews in Food Science and Nutrition, 44*, 57.
Sankoh, F. P., Yan, X., & Conteh, A. M. H. (2012). A situational assessment of socioeconomic factors affecting solid waste generation and composition in Freetown, Sierra Leone. *Journal of Environmental Protection, 3*, 562.
Satyanarayanan, M., Balaram, V., Sawant, S. S., Subramanyam, K. S. V., Krishna, G. V., Dasaram, B., & Manikyamba, C. (2018). Rapid determination of REE, PGE, and other trace elements in geological and environmental materials by high resolution inductively coupled plasma mass spectrometry. *Atomic Spectroscopy, 39*, 1.
Scheinberg, A., Wilson, D. C., & Rodic, L. (2010). *Solid Waste Management in the World's Cities. Water and Sanitation in the World Cities.* Earthscan Published for UN Habitat.
Shaily, & Khan, D. (2026). Sensors and Biosensors for Polycyclic Aromatic Hydrocarbons. In *Ubiquitous Polycyclic Aromatic Hydrocarbon Contamination: Recent Perspectives* (pp. 235-259). Cham: Springer Nature Switzerland.
Sharholy, M., Ahmad, K., Mahmood, G., & Trivedi, R. C. (2008). Municipal solid waste management in Indian cities-A review. *Waste Management, 28*, 459.
Sharma, K. D., & Jain, S. (2018). Overview of municipal solid waste generation, composition, and management in India. *Journal of Environmental Engineering., 145*. https://doi.org/10.1061/(ASCE)EE.1943-7870.000149

Silva, C. P., Carvalho, T., Schneider, R. J., Estevesa, V. I., & Lima, D. L. D. (2020). ELISA as an effective tool to determine spatial and seasonal occurrence of emerging contaminants in the aquatic environment. *Analytical Methods, 12*, 2517.

Simpson, A. J., Simpson, M. J., & Soong, R. (2012). Nuclear magnetic resonance spectroscopy and its key role in environmental research. *Environmental Science and Technology, 46*, 11488.

Singh, S., Gupta, H., Dhiman, S., & Sahu, N. K. (2022). Decontamination of cationic dye brilliant green from the aqueous media. *Applied Water Science, 12*(4), 61.

Singh, S., Kumar, A., & Gupta, H. (2020). Activated banana peel carbon: a potential adsorbent for Rhodamine B decontamination from aqueous system. *Applied Water Science, 10*(8), 185. https://doi.org/10.1007/s13201-020-01274-4

Singh, S., Parveen, N., & Gupta, H. (2018). Adsorptive decontamination of rhodamine-B from water using banana peel powder: a biosorbent. Environmental Technology & Innovation, 12, 189–195.

Sirajudeen, J., & Pravinkumar, J. (2021). Heavy metal analysis of groundwater of Thirukoviloor, Villuppuram, Tamil Nadu, India. *Materials Today: Proceedings, 36*, 828.

Slack, R. J., Gronow, J. R., & Voulvoulis, N. (2004). Hazardous components of household waste. *Critical Reviews in Environmental Science and Technology, 34*, 419.

Slack, R. J., Gronow, J. R., & Voulvoulis, N. (2005). Household hazardous waste in municipal landfills: contaminants in leachate. *Science of the Total Environment, 337*, 119.

Śliwińska, A., Smolinski, A., & Kucharski, P. (2019). Simultaneous Analysis of Heavy Metal Concentration in Soil Samples. *Applied Sciences, 9*, 4705.

Sujauddin, M., Huda, M. S., & Rafiqul, A. T. M. (2008). Household solid waste characteristics and management in Chittagong, Bangladesh. *Journal of Waste Management, 28*, 1688.

Sun, Q., Dong, Y., Wen, X., Zhang, X., Hou, S., Zhao, W., & Yin, D. (2023). A review on recent advances in mass spectrometry analysis of harmful contaminants in food. *Frontiers in Nutrition, 10*, 1244459.

Tadesse, T., Ruijs, A., & Hagos, F. (2008). Household waste disposal in Mekelle city, Northern Ethiopia. *Northern Ethiopia Journal of Waste Management, 28*, 2012.

Thanh, N. P., Matsui, Y., & Fujiwara, T. (2010). Household solid waste generation and characteristic in a Mekong Delta city, Vietnam. *Journal of Environmental Management, 91*, 2307.

Thomaidis, N. S., Asimakopoulos, A. G., & Bletsou, A. A. (2012). Emerging contaminants: a tutorial mini-review. *Global NEST Journal, 14*, 72.

Tomšíková, H., Aufartová, J., Solich, P., Nováková, L., Sosa-Ferrera, Z., & Santana-Rodríguez, J. J. (2012). High-sensitivity analysis of female-steroid hormones in environmental samples. *TrAC Trends in Analytical Chemistry, 34*, 35–58. https://doi.org/10.1016/j.trac.2011.11.008

U.S. Environmental Protection Agency. (2013). Wastes-Non-Hazardous Waste Municipal Solid Waste, 1200 Pennsylvania Ave., N. W. Washington, DC 20460, U.S.A.

Umapathi, R., Ghoreishian, S. M., Rani, G. M., Cho, Y., & Huh, Y. S. (2022). Review-emerging trends in the development of electrochemical devices for the on-site detection of food contaminants. *ECS Sensors Plus, 1*, 044601.

UNEP. (2010). Trends in solid waste management: issues, challenges and opportunities. In *International consultative meeting on expanding waste management services in developing countries 18–19 March 2010 Tokyo, Japan.*

US Code. (1976). Solid Waste Disposal Act, as amended. Resource Conservation and Recovery Act: Subtitle D (Solid Waste Program). US Code (Acts of Congress) Title 42, Chapter 82, Subchapter I (Section 6901); 1976.

Vashisht, D., Kumar, A., Mehta, S. K., & Ibhadon, A. (2020). Analysis of emerging contaminants: A case study of the underground and drinking water samples in Chandigarh, India. *Environmental Advances, 1*, 100002.

Veerasingam, S., Ranjani, M., Venkatachalapathy, R., Bagaev, A., Mukhanov, V., Litvinyuk, D., & Vethamony, P. (2020). Contributions of Fourier transform infrared spectroscopy in microplastic pollution research: A review. *Critical Reviews in Environmental Science and Technology, 51*, 2681.

Vergara, S. E., & Tchobanoglous, G. (2012). Municipal solid waste and the environment: A global perspective. *Annual Review of Environment and Resources, 37*, 277.

Wang, Y., Wang, M., Zhou, L., Geng, X., Xu, Z., & Zhang, H. (2023). Development of a competitive ELISA based on estrogen receptor and weak competitive molecule for the screening of potential estrogens in foods. *Food Chemistry, 401*, 134084.

Weigand, H., Fripan, J., Przybilla, I., & Marb, C. (2003). Composition and contaminant loads of household waste in Bavaria, Germany: investigating effects of settlement structure and waste management practice. In *Proceedings of the 9th International Waste Management and Landfill Symposium, Cagliari.*

Wilkinson, J. L., Hooda, P. S., Barker, J., Barton, S., & Swinden, J. (2016). Ecotoxic pharmaceuticals, personal care products, and other emerging contaminants: a review of environmental, receptor-mediated, developmental, and epigenetic toxicity with discussion of proposed toxicity to humans. *Critical Reviews in Environmental Science and Technology, 46*, 336.

Yay, S. E. (2015). Application of life cycle assessment (LCA) for municipal solid waste management: a case study of Sakarya. *Journal of Cleaner Production, 94*, 284.

Zacarias-Farah, A., & Geyer-Allely, E. (2003). Household consumption patterns in OECD countries: Trends and figures. *Journal of Cleaner Production, 11*, 819.

Zhang, L., Jia, Q., Liao, G., Qian, Y., & Qiu, J. (2022). Multi-Residue determination of 244 chemical contaminants in chicken eggs by liquid chromatography-tandem mass spectrometry after effective lipid clean-up. *Agriculture, 12*, 869.

Zhang, R., El-Mashad, H. M., Hartman, K., Wang, F., Liu, G., Choate, C., & Gamble, P. (2007). Characterization of food waste as feedstock for anaerobic digestion. *Bioresource Technology, 98*, 929.

Zheng, G., & Price, W. S. (2013). Environmental NMR: Diffusion ordered spectroscopy methods. *MagRes, 2*, 561.

Recovery of Precious Metals from Waste Printed Circuit Boards

Shubhangee Agarwal, Amanpreet Kaur, Priyanka Rani, and Himanshu Gupta

1 Introduction

With advancement of technology, the restructuring of society and the economy globally has resulted in a dramatic upswing in e-waste over the past few years. This e-waste is persistently being discarded off in the environment (Agarwal et al., 2024; Dhiman et al., 2024; Gupta et al., 2025; Gupta and Shiju, 2025; Kumar et al., 2025). According to the current analysis from the Global E-waste Monitor (Charpentier et al., 2023), 62 million metric ton of discarded electrical and electronic equipment (WEEE, or e-waste) were produced globally in 2022, but 77.7% of this was not officially recorded to be retrieved and refurbished. Among all types of e-waste, WPCBs are garnering significant attention for their recycling potential, both from an environmental and economic perspective (Baldé et al., 2024). PCBs serve as the essential components in each and every electronic device, facilitating the interconnection between software and hardware. PCBs, comprising 30% ceramics, and 30% plastics, 40% metals, pose substantial challenges in waste reclamation due to their complex framework and compositions (Martelo et al., 2024; Li et al., 2018). Consequently, the recycling of WPCBs represents a significant topic, crucial for environmental preservation and the restoration of valuable metals (Yazici et al., 2010). WPCBs, being abundant in resources, are commonly termed as "urban mines". The composition includes metals such as Zn (approximately 1%), Cu (approximately 16%), Fe (around 3%), Sn (about 4%), and Ni (approximately 2%) (Chen et al., 2013). Furthermore, precious metals such as Ag (0.156%), Au (0.039%), and Pd (0.009%) (Chehade et al., 2012), utilized as interface substances or coating layers due to their conductive properties and chemical resistance (Chen et al., 2013), are found to be ten times (Eygen et al., 2016) more prevalent in WPCBs compared to naturally occurring ores. Consequently, it is evident that the recovering precious metal from WPCBs holds substantial importance. The

S. Agarwal (✉) · A. Kaur · P. Rani · H. Gupta
Department of Chemistry, School of Sciences, IFTM University, Moradabad, India

H. Gupta et al. (eds.), *Advanced Technologies for Assessment, Recycling, and Management of Solid Waste*, Waste as a Resource,
https://doi.org/10.1007/978-3-032-24788-9_8

manufacturing of electronic equipments is increasing, as the utilization of PCBs has become increasingly prevalent. This serves as a crucial source for the recovering of trace and valuable elements. The direct disposal of WPCBs poses a significant risk of causing irreversible damage to the environment. When WPCBs are disposed of in landfills, they can be present alongside municipal waste, leading to the formation of leachate that contains harmful heavy metals (Silva et al., 2021) and bromides (Ning et al., 2017a, 2017b; Kopacek & Kopacek, 2015). Improper treatment of leachate containing heavy metal brominated compounds can lead to their release into aquatic ecosystems, resulting in water pollution (Kumar et al., 2021; Janani et al., 2022). Recycling gives a sound economic opportunity while maintaining a minimal environmental footprint (Silva et al., 2021). Recycling technologies include pyrometallurgical, hydrometallurgical, and biometallurgical processes, all presenting unique advantages and disadvantages. Many investigations are done into foremost mentioned two techniques which focus primarily on the recovery of metals from WPCBs, often neglecting environmental associated impacts (Shokri & Fotovat, 2023; Yan et al., 2022).

2 Composition of WPCBs

Nearly every electrical and electronic device incorporates PCBs. The relationship between hardware and software is crucial. It serves as the core component of every electronic device. Typically, waste PCBs comprises of both metals and nonmetals, approximately 70% are nonmetals and 30% metals (Ning et al., 2017a, 2017b). The composition includes a range of metals, including gold (Au), platinum (Pt), silver (Ag), iron (Fe), nickel (Ni), lead (Pb), copper (Cu), tin (Sn), manganese (Mn), zinc (Zn). The nonmetal components of PCBs are framed of thermosetting and thermoplastic resins, including brominated epoxy resins, which is intricately mixed and often contain embedded metals or other substances. A circuit predominantly composed of copper is affixed to the circuit board using adhesives and protective paints or coatings. The circuit board consists of epoxy resin or fiberglass and is typically finished with a coating of gold. Copper contacts that have been plated with silver are used to build solder joints. The PCBs comprise components made of plastic, metal, and ceramic housing materials. The PCBs contain both precious metals as well as various harmful metals, including Ag, Au, Pd, Zn, Be, Br, Sb, Cd, Ni, Pb, Tl, and As. Certain elements found in discarded PCBs constitute a serious risk to the planet and can lead to a variety of medical issues. Consequently, the recycling of discarded PCBs has emerged as a pressing global concern (Kaya, 2018). Rather than disposing of or sending to landfills, it is advantageous to utilize PCBs as a additional source of metals, given their significant metal percentage (Sankar et al., 2022). Consequently, both environmental pollution and ecosystem will be mitigated and primary natural resources will be safeguarded (Fig. 1).

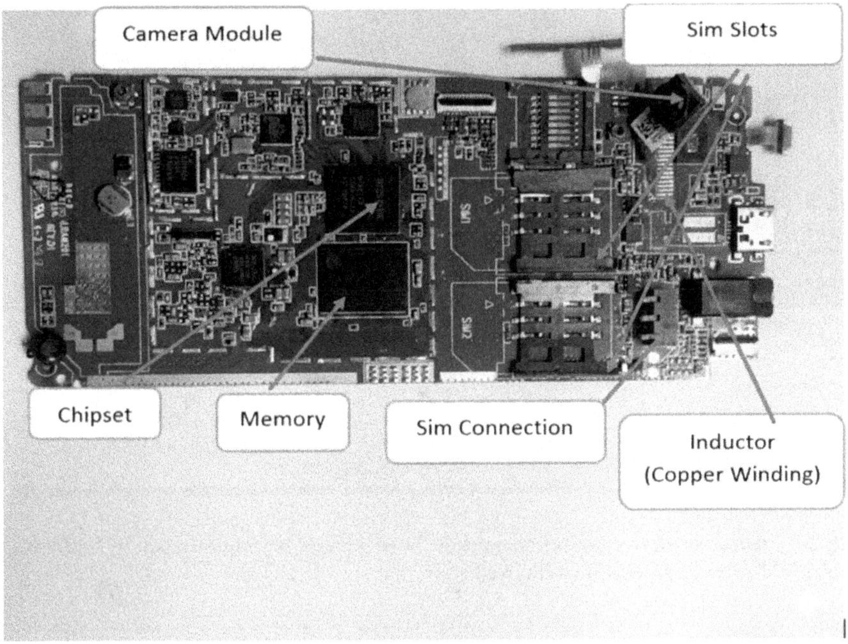

Fig. 1 Components of PCBs (reprinted from original article by Udaykumar et al. (2022) after permission from Elsevier)

3 Techniques of Recycling for Metal Recovery

WPCBs contain high percentage of valuable metals like platinum, palladium, gold and silver often surpassing those present in natural ores. This makes their recovery not only economically beneficial but also crucial for long term environment preservation. The wrong way of discarding WPCBs presents significant environmental risks, primarily due to the pack of toxic metals and persistent organic pollutants. The reclamation of metals from WPCBs holds significant importance, contributing to both environmental sustainability and the metal recovery (Ning et al., 2017a, 2017b). Despite being much lower than base metal concentrations, WPCBs have higher concentrations of precious metals than even the most abundant conventional natural ores (Lin et al., 2022). Recycling technologies can be categorized into traditional methods, including thermal, mechanical, pyrometallurgical, and hydrometallurgical processes, as well as advanced green emerging technologies such as bioleaching, ionic liquids, supercritical fluid extraction, and deep eutectic solvents. Each of these approaches presents its own unique advantages and disadvantages. Many investigations into pyrometallurgical and hydrometallurgical approaches focus primarily on the recovery metals from WPCBs, often neglecting the associated environmental impacts (Guo et al., 2014). Conventional approaches involve the direct handling of waste (Fig. 2).

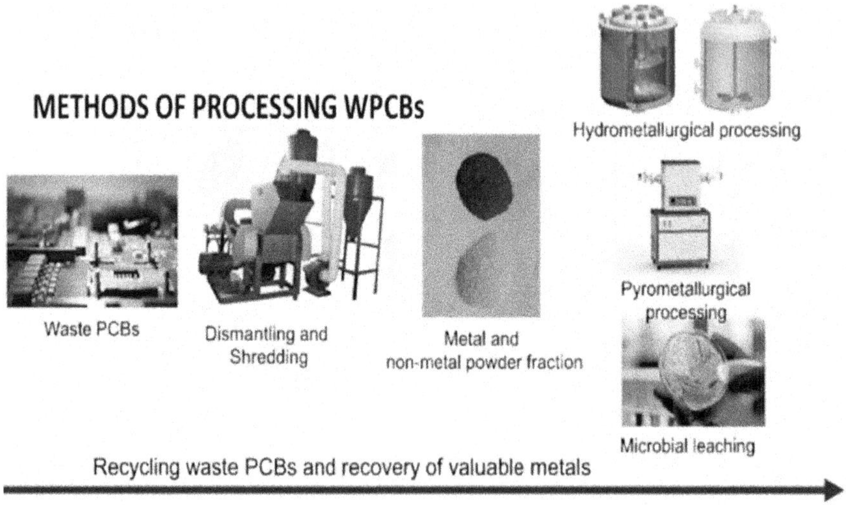

Fig. 2 Flowchart to recover metals from waste PCBs (reprinted from original article by Udaykumar et al. (2022) after permission from Elsevier)

3.1 Physico-Mechanical Separation

This method represents the earliest approach for extracting metals from electronic parts. The initial phase is the dismantling of ECs from the WPCBs. The metallic structure are frequently enclosed with different plastic or ceramic substances on WPCBs (Guo et al., 2015); thus, a preliminary treatment process is essential to reveal the metals for activation (Guo et al., 2015). To streamline the challenges associated with future therapies and enhance treatment efficiency, the electric components of the WPCBs surfaces, including capacitors and resistors, should be disassembled and organized in suitable classifications (Fazari et al., 2024). The mechanical processes are straightforward to execute, requiring minimal chemical usage and demonstrating a greater compatibility with environmental considerations. The process entails the differentiation of metallic and nonmetallic components based on various characteristics such as surface properties, electrical conductivity, size, density, and magnetic properties, among others. The primary obstacle in this process is the reduced recovery rates of valuable metals and base metals (Ning et al., 2017a, 2017b).

3.2 Incineration

Incineration serves as a prevalent disposal technique that entails the combustion of organic compounds in solid waste, primarily because of its effectiveness in reducing volume, sterilizing waste, and facilitating energy recovery. The significant calorific value of WEEE, characterized as a diverse blend of organic and inorganic substances with a notable proportion of organic content, renders its incineration a viable option (Udayakumar et al., 2022). A significant mass reduction of 70% can be attained through the incineration of WEEE. Nonetheless, the incineration of WEEE leads to significant concentrations of heavy metals present in the filter cake, flue gas, and slag. When brominated flame retardants are present in WEEE, they produce furans, dioxins, and brominated compounds, which can cause serious pollution, especially when there is uncontrolled combustion or open burning (Zhu et al., 2025). Open burning represents a rudimentary recycling method that relies on uncontrolled combustion, primarily aimed at extracting valuable metals. Materials like cables are frequently incinerated outdoors to eliminate the plastic insulation and enable the recovery of copper, or chips are extracted from circuit boards post-burning to reclaim noble metals (Au, Ag, Pd, and Pt). In developing countries, small workshops have engaged in open burning of WEEE to recover copper and other valuable metals, driven by cost-effectiveness and simplicity of the process. The efforts in question contribute to environmental issues while simultaneously heightening the health risks faced by workers (Hadi et al., 2015; Cardamone et al., 2021).

3.3 Pyrometallurgy

Pyrometallurgy presents a highly promising, adaptable, and efficient method for the recovery of metallic fractions. These pyrometallurgical technologies can effectively process WEEE with minimal modifications required. The approach involves the direct treatment of electronic components while maintaining their original size. This technique operates at elevated temperatures and is utilized to remove metals from urban garbage dumps. The primary pyrometallurgical techniques utilized for extraction of metals include combustion, smelting, pyrolysis, and molten salt processes (Chancerel & Rotter, 2013). Pyrometallurgical processes have the capacity to process between 1000 and 5000 tons of WPCBs daily, resulting in the yearly production of 100 tons of gold, 2400 tons of silver, 100,000 tons of copper, and 50 tons of PGMs (Kaya, 2018). Consequently, pyrometallurgy presents significant promise as an effective method for the extensive processing of the continuously increasing stream of WPCBs. The primary pyrometallurgical techniques utilized to retrieve metals from WEEEs include combustion, pyrolysis, smelting, and molten salt processes (Chancerel & Rotter, 2013). This technique does have certain limitations.

3.4 Smelting

Smelting involves specific physical and chemical transformations occurring within the furnace at elevated temperatures of 1300–1600 K. It focuses on the concentration of metals into metal alloys. The foremost stage in the handling of this waste typically includes fragmentation and shredding, succeeded by various physical segregation methods (including magnetic separation, air current separation, eddy current separation as detailed by to categorize different categories, each possessing unique chemical properties (Zhang & Xu, 2016). The fraction rich in non-ferrous metals is subsequently subjected to processing using one of the existing smelting technologies. The composition primarily includes prevalent base metals like tin, iron, lead, zinc, aluminum, copper, and nickel along with some residual ferrous elements and plastics, which cannot be completely eliminated during the physical segregation processes. The treatment of the metallic components is contingent upon the primary base metal, whether it is copper or lead. This fraction can be treated in a smelter specific to either copper or lead, functioning as a recycling unit to reclaim these base metals along with inert metals such as silver, gold, and platinum family, which are subsequently extracted through additional hydrometallurgical and electrochemical methods (Chancerel and Rotter, 2013). Copper and lead smelters have demonstrated effectiveness in recycling waste electronics for recovering copper, lead and valuable metals. Electronic scrap is introduced into a furnace with either copper or lead scrap, where the WEEE products undergo melting. In the molten bath, metallic components, primarily copper, lead, and precious metals, persist, while the remaining components aggregate to form the slag. To retrieve primary metals from scrap electronic equipments during the smelting process presents challenges, particularly due to the substantial loss of organic and glass components.

Copper-Smelting Process—Sulfide Method

The process of copper smelting involves the removal of sulfur and iron through the application of oxygen. The investigation of $CuFeS–SiO_2–O_2$ system was done at 1200 °C. Silica is incorporated at a 2:1 molar ratio of $CuFeS_2$ to SiO_2 to facilitate the development of a slag phase during the generation of ferrous oxide (FeO) while blowing O_2-enriched air (Eq. 1) (Schlesinger et al., 2021).

$$2CuFeS_2(s) + 4O_2(g) + SiO_2(s) \rightarrow Fe_2SiO_4(slag) + 3SO_2 + Cu_2S(matte) \quad (1)$$

Lead-Smelting Processes

Lead like copper being chalcophile element occurs naturally in conjunction with sulfur in various other minerals. Lead sulfide ores like galena (PbS) generally include additional minerals that are abundant in zinc, copper, and iron (as sulfates and sulfides), along with valuable metals, particularly silver (Harvey et al., 2022). The concentrate undergoes an initial roasting process with air to facilitate the conversion of sulfide to oxide through subsequent reaction (Eq. 2):

$$2PbS(s) + 3\,O_2(g) \overset{400-900}{\rightarrow} PbO(s/l) + 2SO_2 \quad (2)$$

3.5 *Hydrometallurgy*

Hydrometallurgical leaching technology offers a level of precision that is technically superior, highly predictable, and adaptable, allowing for effective control in the extraction of materials from resources. This method has been utilized in the recycling of WPCBs for nearly thirty years, often in conjunction with mechanical grinding as a preliminary step. It is a specialized area within metallurgy which employs application of aqueous solutions and targeted chemical interactions to recover metal content from both primary ores and additional sources (Ahirwar & Tripathi, 2021). These processes have surfaced as a promising approach for recovering metals, attributed to their capability for significant extraction of metals yields, minimized environmental effect, and comparatively minimum consumption of energy (Tuncuk et al., 2012). Previous investigations have primarily focused on techniques for extracting valuable metals while the leftover base metals are typically deposited as hydroxides and subsequently discarded in landfills (Li et al., 2021). The financial sustainability of the reuse as well as recycling is significantly impacted by the recovery of fundamental metals like copper, tin and lead, as these elements are major components of PCBs (Li et al., 2018). These processes mainly entail the dissolution of metals within an alkaline or acidic environment. Numerous investigations have documented the application of HNO_3, H_2SO_4, HCl, and aqua regia in the extraction of metals from discarded PCBs (Oraby et al., 2020). Cyanide, halide, thiosulfate, and thiourea are often utilized in the extraction process of precious metals. Following the leaching process, the material proceeds to metal recovery methods including electrowinning, electrorefining, cementation, and ion exchange (Ma et al., 2019). For example, Hydrometallurgical technique i.e., bromine leaching can attain the recovery rate exceeding 95% for both gold and silver at optimum conditions (Cui & Anderson, 2020). Furthermore, it is possible to achieve Ag recovery rates as high as 98% through physical segregation methods, particularly direct leaching (Tanısalı et al., 2021). The process parameters in the hydrometallurgical process are more effectively controlled, including the types of acids utilized, their concentrations, the leaching temperature, the duration of leaching, and various other factors. By carefully choosing the right reagents, it is feasible to achieve selective extraction. HNO_3 primarily serves the purpose of leaching copper, palladium, tin and lead for desoldering. Leaching out the tin and lead is crucial, as the soldering alloy used in electronic PCBs primarily consists of these metals. Lead is a harmful substance that if improperly handled or disposed off, can pollute the environment. Nitric acid is effective for extracting copper from PCBs. The productivity of copper recovery increases as the concentration of nitric acid (Birloaga et al., 2014; Bas et al., 2014). Multiple research studies have investigated the interaction of H_2O_2 with acids, demonstrating that the leaching efficiency improves with the addition of oxidizing agent like hydrogen peroxide (H_2O_2). The process of extracting PCBs using a combination of 2 M of H_2SO_4 and 0.2 M H_2O_2 at 80–85 °C for twelve hours resulted in a recovery of approximately 95% of copper. (Birloaga et al., 2013; Jha et al., 2011). The benefit of sulfuric acid lies in its ability to selectively target copper when used with oxidizing agents like ferric iron (Quinet et al., 2005). The drawbacks include its corrosive properties and the requirement for a reactor that can withstand acid while it is being leached. Aqua regia consists of a combination of HNO_3 and HCl is very effective in the reclamation of precious metals like gold, attributed to its strong oxidizing properties and minimal capital investment (Wang et al., 2021). The PCBs undergo a three-stage leaching process, where nitric acid is utilized in the first two stages to extract mainly gold and base metals. In the final stage, combination of HNO_3 and HCl is applied to obtain the gold from the materials obtained in the former stages. As the temperature increases during leaching process, the recovery of gold accelerates. Cyanide leaching represents a conventional technique employed for the extraction of Au and Ag within gold mining treatment facilities and have been utilized in commercial applications. Cyanidation involves the application of cyanide as a leaching agent to extract Au from PCBs. The compound that contains cyanide is applicable for direct use in the electroplating method. The factors that should be taken into account in cyanide leaching for gold extraction include the presence of oxygen, cyanide concentration, pH, temperature, and stirring rate the presence of both cations and anions at the solid–liquid interface (Akcil et al., 2015). KCN is frequently utilized in the cyanide leaching process for Au extraction on a significant scale. The drawbacks include the production of wastewater that is toxic and poses risks to both human health and the environment. Thiourea, recognized for its complexing properties, is a sulfur-containing organic compound that interacts with PCBs to create a cationic soluble complex with the desired metal. This method can extract up to 99% of the gold through leaching. At particular pH values, the ferric ion and thiourea combine to form iron-thiourea complexes, which speed up the rate of suspension (Kaya, 2019). Thiourea can function effectively under both alkaline and acidic conditions and is applicable for the recovery of Au from PCBs. In alkaline environment, thiourea is resistant and is utilized to obtain gold, but in acidic conditions, it shows an efficient capacity to recover precious metals. Chemicals like sodium hyposulfite and ammonium thiosulfate are being used to process gold ore in a more environmentally friendly technique. The solubility of these chemical substances in water is notable, and it is essential for the leaching medium to be alkaline, as the sulfate radical exhibits instability in acidic conditions (Moyo et al., 2020). Halogens like fluorine, chlorine, bromine, iodine and astatine are used in the extraction procedure of precious metals from PCBs. This process is halide leaching process. At the particular conditions of the solutions, this leaching process creates gold (III) and gold (I) complexes with halide ions as a result of the dissolution of gold. The leaching rates utilizing halides are elevated and have the potential to substitute cyanide leaching for gold dissolution, which poses detrimental effects on human well-being and the natural environment. The gold recovery achieved through iodine halide leaching was approximately 98%. In addition to the dissolution of gold, the leaching by iodine can effectively recover Cu in a sulfuric acid medium, utilizing H_2O_2 as an oxidizing

Table 1 Hydrometallurgical extraction of metals from WPCBs

Leaching agent	Reaction involved	Percentage metal recovery	References
HNO_3	$8\ HNO_3 + 3Cu \rightarrow 3Cu(NO_3)_2 + 2\ NO + 4\ H_2O$	95% cu	Hsu et al. (2019)
Aqua regia	$2Au + 11HCl + 3HNO_3 \rightarrow 2HAuCl_4 + 3NOCl + 6\ H_2O$	98% ag & 97% au	Liu et al. (2021)
H_2SO_4	$2\ H^+ + 3Cu \rightarrow Cu^{2+} + 2\ H_2O$	95% cu	Jha et al. (2011)
Halide leaching	$2\ M + 2\ L^- +$ oxidizing agent $\rightarrow 2ML_2^-$ (M = metal; L = halogen)	97% au	Liu et al. (2021)
Thiosulphate leaching	$4Au + 8\ (S_2O_3)^{2-} + 4H_2O \rightarrow 4Au((S_2O_3)^{2-})_2 + 4OH^-$	95% au & 100% ag	Akcil et al. (2015)
Thiourea leaching	$4Au + 2\ CS(NH_2)_2 \rightarrow 4Au(CS(NH_2)_2^- + 4e^-$	71 5 ag & 86% au	Birloaga et al. (2014)
Cyanide leaching	$4Au + 8\ CN^- \rightarrow 4Au(CN)_2 + 4e^-$	90% Pt, Pd & ag	Park et al. (2019)

agent (Ashiq et al., 2019). The use of chlorine/chloride leaching systems has been restricted by the control of Cl_2 gas and the requirement for specialized reactors, while iodide leaching is thought to be extremely selective for gold and environmentally friendly and non-corrosive. One innovative hydrometallurgical method investigated for copper extraction is the supercritical methanol (SCM) process. After a nitric acid pretreatment, printed circuit boards are treated with this supply chain management procedure to create ultrafine copper. In order to separate metals and polymers from ground PCBs, supercritical methanol was applied (Xiu et al., 2013) (Table 1).

3.6 *Bioleaching*

This technique is economically friendly for recovering metals from WPCBs, aligning with the principles of sustainable development. Microbial leaching is another term for this process. This method employs use of microorganisms like fungi and bacteria to help remove metals from solid substrates. For the recovery of metals from e-waste, this technique is gradually becoming a commercially viable strategy (Esamaeli et al., 2020). According to Yken et al. (2021), this technique requires a high level of expertise, involves time taking operations, and has a relatively slow process flow. Recycling has a small environmental impact and offers a good financial opportunity (Ning et al., 2017a, 2017b). Technologies used for recycling include hydrometallurgical, biometallurgical, and pyrometallurgical processes; each has unique advantages as well as limitations. Many investigations into pyrometallurgical and hydrometallurgical technologies focus primarily on the metal extraction from WPCBs, often neglecting the associated environmental effect (Guo et al., 2014). Significant advantages of bioleaching include comparatively high recovery rates, simplicity of use, cost-effectiveness, minimal use of energy, and the absence of specialized industrial tools under particular process optimal environments (Priya & Hait, 2018). Under optimal conditions, various bioleaching methods yield elevated leaching rates to attain enhanced metal recovery. For instance, in a two-phase leaching process, *Pseudomonas aeruginosa* achieves a leaching rate of as much as 91% for silver (Natarajan & Ting, 2014). The application of *C. violaceum* and *P. aeruginosa* during the initial and subsequent leaching phases led to a 74% extraction of gold (Pradhan & Kumar, 2012). Fungi like *Penicillium* spp. and *Aspergillus* spp., along with yeasts such as Yarrowia lipolytica, are utilized in the bioleaching of metals by synthesizing of organic acids (Trivedi et al., 2022).

Selection of Microorganisms

The microbes used in bioleaching in based on the metals to be extracted. These microbes generate organic and inorganic acids which are used in leaching. The bacteria mostly used for leaching are *Pseudomonas aeruginosa, C. violaceum, Acidio thiobacillusc ferrooxidans*, etc. And the fungi commonly used are *Aspergillus* spp., *Penicillium* spp.etc.

Types of Bioleaching

Various leaching methods employ different microorganisms. The process of bioleaching is contingent upon the specific species of microorganisms that are engaged. The technical approach employed for the reclamation of valuable metals, also demands the management of cyanide disposal, is of paramount importance. A Cynogenic bacterium generates cyanide for the leaching of precious metals and can also self-digest excess cyanide (Arshadi et al., 2020; Van Yken et al., 2021). This technique is mainly categorized into two approaches: Direct and indirect leaching. Direct leaching is further categorized into one-step and two-step bioleaching methods, whereas indirect leaching encompasses spent medium bioleaching.

One-Step Bioleaching

In this technique e-waste samples are directly added to microorganisms that have been injected into sterile conditioned media as part of this bioleaching process (Arya & Kumar, 2020). This technique demonstrates how the complex structure and composition of WPCBs may cause the reaction to release toxic chemicals and

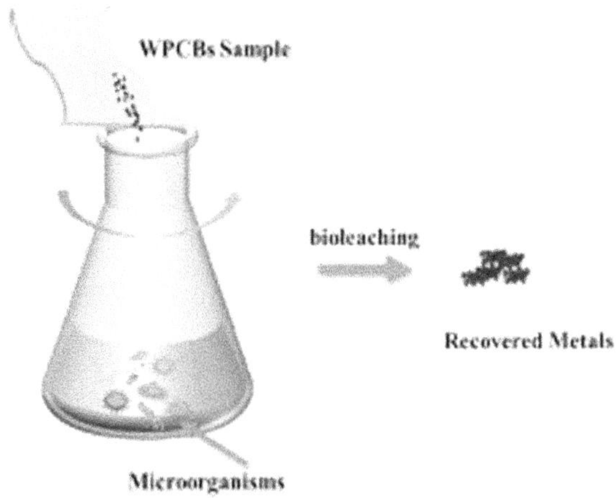

Fig. 3 One-step bioleaching (reprinted from original article by Dong et al. (2023) after permission from Elsevier)

other detrimental components, which could hinder the growth of microorganisms and ultimately lead to suboptimal bioleaching capacity (Faramarzi et al., 2020) (Fig. 3).

Two-Step Bioleaching

This process is mainly employed to promote growth of microorganisms while minimizing any hindrances or toxic impacts. This method involves cultivating particular microorganisms in the appropriate nutrient medium while maintaining ideal conditions. The subsequent stage of the bioleaching process is adding e-waste samples to the microbial environment once the bacteria have reached the logarithmic growth phase (Shah et al., 2015) (Fig. 4).

Spent Medium Bioleaching

This process entails the cultivation of microorganisms within a environment until reaching the phase of fixation, at which point metabolite are generated. Following centrifugation, the suspended particles and micro living organisms present in this medium undergo filtration to yield cellular-free medium, which is subsequently utilized for bioleaching (Arya & Kumar, 2020). This process presents multiple benefits, such as diminishing the toxic effects of WPCBs to microbes and enhancing the regulation of leaching conditions. This method is characterized by a brief processing duration, a simple operational procedure, and the capacity to attain enhanced metal retrieval under optimal leaching conditions, in the absence of microorganisms (Moazzam et al., 2021). For example, the two-step bioleaching of discarded PCBs utilizing C. violaceum under comparable environment yielded only 11.3% recovery of Au, which is significantly lesser than 30% recovery obtained through

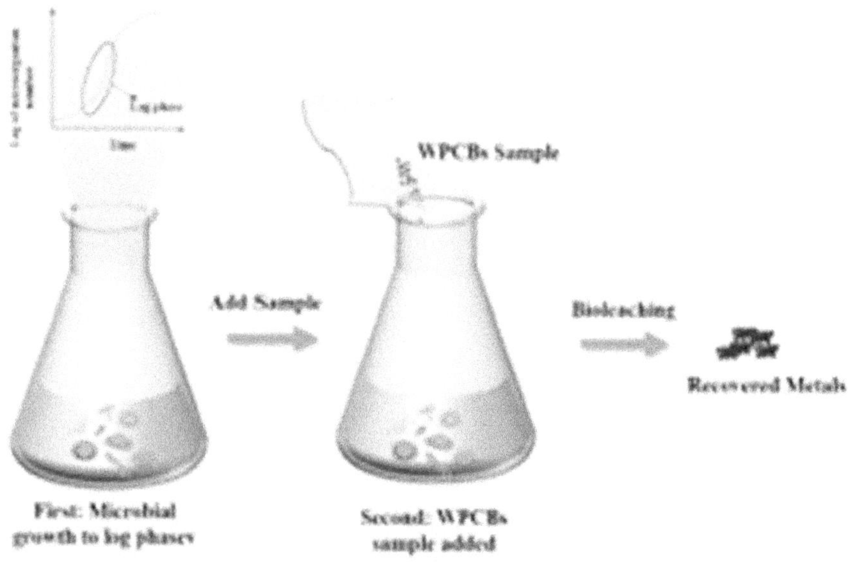

Fig. 4 Two-step bioleaching (reprinted from original article by Dong et al. (2023) after permission from Elsevier)

spent medium bioleaching (Natarajan & Ting, 2015). Using a separate investigation, C. violaceum demonstrated 11.3% recovery of Au in 0.5% w/w slurry. Conversely, in this bioleaching process, a 30% recovery of gold was attained at pH level 10 by increasing the pH of the microbial-free bio-culture prior to the infusion of WPCB (Natarajan & Ting, 2014) (Fig. 5).

Mechanism of Bioleaching

The bioleaching process involves various mechanisms, including acidolysis, redoxolysis and complexolysis (Jia et al., 2021). The mechanisms are contingent upon the specific species of microorganisms present. There are two types of bioleaching mechanisms: those involving autotrophic microorganisms and those involving heterotrophic microorganisms (Glombitza & Reichel, 2014).

Autotrophic Bioleaching

Autotrophic bioleaching mainly employs a metabolic pathway that involves thiosulfate and polysulfide for the process of leaching of metals (Mishra et al., 2005). The autotrophic microorganisms, such as *Thiobacillus* and *Aspergillus*, contributes role in metal dissolution. The bioleaching procedure by these microorganisms generally involves the utilization of thiosulfate and polysulfide pathways (Trivedi et al., 2022).

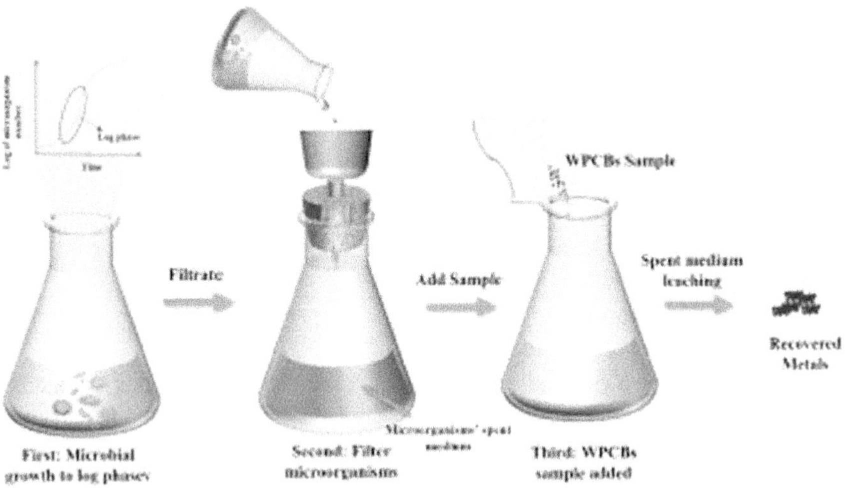

Fig. 5 Spent medium bioleaching (reprinted from original article by Dong et al. (2023) after permission from Elsevier)

Microorganisms facilitate oxidation of Fe^{2+} to Fe^{3+} in the thiosulfate pathway (Eq. (3)), and subsequent modifications in the metal oxidation states in WPCBs cause Fe^{3+} to be reduced to its initial form Fe^{2+} (Eqs. (4 and 6)), in which M symbolizes base metals like copper and iron in WPCBs. The polysulfide route employs synergistic effects of Fe^{3+} and H^+ to facilitate the extraction of metals from metallic frameworks like WPCBs. The existence of base metals like copper and iron in WPCBs can disrupt the leaching process of valuable metals. *A. thiooxidans* demonstrated an effective capacity to eliminate over 80% of the Cu from WPCBs, leading to a reduction in the copper to gold ratio from 5.7 to 3.0 (Pham & Ting, 2009).

$$\text{Microbial}: 4Fe^{2+} + O_2 + 2\,H^+ \rightarrow 4Fe^{3+} + 2OH^- \qquad (3)$$

$$\text{Chemical}: M^0 + 2Fe^{3+} \rightarrow M^{2+} + 2\,Fe^{2+} \qquad (4)$$

In the leaching of gold by *A. thiooxidans*, thiosulfate ($S_2O_32^{2-}$) serves as an intermediate metabolite, exhibiting significant instability in acidic conditions (Eq. (5)). Nonetheless, stabilization can be achieved by adjusting the pH and adding inhibitors. Metals engage with microbes within the exopolysaccharide layer, where gold easily infiltrates peptidoglycan and attaches to glutamic acid (–COO–) which is highly reactive. As shown in reaction Eqs. (7 and 8), this produces a significant amount of precipitate, which in turn causes the formation of crosslinked chains (Jaiswal & Srivastava, 2024).

$$\text{Microbial}: 2S^o + O_2 + 2\,H^+ \rightarrow 2H^+ + 2S_2O_3{}^{2-} \qquad (5)$$

$$\text{Chemical} : M^0 + 2Fe^{3+} \rightarrow M^{2+} + 2Fe^{2+} \quad (6)$$

$$2Au^{3+} + R - CHO + H_2O \rightarrow 2Au^{+} + R - COOH + 2H^{+} \quad (7)$$

$$2Au^{+} + R - COOH + H_2O \rightarrow 2Au^{o} + R - COOH + 2H^{+} \quad (8)$$

Heterotrophic Bioleaching

In this leaching process especially fungi, facilitates dissolution of metals in concrete substances through a synergistic action of chemical compounds, such as enzymes and acids that are organic in nature. Fungi like *Aspergillus niger* are crucial for bioleaching through the biosynthesis of metallic soluble organic acids, including oxalic acid, acetic acid, citric acid. These organic acids contribute ligands and proton that facilitate the dispersal of metals present in WPCBs, as demonstrated in Eq. (9). Furthermore, in complexation reactions, organic acids act as ligands, forming stable complexes with the metals found in WPCBs. Eq. (10) shows how this process enhances the dissolution of metals in the resulting solution (Saidan et al., 2012).

$$M + H^{+} + O_2 + H_2O \rightarrow M^{n+} \quad (9)$$

$$RCOO^{-} + M^{H_3}{}^{O+} \rightarrow RCOOM + H_2O \quad (10)$$

Among heterotrophic bacteria, cyanogenic bacteria are thought to be most significant. The bioleaching of metals can be exemplified by the cyanogenic bacteria *C. violaceum* involves the utilization of cyanide, generated by the oxidatively decarboxylating glycine, for leaching metals while bacterium is growing in its early stationary stages (Liu et al., 2021). Furthermore, *C. violaceum* has the capacity to leach metals that form complexes through the chelation of the metabolically generated cyanide ion. Eqs. (11 and 12) provide more information through the specific reaction equations (Faramarzi et al., 2020) (Table 2).

$$4Au + 8CN^{-} + O_2 + H_2O \rightarrow 4Au(CN)_2{}^{-} + 4OH^{-} \quad (11)$$

$$4Ag + 8CN^{-} + O_2 + H_2O \rightarrow 4Au(CN)_2{}^{-} + 4OH^{-} \quad (12)$$

Table 2 Comaparative analysis of WPCBs processing techniques

Pyrometallurgical process	Hydrometallurgical process	Bioleaching process
This technique encompasses burning at high temperature and smelting in a blast furnace at elevated temperatures to primarily recover non-ferrous metals	This technique demonstrates higher selectivity for metal extraction and presents minimal environmental risks in comparison	This method utilizes bacteria to extract metals from discarded PCBs
The configuration incurs excessive expenses and must include an air pollution control system	The process entails utilizing a range of organic and inorganic acids under different reaction conditions for the recovery of valuable and base metals	Bacteria are employed in the recovery of metals like Cr, Zn, Ni, and cu, as well as precious metals, from WPCBs
Takes scrap of any kind from any source	Mechanical preliminary treatment of the scrap is done prior to leaching	Takes scrap of any kind from any source
Pyrolysis occurs in an environment with inert gases and elevated temperatures	The usage of reagents is excessively high, resulting in waste being disposed of in the water stream	Bioleaching is selective to specific metals and specific microorganisms are used to culture

4 Global Impact of WPCB Recycling

WPCBs are significant waste that is expanding quickly, harming the environment, and being challenging to handle. The multi-metal character of the waste leads to challenges in the processing of WPCBs. Because of this, creating a common and effective procedure to handle WPCBs is challenging. WPCB recycling has an influence on the environment and the economy. Waste recycling is essential for the development of a sustainable ecosystem. Lead, mercury, nickel, tin, polychlorinated biphenyls, polybrominated diphenyl ethers, and other metals are all present in WPCBs. Hazardous gases are released when these wastes are burned (Ahmad et al., 2025). Recycling these wastes saves natural resources and lessens the quantity of waste that is landfilled. Additionally, it lessens the laborious process of extracting natural resources. At this moment, e-waste is expanding drastically than any other types of municipal waste (Nithya et al., 2021). The enhanced use of new technologies has raised the demand for latest electronic equipments which in turn and has decreased the lifespan of these equipments. Currently, 53.6 million metric tons of e-waste are created yearly worldwide; by 2050, this rate is predicted to treble (Farazi et al., 2022). This rise in the amount of e-waste accumulation is a result of both developed and emerging nations. Because of their complexity and heterogeneity, WPCBs make about 3–5% of all e-waste. It is regarded as one of the most difficult waste to recycle (Golmohammaszadeh et al., 2018).

5 Conclusion and Future Perspectives

Currently, the rapid pace of economic development, ongoing innovations in technology, and enhancements in living standards has lead to significant quantities of waste electric and electronic equipments (WEEE). Globally almost 40 million metric tons of e-waste are produces annually, but unfortunately a little portion of this waste is recycled. The majority is either incinerated or dumped in landfills. Within the category of WEEE the WPCBs have been reported as useful component as it contains precious metals in them. Recycling of WPCBs offers a viable and affordable way to extract valuable metals, offering a substitute for conventional mining methods. The intricate composition and presence of hazardous materials render the reuse and recycling methodology complicated and necessitate further improvements. The rising occurrence of WPCBs and the related dangers of inadequate management have attracted considerable focus. Because of this, recycling precious metals from WPCBs has become a profitable industry, even though developing countries frequently lack proper pollution controls. Recovery of precious metals from WPCBs presents a extremely difficult task due to their dual nature of being both valuable and harmful, a situation arising from their complex material composition. Therefore, it is crucial and imperative to develop appropriate technologies for recycling metals from WPCBs that have not adverse impact on the atmosphere and human health. Bioleaching technology presents significant potential due to its economical viability, ecological benefits, and minimal risk nature. Nonetheless, this technique is marked by constraints including bacterial adaptations and elevated pulp volumes, resulting in most prior investigations being carried out mainly at the laboratory scale. Consequently, WPCB recycling is an essential and significant procedure for both the retrieval of precious metals and the elimination of hazardous waste. The current technologies for recycling metals from WPCBs are thoroughly reviewed in this article. An comprehensive technological approach is proposed, encompassing metals enrichment and the precious metal extraction. In conclusion, a number of enhancements and suggestions pertaining to methods and future developments are put forth in order to enhance the retrieval of precious metals from WPCBs. Thus, the establishment of hydrometallurgical processes that overcome existing constraints is necessary for effective and sustainable recovery of precious metals, providing a novel method for the recovery and disposal of WPCBs.

References

Agarwal, S., Dhiman, S., & Gupta, H. (2024). Recovery of lithium and copper from anode electrode materials of spent LIBs by acidic leaching. *Environmental Science and Pollution Research, 31*, 34249–34257. https://doi.org/10.1007/s11356-024-33537-x

Ahirwar, R., & Tripathi, A. K. (2021). E-waste management: A review of recycling process, environmental and occupational health hazards, and potential solutions. *Environmental Nanotechnology, Monitoring & Management, 15*, 100409.

Ahmad, F., Qureshi, M. I., Rawat, S., Alkharisi, M. K., & Alturki, M. (2025). E-waste in concrete construction: recycling, applications, and impact on mechanical, durability, and thermal properties—A review. *Innovative Infrastructure Solutions, 10*(6), 246.

Akcil, A., Erust, C., Gahan, C. S., Ozgun, M., Sahin, M., & Tuncuk, A. (2015). Precious metal recovery from waste printed circuit boards using cyanide and non-cyanide lixiviants–a review. *Waste Management, 45*, 258–271.

Arshadi, M., Yaghmaei, S., & Esmaeili, A. (2020). Evaluating the optimal digestion method and value distribution of precious metals from different waste printed circuit boards. *Journal of Material Cycles and Waste Management, 22*(5), 1690–1698.

Arya, S., & Kumar, S. (2020). Bioleaching: urban mining option to curb the menace of E-waste challenge. *Bioengineered, 11*(1), 640–660.

Ashiq, A., Kulkarni, J., & Vithanage, M. (2019). Hydrometallurgical recovery of metals from e-waste. In *Electronic waste management and treatment technology* (pp. 225–246). Butterworth-Heinemann.

Baldé, C. P., Kuehr, R., Yamamoto, T., McDonald, R., D'Angelo, E., Althaf, S., Bel, G., & Wagner, M. (2024). Global e-waste monitor 2024.

Bas, A. D., Deveci, H., & Yazici, E. Y. (2014). Treatment of manufacturing scrap TV boards by nitric acid leaching. *Separation and Purification Technology, 130*, 151–159.

Birloaga, I., Coman, V., Kopacek, B., & Vegliò, F. (2014). An advanced study on the hydrometallurgical processing of waste computer printed circuit boards to extract their valuable content of metals. *Waste Management, 34*(12), 2581–2586.

Birloaga, I., De Michelis, I., Ferella, F., Buzatu, M., & Vegliò, F. (2013). Study on the influence of various factors in the hydrometallurgical processing of waste printed circuit boards for copper and gold recovery. *Waste Management, 33*(4), 935–941.

Cardamone, G. F., Ardolino, F., & Arena, U. (2021). About the environmental sustainability of the European management of WEEE plastics. *Waste Management, 126*, 119–132.

Chancerel, P., Rotter, V. S., Ueberschaar, M., Marwede, M., Nissen, N. F., & Lang, K. D. (2013). Data availability and the need for research to localize, quantify and recycle critical metals in information technology, telecommunication and consumer equipment. *Waste Management & Research, 31*(10_suppl), 3–16.

Charpentier, N. M., Maurice, A. A., Xia, D., Li, W. J., Chua, C. S., Brambilla, A., & Gabriel, J. C. P. (2023). Urban mining of unexploited spent critical metals from E-waste made possible using advanced sorting. *Resources, Conservation and Recycling, 196*, 107033.

Chehade, Y., Siddique, A., Alayan, H., Sadasivam, N., Nusri, S., Ibrahim, T.. (2012). Recovery of gold, silver, palladium, and copper from waste printed circuitboards. In: International conference on chemical, civil and environment engineering (ICCEE2012), March 24–25, Dubai.

Chen, M., Wang, J., Chen, H., Ogunseitan, O. A., Zhang, M., Zang, H., & Hu, J. (2013). Electronic waste disassembly with industrial waste heat. *Environmental Science & Technology, 47*(21), 12409–12416.

Cui, H., & Anderson, C. (2020). Hydrometallurgical treatment of waste printed circuit boards: bromine leaching. *Metals, 10*(4), 462.

Dhiman, S., Agarwal, S., & Gupta, H. (2024). Application of phosphonium ionic liquids to separate Ga, Ge and In utilizing solvent extraction: A review. *Journal of Ionic Liquids, 4*(1), 100080.

Dong, Y., Zan, J., & Lin, H. (2023). Recovery of precious metals from waste printed circuit boards though bioleaching route: A review of the recent progress and perspective. *Journal of Environmental Management, 348*, 119354.

Esmaeili, M., Rastegar, S. O., Beigzadeh, R., & Gu, T. (2020). Ultrasound-assisted leaching of spent lithium ion batteries by natural organic acids and H2O2. *Chemosphere, 254*, 126670.

Faraji, F., Golmohammadzadeh, R., & Pickles, C. A. (2022). Potential and current practices of recycling waste printed circuit boards: a review of the recent progress in pyrometallurgy. *Journal of environmental management, 316*, 115242.

Faramarzi, M. A., Mogharabi-Manzari, M., & Brandl, H. (2020). Bioleaching of metals from wastes and low-grade sources by HCN-forming microorganisms. *Hydrometallurgy, 191*, 105228.

Fazari, J., Hossain, M. Z., & Charpentier, P. (2024). A review on metal extraction from waste printed circuit boards (wPCBs). *Journal of Materials Science, 59*(27), 12257–12284.

Glombitza, F., & Reichel, S. (2014). Metal-containing residues from industry and in the environment: Geobiotechnological urban mining. *Geobiotechnology I: Metal-related Issues*, 49–107.

Golmohammadzadeh, R., Faraji, F., & Rashchi, F. (2018). Recovery of lithium and cobalt from spent lithium ion batteries (LIBs) using organic acids as leaching reagents: A review. *Resources, Conservation and Recycling, 136*, 418–435.

Guo, J., Zhang, R., & Xu, Z. (2015). PBDEs emission from waste printed wiring boards during thermal process. *Environmental Science & Technology, 49*(5), 2716–2723.

Guo, X., Qin, F. G., Yang, X., & Jiang, R. (2014). Study on low-temperature pyrolysis of large-size printed circuit boards. *Journal of Analytical and Applied Pyrolysis, 105*, 151–156.

Gupta, H., & Shiju, N. R. (2025). Recycling and reuse of olefinic separators in lithium-ion batteries: A mini review. *Current Opinion in Green and Sustainable Chemistry, 56*, 101054. https://doi.org/10.1016/j.cogsc.2025.101054

Gupta, H., Dhiman, S., Stam, P. G., Chaudhary, S., Murugavel, R., & Shiju, N. R. (2025). Sustainable approach to cobalt recycling: Levulinic acid leaching of waste lithium-ion battery cathode materials. *Environmental Technology & Innovation*, 104675. https://doi.org/10.1016/j.eti.2025.104675

Hadi, P., Xu, M., Lin, C. S., Hui, C. W., & McKay, G. (2015). Waste printed circuit board recycling techniques and product utilization. *Journal of Hazardous Materials, 283*, 234–243.

Harvey, J. P., Khalil, M., & Chaouki, J. (2022). Pyrometallurgical processes for recycling waste electrical and electronic equipment. *Electronic Waste: Recycling and Reprocessing for a Sustainable Future*, 135–164.

Hsu, E., Barmak, K., West, A. C., & Park, A. H. A. (2019). Advancements in the treatment and processing of electronic waste with sustainability: a review of metal extraction and recovery technologies. *Green Chemistry, 21*(5), 919–936.

Jaiswal, M., & Srivastava, S. (2024). A review on sustainable approach of bioleaching of precious metals from electronic wastes. *Journal of Hazardous Materials Advances, 14*, 100435.

Janani, R., Gurunathan, B., Sivakumar, K., Varjani, S., Ngo, H. H., & Gnansounou, E. (2022). Advancements in heavy metals removal from effluents employing nano-adsorbents: Way towards cleaner production. *Environmental Research, 203*, 111815.

Jha, M. K., Lee, J. C., Kumari, A., Choubey, P. K., Kumar, V., & Jeong, J. (2011). Pressure leaching of metals from waste printed circuit boards using sulfuric acid. *Jom, 63*(8), 29–32.

Jia, Y., Sun, H., Tan, Q., Xu, J., Feng, X., & Ruan, R. (2021). Industrial heap bioleaching of copper sulfide ore started with only water irrigation. *Minerals, 11*(11), 1299.

Kaya, M. (2018). Current WEEE recycling solutions. In *Waste electrical and electronic equipment recycling* (pp. 33–93). Woodhead Publishing.

Kaya, M. (2019). Electronic waste and printed circuit board recycling technologies.

Kopacek, B., & Kopacek, P. (2015). Extracting rare materials from electr (on) ic scrap. *IFAC-PapersOnLine, 48*(24), 157–161.

Kumar, A., Dhiman, S., Kumar, R., & Gupta, H. (2025). Recovery of germanium from waste optical fibres using phosphonium ionic liquid. *Journal of Cleaner Production, 505*, 145466. https://doi.org/10.1016/j.jclepro.2025.145466

Kumar, A., Saini, H. S., Şengör, S., Sani, R. K., & Kumar, S. (2021). Bioleaching of metals from waste printed circuit boards using bacterial isolates native to abandoned gold mine. *BioMetals, 34*(5), 1043–1058.

Li, H., Eksteen, J., & Oraby, E. (2018). Hydrometallurgical recovery of metals from waste printed circuit boards (WPCBs): Current status and perspectives–A review. *Resources, Conservation and Recycling, 139*, 122–139.

Li, H., Oraby, E., & Eksteen, J. (2021). Recovery of copper and the deportment of other base metals from alkaline glycine leachates derived from waste printed circuit boards (WPCBs). *Hydrometallurgy, 199*, 105540.

Lin, C., Zeng, Z., Xu, R., Liang, W., Guo, Y., & Huo, X. (2022). Risk assessment of PBDEs and PCBs in dust from an e-waste recycling area of China. *Science of the Total Environment, 803*, 150016.

Liu, G., Pan, D. A., Wu, Y., Yuan, H., Yu, L., & Wang, W. (2021). An integrated and sustainable hydrometallurgical process for enrichment of precious metals and selective separation of copper, zinc, and lead from a roasted sand. *Waste Management, 132*, 133–141.

Ma, E. (2019). Recovery of waste printed circuit boards through pyrometallurgy. *Electronic Waste Management and Treatment Technology*, 247–267.

Martelo, L. M., Bastos, M. M., & Soares, H. M. (2024). Separation of the metallic and non-metallic fractions of waste printed circuit boards–A review focused on the organic swelling. *Minerals Engineering, 206*, 108529.

Mishra, D., Kim, D. J., Ahn, J. G., & Rhee, Y. H. (2005). Bioleaching: a microbial process of metal recovery; a review. *Metals and Materials International, 11*(3), 249–256.

Moazzam, P., Boroumand, Y., Rabiei, P., Baghbaderani, S. S., Mokarian, P., Mohagheghian, F., Mohammad, L. J., & Razmjou, A. (2021). Lithium bioleaching: An emerging approach for the recovery of Li from spent lithium ion batteries. *Chemosphere, 277*, 130196.

Moyo, T., Chirume, B. H., & Petersen, J. (2020). Assessing alternative pre-treatment methods to promote metal recovery in the leaching of printed circuit boards. *Resources, Conservation and Recycling, 152*, 104545.

Natarajan, G., & Ting, Y. P. (2014). Pretreatment of e-waste and mutation of alkali-tolerant cyanogenic bacteria promote gold biorecovery. *Bioresource Technology, 152*, 80–85.

Natarajan, G., & Ting, Y. P. (2015). Gold biorecovery from e-waste: an improved strategy through spent medium leaching with pH modification. *Chemosphere, 136*, 232–238.

Ning, C., Lin, C. S. K., Hui, D. C. W., & McKay, G. (2017a). Waste printed circuit board (PCB) recycling techniques. *Chemistry and Chemical Technologies in Waste Valorization*, 21–56.

Ning, C., Lin, C. S. K., Hui, D. C. W., & McKay, G. (2017b). Waste printed circuit board (PCB) recycling techniques. *Chemistry and Chemical Technologies in Waste Valorization*, 21–56.

Nithya, R., Sivasankari, C., & Thirunavukkarasu, A. (2021). Electronic waste generation, regulation and metal recovery: a review. *Environmental Chemistry Letters, 19*(2), 1347–1368.

Oraby, E. A., Li, H., & Eksteen, J. J. (2020). An alkaline glycine-based leach process of base and precious metals from powdered waste printed circuit boards. *Waste and Biomass Valorization, 11*(8), 3897–3909.

Park, H. S., Han, Y. S., & Park, J. H. (2019). Massive recycling of waste mobile phones: Pyrolysis, physical treatment, and pyrometallurgical processing of insoluble residue. *ACS Sustainable Chemistry & Engineering, 7*(16), 14119–14125.

Pham, V. A., & Ting, Y. P. (2009). Gold bioleaching of electronic waste by cyanogenic bacteria and its enhancement with bio-oxidation. *Advanced Materials Research, 71*, 661–664.

Pradhan, J. K., & Kumar, S. (2012). Metals bioleaching from electronic waste by *Chromobacteri-umviolaceum* and *Pseudomonads* sp. *Waste Management & Research, 30*(11), 1151–1159.

Priya, A., & Hait, S. (2018). Feasibility of bioleaching of selected metals from electronic waste by *Acidiphiliumacidophilum. Waste and Biomass Valorization, 9*(5), 871–877.

Quinet, P., Proost, J., & Van Lierde, A. (2005). Recovery of precious metals from electronic scrap by hydrometallurgical processing routes. *Mining, Metallurgy & Exploration, 22*(1), 17–22.

Saidan, M., Brown, B., & Valix, M. (2012). Leaching of electronic waste using biometabolised acids. *Chinese Journal of Chemical Engineering, 20*(3), 530–534.

Sankar, V. U., Lakshmi, G., & Sankar, Y. S. (2022). A review of various defects in PCB. *Journal of Electronic Testing, 38*(5), –491.

Schlesinger, M. E., Sole, K. C., Davenport, W. G., & Flores, G. R. A. (2021). *Extractive metallurgy of copper*. Elsevier.

Shah, M. B., Tipre, D. R., Purohit, M. S., & Dave, S. R. (2015). Development of two-step process for enhanced biorecovery of Cu–Zn–Ni from computer printed circuit boards. *Journal of Bioscience and Bioengineering, 120*(2), 167–173.

Shokri, A., & Fotovat, F. (2023). Kinetic analysis and modeling of paper-laminated phenolic printed circuit board (PLP-PCB) pyrolysis using distributed activation energy models (DAEMs). *Thermochimica Acta, 724*, 179513.

Silva, L. H. D. S., Júnior, A. A., Azevedo, G. O., Oliveira, S. C., & Fernandes, B. J. (2021). Estimating recycling return of integrated circuits using computer vision on printed circuit boards. *Applied Sciences, 11*(6), 2808.

Tanısalı, E., Özer, M., & Burat, F. (2021). Precious metals recovery from waste printed circuit boards by gravity separation and leaching. *Mineral Processing and Extractive Metallurgy Review, 42*(1), 24–37.

Trivedi, A., Vishwakarma, A., Saawarn, B., Mahanty, B., & Hait, S. (2022). Fungal biotechnology for urban mining of metals from waste printed circuit boards: a review. *Journal of Environmental Management, 323*, 116133.

Tuncuk, A., Stazi, V., Akcil, A., Yazici, E. Y., & Deveci, H. (2012). Aqueous metal recovery techniques from e-scrap: Hydrometallurgy in recycling. *Minerals Engineering, 25*(1), 28–37.

Udayakumar, S., Abd Razak, M. I. B., & Ismail, S. (2022). Recovering valuable metals from Waste Printed Circuit Boards (WPCB): A short review. *Materials Today: Proceedings, 66*, 3062–3070.

Van Eygen, E., De Meester, S., Tran, H. P., & Dewulf, J. (2016). Resource savings by urban mining: The case of desktop and laptop computers in Belgium. *Resources, Conservation and Recycling, 107*, 53–64.

Van Yken, J., Boxall, N. J., Cheng, K. Y., Nikoloski, A. N., Moheimani, N. R., & Kaksonen, A. H. (2021). E-waste recycling and resource recovery: A review on technologies, barriers and enablers with a focus on oceania. *Metals, 11*(8), 1313.

Wang, R., Zhang, C., Zhao, Y., Zhou, Y., Ma, E., Bai, J., & Wang, J. (2021). Recycling gold from printed circuit boards gold-plated layer of waste mobile phones in "mild aqua regia" system. *Journal of Cleaner Production, 278*, 123597.

Xiu, F. R., Qi, Y., & Zhang, F. S. (2013). Recovery of metals from waste printed circuit boards by supercritical water pre-treatment combined with acid leaching process. *Waste Management, 33*(5), 1251–1257.

Yan, K., Liu, C., Liu, L., Xiong, M., Chen, J., Zhang, Z., Zhong, S., Xu, Z., & Huang, J. (2022). Pyrolysis behaviour and combustion kinetics of waste printed circuit boards. *International Journal of Minerals, Metallurgy and Materials, 29*(9), 1722–1732.

Yazıci, E., Deveci, H., Alp, I., Akcil, A., & Yazıcı, R. (2010). Characterisation of computer printed circuit boards for hazardous properties and beneficiation studies. In *XXV international mineral processing congress (IMPC)* (pp. 4009–4015).

Zhang, L., & Xu, Z. (2016). A review of current progress of recycling technologies for metals from waste electrical and electronic equipment. *Journal of Cleaner Production, 127*, 19–36.

Zhu, Y., Wen, X., Zhang, W., Shen, F., & Cao, Y. (2025). Pilot study on pyrolytic utilization of WPCB based on rotary kilns. *Fuel, 400*, 135717.

Separation and Recovery of Valuable Metals from Waste Lithium-Ion Batteries

Mili Agrawala, Sibananda Sahu, and Niharbala Devi

1 Introduction

The rising global demand for energy, paired with urgent concerns over climate change and resource depletion, has prompted a profound shift in how energy is produced, stored, and consumed. Traditional energy systems, particularly those based on fossil fuels, have historically fuelled economic development but are now recognised as significant contributors to greenhouse gas emissions, environmental degradation, and economic instability due to fluctuating resource availability and market prices (Harper et al., 2019; International Energy Agency, 2023). In response, governments and industries worldwide have intensified efforts to transition towards cleaner energy alternatives, for instance solar, wind, and energy storage systems (ESS). Central to this transformation is the adoption of efficient, high-capacity battery technologies that can support both stationary storage for grid stability and consumer electronics like those adopted in electric vehicles (EVs). Among these, lithium-ion batteries (LIBs) have procured global prominence owing to favourable electrochemical behaviour, lightweight structure, and adaptability across multiple sectors (Kumar et al., 2021; Mishra et al., 2022). These developments aim to reduce reliance on conventional combustion engines and support broader carbon neutrality goals. The evolution of LIB technology traces back to the surging necessity for efficient, lightweight, and rechargeable ESS in the late twentieth century (Choi & Kim, 2021). Lithium's small ionic radius and high electrochemical potential enable efficient energy storage

M. Agrawala · N. Devi (✉)
Department of Chemistry, Institute of Technical Education and Research, Siksha 'O' Anusandhan Deemed to be University, Bhubaneswar, Odisha, India
e-mail: niharbaladevi@soa.ac.in

S. Sahu
International PranaGraf Mintech Research Centre, a Unit of PranaGraf Materials & Technologies Pvt. Ltd., Bhubaneswar, Odisha, India

H. Gupta et al. (eds.), *Advanced Technologies for Assessment, Recycling, and Management of Solid Waste*, Waste as a Resource,
https://doi.org/10.1007/978-3-032-24788-9_9

in a compact form. This led to immediate adoption in portable consumer electronics, including mobile phones, camcorders, and laptops, where weight and energy density were critical design constraints (Goodenough & Kim, 2009; Scrosati, 2011). The early success of LIBs was closely tied to the exponential growth of portable electronics. Over the past two decades, LIBs have also become the cornerstone of consumer electronics standard, grid-scale energy storage solutions, and electric vehicle (EV) propulsion, driven by decarbonisation goals and global energy transitions (Nitta et al., 2014). Transition metals present in cathode chemistries are not only essential to the electrochemical performance of LIBs but also represent finite resources, mined at significant environmental cost. The use of critical raw materials (CRMs) has prompted concern over the sustainability of the LIB value chain, making end-of-life battery management a global research priority (Ciez & Whitacre, 2019).

1.1 Current Trends in the LIB Market

The global LIB market has grown at an annual rate of over 15%, and it is expected to exceed USD 135 billion by 2031, driven by EV adoption, grid storage, and consumer electronics (International Energy Agency, 2023). Cathode materials dominate the cost and environmental footprint of these batteries. Nickel-rich chemistries like NMC811 and NCMA are benefiting popularity owing to their higher energy density and performance. LFP (lithium iron phosphate) batteries are gaining prominence due to thermal stability and cost-effectiveness (Koech et al., 2024). China currently dominates the LIB supply chain, accounting for over 70% of global cell production and significant control over cathode and anode material processing (BloombergNEF, 2021).

Beyond LIBs, numerous alternative energy storage systems are being explored to address the diverse needs of grid-scale storage, mobility, and short-duration power applications (Hamid et al., 2024; Jafarizadeh et al., 2024). Figure 1 summarises a normalised value of cost analysis of major energy storage systems, such as compressed air, flywheel, pumped hydro, super capacitors, and battery materials, covering capital investment, energy cost per kWh, and operational expenditures. While the operational and maintenance (O&M) cost of major battery materials is high, the power investment and energy capital cost in comparison with other energy storage devices seems economically accessible. However, to further comprehend the effectiveness of varied battery storage materials, Fig. 2 details a comprehensive analysis of normalised value of major operational characteristics such as lifespan, cycle life, energy and power density, and efficiency. It suggests the operational characteristics of LIBs are favoured over other battery materials, demonstrating their promising utilisation and applications worldwide. Further LIBs serve an advantage owing to their scalability, compactness, and optimal performance, still they remain comparatively expensive due to feedstock and manufacturing costs (Hadoune et al., 2023; Loulijat et al., 2024; Pilali et al., 2025).

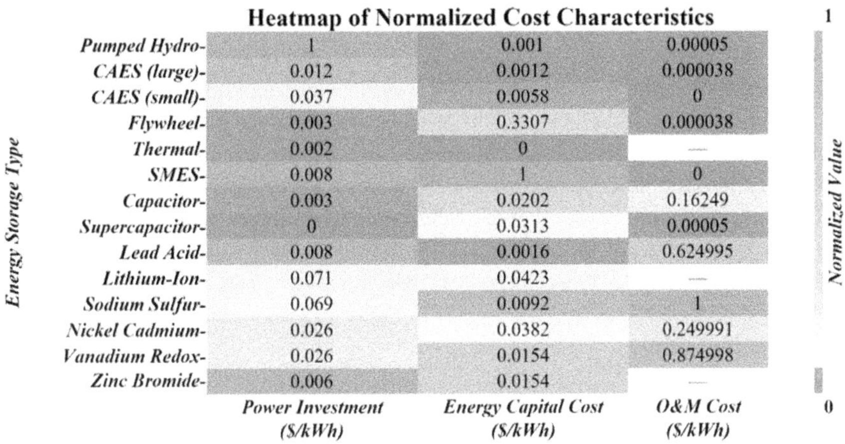

Fig. 1 Cost analysis of prominent energy storage systems, detailing power investment, energy capital cost, and operation and maintenance (O&M) cost per kilowatt-hour (Pilali et al., 2025)

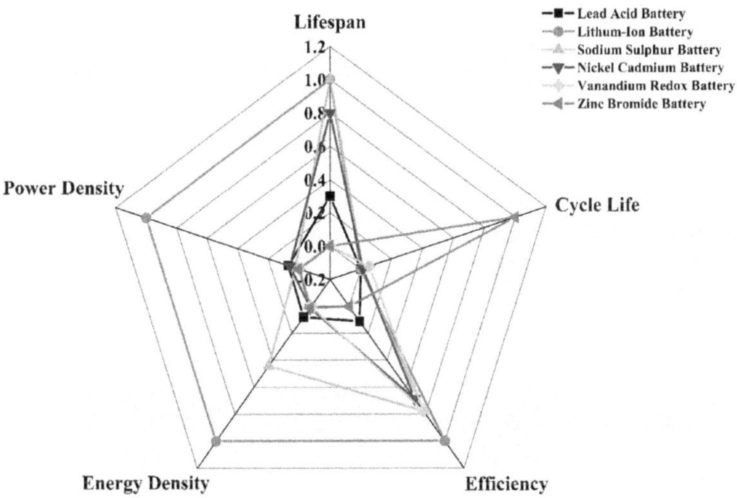

Fig. 2 Comparative operational characteristics of major energy storage technologies (Pilali et al., 2025)

1.2 Environmental Impact of Improper LIB Disposal

Spent LIBs, when not handled through formal recycling systems, pose serious environmental and health risks (Roy et al., 2021; Winslow et al., 2017). They contain reactive electrolytes (e.g. $LiPF_6$ inorganic solvents), flammable components, and leachable toxic metals, including cobalt, nickel, and manganese (Winslow et al.,

2017). Landfilling or incineration can lead to environmental concerns pertaining to heavy metal leaching into groundwater, decomposition of organic electrolyte salts, thermal runaways, fires, and explosions. Furthermore, landfilling or incineration practices in informal recycling sectors can cause occupational hazards. A 2022 report by the UNEP (United Nations Environment Programme) emphasised that informal battery dismantling practices in developing countries contribute significantly to soil and water contamination, often in densely populated urban environments (Kendall et al., 2023). Moreover, lithium extraction itself is associated with water depletion and ecosystem degradation, particularly in South America's "lithium triangle" (Parker et al., 2023; Vera et al., 2023).

1.3 Necessity for LIB Recycling

Given the increasing LIB usage and finite global reserves of critical metals, efficient recycling is both a sustainability imperative and a strategic economic opportunity (Martin et al., 2016). Removal of lithium (Li), cobalt (Co), manganese (Mn), and nickel (Ni) from retired LIBs could reduce reliance on primary mining, lower environmental impact, and promote circularity (Sobianowska-Turek et al., 2021). Also, unregulatory disposal of spent battery material contributes to the accumulation of hazardous e-waste (Golmohammadzadeh et al., 2018). If not handled responsibly, spent batteries can release toxic substances, including heavy metals and electrolyte residues, into the environment, thereby contaminating soil and water. Recognising these critical concerns, governments and regulatory bodies have begun implementing targeted policies. In India, for example, the battery waste management rules (2022), formulated by the MoEFCC, Ministry of Environment, Forest and Climate Change, in collaboration with NITI Aayog, outline responsibilities for producers, importers, and recyclers under an Extended Producer Responsibility (EPR) framework (Kumar & Rizvi, 2024; Rakshit et al., 2025). Therefore, the necessity for LIB recycling extends beyond environmental stewardship; it is a strategic imperative to ensure material security, promote green innovation, and zero waste creation, which align with global sustainability goals.

2 Composition and Classification of LIBs

LIBs are built around four fundamental components: cathode, anode, electrolyte, and separator. Each part plays a distinct electrochemical role, and the collective choice of materials determines crucial performance attributes such as specific energy, power capability, thermal stability, and service life (Sobianowska-Turek et al., 2021; Tang et al., 2015).

2.1 Cathode

The cathode material acts as the primary source of Li^+ ions during discharge and offers a decisive role in defining a LIB's overall performance including energy density as well as voltage. Structurally, it consists of a thin aluminium current collector coated with active lithium metal oxide compounds. These compounds include a combination of important elements for instance iron (Fe), cobalt (Co), manganese (Mn), lithium (Li), and nickel (Ni) all of which contribute to the electrode's redox behaviour and stability under varying operational conditions (Li et al., 2017a, 2017b).

Several cathode chemistries are in widespread commercial use, each offering a unique balance of cost, energy capacity, thermal stability, and cycling performance (Koech et al., 2024). Commonly adopted battery chemistries include lithium cobalt oxide ($LiCoO_2$; LCO), nickel manganese cobalt oxides ($LiNi_{1/3}Mn_{1/3}Co_{1/3}O_2$; NMC), nickel cobalt aluminium oxide (NCA; $LiNiCoAlO_2$), lithium manganese oxide (LMO; $LiMn_2O_4$), and lithium iron phosphate ($LiFePO_4$; LFP). LCO, a common battery material in mobile electronics, serves as potent energy-dense material yet suffers disadvantages from high cost and cobalt scarcity. In contrast, NMC, widely used in electric vehicles, balances energy, cost, and safety through variants like NMC622 and NMC811. NCA, primarily used by manufacturers like Tesla, provides high energy density and excellent performance. LMO, with moderate energy density and cycle life, serves as a cost-efficient and safer alternative. Furthermore, LFP, with its long lifecycle and high thermal stability, is preferred in stationary and public transport applications.

Recent advancements have focussed on nickel-rich cathodes to enhance energy density of the cathode material and even reduce reliance on cobalt as key composition. However, these materials can introduce challenges related to thermal stability and structural degradation during cycling (Saaid et al., 2023). Modern cathodes are engineered for improved specific capacities (typically ranging from 140 to 250 mAh/g) and optimised lithium-ion diffusion. Moreover, the push towards cobalt-free or low-cobalt cathodes, for instance lithium-rich layered oxides and lithium manganese-rich variants, is gaining momentum due to ethical sourcing and supply chain risks (Hasan et al., 2025). While all LIB types share a baseline lithium content (~ 11–12 wt.%), the ratios of Ni, Co, and Mn vary substantially (Fig. 3), influencing thermal behaviour, cost, and recyclability (Buchmann, 2021).

2.2 Anode

The anode in LIB functions as the host structure for lithium-ion insertion during charging and portrays a critical role in governing battery's voltage window, capacity, and cycling performance (Yoshino, 2014). Typically, the anode material is coated onto a copper foil current collector, which offers good electrical conductivity and compatibility with the low operating potential of most anode chemistries. Graphite's

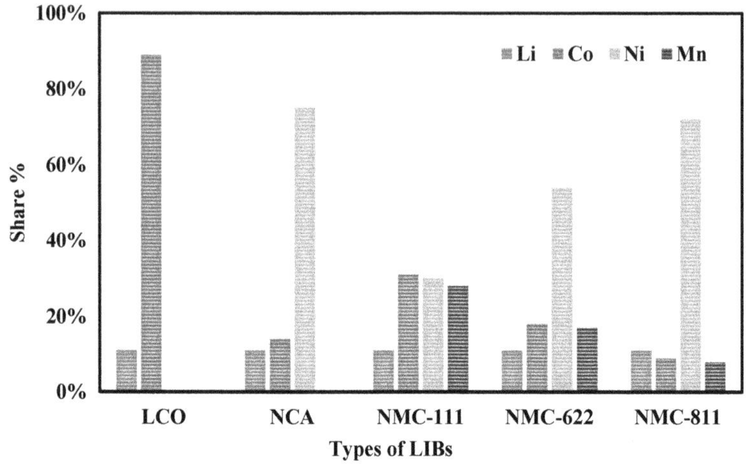

Fig. 3 Elemental composition across different LIB chemistries (Buchmann, 2021)

predominance as anode material in commercial LIBs since the 1990s is primarily because of its structural stability, relatively low lithiation potential (~ 0.1 V vs. Li/Li^+), and excellent reversibility over repeated cycles (Liu et al., 2023a, 2023b). However, the theoretical specific capacity of graphite is limited to approximately 372 mAh/g. Under high charging rates or deep cycling, graphite can experience lithium plating on the surface, leading to dendrite formation—an issue that can compromise both performance and safety (Shen et al., 2017). To overcome these limitations, several alternative anode materials have been explored. Silicon-based materials offer an elevated theoretical capacity (~ 4200 mAh/g); however, poor cycling stability and large volume expansion proposes major disadvantage. Tin (Sn)-based compounds provide intermediate capacity but face structural challenges. Lithium titanate ($Li_4T_5O_{12}$, LTO), a spinel-structured oxide, provides an excellent safety profile. Functioning at relatively higher potential of ~ 1.55 V vs. Li/Li^+, LTO completely avoids lithium plating, thereby eliminating dendrite formation. Additionally, it delivers exceptional rate capability and long cycle life, although it sacrifices energy density compared to graphite-based systems (Lim et al., 2018; Tang et al., 2015).

2.3 Electrolyte

The electrolyte forms the ionic conduit between the cathode and anode, allowing lithium-ion migration during charge and discharge. In conventional cells this medium is a solution of a lithium salt, most often $LiPF_6$, though $LiBF_4$, $LiClO_4$, and $LiAsF_6$ are also used, dissolved in a blend of low-viscosity organic carbonates, for instance ethyl-methyl carbonate (EMC), dimethyl carbonate (DMC), diethyl carbonate (DEC), propylene carbonate (PC), or, less commonly, dimethyl sulfoxide (DMSO) (Aravindan et al., 2011; Xu et al., 2023; Zheng et al., 2018).

3 Recycling Methods

Recycling spent LIBs is essential for conserving critical mineral resources such as Li, Mn, Co, and Ni, while limiting the environmental and safety hazards associated with improper disposal (Zeng et al., 2013). A robust LIB recycling framework typically involves two broad stages: physical (or mechanical) processes, which serve as pretreatment to concentrate and segregate active materials, and chemical processes, which extract and purify the valuable metals. Due to the intricate layered LIB's structure and the presence of flammable electrolytes, it is imperative to safely discharge batteries before initiating any treatment process (Doerffel & Sharkh, 2005). Physical processes thus play a foundational role in preparing battery materials for efficient downstream recovery (Diekmann et al., 2016).

3.1 Physical and Pretreatment Processes

Physical processes aim to separate valuable and hazardous components of LIBs based on mechanical, thermal, and solvent-assisted principles (Li et al., 2018). These approaches help isolate black mass (a mixture of cathode/anode active materials) from casings, current collectors, binders, and electrolytes, thereby improving the purity and efficiency of subsequent chemical treatments (Ordoñez et al., 2016). The major physical operations are categorised as follows.

Discharging and Safety Handling

Spent LIBs retain residual charge, and improper handling during recycling may lead to short-circuiting, fire, or toxic gas emissions. Therefore, controlled discharge is a critical safety step. Various studies recommend immersing battery materials in electrolytic solutions namely NaCl or Na_2SO_4 to facilitate slow and safe discharge (Nie et al., 2014). Li et al. (2016) examined the effects of variation in discharge time and electrolyte concentration, reporting an optimal 72% efficiency using 10 wt% NaCl over 358 min. Immersion methods, while simple and scalable, must consider hydrogen evolution and corrosion risks. Salt solution discharging is favoured in industrial setups due to its economical advantage and minimal ecological impact (Diekmann et al., 2016).

Mechanical Separation

Mechanical operations, including dismantling, crushing, sieving, flotation, gravity, and magnetic separation, are widely applied to isolate valuable components (Da Costa et al., 2015; He et al., 2016; Wang et al., 2016; Yu et al., 2017; Zhu et al., 2011). The LIB is typically disassembled manually or robotically to separate modules, followed by crushing to liberate embedded materials. The process must be conducted under inert conditions to prevent flammable gas ignition (Dorella & Mansur, 2007). Diekmann et al. (2016) identified dimethyl carbonate (DMC), ethyl methyl carbonate (EMC), and CO_2 as major volatiles released during mechanical crushing. After crushing, the particle mixture undergoes sieving and classification. Wang et al. (2016) observed that components like plastics and foils (Al, Cu) accumulate in coarser fractions (> 1 mm), while cathode and anode active materials concentrate in finer fractions (< 1 mm). In this study, cobalt was found to comprise 85% of the < 1 mm fraction. Such presorting significantly improves flotation efficiency and the purity of recovered black mass. Flotation techniques exploit surface wettability differences between cathode and anode materials. Graphite, being naturally hydrophobic, can be separated from hydrophilic oxides using suitable frothers (Yu et al., 2017). Gravity separation enhances liberation by leveraging density differences, while magnetic separation targets ferrous impurities. Shin et al. (2005) proposed a novel mechanical–hydrometallurgical hybrid route for recovering cobalt and lithium using flotation followed by acid leaching, yielding near-complete recovery with a low environmental footprint. Nevertheless, the complexity of LIB architecture and the tight integration of materials often limit the efficiency of mechanical separation (Li et al., 2009). Residual cross-contamination between Cu, Al, and black mass remains a concern, which may affect downstream leaching and purification steps.

Thermal Pretreatment

Thermal processes are employed to decompose binders (PVDF), remove electrolyte residues, and weaken adhesion between components. Low-temperature baking (100–150 °C) under nitrogen or vacuum helps evaporate volatile solvents and soften polymers. Thermal treatment may also be integrated between leaching steps to regenerate surfaces or improve leaching kinetics (Xu et al., 2007). Bahgat et al. (2006) efficiently synthesised Li/Co ferrite microcrystalline by calcining spent cathode and Fe_2O_3 at over 1000 °C. Similarly, Castillo (2002) treated mixed hydroxide residue for 2 h at 500 °C to remove residual carbon and improve purity. However, these treatments must account for emissions of dioxins, heavy metal vapours, and corrosive gases, necessitating the installation of gas scrubbing units and emission control systems. Lee and Rhee (2002) developed a combined thermal–hydrometallurgical–sol–gel route, recovering Li and Co and re-synthesising $LiCoO_2$. Though effective, thermal treatment consumes high energy and raises environmental challenges, limiting its sustainability.

Mechanochemical Activation

Mechanochemical processes have garnered significant attention as an environmentally friendly and energy-efficient method to enhance LIB recycling. These processes rely on high-energy ball milling to disrupt the crystal lattice of electrode materials and break down polymeric binders such as PVDF (Wang et al., 2022a, 2022b; Zhang et al., 2000). Mechanochemistry facilitates increased surface area and reactivity of cathode materials, which significantly improves metal leaching efficiency at ambient conditions. A specialised grinding technique for $LiCoO_2$, for example, effectively exposes lithium and cobalt-bearing phases for subsequent acid leaching. In an experimental work performed by Zhang et al. (2000), the recovery and separation of essential elements from scrap $LiCo_{0.2}Ni_{0.8}O_2$ was achieved at room temperature through planetary milling, both with and without the addition of Al_2O_3 powder. The inclusion of Al_2O_3 significantly enhanced the breakdown of PVDF and increased the leaching yield. Co, Ni, and Li were recovered with yields exceeding 90%, and around 1% fluorine was found in the filtrate, indicating partial PVDF dissolution only in samples ground with Al_2O_3. Saeki et al. (2004) developed a similar procedure where $LiCoO_2$ was ground with PVC powder to induce mechanochemical formation of Li- and Co-chlorides, which are water-soluble. These chlorides were then easily recovered through simple water leaching. This study also highlighted the dual benefit of mechanochemically neutralising halogenated plastics, such as PVC, which are often problematic waste streams in LIB recycling. Moreover, Saito et al. (2024) and Tongamp et al. (2006) reported a two-step dry milling/grinding and leaching method for dechlorinating PVC waste and extracting Co and Li. Wang et al. (2017) co-processed waste PVC with $LiCoO_2$ from spent LIBs, where chloride ions from PVC reacted with Li to form LiCl and reorganised Co-Fe crystals into magnetic $CoFe_4O_6$. The process achieved 100% and 96.4% conversion of Li and Cl, respectively, with 91.9% of Co retained as magnetic $CoFe_xO_y$. Lee et al. (2000) supported this by demonstrating that mechanochemically treated samples exhibit enhanced leachability due to lattice distortion and defect formation. Mechanochemical activation offers multiple advantages: it eliminates the need for aggressive chemicals, operates at room temperature, and reduces emissions. The mechanochemical conversion of electrode materials into reactive intermediates such as chlorides, oxides, or sulphates also shortens the reaction time during hydrometallurgical treatments. However, scale-up of this technology poses challenges due to the energy demand of milling and difficulties in uniformly processing large volumes of black mass. Nevertheless, the integration of mechanochemical pretreatment with downstream green leaching methods is emerging as a viable and sustainable pathway for next-generation LIB recycling.

Solvent-Based Dissolution and Delamination

To effectively remove PVDF or PTFE binders that hold electrode layers on metal foils, solvent-based dissolution is widely practised. Common solvents include dimethyl sulfoxide (DMSO), dimethylacetamide (DMAc), N-methylpyrrolidone (NMP), and dimethylformamide (DMF) (Zhou et al., 2010). Bai et al. (2021) evidenced the dissolution of PVDF binder in NMC622 at 100 °C using triethyl phosphate (TEP) as a green substituent for NMP and DMF, along with separation of cathode materials under refined recovery parameters, thereby demonstrating improved environmental benefits. Bankole and Lei (2014) explored the recovery of $LiPF_6$ electrolyte using ethanol and conversion to Li_2SiF_6, highlighting solvent-based strategies for detoxifying spent LIBs. Recent innovations explore green sustainable solvating medium, including those of ionic liquids (ILs) and deep eutectic solvents (DES) for binder breakdown. These solvents offer lower volatility and greater recyclability, but high cost and recovery complexity limit commercial adoption (Bankole et al., 2013). Solvent-based delamination also facilitates foil recovery (Al, Cu) and enhances cathode material purity, especially when followed by gentle washing and sieving. Integration of solvent extraction with mechanochemistry has shown synergistic effects, improving both efficiency and selectivity.

3.2 *Chemical Recycling Processes*

The employment of pretreatment procedures on spent battery materials facilitates the adoption of subsequent/successive extraction strategies by segregating the intricate battery components. Current recycling practices to recover essential cathode materials involve the utilisation of extractive metallurgical pathways such as pyrometallurgy, hydrometallurgy, and bioleaching techniques. Figure 4 demonstrates common extractive strategies adopted for LIB recycling.

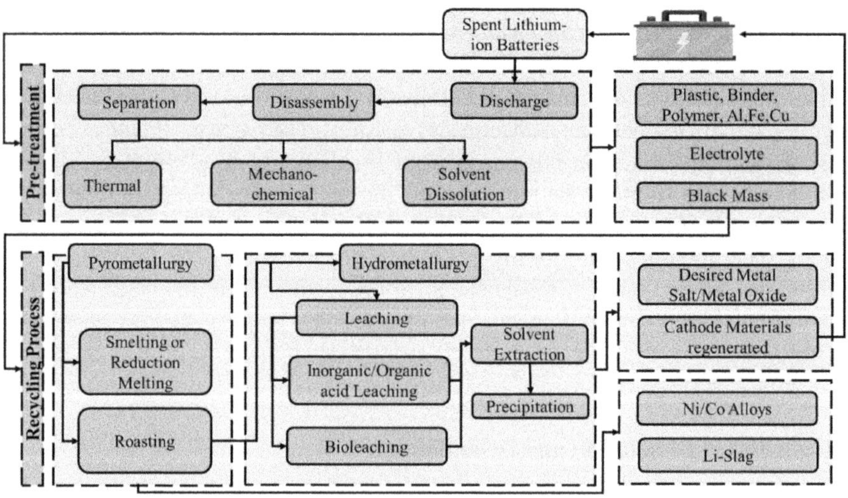

Fig. 4 Extractive strategies adopted for LIB recycling

4 Pyrometallurgical Aspect of Recycling Spent LIBs

The pyrometallurgical recycling strategy, widely adopted among commercial companies, comprises 17% of overall global battery processing and recovery procedures. It is widely adopted owing to the advantages in extracting high-value essential metals like Ni, Cu, and Co and partial recovery of Li, featuring diverse melting points (MPs) and boiling points (BPs) of varied metals, minimal raw materials utilisation, and minimal waste liquid discharge (Shi et al., 2025). Though it is characterised by high-temperature thermal reaction, it offers a scalability advantage, operational robustness, and ability to handle unsorted heterogeneous battery waste streams (Cornelio et al., 2024). Pyrometallurgical recycling typically begins with a thermal pretreatment step to remove organic components such as binders, separators, and electrolytes using pyrolysis or incineration approaches (Tong et al., 2024). Pyrolysis pretreatment decomposes organic components to isolate the active materials from metal foils by thermally treating the battery materials in a rotary roaster, maintaining an inert environment at a temperature range of 600–800 °C (Liu et al., 2019). It ascertains the structural integrity of cathode materials and aids mechanical separation by rendering Al/Cu foils (Kwade et al., 2022). Incineration pretreatment, on the other hand, involves combusting organics in a drum furnace at an operational temperature of 700–800 °C in an oxygen-rich environment to isolate black mass rich in metals (Lombardo et al., 2020; Neumann et al., 2022). This process, although it offers operational simplicity, results in higher energy and Li-loss when compared to pyrolysis pretreatment (Tong et al., 2024).

The pyrometallurgically pretreated organic-free feedstock then undergoes primary extractive pyrometallurgical treatment through smelting and reduction roasting, which majorly differ in reagent adaptation, temperature profiles, and recovery efficiencies.

4.1 Smelting and Reduction Melting

Smelting, a high-temperature ($\geq$ 1400 °C) process wherein metals with low melting points coalesce into an immiscible molten alloy phase, while non-metallic impurities and less reactive metals are incorporated into the slag phase or emitted as gas (Ren et al., 2017; Tong et al., 2024; Zheng et al., 2024). Smelting procedure for spent battery material, having industrial implications in Umicore's Ultra-High-Temperature (UHT) process and Inmetco's High-Temperature Metal Recovery (HTMR) system, involves feeding the battery material directly into a furnace with common reductant (graphite) or existing anode materials to recover Fe, Cu, Ni, and Co as alloys and Al, Mn, and Li as slag (Ekberg & Petranikova, 2015; Shi et al., 2019; Zheng et al., 2024). The smelting procedure adapted by Hu et al. (2020) ascertains recovery of 91.5% Mn, 98.2% Co, and 98.4% Ni as Co-Ni-Mn alloy from NMC622-type battery material and recovering 68.3% Li from Li_2CO_3-flue dust using a carbonated water leaching procedure. Similarly, Qu et al. (2022) demonstrate the efficiency and applicability of co-smelting copper slag along LCNO-type battery materials, resulting in the recovery of 99.30% Ni, 98.04% Cu, 99.13% Co, and 97.98% Fe. However, despite its scalability and ability to process mixed feedstock, smelting results in additional lithium treatment, high operation costs and energy input, poor selectivity, and release of harmful gases and toxic halogens, necessitating the utilisation and installation of advanced purification systems (Zheng et al., 2024).

4.2 Roasting-Based Recovery Processes

Roasting-based pyrometallurgical recovery enables recovery of valuable metals by conducting gas–solid reactions at moderate temperature (600–1000 °C) using reducing agents, resulting in the production of transition metal, metal oxide, and Li-salt. This method works in conjunction with hydrometallurgical techniques and offers improved lithium retention when compared to smelting (Li et al., 2025). Depending on the type of reactants used, reduction roasting can be categorised as carbothermal reduction roasting and salt-assisted reduction roasting.

Carbothermal Reduction Roasting

Carbothermic reactions on spent battery materials utilise anodic graphite and other carbonaceous materials as reducing agents to thermally break down the oxygen octahedron in lithiated metal oxides and convert them into lower valence state metal oxide forms like CoO, NiO, and MnO, more amenable to subsequent acid leaching, and Li_2CO_3 phase (Tong et al., 2024). Here, the synergistic effect arises from the liberation of molecular oxygen (O_2) as well as the reduction chemical reactions of carbon with oxygen and transition metal oxide (Li et al., 2025). Liu et al. (2018) demonstrate efficient carbothermal reduction roasting at 650 °C using 10% coke dosage followed by water leaching to recover 93.33% Ni, 98.68% Mn, 98.08% Co, and 93.67% Li.

Current research work focusses on utilising waste resources as a common reductant for the carbothermal roasting of spent batteries. Ma et al. (2020) proposed reduction roasting using 5% spent anode powder at an optimised condition of 650 °C for 1 h, followed by NH_3 leaching to extract 82.1% Li, 97.7% Ni and 99.1% Co. Zhang et al. (2019) utilised a 6:1 mass ratio of NMC: anodic graphite at 600 °C for 3 h to carbothermally reduce spent battery material into a metal oxide phase amenable for subsequent leaching. Feng et al. (2025) proposed biomass-assisted carbothermal reduction roasting at a 1:2 mass ratio at roasting temperature of 650 °C and time interval of 20 min, followed by selective leaching of NCM811 battery material to recover 99.6% Li. However, the introduction of carbon in carbothermal reduction bears reactions at higher temperatures and results in huge emissions of greenhouse gases along with the inclusion of waste carbon (Tong et al., 2024).

Salt-Assisted Roasting

Salt-assisted roasting, an energy-efficient and selective pyrometallurgical approach, involves the introduction of reactive salts (chlorides, nitrates, and sulphates) during roasting to stabilise the transition metal oxides/salts and promote water-soluble lithium salts formation (Zhu et al., 2024). Zhang et al. (2021) conducted sulphate-assisted reduction roasting, which involves the utilisation of ammonium sulphate $(NH_4)_2SO_4$, of NMC111 with a salt-to-material mass ratio of 3.5:1 under an optimised reaction parameter of 400 °C and 20 min to recover 97.30% Mn, 97.32% Co, 98.31% Li, and 96.29% Ni. Similarly, Tang et al. (2020) demonstrated analogous results with recovery percentages exceeding 98% by conducting the reaction at 400 °C in an inert environment using $(NH_4)_2SO_4$, demonstrating the efficacy of sulphate-based roasting in conducting the reaction under lower temperature conditions as well as reducing reliance on harsh chemicals. Using a similar approach, Li et al. (2010) demonstrated that chlorination roasting utilised chlorinating agents such as NH_4Cl and NaCl to facilitate the formation of soluble LiCl with a recovery yield exceeding 99% at 450 °C. Similarly, Barrios et al. (2021) proclaimed near-complete conversion of battery metals as metal chlorides at 900 °C within 90 min. Further to extend the studies on salt-assisted reduction roasting, Li et al. (2011) explored nitrate-assisted roasting on $LiCoO_2$ using nitric acid (HNO_3) at temperatures as low as 250 °C to recover ~ 93% Li with $LiNO_3$ and Co_3O_4 as major product phases. Although salt-assisted roasting shows promise in lithium recovery with limited reagent consumption and minimal reaction temperature, studies conducted on sulphate-assisted roasting suggest poor selectivity under non-optimised sulphate roasting conditions and demonstrate concerns pertaining to SO_x emission. Similarly, chlorination roasting as well as nitrate-assisted roasting also raise environmental constraints and equipment concerns due to the corrosive effect of HCl and release of hazardous Cl_2 gases and NO_X gases from chloride-based and nitrate-based systems, respectively (Tong et al., 2024). Thus, though the process suggests high lithium recovery with reactions at low thermal conditions, it mandates precision in salt selection, necessitates gas management, and recommends product-purity concerns (Bresolin et al., 2025).

Microwave-Assisted Roasting

Microwave-assisted reduction roasting suggests a sustainable replacement of the traditional thermal-based pyrometallurgical approach, which typically relies on high-temperature reactors that utilise heat transfer mechanisms involving conduction, convection, and radiation. Unlike traditional systems, microwave treatment involves the irradiation of electromagnetic waves directly onto the battery materials, thereby minimising energy loss to the surrounding environment (Bresolin et al., 2025). The underlying heating mechanism arises from Joule heating or dipolar and ionic interactions, which is a result of rapid oscillations and collisions of electrons, free ions, and polar molecules, resulting in enhanced heating rates, uniform energy distribution, safe operation, and a cost-efficient process (Men et al., 2023; Sun et al., 2023). Additionally, the anodic graphite present in spent battery materials not only serves as a reductant but also acts as an active microwave absorber, resulting in enhanced metal recovery at mild reaction conditions (Fahimi et al., 2023). Fu et al. (2020a) demonstrated microwave-assisted roasting of NMC111 in a nitrogen atmosphere under optimised conditions of 500 W and 30 min, resulting in disordered crystal structures with 850 induced-lattice cracks and a leaching efficacy of 99% Li and 97% for other transition metals (Co, Mn, and Ni). Further trials on reduction of processing time from 30 to 15 min, followed by water leaching, result in 99% recovery of Li as Li_2CO_3 and > 96% recovery of transition metals from fumaric acid leaching of the obtained leached residues (Fu et al., 2020b). Similarly, Liu et al. (2023a, 2023b) recovered 96.12% Li by inducing 700W microwave irradiation for an hour, followed by water and alkaline leaching. Zhao et al. (2020) replaced the anodic graphite with biomass-assisted microwave-roasting using 1500 W power of microwaves for 25 min to recover 93.4% Li from NMC111. Similarly, You et al. (2023) utilised microwave-assisted hydrogen roasting at a temperature of 500 °C and 60-min time interval to recover 97.7% Li, 99.5% Ni, 97.7% Co, and 88.4% Mn. Similarly, Guan et al. (2022) as well as Liang et al. (2023) conducted microwave-assisted chlorination as well as sulphation roasting to efficiently recover critical metals from spent battery materials. Despite these advantages, technical challenges remain, including limited microwave penetration, difficulties in scaling, and variability in dielectric properties across LIB feedstocks. Nevertheless, its potential for lower environmental impact and compatibility with renewable energy sources positions microwave-assisted roasting as a viable direction for sustainable LIB recycling (Bresolin et al., 2025).

5 Hydrometallurgical Recycling

Hydrometallurgy, widely adopted as a dominant recycling strategy, uses aqueous dissolution and separation mechanisms to recover essential critical and strategic minerals from spent battery materials. Hydrometallurgical recycling strategy promises low-energy treatment of spent battery materials, followed by high recovery purity with reduced gas emissions to recycle spent battery materials (Lee et al., 2025). It involves the utilisation of leaching and separations through solvent extraction and precipitation as a general LIB recycling pathway.

5.1 Leaching Strategy in LIB Recycling

Leaching involves mobilisation, extraction, and retraction of soluble elements from the waste LIB matrix by dissolving them in a solvent (leachate) phase, allowing the major components (Li, Co, Ni, Mn, etc.) to be regenerated as a metal-rich leachate phase through dissolution, desorption, and complexation strategy and enabling their subsequent separation stages (Kulkarni, 2015). However, with the diversification of LIB technologies, treatment of different cathode materials requires tailored leaching strategies due to their varied chemical composition and phase stability. Industrial practices on LIB recycling focus on chemical leaching using conventional solvents, including both organic and inorganic solvents, along with a reductant phase, owing to their relatively short processing time, ease of scalability, and high metal recovery (Lee et al., 2025; Srinivasan et al., 2024). However, current research focusses on maximising recycling–recovery through leaching systems involving varied chemical environments. However, acid leaching is commonly adopted as a solvent phase for LIB recycling regardless of the cathode chemistry. Thus, the current paper focusses on introducing an in-depth literature review of acid leaching processes, such as inorganic acid-based leaching systems and organic-acid-based leaching systems, and highlights the role of green solvents in the leaching strategy of spent battery material.

Inorganic Acid Leaching

Studies have shown the utilisation of traditional inorganic acids, for instance, sulphuric acid (H_2SO_4), hydrochloric acid (HCl), nitric acid (HNO_3), and phosphoric acid (H_3PO_4) as common leachants for extraction of critical minerals from retired battery chemistries (Aikawa et al., 2017; Chen et al., 2016; Peng et al., 2018; Takacova et al., 2016). Vieceli et al. (2021) reported efficient leaching of NMC using sulphuric acid (1.5 mol/L) at 60 °C temperature, 300 rpm stirring speed, and S/L ratio of 100 g/L at a time duration of 60 min to recover 84% Mn, 99% Ni, 90% Co, and 91% Li. Similarly, Takahashi et al. (2020) leached mixed-type battery material using hydrochloric acid, maintaining a S/L ratio of 200 g/L at pH 3.0 at 50 °C for 240 min to selectively leach 98.4% Co. Yang et al. (2021) leached LFP-type battery materials using H_3PO_4 (1.5 mol/L) and HCl (0.5 mol/L) with 100 g/L solid–liquid ratio to leach 97% Li after 180 min at 90 °C. Similar results were also obtained with nitric acid leaching of NCA-type batteries with 4 mol/L nitric acid by Muzayanha et al. (2019) to leach 63% Co and 80% Ni at 80 °C within a time interval of 60 min. Further studies on the leaching behaviour of LIBs demonstrate the utilisation of reductants, commonly H_2O_2, because of high standard reduction potential and limited dissolution in the leaching system (Cerrillo-Gonzalez et al., 2021). Current studies showed utilisation of varied reducing agents such as inorganic compounds (sodium sulphite (Na_2SO_3), ammonium chloride (NH_4Cl), sodium chloride (NaCl), sodium bisulphite ($NaHSO_3$)), organic acids (ascorbic acid, formic acid, citric acid), biomass derived reducing agents (glucose, waste areca powder, grape skin), and metals and metal-scrap (Cu-Al, iron scrap, Cu-Al waste), etc. (Lee et al., 2025). These reducing agents are used along with the solvent phase in a particular proportion to facilitate the leachability of spent battery material by converting higher-valent transition metals into leachable or more soluble lower-valent transition states, more amenable to leaching (Davis & Demopoulos, 2023). Table 1 displays the effect of various reducing agents on the leaching efficiency of critical metals of varied cathode materials.

Cerrillo-Gonzalez et al. (2021) studied the acid-leaching of spent $LiCoO_2$ battery material through mathematical modelling and validated the extractive recovery of both Co and Li species from $LiCoO_2$ powder through a reduction mechanism. Equations 1 and 2 demonstrate that the acid-leaching behaviour of $LiCoO_2$ battery material pertains through the water oxidation half-reaction to complete the redox pair, without the presence of reducing agent. It results in the extraction of Co and Li from the battery material, leaving behind Co_3O_4 residue, which pertains as a crust around the unreacted core of $LiCoO_2$ and limits the dissolution in acidic media. However, in the presence of a reductant, the leaching behaviour gets enhanced and undergoes a reaction mechanism following Eqs. (3) and (4) (Cerrillo-Gonzalez et al., 2021). This leaching behaviour was also observed by Nayl et al. (2017), demonstrating enhanced leaching efficiency with the use of reducing agents.

$$\mathrm{LiCoO_2(s) + 2H^+ \rightarrow Li^+ + \frac{1}{2}Co^{2+} + \frac{1}{6}Co_3O_4(s) + H_2O + \frac{1}{6}O_2} \tag{1}$$

$$\mathrm{Co_3O_4(s) + 6H^+ \rightarrow 3Co^{2+} + 3H_2O + \frac{1}{2}O_2} \tag{2}$$

$$\mathrm{LiCoO_2}(s) + \mathrm{3H^+ + \frac{1}{2}H_2O_2 \rightarrow Li^+ + Co^{2+} + 2H_2O + \frac{1}{2}O_2} \tag{3}$$

$$\mathrm{Co_3O_4(s) + H_2O_2 + 6H^+ \rightarrow 3Co^{2+} + O_2 + 4H_2O} \tag{4}$$

Surveys on leaching behaviour have also demonstrated the role of varied variables for instance, time, acid concentrate, temperature, and reductant concentrations on the leachability of critical metals. Studies pertaining to optimal inorganic acid concentrations exhibit maximum concentration of H_2SO_4 (1.5–2 mol/L), HCl (4 mol/L), H_3PO_4 (2 mol/L), and HNO_3 (1 mol/L) to be utilised to gain maximum efficiency in leaching. Increased acid concentration increases the concentration of free-H^+ ions, which facilitates faster displacement of Li^+ ions with H^+ ions in the crystal lattice.

Further, higher acidic conditions enhance the dissolution and leaching kinetics of the metals, as the reaction pertains majorly at the solid–liquid (S/L) interface, resulting in aiding maximum leachability due to stoichiometric, kinetic, and thermodynamic limitations (Lee et al., 2025). Similarly, Cheng et al. (2019) investigated optimal hydrogen peroxide concentration as 4% owing to its accelerated decomposition at high-temperature conditions. Likewise, temperature also considerably affects the leaching behaviour of battery materials. Enhanced reaction temperature boosts

Table 1 Effect of reductant on leaching varied cathode material in inorganic acid media

Battery type	[Acid] (mol/L)	[Reductant], %	Reaction condition	Mechanical or other external aid	Leaching efficiency (in %)				References
					Co	Ni	Mn	Li	
NMC	H_2SO_4 (2 mol/L)	Fe^{2+} (10%)	30 °C, 100 g/L, 360 min	60 W ultrasonic power	90.4	85.5	91.8	89.9	Do et al. (2022)
Mixed	H_2SO_4 (4 mol/L)	Acetylene black (10%)	85 °C, 100 g/L, 60 min	–	97.8	95.3	98	95.7	Jiang et al. (2022)
Mixed	HCl (0.5 mol/L)	L-ascorbic acid	90 °C, 10 g/L, 10 min	–	99.9	99.9	99.9	99.9	Gu et al. (2022)
LCO	H_3PO_4 (1.5 mol/L)	Glucose (0.02 mol/L)	80 °C, 120 min	–	95.3	–	–	100	Meng et al. (2017)
LFP	HCl (0.5 mol/L) + H_3PO_4 (1.5 mol/L)	–	90 °C, 100 g/L, 180 min	–	–	–	–	97	Yang et al. (2021)

the kinetic diffusion of the particles in the solvated medium and thus enhances diffusion and dissolution mechanisms, leading to enhanced leaching. In the same way, leaching behaviour also enhances with the rise in reaction time but remains constant after the optimal condition is reached. Correspondingly, pulp density also affects the leaching behaviour by regulating the accessible surface area for leaching, controlled by the stoichiometry of the reaction. Generally, leaching efficiency increases with an increase in S/L ratio to an extent, reaching a maximum, and on further increase in S/L ratio, leaching efficiency then decreases (Faraji et al., 2020). Thus, optimised recovery of spent battery materials necessitates optimising the parameters and conducting reactions, maintaining a particular reaction condition.

Organic Acid Leaching

The growing weightage on green chemistry has stimulated the adoption of biodegradable, microorganism-derived organic acids as leaching agents (Theron & Lues, 2007). Owing to their carboxylic and hydroxyl functional groups, these acids (i) behave as proton donors, (ii) exhibit strong chelation with metal ions, and (iii) help to minimise secondary pollution and wastewater generation (Sahu & Devi, 2022). The number of dissociation steps in solution correlates directly with an acid's leaching ability: each released proton increases metal solubility, thereby enhancing overall extraction (Lee et al., 2025).

Song et al. (2021) reported that 5 mol/L acetic acid exhibited lower leaching efficiency (< 80%) than 2 mol/L inorganic acid (H_3PO_4) for extracting critical metals from retired battery materials. Furthermore, the superior leaching performance of lactic acid, another monobasic acid, could be due to the presence of both carboxylic and hydroxyl groups. Roshanfar et al. (2018) observed an increase in leaching efficiency from 79 to 98% as the lactic acid concentrate rose from 0.5 to 1.75 mol/L, respectively. Li et al. (2017a, 2017b) demonstrated that citric acid, a tricarboxylic acid having strong chelating ability, achieved > 94% leaching efficiency for Co, Ni, Li, and Mn within 1 h at 90 °C and 20 g/L pulp density by dissolving mixed battery materials in citric acid (0.5 mol/L) with H_2O_2 (1.5 vol%). Other organic acids, for instance oxalic, malic, and formic acids, also demonstrate comparable leaching efficiency with inorganic acid, with values exceeding 90%, when paired with a suitable reductant (Lee et al., 2025). For instance, leaching spent batteries with gluconic acid in the presence of 0.5 vol% H_2O_2 yields values exceeding 90% while those below 50% when paired without the reductant (Fan et al., 2020a, 2020b). Table 2 demonstrates the leaching efficiency of organic acids with varied reductants. Leaching spent battery materials with ascorbic acid often delivers synergistic enhancement in metal dissolution owing to the acidic functionality as well as reductant property (Li et al., 2012). Consequently, the stable metal–ligand complexation formed with an organic acid exhibits efficient dissolution when operated at a particular leaching condition. This, however, requires optimisation of the operating parameters to achieve maximal or selective leaching efficiency.

Table 2 Effect of reductant on leaching varied cathode material in organic acid media

Battery type	[Acid], mol/L	[Reductant]	Reaction condition	Mechanical or other external aid	Leaching efficiency (in %)				References
					Co	Ni	Mn	Li	
LCO	Citric acid (1.5 mol/L)	Tea waste (0.4 g/g)	90 °C, 30 g/L, 120 min	–	96	–	–	98	Chen et al. (2015)
LCO	Citric acid:pyrine—3:1 (148 g/L)	H_2O_2 (0.14 mol/L)	65 °C, 30 g/L, 40–50 min	120 W ultrasonic power	94	–	–	99	Lin et al. (2021)
Mixed	Ascorbic acid (0.5 N)	–	10 g/L, 40–5 min	300 W microwave power	100	–	100	100	Chu et al. (2021)
LCO	Tannic acid (1 mol/L)	Acetic acid (1 mol/L)	80 °C, 20 g/L, 80 min	Stirring (250 rpm)	86	–	–	100	Prasetyo et al. (2022)
NMC111	DL-malic acid (0.3 mol/L)	Glucose (0.3 g/g)	80 °C, 20 g/L, 30 min	–	95	95	95	95	Sidiq et al. (2022)

Furthermore, current leaching aspects involve the use of DESs as an emerging solvent and green strategy exhibiting tuneable acidity, low volatility, and biodegradability, and can serve both as a leachant and reducing medium (Abbott et al., 2002; Tran et al., 2020) Moreover, the integration of biomass-derived reductants like tea waste, grape seed extract, and phytochemicals has attracted attention for it (Lee et al., 2025). These materials can support low-cost and low-toxicity leaching, aligning with circular economy principles.

Bioleaching

Bioleaching is a time-consuming process, but by modifications, the leaching kinetics can be improved. Horeh et al. (2016) used *Aspergillus niger* to leach the metal ions present in the mobile phone batteries. Bioleaching experiments of one-step, two-step, and spent medium were carried out. In the spent medium, the culture of the fungi attended the stationary phase first, and then battery material was mixed. The experimental period for all cases was approximately 30 days. The percentage leaching of different metal ions from three experiments is shown in Fig. 5a.

Maximum efficiency of the critical metal ions was attained from the spent medium (Fig. 5a). The reason provided by the authors was that these fungi produce organic acids, which are responsible for metal leaching. Figure 5b shows the amount of different organic acids produced during the leaching process. Oxalic acid was solely responsible for one-step leaching; that's why the Co and Ni oxalates remained in the residue (Fig. 5b). But in the spent medium, the concentration of citric acid is higher and portrays an essential role in metal leaching, which is also confirmed by the percentage leaching (%L) of Co and Ni.

In another approach, Biswal et al. (2018) demonstrated a bioleaching study on spent LIBs using *Acidithiobacillus thiooxidans* strain 80191, and *Aspergillus niger* strains MM1 and SG1, succeeded by chemical precipitation to retract the cations from retired LIBs. In both cases, the dissolution of metal ions was better in Type 2 (after the biomass growth, LIB was added) than in Type I. Comparing fungi and bacteria, the leaching efficacy of metal cations was more pronounced in the case of two *Aspergillus niger* strains.

Combining organic acid leaching and electrochemically assisted leaching, Mukhopadhyay et al. (2024) reported a novel recycling route for critical metal removal from the EOL (end-of-life) of LIBs. Gluconic acid produced from *Gluconobacter oxydans* was used as the lixiviant and electrolyte for the electrochemical process. Fe(II) is used as the reducing agent. Optimisation approaches of Design of Experiments (DOE) and Response Surface Methodology (RSM) were applied to maximise the metal extraction efficiencies. Overall, the gluconic acid concentration plays a significant role in metal leaching efficiency.

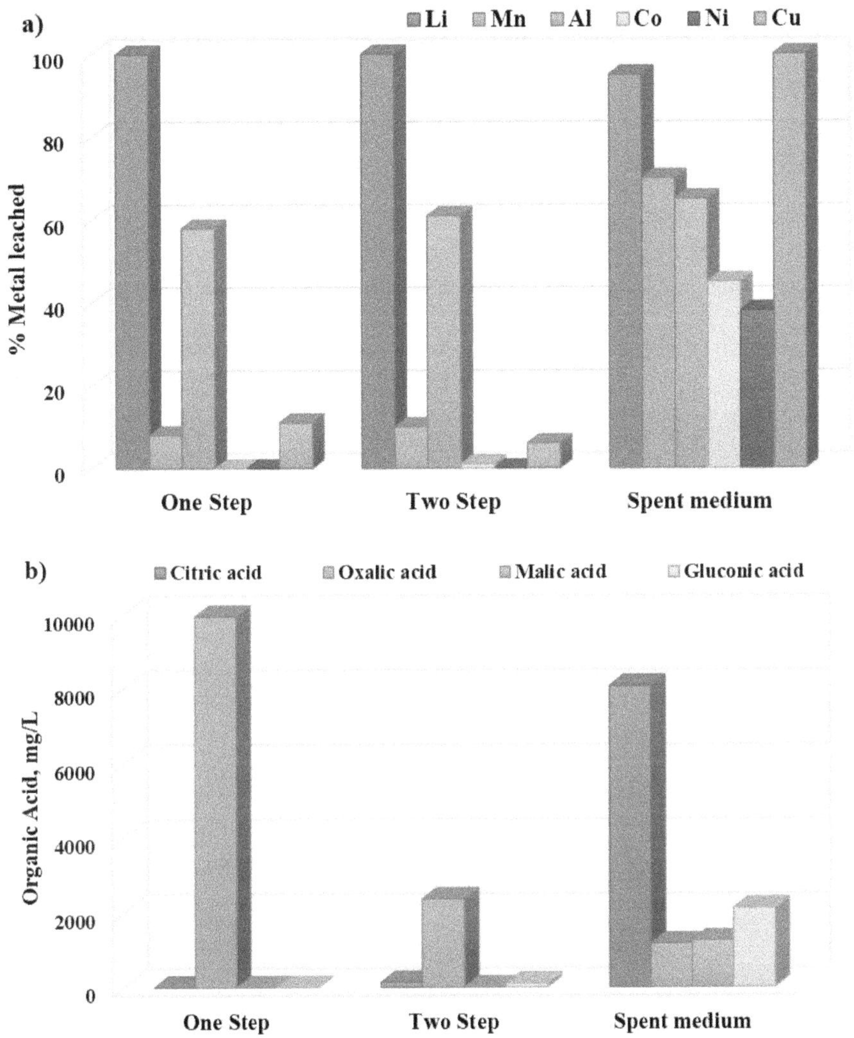

Fig. 5 Effect of bioleaching one-step, two-step, and spent medium on **a** leaching efficacy of varied metal ions, and **b** concentration of organic acid produced (Horeh et al., 2016)

5.2 *Solvent Extraction and Chemical Precipitation*

Liquid–liquid extraction or solvent extraction involves the selective transfer of a solute, usually metal ions, from an aqueous phase (AP) into an organic phase (OP), facilitated using specific extractants. It is mostly used after the leaching step, where the metals from the ore/waste come into the AP as dissolved ions. This technique plays a vital role in sustainable resource management, efficiently reclaiming valuable metals from LIBs. The most used extractants for metal ions recovery from LIBs were ionic liquids and organophosphorus molecular extractants.

Using Ionic Liquids (ILs)

ILs are distinguished green solvents owing to their unique properties, and they are liquid salts comprised of asymmetric organic species (cations) of high molecular weight and anions species (inorganic or organic) (Marsh et al., 2004; Singh & Savoy, 2019; Welton, 2018). Due to the large ion size of the IL and conformational flexibility, they have high entropy changes and low lattice enthalpies, favouring the liquid state. Some of the ionic liquids, including their chemical name, molecular formula, mass, and appearance, are shown in Table 3.

Xu et al. (2020) used Cyphos IL 101 for selective recovery of critical minerals such as Co and Li from waste LIBs. The IL was first mixed with 14% water to reduce its viscosity, and this was mixed with the LIB leach liquor. The experiments were conducted by varying HCl concentration (0.5–4.0 mol/L) in the aqueous solution, and the maintaining temperature in the range of 50–90 °C. After mixing, followed by centrifugation, the raffinate was analysed. The optimum conditions for Co/Li separation were investigated by studying different influential parameters (Fig. 6a and b) (Xu et al., 2020). Figure 6a and b shows that the highest separation factor, 691.4, was observed at 4 mol/L HCl. At high HCl concentration, cobalt forms $CoCl_4^{2-}$ anionic complex, which can be easily extracted to Cyphos IL 101 through an anion exchange mechanism. The authors also studied the extraction cycle of the extractant (Xu et al., 2020) and, after successive use for 10 cycles, without regeneration by

Table 3 Chemical properties of ionic liquid

Commercial name	Chemical name	Formula	Mol. wt., g/mol	Appearance
Cyphos IL 101	Trihexyl(tetradecyl)phosphonium chloride	$C_{32}H_{68}ClP$	519.3	Colourless viscous liquid
Cyphos IL 102	Trihexyl(tetradecyl)phosphonium bromide	$C_{32}H_{68}PBr$	563.8	Colourless-pale yellow liquid
$[P_{6,6,6,14}][SCN]$	Trihexyl(tetradecyl)phosphonium thiocyanate	$C_{33}H_{68}NPS$	541.9	Colourless viscous liquid
$[C_6mim][PF_6]$	1-Hexyl-3-methylimidazolium hexafluorophosphate	$C_{10}H_{19}F_6N_2P$	312.3	Colourless-pale yellow crystalline solid or viscous liquid
Cyphos IL 104	Trihexyl(tetradecyl)phosphonium bis(2,4,4-trimethylpentyl)phosphinate	$C_{48}H_{102}O_2P_2$	773.3	Colourless-yellow solid
Aliquat 336	Trioctylmethylammonium chloride	$C_{25}H_{54}NCl$	404.2	Clear-pale yellow liquid

stripping, calculated the cumulative extraction capacity, which reached 35.25 g/L for cobalt.

Another strategy for the selective extractive removal of Li from waste NMC battery leach liquor was reported by Zheng et al. (2021) using an IL carboxymethyl trimethylammonium bis(trifluoromethyl) sulphonimide in TBP. This functional IL was synthesised in the laboratory, taking carboxymethyl trimethylammonium chloride and lithium bis(trifluoromethane sulphonyl) imide. The synthesised IL was characterised by FTIR, and Mass Spectroscopy. 20% IL in 80% TBP was used for the

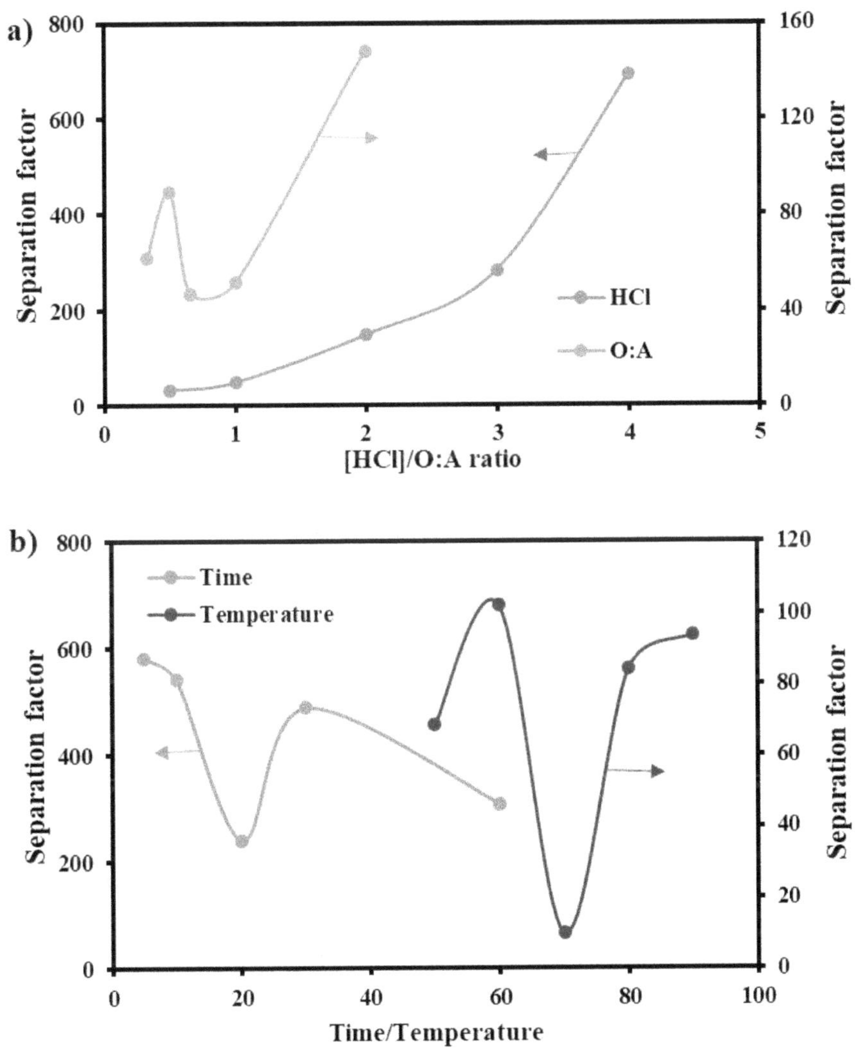

Fig. 6 Effect of **a** hydrochloric acid concentration and O:A ratio, **b** temperature and time on the separation factor of cobalt and lithium. (Xu et al., 2020)

extraction study of NMC sulphate leach liquor at different acidic pH ranges (Zheng et al., 2021). The result showed that the extraction of Li (~ 60%) barely changed with the change in pH. The construction of the McCabe–Thiele plot suggested that the five-stage extraction removed 99.88% Li from other metal ions from the leach liquor at an A/O of 1/2. Multistage extractions were carried out by the authors, and 0.128 g/L Li was left over in the AP. The Li-selective extraction had been demonstrated by the following mechanism (Zheng et al., 2021).

$$[(CH_3)_3NCH_2COOH]_{org} + Li^+_{aq} + 3TBP \rightarrow [(CH_3)_3NCH_2COOLi \cdot 3TBP]_{org} + H^+ \quad (5)$$

Dhiman and Gupta (2019) demonstrated selective extraction of cobalt from waste lithium-ion battery using Cyphos IL 102. Spent LIB was leached by adjusting the acidity to 5 mol/L HCl, resulting in a leach liquor with 1.406, 0.273, and 0.181 g/L Co, Li, and Ni, respectively. Extraction experiment demonstrated selective extraction of 99.9% cobalt at 0.3 mol/L using Cyphos IL 102. However, for practical recovery, 0.2 mol/L of extractant was used in a two-stage counter-current extraction procedure maintaining an O:A ratio of 1, yielding a loaded OP with nearly all cobalt transferred to the OP as $[P_{66614}{}^{+}]_2[CoCl_4{}^{2-}]$. Following this regeneration of cobalt involved stripping the OP with HCl (0.05 mol/L) in a single stage, resulting in 99.8% recovery of cobalt with 100% purity.

Wang et al. (2023) used IL (1-Butyl-3-MethylImidazolium bis (trifluoromethyl sulfonyl) imide) as the diluent for the isolated extraction system, as well as a synergistic mixture of TOPO and HTTA, to extract critical minerals from the acetic acid leachate of spent LIBs. Experiments on the single extraction system (HTTA in IL and TOPO in IL) manifested that 86.6%, 47.5%, and 96.3% of Co, Al, and Ni were extracted using 0.5 mol/L HTTA in IL whilst using TOPO (0.5 mol/L), the recovery percentage of Co, Al and Ni was 37.9%, 54.9%, and 23.6%, respectively. An enormous difference in Mn extraction has been observed between the two extractants. The extraction efficiency of TOPO for Mn was 93.7%, but it was only 2–3% for HTTA. A very small amount of Li extraction has been reported for both HTTA and TOPO. When a synergistic approach has been adopted (mixing equal concentrations of TOPO and HTTA in IL) by the authors, the recovery rate of all metal ions increased drastically, except for Li. The extraction efficacy of metal ions Co^{2+}, Mn^{2+}, Al^{3+}, and Ni^{2+} was 99.54%, 99.85%, 89.81%, and 99.68%, respectively (Wang et al., 2023). The co-extraction efficacy of Li was merely 2.84% at the highest conditions: O/A = 1/1, 20 min, 0.4 mol/L HTTA and TOPO concentration, temperature 25 °C. Based on UV spectroscopy and slope methods, the authors deduced the following extraction mechanism (Wang et al., 2023).

$$2TOPO_{org} + HTTA_{org} + Co^{2+}_{aq} + CH_3COO^-_{aq} \rightarrow H^+_{aq} + [Co(CH_3COO)(TTA) \cdot 2TOPO]_{org} \quad (6)$$

Using synthesised ionic liquids, Tran et al. (2021) separated the metal ions Cu, Fe, Co, Ni, and Mn from the sulphate medium of spent LIB. The IL ALi-Cy301 was prepared using an equimolar mixture of bis(2,4,4-trimethylpentyl)dithiophosphinic acid (Cyanex 301) and *N*-methyl-*N,N, N*-trioctylammonium chloride (Aliquat 336) in kerosene diluent, followed by mixing them vigorously. After mixing, the solution was washed with $NaHCO_3$ to remove H^+ and Cl^- present in it. Similarly, the authors also prepared ALi-SCN (1.0 mol/L) by treating Aliquat 336 several times with NH_4SCN for anion exchange (Tran et al., 2021). After leaching with sulphuric acid under maximal leaching conditions (20% (v/v) H_2O_2, 2.0 mol/L H_2SO_4, 25 g/L pulp density at 60 °C), the Cu^{2+} and Si(IV) removal from the rest of the metal ions was tried with Cyanex 301. The authors got good selectivity of Cu^{2+} over Co^{2+} and Ni^{2+}, but due to the liberation of H^+ ions to the AP (exchange of metal and H^+ ions), the acidity of the AP increased, which created a problem for other metal extractions. Hence, to overcome the difficulty, ALi-Cy301 was used to separate Cu(II) according to the reaction proposed by the authors (Tran et al., 2021)

$$\begin{aligned} & Cu^{2+}_{(aq)} + 3HSO^-_{4(aq)} + H^+_{(aq)} + 3\left[ALi - Cy301\right]_{(org)} \\ & \quad = CuA_{2(org)} + HA_{(org)} + 3\left[ALi^+ - HSO_4^-\right]_{(org)} \end{aligned} \tag{7}$$

After Cu(II) removal, the concentration of other metal ions was 0.126 g/L Fe, 15.5 g/L Ni, 3.23 g/L Co, and 0.395 g/L Mn with H^+ of 2 mol/L. Then, employing ALi-SCN (0.5 mol/L), both Fe^{3+} and Co^{2+} were selectively separated from Mn and Ni. Further, selective precipitation of Mn as MnO_2 was proposed because of the concentration gap between Mn (low) and Ni (Tran et al., 2021), and it was carried out using an oxidising agent NaClO with the following mechanism demonstrated in Eq. (8). The loaded organics were selectively stripped using HNO_3 for Cu(II) and NH_3 for Co(II).

$$Mn^{2+}_{(aq)} + NaClO_{(aq)} + H_2O_{(l)} \rightarrow MnO_{2(s)} + NaCl_{(aq)} + 2H^+_{(aq)} \tag{8}$$

In another investigation (Guo et al., 2024), the authors used different ionic liquids and molecular extractants to prepare a synergistic extractant system. 1-aminopropyl-3-methylimidazolium bis((trifluoromethyl)sulfonyl)imide ([APMIM][Tf_2N]), 1-butyl-3-methylimidazolium bis((trifluoromethyl)sulfonyl) imide ([BMIM][Tf_2N]), 1-ethyl-3-methylimidazolium bis((trifluoromethyl)sulfonyl) imide ([EMIM][Tf_2N]), 1-allyl-3-methylimidazolium bis((trifluoromethyl) sulfonyl)imide ([AMIM][Tf_2N]), 1-ethyl-3-methylimidazolium propionate ([EMIM][OPr]), 1-ethyl-3-methylimidazolium diethyl phosphate [EMIM][Et_2PO_4]), ionic liquids, and molecular extractants such as Cyanex 923 (C923, a commercial mixture of trialkyl phosphine oxides with n-hexyl, n-butyl, and n-octyl), di-(2-ethylhexyl)phosphoric acid (P-204), tributyl phosphate (TBP), and 2-ethylhexyl hydrogen-2-ethylhexylphosphonate (P-507). After rigorous effort, the authors adopted [APMIM][Tf_2N] + Cyanex 923 extraction system to separate essential metal from the waste NMC (Guo et al., 2024). To get the optimum conditions, the aqueous solution pH, mass fraction of the IL, and A/O ratio have been varied, and the optimised conditions were at A/O = 1:4, pH = 4, and mIL = 30 wt% where quantitative separation of Mn, Ni, and Co from Li (<10%) in the AP was obtained. Furthermore, the OP reloading of the collaborative extractant [APMIM][Tf_2N] + Cyanex 923 was a challenge to the authors because of the complex system. They applied a mixture of HCl and 2 wt% ascorbic acid to bring the metal ions to an aqueous solution. The maximum recovery (95.61% Co, 94.37% Ni, and 99.15% Mn) was noticed at HCl (3 mol/L) and an O:A ratio of 1:2. The extractive mechanism of the system was analysed using spectroscopic analysis coupled with slope experiment before and after extraction of the OP. The authors applied IGMH (Independent Gradient Model based on Hirshfeld partition) and ESP (Electrostatic Potential) to support the coordination of C923 with the metal ions. They also confirmed that the $[Tf_2N]^-$ anion of the IL is responsible for the stabilisation of [Metal ion-[Tf_2N]-C923] complexation in the OP (Guo et al., 2024).

Not always is an additional ligand necessary to extract metal ions. For example, only replacing the anion of the Cyphos IL 101 (Cl^- to SCN^-), the selectivity of Li and Co extraction increased by equilibrating it with ascorbic acid leach liquor of spent calcined $LiCoO_2$ (Sahu & Devi, 2023). Selective recovery of Co (99.9%) with no co-extraction of Li was obtained by performing 2 stages counter-current extraction of Aliquat 336-SCN at a phase ratio of 1:1. A similar trend was also observed when it was applied to the chloride leach liquor of $LiCoO_2$ battery waste (Mohanty & Devi, 2023).

Using Organophosphorus Acids

Organophosphorus extractants are widely applicable to many metal extraction processes, and to LIBs. The three most commonly used extractants are outlined in Table 4.

Wang et al. (2022a, 2022b) studied the extraction and separation behaviour of Co, Mn, and Al from acetic acid leached liquor using TBP (10%), Cyanex 272 (15%), and kerosene (75% v/v). From the pH variation study, it has been found that Al is independent of pH, and the extraction of Mn and Co escalated with rise in pH. The

Table 4 Chemical properties of molecular extractants

Commercial name	Chemical name	Formula	Mol. wt. (in g/mol)	Appearance
D2EHPA	Di-(2-ethylhexyl)phosphoric acid	$C_{16}H_{35}O_4P$	322.4	Pale yellow liquid
PC 88A	(2-Ethylhexyl)phosphonic acid mono-2-ethylhexyl ester	$C_{16}H_{35}O_3P$	306.4	Light yellow liquid
Cyanex 272	Bis(2,4,4-trimethylpentyl) phosphinic acid	$C_{16}H_{35}O_2P$	290.4	Colourless to light yellow liquid

order of the metal ions at pH 6.0 was Mn > Co > Al. The aqueous medium is the acetic acid leach liquor, so the extraction mechanism for the metal ions using Cyanex 272 has been proposed based on slope analysis. The extraction of Mn^{2+} and Al^{3+} followed a cation exchange mechanism with non-involvement of the acetate ion.

$$Al^{3+}_{(aq)} + 3(HA)_{2(org)} \rightleftarrows Al(HA_2)_{3(org)} + 3H^{+}_{(aq)} \tag{9}$$

$$Mn^{2+}_{(aq)} + 2(HA)_{2(org)} \rightleftarrows Mn(HA_2)_{2(org)} + 2H^{+}_{(aq)} \tag{10}$$

But for Co^{2+}, there is involvement of acetate ions in the coordination mechanism, and the slope value is ~ 1.0 (Fig. 7) for log[CH_3COOH] versus log D. The authors proposed that the mechanism for cobalt extraction is

$$Co^{2+}_{(aq)} + CH_3COO^{-}_{(aq)} + (HA)_{2(org)} \rightleftarrows Co(CH_3COO)(HA_2)_{(org)} + H^{+}_{(aq)} \tag{11}$$

After complete extraction of the three metal ions at pH 6.0 with the desired Cyanex 272 concentration from leach solution, the selective separation of Mn and Co from Al has been carried out using 1 mol/L NH_4HSO_4. A high concentration of acid (5 mol/L HNO_3) was needed to strip 9.51% from the Co and Mn-free loaded organic maintained at O:A ratio of 5:1.

Zhang et al. (2022) used both precipitation and solvent extraction procedure to separate the transition metal cations (Mn, Co, Ni) and Li from waste ternary LIBs leached in sulphate medium. The composition of the leach liquor was quantitatively analysed to be 12.65 g/L Ni, 5.65 g/L Co, 3.02 g/L Li, and 5.88 g/L Mn, along with 2.91 g/L of Al. The unwanted Al from the leach liquor was separated using phosphate precipitation by controlling the overall pH of the solution. Further, the neutralisation of the leach liquor was tested using different bases such as $NaHCO_3$, $Na_2C_2O_3$, and NaOH. These bases have no impact on Al precipitation, but the co-precipitation of Mn escalated with rise in basic strength which follows the order, i.e. NaOH > NaC_2O_3 > $NaHCO_3$. So, $NaHCO_3$ was used for pH adjustment, maintaining the

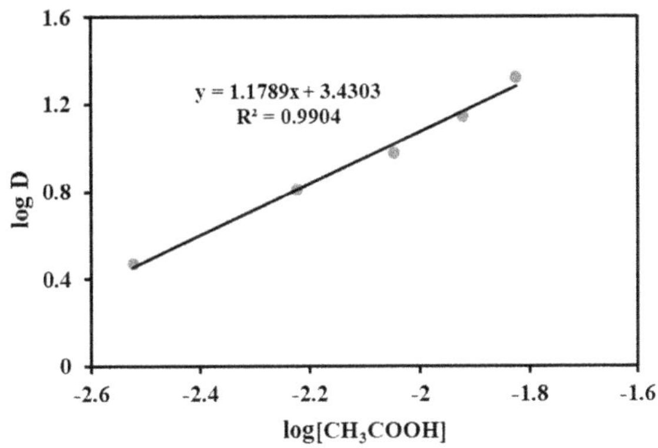

Fig. 7 Log–log plot of distribution coefficient and acetic acid concentration (Wang et al., 2022a; Wang et al., 2022b)

temperature at 80 °C to ensure complete precipitation using a 1:1 ratio of Na_3PO_4. The formation of $AlPO_4$ was confirmed by analysing through XRD, SEM–EDS, and XPS. The aluminium-free leach liquor was treated with P507 for the separation of Mn, Ni, and Co from lithium (Zhang et al., 2022). First, the organophosphorus extractant was saponified 60% to acquire better extraction ability. If we analyse the extraction mechanism for acidic extractants ($M^{n+} + nHA \rightleftharpoons MA_n.HA + nH^+$), the release of H^+ ions to aqueous solution inhibits the metal extraction by decreasing the equilibrium pH. Once saponified, the release of Na^+ ions does not alter the equilibrium pH, so metal extraction is enhanced. The authors performed five-stage counter-current extraction using P507 (1 mol/L) in kerosene at an A:O ratio of 1:1.4. The strip solutions were enriched using H_2SO_4 (1.5 mol/L) by conducting three-stage counter-current back-extraction with an O:A ratio of 5:1.

Further the application of both solvent extraction and solvoleaching was employed to efficiently recover Co species from LCO using the current collectors (Cu and Al) as the reductant (Peeters et al., 2022). The organophosphorus extractants were tested for solvoleaching, and from Fig. 8, it has been observed that D2EHPA could leach and extract Co more efficiently than P507 and Cyanex 272.

Another point of view is represented by Mohanty et al. (2022) using D2EHPA as extractant to selectively recover manganese from the sulphate leach medium of $LiMn_2O_4$. The extraction of the metal cations was conducted using NaD2EHPA (0.3 mol/L) in aliphatic kerosene, maintaining an equilibrium pH from 2.0 to 4.43. With increasing equilibrium pH, the separation factor increased (Fig. 9).

From Fig. 9, the highest separation factor ($\beta = D_{Co}/D_{Li}$) was found to be 1462.79. The Mn extraction increased from 92 to 99.8% but the co-extraction of Li increased up to 30.2%. In case of acidic extractants, the solution pH plays a major role (Pilali et al., 2025). Hence, selecting the pH is very important. So, the McCabe–Thiele plot was constructed at equilibrium pH 2.0 and two-stage counter-current extraction at

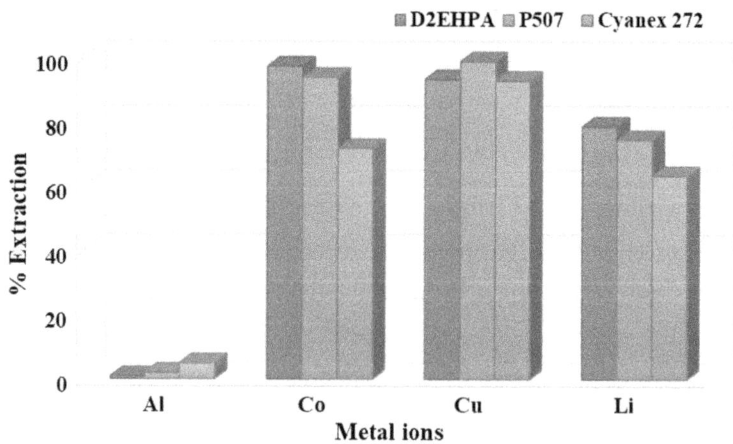

Fig. 8 Percentage extraction of different metal cations using D2EHPA, P507, and Cyanex 272 (Peeters et al., 2022)

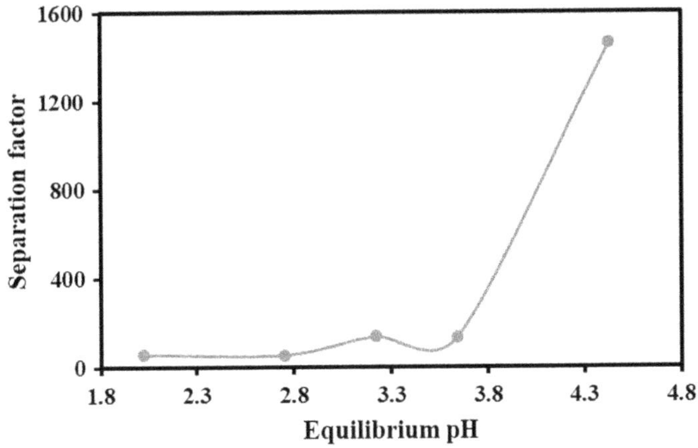

Fig. 9 Effect of separation factor as a function of equilibrium pH (Mohanty et al., 2022)

O:A ratio of unity, followed by Li scrubbing. Pure Mn was left in the loaded OP. Then two stripping stages at an O:A ratio of 1.5:1 using 10% H_2SO_4 demonstrated 1.5 times enrichment of Mn ion as compared to the initial loaded OP.

Another study has been devoted to recovering the metal ions from LIBs of used mobile phones (Shih et al., 2019). After preliminary treatment followed by leaching with suitable acids, the leach liquor of the LIBs was generated. This leached solution was further treated with $KMnO_4$ to precipitate Mn^{2+} as MnO_2. For the precipitation of Mn, pH has a major role. Over 99.5% Mn removal occurred in 60 min at pH 3.0; however, Ni and Co removal only approached 8% and 3%, respectively (Shih et al., 2019). The concentrations of the metal ions after Mn removal were 0.945 g/L Ni, 17.4 g/L Co, 2.75 g/L Li, 0.065 g/L Al, and 0.0431 g/L Cu, respectively. The Mn-free leach liquor was extracted with D2EHPA diluted in Isopar-L. The order of extraction of the metal ions was 99.6% Co, 98.7% Ni, and 95.4% Cu using 0.25 mol/L D2EHPA at an AP pH of 6.0 and 1.5 O:A ratio. After separating these metal ions, the raffinate contains Li and Al. Selective back extraction of Co from loaded organic was conducted using HCl (2 mol/L) followed by oxalate precipitation to form $CoC_2O_4 \cdot 2H_2O$.

6 Global Impact for LIB Recycling

Recycling of retired LIBs has significant implications for resource reliability, circular economy, sustainability, and decarbonisation concerns. From an environmental perspective, an industrial-scale life cycle assessment (LCA) conducted by Machala et al. (2025) demonstrated that LIB recycling reduces environmental impacts, primarily energy consumption, water usage, and greenhouse gas (GHG) emissions, to a minimum of 58% compared with conventional mining-based supply chains

(Machala et al., 2025; Rezaei et al., 2025). In circular supply chains, upstream stages such as collection and transport contribute less than 4% of the total impact, whereas conventional mining accounts for up to 30%. The study also identified electricity usage as the largest contributor to recycling impacts, with sourcing from low-carbon grids reducing emissions by up to 93%, although grids reliant on hydro or geothermal power may increase water use (Machala et al., 2025). From an economic and supply chain perspective, recycling demonstrated a 25% decline in essential mineral resource inadequacy potential, thereby reducing shortages of critical minerals while also stabilising supply chains by avoiding occupational hazards, ecological degradation, and geopolitical risks associated with mining (Rezaei et al., 2025; Winslow et al., 2017). Recycling further provides macroeconomic advantages by substantially lowering the cost of recovered metals; for example, recycled copper (~$3000/ton) and aluminium (~ $1660/ton) are considerably cheaper than their virgin mining counterparts, which cost ~ $5500/ton and ~ $2500/ton, respectively (Zeng et al., 2021). Additionally, battery recycling has been shown to reduce fine particulate matter formation from EVs by 17%, refining air quality by decreasing waste landfilling and incineration (Rezaei et al., 2025). According to ReCiPe 2016, recycling also lowers the toxicity effects of noncarcinogens and carcinogens in humans by 22% and 14%, respectively, further underscoring its environmental benefits (Koroma et al., 2022).

Recycling creates opportunities to recover the intrinsic value of cathode materials, generating an extra revenue source and creating a circular supply network that can withstand shift in battery component costs (Hagelüken & Goldmann, 2022). Overall, LIB recycling not only reduces environmental burdens compared with virgin production but also outperforms older battery technologies in both economic viability and critical material recovery potential.

7 Conclusions and Outlook

The diversification of LIBs, driven by amplified usage, requires the utilisation of varied cathode materials, which, after attaining their EOL, serve as a potent source of critical minerals. Adaptation of efficient recycling strategies maintains the material criticality and ensures a smooth supply chain of these minerals demonstrating 25% decline in mineral resource scarcity potential globally. This chapter summarises separation and recycling tools for effective recovery by detailing pretreatment and extractive metallurgical recycling aspects, covering physical as well as chemical aspects. Pretreatment seems to be a crucial step for subsequent material separation and metal recovery. It involves discharging followed by mechanical separation and pretreatment via thermal, mechanochemical, or solvent-based dissolution mechanisms to separate the cathode materials as black mass. Pyrometallurgical technique, having wide industrial application, generally requires no such pretreatment steps and conducts critical metal recovery through smelting or reduction roasting processes often at temperature above 1000 °C. While recovery efficiencies for these metals can exceed 90% using optimised reducing melting or roasting conditions, the potential loss of Li in slag mandates utilisation of additional processing steps for recovery. The approach benefits from operational simplicity and is well suited for mixed battery streams; however, the high energy input, elevated reaction temperature, emissions of toxic gases, and hazardous by-product demonstrate utilisation of greener aspects of the conventional methods through salt-assisted roasting and microwave-assisted strategies followed by leaching. These greener aspects of the pyrometallurgical treatment process ensure conducting the reaction at milder reaction conditions with minimal loss of critical metals. However, the difficulty in scaling up the process in microwave-assisted roasting and emission of harmful gases in the salt-assisted roasting process still decreases the green recycling aspect of pyrometallurgical methods and necessitates the utilisation of hydrometallurgical strategies in the R&D sector owing to its minimal-energy redemption and limited pollution advantage.

The hydrometallurgical sector involves the dissolution of critical metals through leaching using inorganic, organic, and bioleaching processes, followed by conducting metal recovery and separation procedures, mainly following solvent extraction and precipitation techniques. Varied inorganic acids such as sulphuric acid, phosphoric acid, and nitric acid commonly demonstrate higher leachability than organic acids, making enhanced industrial applications. With optimal acid concentration in the range of (0.5–2 mol/L) in the presence of reducing agent at temperature range of 30–70 °C, inorganic acids show leachability value exceeding 90%; however, it provides disadvantage due to huge operating expenses and their explicit corrosive nature. These mandate the adoption of hydrometallurgical leaching using mild leaching conditions and hence using organic acid, bioleaching, and even utilising reductants for efficient recovery of critical minerals from spent battery materials. Studies demonstrate organic acids, such as citric acid, tannic acid, and malic acid, when used with organic or green reductants at elevated temperature not exceeding 90 °C or with ultrasonic power, and demonstrate recovery of critical minerals similar to that of inorganic acids. Then again, current research proposes integration of ultrasonic applications along with greener solvents, such as the use of DES and green reductant via the use of biomass-based reductant in leaching aspects. The downstream process is mainly focussed on adapting solvent extraction, followed by the subsequent application of the precipitation technique to extract the metal ions from waste LIBs. Solvent extraction involves the use of ionic liquids and molecular extractants as an effective strategy to selectively recover critical metals from leached liquor. The study demonstrates a detailed comparison among various recycling methods and proposes a closed-loop recycling pathway for recycling and recovering critical metals from spent battery materials.

References

Abbott, A. P., et al. (2002). Novel solvent properties of choline chloride/urea. *Chemical Communications, 1*, 70–71. https://doi.org/10.1039/b210714g

Aikawa, T., et al. (2017). Zydrothermal leaching of $LiCoO_2$ with Sulfuric acid, nitric acid, and citric acid. *Kagaku Kogaku Ronbunshu, 43*(4), 313–318. https://doi.org/10.1252/kakoronbunshu.43.313

Aravindan, V., et al. (2011). Lithium-Ion conducting electrolyte salts for lithium batteries. *Chemistry—A European Journal, 17*(51), 14326–14346. https://doi.org/10.1002/chem.201101486

Bahgat, M., et al. (2006). Synthesis, characterization and magnetic properties of microcrystalline lithium cobalt ferrite from spent lithium-ion batteries. *Journal of Materials Processing Technology, 183*(1), 117–121. https://doi.org/10.1016/j.jmatprotec.2006.10.005

Bai, Y., et al. (2021). Recovery of cathode materials and aluminum foil using a green solvent. *ACS Sustainable Chemistry & Engineering, 9*(17), 6048–6055. https://doi.org/10.1021/acssuschemeng.1c01293

Bankole, O. E., Gong, C., & Lei, L. (2013). Battery recycling technologies: Recycling waste lithium ion batteries with the impact on the environment in-view. *Journal of Environment and Ecology, 4*(1), 14. https://doi.org/10.5296/jee.v4i1.3257

Bankole, O. E., & Lei, L. (2014). Silicon exchange effects of glassware on the recovery of LiPF6: Alternative route to preparation of Li2SiF6. *The Journal of Solid Waste Technology and Management, 39*(4), 254–259. https://doi.org/10.5276/jswtm.2013.254

Barrios, O. C., et al. (2021). Chlorination roasting of the cathode material contained in spent lithium-ion batteries to recover lithium, manganese, nickel and cobalt. *Minerals Engineering, 176*, Article 107321. https://doi.org/10.1016/j.mineng.2021.107321

Biswal, B. K., et al. (2018). Biological leaching and chemical precipitation methods for recovery of Co and Li from spent lithium-ion batteries. *ACS Sustainable Chemistry & Engineering, 6*(9), 12343–12352. https://doi.org/10.1021/acssuschemeng.8b02810

BloombergNEF. (2021). Electric vehicle outlook. https://about.bnef.com/electric-vehicle-outlook/

Bresolin, B. M., Zanoletti, A., & Bontempi, E. (2025). Recent improvements in salt-assisted and microwave-assisted recovery methods for sustainable metal extraction from NCM cathodes in spent lithium-ion batteries: A review. *Separation and Purification Technology*, 131918. https://doi.org/10.1016/j.seppur.2025.131918

Buchmann, I. (2021). *BU-216: Summary table of lithium-based batteries*. Battery University. https://batteryuniversity.com/article/bu-216-summary-table-of-lithium-based-batteries

Castillo, S. (2002). Advances in the recovering of spent lithium battery compounds. *Journal of Power Sources, 112*(1), 247–254. https://doi.org/10.1016/s0378-7753(02)00361-0

Cerrillo-Gonzalez, M. M., et al. (2021). Acid leaching of $LiCoO_2$ enhanced by reducing agent. Model formulation and validation. *Chemosphere, 287*, Article 132020. https://doi.org/10.1016/j.chemosphere.2021.132020

Chen, X., et al. (2015). Sustainable recovery of metals from spent lithium-ion batteries: A green process. *ACS Sustainable Chemistry & Engineering, 3*(12), 3104–3113. https://doi.org/10.1021/acssuschemeng.5b01000

Chen, X., et al. (2016). Recovery of valuable metals from waste cathode materials of spent lithium-ion batteries using mild phosphoric acid. *Journal of Hazardous Materials, 326*, 77–86. https://doi.org/10.1016/j.jhazmat.2016.12.021

Cheng, Q., et al. (2019). Characterization, modeling, and optimization of a single-step process for leaching metallic ions from $LiNi_{1/3}Co_{1/3}Mn_{1/3}O_2$ cathodes for the recycling of spent lithium-ion batteries. *Hydrometallurgy, 185*, 1–11. https://doi.org/10.1016/j.hydromet.2019.01.003

Choi, J., & Kim, P. J. (2021). A roadmap of battery separator development: Past and future. *Current Opinion in Electrochemistry, 31*, Article 100858. https://doi.org/10.1016/j.coelec.2021.100858

Chu, H., Lie, J., & Liu, J.-C. (2021). Rapid leaching of valuable metals from spent lithium-ion batteries with microwave irradiation using organic and inorganic acid. *Journal of Sustainable Metallurgy, 7*(2), 630–641. https://doi.org/10.1007/s40831-021-00362-2

Ciez, R. E., & Whitacre, J. F. (2019). Examining different recycling processes for lithium-ion batteries. *Nature Sustainability, 2*(2), 148–156. https://doi.org/10.1038/s41893-019-0222-5

Cornelio, A., Zanoletti, A., & Bontempi, E. (2024). Recent progress in pyrometallurgy for the recovery of spent lithium-ion batteries: A review of state-of-the-art developments. *Current*

Opinion in Green and Sustainable Chemistry, 46, Article 100881. https://doi.org/10.1016/j.cogsc.2024.100881

Da Costa, A. J., et al. (2015). Beneficiation of cobalt, copper and aluminum from wasted lithium-ion batteries by mechanical processing. *International Journal of Mineral Processing, 145*, 77–82. https://doi.org/10.1016/j.minpro.2015.06.015

Davis, K., & Demopoulos, G. P. (2023). Hydrometallurgical recycling technologies for NMC Li-ion battery cathodes: Current industrial practice and new R&D trends. *RSC Sustainability, 1*(8), 1932–1951. https://doi.org/10.1039/d3su00142c

Dhiman, S., & Gupta, B. (2019). Partition studies on cobalt and recycling of valuable metals from waste Li-ion batteries via solvent extraction and chemical precipitation. *Journal of Cleaner Production, 225*, 820–832. https://doi.org/10.1016/j.jclepro.2019.04.004

Diekmann, J., et al. (2016). Ecological recycling of lithium-ion batteries from electric vehicles with focus on mechanical processes. *Journal of the Electrochemical Society, 164*(1), A6184–A6191. https://doi.org/10.1149/2.0271701jes

Do, M. P., et al. (2022). Green closed-loop cathode regeneration from spent NMC-based lithium-ion batteries through bioleaching. *ACS Sustainable Chemistry & Engineering, 10*(8), 2634–2644. https://doi.org/10.1021/acssuschemeng.1c06885

Doerffel, D., & Sharkh, S. A. (2005). A critical review of using the Peukert equation for determining the remaining capacity of lead-acid and lithium-ion batteries. *Journal of Power Sources, 155*(2), 395–400. https://doi.org/10.1016/j.jpowsour.2005.04.030

Dorella, G., & Mansur, M. B. (2007). A study of the separation of cobalt from spent Li-ion battery residues. *Journal of Power Sources, 170*(1), 210–215. https://doi.org/10.1016/j.jpowsour.2007.04.025

Ekberg, C., & Petranikova, M. (2015). Lithium batteries recycling. In *Elsevier eBooks* (pp. 233–267). https://doi.org/10.1016/b978-0-12-801417-2.00007-4

Fahimi, A., et al. (2023). A microwave-enhanced method able to substitute traditional pyrometallurgy for the future of metals supply from spent lithium-ion batteries. *Resources Conservation and Recycling, 194*, Article 106989. https://doi.org/10.1016/j.resconrec.2023.106989

Fan, E., et al. (2020a). Glucose oxidase-based biocatalytic acid-leaching process for recovering valuable metals from spent lithium-ion batteries. *Waste Management, 114*, 166–173. https://doi.org/10.1016/j.wasman.2020.06.047

Fan, E., Yang, J., et al. (2020b). Leaching mechanisms of recycling valuable metals from spent lithium-ion batteries by a malonic acid-based leaching system. *ACS Applied Energy Materials, 3*(9), 8532–8542. https://doi.org/10.1021/acsaem.0c01166

Faraji, F., et al. (2020). Kinetics of leaching: A review. *Reviews in Chemical Engineering, 38*(2), 113–148. https://doi.org/10.1515/revce-2019-0073

Feng, S., et al. (2025). Efficient synergetic recycling of spent Lithium-ion batteries and waste biomass based on segregated reduction roasting strategy. *Separation and Purification Technology, 373*, Article 133648. https://doi.org/10.1016/j.seppur.2025.133648

Fu, Y., et al. (2020a). Microwave reduction enhanced leaching of valuable metals from spent lithium-ion batteries. *Journal of Alloys and Compounds, 832*, Article 154920. https://doi.org/10.1016/j.jallcom.2020.154920

Fu, Y., He, Y., Li, J., et al. (2020b). Improved hydrometallurgical extraction of valuable metals from spent lithium-ion batteries via a closed-loop process. *Journal of Alloys and Compounds, 847*, Article 156489. https://doi.org/10.1016/j.jallcom.2020.156489

Golmohammadzadeh, R., Faraji, F., & Rashchi, F. (2018). Recovery of lithium and cobalt from spent lithium ion batteries (LIBs) using organic acids as leaching reagents: A review. *Resources Conservation and Recycling, 136*, 418–435. https://doi.org/10.1016/j.resconrec.2018.04.024

Goodenough, J. B., & Kim, Y. (2009). Challenges for rechargeable Li batteries. *Chemistry of Materials, 22*(3), 587–603. https://doi.org/10.1021/cm901452z

Gu, S., et al. (2022). Insights into the coordination enhanced leaching mechanism of spent lithium-ion batteries cathode materials. *Journal of Environmental Chemical Engineering, 10*(3), Article 107745. https://doi.org/10.1016/j.jece.2022.107745

Guan, J., et al. (2022). Microwave-assisted chlorination extraction of valuable metals from spent power ternary lithium-ion batteries. *International Journal of Environmental & Analytical Chemistry*, 1–14. https://doi.org/10.1080/03067319.2022.2098482

Guo, Y., et al. (2024). Experimental and molecular insights into ionic liquid-based recovery of valuable metals from spent lithium-ion batteries. *Chemical Engineering Science, 302*, Article 120895. https://doi.org/10.1016/j.ces.2024.120895

Hadoune, A., et al. (2023). Optimizing direct power control of DFIG-based WECS using super-twisting algorithm under real wind profile. *Frontiers in Energy Research, 11*. https://doi.org/10.3389/fenrg.2023.1261902

Hagelüken, C., & Goldmann, D. (2022). Recycling and circular economy XE "Circular economy"—Towards a closed loop for metals in emerging clean technologies. *Mineral Economics, 35*(3–4), 539–562. https://doi.org/10.1007/s13563-022-00319-1

Hamid, C., et al. (2024). Robust control system for DFIG-based WECS and energy storage in reel wind conditions. *EAI Endorsed Transactions on Energy Web, 11*. https://doi.org/10.4108/ew.4856

Harper, G., et al. (2019). Recycling lithium-ion batteries from electric vehicles. *Nature, 575*(7781), 75–86. https://doi.org/10.1038/s41586-019-1682-5

Hasan, M. M., et al. (2025). Advancing energy storage: The future trajectory of lithium-ion battery technologies. *Journal of Energy Storage, 120*, Article 116511. https://doi.org/10.1016/j.est.2025.116511

He, Y., et al. (2016). Recovery of $LiCoO_2$ and graphite from spent lithium-ion batteries by Fenton reagent-assisted flotation. *Journal of Cleaner Production, 143*, 319–325. https://doi.org/10.1016/j.jclepro.2016.12.106

Horeh, N. B., Mousavi, S. M., & Shojaosadati, S. A. (2016). Bioleaching of valuable metals from spent lithium-ion mobile phone batteries using *Aspergillus niger*. *Journal of Power Sources, 320*, 257–266. https://doi.org/10.1016/j.jpowsour.2016.04.104

Hu, X., Mousa, E., & Ye, G. (2020). Recovery of Co, Ni, Mn, and Li from Li-ion batteries by smelting reduction—Part II: A pilot-scale demonstration. *Journal of Power Sources, 483*, Article 229089. https://doi.org/10.1016/j.jpowsour.2020.229089

International Energy Agency (IEA). (2023). World energy outlook 2023. https://www.iea.org/reports/world-energy-outlook-2023

Jafarizadeh, H., et al. (2024). Navigating challenges in large-scale renewable energy storage: Barriers, solutions, and innovations. *Energy Reports, 12*, 2179–2192. https://doi.org/10.1016/j.egyr.2024.02.123

Jiang, H., et al. (2022). Study on the thermal reduction effect of organic components in spent ternary lithium battery cathode active materials. *Waste Management, 148*, 33–42. https://doi.org/10.1016/j.wasman.2022.05.011

Kendall, A., et al. (2023). Electric vehicle lithium-ion batteries in lower- and middle-income countries: Life cycle impacts and issues. https://escholarship.org/uc/item/7m2536mp

Koech, A. K., et al. (2024). Lithium-ion battery fundamentals and exploration of cathode materials: A review. *South African Journal of Chemical Engineering, 50*, 321–339. https://doi.org/10.1016/j.sajce.2024.09.008

Koroma, M. S., et al. (2022). Life cycle assessment of battery electric vehicles: Implications of future electricity mix and different battery end-of-life management. *The Science of the Total Environment, 831*, Article 154859. https://doi.org/10.1016/j.scitotenv.2022.154859

Kulkarni, S. J. (2015). A review on studies and research on various aspects of leaching. *International Journal of Research and Review, 2*(9), 579–583. https://ijrrjournal.com/IJRR_Vol.2_Issue.9_Sep2015/Kulkarni.html

Kumar, A., & Rizvi, F. F. (2024)/ Hazardous effects of battery waste and role of the Battery Waste Management Rule 2022 in enhancing energy efficiency. *International Journal of Current Science Research and Review, 07*(11). https://doi.org/10.47191/ijcsrr/v7-i11-14

Kumar, R., Lamba, K., & Raman, A. (2021). Role of zero emission vehicles in sustainable transformation of the Indian automobile industry. *Research in Transportation Economics, 90*, Article 101064. https://doi.org/10.1016/j.retrec.2021.101064

Kwade, A., et al. (2022). Comminution and classification as important process steps for the circular production of lithium batteries. *KONA Powder and Particle Journal, 40*(0), 50–73. https://doi.org/10.14356/kona.2023006

Lee, C., et al. (2025). Hydrometallurgical process of spent lithium-ion battery recycling Part. 1 Chemical leaching of valuable metals from cathode active materials: Review and case study. *Hydrometallurgy*, 106494. https://doi.org/10.1016/j.hydromet.2025.106494

Lee, C. K., & Rhee, K.-I. (2002). Preparation of $LiCoO_2$ from spent lithium-ion batteries. *Journal of Power Sources, 109*(1), 17–21. https://doi.org/10.1016/s0378-7753(02)00037-x

Lee, J., Zhang, Q., & Saito, F. (2000). Room temperature extraction of Co and Li from ground lithium-ion secondary battery scrap. *Shigen-to-Sozai, 116*(11), 919–922. https://doi.org/10.2473/shigentosozai.116.919

Li, J., et al. (2025). A critical review on the advances in thermal reduction technology for recycling spent lithium-ion batteries. *Next Materials, 8*, Article 100641. https://doi.org/10.1016/j.nxmate.2025.100641

Li, J., Wang, G., & Xu, Z. (2016). Generation and detection of metal ions and volatile organic compounds (VOCs) emissions from the pretreatment processes for recycling spent lithium-ion batteries. *Waste Management, 52*, 221–227. https://doi.org/10.1016/j.wasman.2016.03.011

Li, L., et al. (2009). Recovery of cobalt and lithium from spent lithium ion batteries using organic citric acid as leachant. *Journal of Hazardous Materials, 176*(1–3), 288–293. https://doi.org/10.1016/j.jhazmat.2009.11.026

Li, L., et al. (2012). Ascorbic-acid-assisted recovery of cobalt and lithium from spent Li-ion batteries. *Journal of Power Sources, 218*, 21–27. https://doi.org/10.1016/j.jpowsour.2012.06.068

Li, L., et al. (2017a). Process for recycling mixed-cathode materials from spent lithium-ion batteries and kinetics of leaching. *Waste Management, 71*, 362–371. https://doi.org/10.1016/j.wasman.2017.10.028

Li, L., et al. (2018). The recycling of spent lithium-ion batteries: A review of current processes and technologies. *Electrochemical Energy Reviews, 1*(4), 461–482. https://doi.org/10.1007/s41918-018-0012-1

Li, M., et al. (2010). Solubilities of NACL, KCL, LICL, and LIBR in methanol, ethanol, acetone, and mixed solvents and correlation using the LIQUAC model. *Industrial & Engineering Chemistry Research, 49*(10), 4981–4988. https://doi.org/10.1021/ie100027c

Li, M.-Y., et al. (2011). Experimental measurement and modeling of solubility of LiBr and $LiNO_3$ in methanol, ethanol, 1-propanol, 2-propanol and 1-butanol. *Fluid Phase Equilibria, 307*(1), 104–109. https://doi.org/10.1016/j.fluid.2011.03.017

Li, W., Song, B., & Manthiram, A. (2017b). High-voltage positive electrode materials for lithium-ion batteries. *Chemical Society Reviews, 46*(10), 3006–3059. https://doi.org/10.1039/c6cs00875e

Liang, G., et al. (2023). Microwave-assisted sulfation method for lithium recovery from spent $LiNixCoyMnzO_2$ cathode material: Process intensification and conversion mechanism. *ACS Sustainable Chemistry & Engineering, 11*(34), 12484–12493. https://doi.org/10.1021/acssuschemeng.3c00343

Lim, J.-E., et al. (2018). Binder-free hybrid $Li_4Ti_5O_{12}$ anode for high performance lithium-ion batteries. *Electrochimica Acta, 282*, 270–275. https://doi.org/10.1016/j.electacta.2018.06.057

Lin, L., Lu, Z., & Zhang, W. (2021). Recovery of lithium and cobalt from spent Lithium- Ion batteries using organic aqua regia (OAR): Assessment of leaching kinetics and global warming potentials. *Resources Conservation and Recycling, 167*, Article 105416. https://doi.org/10.1016/j.resconrec.2021.105416

Liu, C., et al. (2019). Recycling of spent lithium-ion batteries in view of lithium recovery: A critical review. *Journal of Cleaner Production, 228*, 801–813. https://doi.org/10.1016/j.jclepro.2019.04.304

Liu, C., et al. (2023a). Microwave low-temperature treatment—Step leaching process for recovering black mass from spent lithium-ion batteries. *Journal of Environmental Chemical Engineering, 11*(3), Article 109759. https://doi.org/10.1016/j.jece.2023.109759

Liu, P., et al. (2018). Study on the reduction roasting of spent LiNixCoyMnzO2 lithium-ion battery cathode materials. *Journal of Thermal Analysis and Calorimetry, 136*(3), 1323–1332. https://doi.org/10.1007/s10973-018-7732-7

Liu, Y., Shi, H., & Wu, Z.-S. (2023b). Recent status, key strategies and challenging perspectives of fast-charging graphite anodes for lithium-ion batteries. *Energy & Environmental Science, 16*(11), 4834–4871. https://doi.org/10.1039/d3ee02213g

Lombardo, G., et al. (2020). Incineration of EV Lithium-ion batteries as a pretreatment for recycling—Determination of the potential formation of hazardous by-products and effects on metal compounds. *Journal of Hazardous Materials, 393*, Article 122372. https://doi.org/10.1016/j.jhazmat.2020.122372

Loulijat, A., et al. (2024). Enhancing fault ride-through capacity of DFIG-based WPs by adaptive backstepping command using parametric estimation in non-linear forward power controller design. *IEEE Access, 12*, 58675–58689. https://doi.org/10.1109/access.2024.3381613

Ma, Y., et al. (2020). A promising selective recovery process of valuable metals from spent lithium ion batteries via reduction roasting and ammonia leaching. *Journal of Hazardous Materials, 402*, Article 123491. https://doi.org/10.1016/j.jhazmat.2020.123491

Machala, M. L., et al. (2025). Life cycle comparison of industrial-scale lithium-ion battery recycling and mining supply chains. *Nature Communications, 16*(1). https://doi.org/10.1038/s41467-025-56063-x

Marsh, K. N., Boxall, J. A., & Lichtenthaler, R. (2004). Room temperature ionic liquids and their mixtures—A review. *Fluid Phase Equilibria, 219*(1), 93–98. https://doi.org/10.1016/j.fluid.2004.02.003

Martin, G., et al. (2016). Lithium market research—Global supply, future demand and price development. *Energy Storage Materials, 6*, 171–179. https://doi.org/10.1016/j.ensm.2016.11.004

Men, L., et al. (2023). A systematic review of efficient recycling for the cathode materials of spent lithium-ion batteries: Process intensification technologies beyond traditional methods. *Green Chemistry, 26*(3), 1170–1193. https://doi.org/10.1039/d3gc04088g

Meng, Q., Zhang, Y., & Dong, P. (2017) Use of glucose as reductant to recover Co from spent lithium ions batteries. *Waste Management, 64*, 214–218. https://doi.org/10.1016/j.wasman.2017.03.017

Mishra, G., et al. (2022). A review on recycling of lithium-ion batteries to recover critical metals. *Journal of Environmental Chemical Engineering, 10*(6), Article 108534. https://doi.org/10.1016/j.jece.2022.108534

Mohanty, A., & Devi, N. (2023). Separation of critical metals using trihexyl(tetradecyl) phosphonium thiocyanate from spent lithium-ion battery. *Transactions of the Indian Institute of Metals, 77*(1), 119–125. https://doi.org/10.1007/s12666-023-03065-2

Mohanty, A., et al. (2022). Selective recovery and intensification of Mn from spent $LiMn_2O_4$ using sulfuric acid as lixiviant and Na-D2EHPA as extractant. *Geosystem Engineering, 25*(5–6), 246–255. https://doi.org/10.1080/12269328.2022.2127426

Mukhopadhyay, A., et al. (2024). Electrochemically assisted (bio)leaching of end-of-life lithium-ion batteries for critical metals recovery. *ACS Sustainable Chemistry & Engineering, 12*(37), 14119–14127. https://doi.org/10.1021/acssuschemeng.4c06090

Muzayanha, S. U., et al. (2019). A fast metals recovery method for the synthesis of lithium nickel cobalt aluminum oxide material from cathode waste. *Metals, 9*(5), 615. https://doi.org/10.3390/met9050615

Nayl, A. A., et al. (2017). Acid leaching of mixed spent Li-ion batteries. *Arabian Journal of Chemistry, 10*, S3632–S3639. https://doi.org/10.1016/j.arabjc.2014.04.001

Neumann, J., et al. (2022). Recycling of lithium-ion batteries—Current state of the art, circular economy, and next generation recycling. *Advanced Energy Materials, 12*(17), 2102917. https://doi.org/10.1002/aenm.202102917

Nie, H., et al. (2014). $LiCoO_2$: Recycling from spent batteries and regeneration with solid state synthesis. *Green Chemistry, 17*(2), 1276–1280. https://doi.org/10.1039/c4gc01951b

Nitta, N., et al. (2014). Li-ion battery materials: Present and future. *Materials Today, 18*(5), 252–264. https://doi.org/10.1016/j.mattod.2014.10.040

Ordoñez, J., Gago, E. J., & Girard, A. (2016). Processes and technologies for the recycling and recovery of spent lithium-ion batteries. *Renewable and Sustainable Energy Reviews, 60*, 195–205. https://doi.org/10.1016/j.rser.2015.12.363

Parker, S. S., Clifford, M. J., & Cohen, B. S. (2023). Potential impacts of proposed lithium extraction on biodiversity and conservation in the contiguous United States. *The Science of the Total Environment, 911*, Article 168639. https://doi.org/10.1016/j.scitotenv.2023.168639

Peeters, N., Binnemans, K., & Riaño, S. (2022). Recovery of cobalt from lithium-ion battery cathode material by combining solvoleaching and solvent extraction. *Green Chemistry, 24*(7), 2839–2852. https://doi.org/10.1039/d1gc03776e

Peng, C., et al. (2018). Selective reductive leaching of cobalt and lithium from industrially crushed waste Li-ion batteries in sulfuric acid system. *Waste Management, 76*, 582–590. https://doi.org/10.1016/j.wasman.2018.02.052

Pilali, E., et al. (2025). SWOT analysis on the transition from Lithium-Ion batteries to Sodium-Ion batteries. *Sustainable Energy Technologies and Assessments, 80*, Article 104371. https://doi.org/10.1016/j.seta.2025.104371

Prasetyo, E., et al. (2022). Tannic acid as a novel and green leaching reagent for cobalt and lithium recycling from spent lithium-ion batteries. *Journal of Material Cycles and Waste Management, 24*(3), 927–938. https://doi.org/10.1007/s10163-022-01368-y

Qu, G., Li, B., & Wei, Y. (2022). A novel approach for the recovery and cyclic utilization of valuable metals by co-smelting spent lithium-ion batteries with copper slag. *Chemical Engineering Journal, 451*, Article 138897. https://doi.org/10.1016/j.cej.2022.138897

Rakshit, P. S., et al. (2025). End-of-life battery management in India's EV sector: Environmental law, biodiversity risk, and regulatory gaps. *International Journal of Environmental Sciences*, 1048–1060. https://doi.org/10.64252/r00xbb81

Ren, G.-X., et al. (2017). Recovery of valuable metals from spent lithium ion batteries by smelting reduction process based on $FeO–SiO_2–Al_2O_3$ slag system. *Transactions of Nonferrous Metals Society of China, 27*(2), 450–456. https://doi.org/10.1016/s1003-6326(17)60051-7

Rezaei, M., et al. (2025). A review of lithium-ion battery recycling for enabling a circular economy. *Journal of Power Sources, 630*, Article 236157. https://doi.org/10.1016/j.jpowsour.2024.236157

Roshanfar, M., Golmohammadzadeh, R., & Rashchi, F. (2018). An environmentally friendly method for recovery of lithium and cobalt from spent lithium-ion batteries using gluconic and lactic acids. *Journal of Environmental Chemical Engineering, 7*(1), Article 102794. https://doi.org/10.1016/j.jece.2018.11.039

Roy, J. J., et al. (2021). Green recycling methods to treat lithium-ion batteries e-waste: A circular approach to sustainability. *Advanced Materials, 34*(25). https://doi.org/10.1002/adma.202103346

Saaid, F. I., et al. (2023). Ni-rich lithium nickel manganese cobalt oxide cathode materials: A review on the synthesis methods and their electrochemical performances. *Heliyon, 10*(1), Article e23968. https://doi.org/10.1016/j.heliyon.2023.e23968

Saeki, S., et al. (2004). Co-grinding $LiCoO_2$ with PVC and water leaching of metal chlorides formed in ground product. *International Journal of Mineral Processing, 74*, S373–S378. https://doi.org/10.1016/j.minpro.2004.08.002

Sahu, S., & Devi, N. (2022). Effective leaching of spent lithium-ion batteries using DL-lactic acid as lixiviant and selective separation of metals through precipitation and solvent extraction. *Environmental Science and Pollution Research, 30*(39), 90152–90167. https://doi.org/10.1007/s11356-022-24560-x

Sahu, S., & Devi, N. (2023). Hydrometallurgical treatment of spent lithium ion batteries using environmentally friendly leachant and extractant. *Journal of Material Cycles and Waste Management, 25*(6), 3303–3315. https://doi.org/10.1007/s10163-023-01754-0

Saito, F., Zhang, Q., & Kano, J. (2024). Mechanochemical dechlorination of waste PVC resin and feedstock recycling. *Powder Technology*, 120330. https://doi.org/10.1016/j.powtec.2024.120330

Scrosati, B. (2011). History of lithium batteries. *Journal of Solid State Electrochemistry, 15*(7–8), 1623–1630. https://doi.org/10.1007/s10008-011-1386-8

Shen, X., et al. (2017). Research progress on silicon/carbon composite anode materials for lithium-ion battery. *Journal of Energy Chemistry, 27*(4), 1067–1090. https://doi.org/10.1016/j.jechem.2017.12.012

Shi, H., Zhang, J., & Ou, L. (2025). A comprehensive review: Evaluating emerging green leaching technologies for recycling spent lithium-ion batteries. *Chemical Engineering Journal*, 160006. https://doi.org/10.1016/j.cej.2025.160006

Shi, J., et al. (2019). Sulfation roasting mechanism for spent Lithium-Ion battery metal oxides under SO2-O2-AR atmosphere. *JOM Journal of the Minerals Metals and Materials Society, 71*(12), 4473–4482. https://doi.org/10.1007/s11837-019-03800-5

Shih, Y.-J., et al. (2019). Chemical leaching, precipitation and solvent extraction for sequential separation of valuable metals in cathode material of spent lithium ion batteries. *Journal of the Taiwan Institute of Chemical Engineers, 100*, 151–159. https://doi.org/10.1016/j.jtice.2019.04.017

Shin, S. M., et al. (2005). Development of a metal recovery process from Li-ion battery wastes. *Hydrometallurgy, 79*(3–4), 172–181. https://doi.org/10.1016/j.hydromet.2005.06.004

Sidiq, A. L., et al. (2022). NCM cathode active materials reproduced from end-of-life Li-ion batteries using a simple and green hydrometallurgical recycling process. *Materials Research Bulletin, 153*, Article 111901. https://doi.org/10.1016/j.materresbull.2022.111901

Singh, S. K., & Savoy, A. W. (2019). Ionic liquids synthesis and applications: An overview. *Journal of Molecular Liquids, 297*, Article 112038. https://doi.org/10.1016/j.molliq.2019.112038

Sobianowska-Turek, A., et al. (2021). The necessity of recycling of waste Li-Ion batteries used in electric vehicles as objects posing a threat to human health and the environment. *Recycling, 6*(2), 35. https://doi.org/10.3390/recycling6020035

Song, D., et al. (2021). Characteristic comparison of leaching valuable metals from spent power Li-ion batteries for vehicles using the inorganic and organic acid system. *Journal of Environmental Chemical Engineering, 10*(1), Article 107102. https://doi.org/10.1016/j.jece.2021.107102

Srinivasan, S., Shanthakumar, S., & Ashok, B. (2024). Sustainable lithium-ion battery recycling: A review on technologies, regulatory approaches and future trends. *Energy Reports, 13*, 789–812. https://doi.org/10.1016/j.egyr.2024.12.043

Sun, J., et al. (2023). Progress on the microwave-assisted recycling of spent lithium battery graphite. *Processes, 11*(5), 1451. https://doi.org/10.3390/pr11051451

Takacova, Z., et al. (2016). Cobalt and lithium recovery from active mass of spent Li-ion batteries: Theoretical and experimental approach. *Hydrometallurgy, 163*, 9–17. https://doi.org/10.1016/j.hydromet.2016.03.007

Takahashi, V. C. I., et al. (2020). Enhancing cobalt recovery from Li-ion batteries using grinding treatment prior to the leaching and solvent extraction process. *Journal of Environmental Chemical Engineering, 8*(3), Article 103801. https://doi.org/10.1016/j.jece.2020.103801

Tang, Y., et al. (2015). Rational material design for ultrafast rechargeable lithium-ion batteries. *Chemical Society Reviews, 44*(17), 5926–5940. https://doi.org/10.1039/c4cs00442f

Tang, Y., et al. (2020). Recovery and regeneration of lithium cobalt oxide from spent lithium-ion batteries through a low-temperature ammonium sulfate roasting approach. *Journal of Power Sources, 474*, Article 228596. https://doi.org/10.1016/j.jpowsour.2020.228596

Theron, M. M., & Lues, J. F. R. (2007). Organic acids and meat preservation: A review. *Food Reviews International, 23*(2), 141–158. https://doi.org/10.1080/87559120701224964

Tong, Z., et al. (2024). Advances in lithium-ion battery recycling: Strategies, pathways, and technologies. ChemPhysMater [Preprint]. https://doi.org/10.1016/j.chphma.2024.05.005

Tongamp, W., Zhang, Q., & Saito, F. (2006). Mechanochemical decomposition of PVC by using La_2O_3 as additive. *Journal of Hazardous Materials, 137*(2), 1226–1230. https://doi.org/10.1016/j.jhazmat.2006.04.013

Tran, T. T., Liu, Y., & Lee, M. S. (2020). Recovery of pure molybdenum and vanadium compounds from spent petroleum catalysts by treatment with ionic liquid solution in the presence of oxidizing agent. *Separation and Purification Technology, 255*, Article 117734. https://doi.org/10.1016/j.seppur.2020.117734

Tran, T. T., Moon, H. S., & Lee, M. S. (2021). Co, Ni, Cu, Fe, and Mn integrated recovery process via sulfuric acid leaching from spent lithium-ion batteries smelted reduction metallic alloys. *Mineral Processing and Extractive Metallurgy Review, 43*(8), 954–968. https://doi.org/10.1080/08827508.2021.1979541

Vera, M. L., et al. (2023). Environmental impact of direct lithium extraction from brines. *Nature Reviews Earth & Environment, 4*(3), 149–165. https://doi.org/10.1038/s43017-022-00387-5

Vieceli, N., et al. (2021). Hydrometallurgical recycling of EV lithium-ion batteries: Effects of incineration on the leaching efficiency of metals using sulfuric acid. *Waste Management, 125*, 192–203. https://doi.org/10.1016/j.wasman.2021.02.039

Wang, K., et al. (2022a). Separation of Co and Mn from acetic acid leaching solution of spent lithium-ion battery by Cyanex272. *Journal of Environmental Chemical Engineering, 10*(5), Article 108250. https://doi.org/10.1016/j.jece.2022.108250

Wang, K., et al. (2023). Separation of metals from acetic acid leaching solution of spent lithium-ion batteries by ionic liquid system. *Chemical Engineering Journal, 472*, Article 145006. https://doi.org/10.1016/j.cej.2023.145006

Wang, M., et al. (2022b). Recycling spent lithium-ion batteries using a mechanochemical approach. *Circular Economy, 1*(2), Article 100012. https://doi.org/10.1016/j.cec.2022.100012

Wang, X., Gaustad, G., & Babbitt, C. W. (2016). Targeting high value metals in lithium-ion battery recycling via shredding and size-based separation. *Waste Management, 51*, 204–213. https://doi.org/10.1016/j.wasman.2015.10.026

Wang, M. M., Zhang, C. C., & Zhang, F. S. (2017). Recycling of spent lithium-ion battery with polyvinyl chloride by mechanochemical process. *Waste Management, 67*, 232–239. https://doi.org/10.1016/j.wasman.2017.05.013

Welton, T. (2018). Ionic liquids: A brief history. *Biophysical Reviews, 10*(3), 691–706. https://doi.org/10.1007/s12551-018-0419-2

Winslow, K. M., Laux, S. J., & Townsend, T. G. (2017). A review on the growing concern and potential management strategies of waste lithium-ion batteries. *Resources Conservation and Recycling, 129*, 263–277. https://doi.org/10.1016/j.resconrec.2017.11.001

Xu, J., et al. (2007). A review of processes and technologies for the recycling of lithium-ion secondary batteries. *Journal of Power Sources, 177*(2), 512–527. https://doi.org/10.1016/j.jpowsour.2007.11.074

Xu, J., et al. (2023). Electrolyte design for Li-ion batteries under extreme operating conditions. *Nature, 614*(7949), 694–700. https://doi.org/10.1038/s41586-022-05627-8

Xu, L., Chen, C., & Fu, M.-L. (2020). Separation of cobalt and lithium from spent lithium-ion battery leach liquors by ionic liquid extraction using Cyphos IL-101. *Hydrometallurgy, 197*, Article 105439. https://doi.org/10.1016/j.hydromet.2020.105439

Yang, L., et al. (2021). Closed-loop regeneration of battery-grade FePO4 from lithium extraction slag of spent Li-ion batteries via phosphoric acid mixture selective leaching. *Chemical Engineering Journal, 431*, Article 133232. https://doi.org/10.1016/j.cej.2021.133232

Yoshino, A. (2014). *Development of the Lithium-Ion battery and recent technological trends* (pp. 1–20). Elsevier eBooks. https://doi.org/10.1016/b978-0-444-59513-3.00001-7

You, J., et al. (2023). Recycling valuable metals from spent lithium-ion battery cathode materials based on microwave-assisted hydrogen reduction followed by grind-leaching and magnetic separation. *Journal of Cleaner Production, 428*, Article 139488. https://doi.org/10.1016/j.jclepro.2023.139488

Yu, J., et al. (2017). A promising physical method for recovery of $LiCoO_2$ and graphite from spent lithium-ion batteries: Grinding flotation. *Separation and Purification Technology, 190*, 45–52. https://doi.org/10.1016/j.seppur.2017.08.049

Zeng, X., et al. (2021). Comparing the costs and benefits of virgin and urban mining. *Journal of Management Science and Engineering, 7*(1), 98–106. https://doi.org/10.1016/j.jmse.2021.05.002

Zeng, X., Li, J., & Singh, N. (2013). Recycling of spent Lithium-Ion battery: A critical review. *Critical Reviews in Environmental Science and Technology, 44*(10), 1129–1165. https://doi.org/10.1080/10643389.2013.763578

Zhang, K., et al. (2022). Recovery of metals from sulfate leach solutions of spent ternary lithium-ion batteries by precipitation with phosphate and solvent extraction with P507. *Hydrometallurgy, 210*, Article 105861. https://doi.org/10.1016/j.hydromet.2022.105861

Zhang, Q., et al. (2000). Room temperature acid extraction of Co from $LiCo_{0.2}Ni_{0.8}O_2$ scrap by a mechanochemical treatment. *Advanced Powder Technology, 11*(3), 353–359. https://doi.org/10.1163/156855200750172222

Zhang, X., et al. (2021). Recovery valuable metals from spent lithium-ion batteries via a low-temperature roasting approach: Thermodynamics and conversion mechanism. *Journal of Hazardous Materials Advances, 1*, Article 100003. https://doi.org/10.1016/j.hazadv.2021.100003

Zhang, Y., et al. (2019). Improved recovery of valuable metals from spent lithium-ion batteries by efficient reduction roasting and facile acid leaching. *Waste Management, 102*, 847–855. https://doi.org/10.1016/j.wasman.2019.11.045

Zhao, Y., et al. (2020). Microwave pyrolysis of macadamia shells for efficiently recycling lithium from spent lithium-ion batteries. *Journal of Hazardous Materials, 396*, Article 122740.

Zheng, H., et al. (2021). Selective extraction of lithium from spent lithium batteries by functional ionic liquid. *ACS Sustainable Chemistry & Engineering, 9*(20), 7022–7029. https://doi.org/10.1021/acssuschemeng.1c00718

Zheng, S., et al. (2024). A review of cathode and electrolyte recovery from spent lithium-ion batteries: Recent technologies, processes and policies. *Resources Chemicals and Materials, 3*(3), 188–229.

Zheng, X., et al. (2018). A mini-review on metal recycling from spent lithium ion batteries. *Engineering, 4*(3), 361–370. https://doi.org/10.1016/j.eng.2018.05.018

Zhou, X., et al. (2010). Recycling of electrode materials from spent lithium-ion batteries. In *4th International Conference on Bioinformatics and Biomedical Engineering (ICBBE)* (pp. 1–4). IEEE. https://doi.org/10.1109/ICBBE.2010.5518015

Zhu, S., et al. (2011). Recovering copper from spent lithium ion battery by a mechanical separation process. In *International Conference on Materials for Renewable Energy & Environment (ICMREE)* (Vol. 1, pp. 1008–1012). IEEE. https://doi.org/10.1109/ICMREE.2011.5930972

Zhu, X., et al. (2024). Improved recovery of cathode materials and enhanced lithium selective extraction from spent $LiNi_{0.5}Co_{0.2}Mn_{0.3}O_2$ batteries via $CaCl_2$-assisted microwave roasting. J. Environ. Chem. Eng. **12**(2), 112037. https://doi.org/10.1016/j.jece.2024.112037

Management of Solid Waste in the Construction Materials for the Circular Economy

Deepak Chandra Joshi, Anurag Sharma, Lavkush Verma, and Keshav Raj Paudel

1 Introduction

In 2022, the construction and demolition activities contributed 30–40% of the total solid waste in the world, with China making the most with more than 2360 million metric tonnes per year (Wang et al., 2021) and the United States producing about 600 million tonnes per year (Patil et al., 2024). Nevertheless, 1% of this waste is recycled, which indicates that the waste is located in landfills, contaminated the air and water, and into the air and water released toxins and particulate matter (Bonifazi et al., 2025). When the disposal of resources is done in a manner that does not preserve the resources, there is an increase in the emission of the greenhouse gases and there is an accelerated consumption of the resources since raw materials are constantly being extracted. Higher landfill costs, loss of material recovery opportunities, and wastage of valuable resources such as metals and concrete that can be recycled are bad to the economy as they result in poor health and lifestyle among people living near the disposal sites and poorer cities (Purchase et al., 2021). There is a tendency to engage in illegal dumping and burning, the problem is aggravated by the deficiencies of the policy framework and the unawareness, particularly in the developing world. The circular economy (CE) in construction attempts to restructure

D. C. Joshi (✉)
Department of Pharmacy, Faculty of Pharmaceutical sciences and Nursing, Vivekananda Global University, Jaipur, Rajasthan, India

A. Sharma
Invertis Institute of Pharmacy, Invertis University, Bareilly, Uttar Pradesh, India

L. Verma
School of Pharmacy, Chandigarh University, Unnao, Uttar Pradesh, India

K. R. Paudel
School of Mechanical and Mechatronic Engineering, University of Technology, 15 Broadway, Sydney, Ultimo, NSW 2007, Australia

H. Gupta et al. (eds.), *Advanced Technologies for Assessment, Recycling, and Management of Solid Waste*, Waste as a Resource,
https://doi.org/10.1007/978-3-032-24788-9_10

the conventional linear approaches with the aim of reducing, reusing, and recycling of resources, which utilises resources more efficiently and reduces the production of waste. This approach puts sustainable design, new materials, and closed-loop systems keeping construction materials in operation as long as possible in the centre of the list (Siddiqua et al., 2022). It promotes cradle-to-cradle policies and application of new recycling technologies. To expand in the industry, CE principles require regulation backing and stakeholder determination (Gamage et al., 2024). This review gives a structured look at the diverse types of waste, where they come from, and the problems that come with managing them. It also looks at innovative ideas, technological tools, and policy frameworks that can help the construction industry use resources more efficiently and protect the environment.

2 Types and Sources of Solid Waste in Construction

The construction sector generates vast amounts of solid waste that pose significant environmental, economic, and social challenges. To strategically manage this waste and transition towards sustainable practices like the circular economy, it is essential to understand the detailed classification, major sources, components, and the physical and chemical characteristics of construction and demolition (C&DW) waste. There are three main types of construction waste: waste from new construction, waste from demolition, and waste from renovation or remodelling. Each category has its own types of waste that are specific to the stage of the construction lifecycle and need to be managed in diverse ways.

1. **Construction Waste (New Builds)**: Construction waste refers to waste that is produced whenever new buildings are being constructed. This incorporates additional materials, scraps, packaging wastes, and dirt during digging and grading. Wastes such as concrete debris, scraps of wood, scraps of metal, packaging materials, waste plaster, and waste plasterboard are common. It occurs when individuals order excessively, break items during their management or storage, fix the design, or when items are not performed correctly at the location (MATERIAL WASTE IN CONSTRUCTION PROJECTS: CAUSES and EFFECTS, 2023). According to the research, construction waste constitutes approximately 1015% of the overall C&DW stream of waste. Particular attention should be paid to addressing this kind of waste since certain materials may be recycled provided that they are segregated on-site. People are attempting to reduce this waste by just-in-time delivery, designing that optimisation, modular construction, and sorting protocols that take place on the site.
2. **Demolition Waste**: In the world, demolition debris represents a huge percentage of construction and demolition (C&D) waste. It is produced by demolishing structures like buildings, roads, bridges, and other infrastructure. The demolition waste is normally comprised of a combination of diverse (differing) materials, such as concrete, bricks, metals, wood, asphalt, and other building materials.

This waste also tends to be more polluted and is more of a blend of various types of waste, and thus, it becomes difficult to separate and recycle. Demolition work is a source of about 60–70% C&DW and is increasingly growing in the world due to redevelopment and renewal of infrastructures. It is possible to recycle concrete and masonry and reuse them as aggregate, recycle metals, and recycle or convert wood to energy, unless it has been contaminated. However, you have to be extremely cautious of such pollutants as asbestos, lead paint, and chemical sealants to prevent damaging the environment (Abedin Khan et al., 2024).

3. **Renovation and Remodelling Waste**: Wastes that are generated during the renovation and remodelling are the garbage that results during the process of renovating and upgrading old buildings. It is not similar to new building and demolition trash. It is mostly made up of small fragments of demolition waste, wrapping paper, old fixtures, tiles, and materials that have been removed to make room to install new systems (Ponnada & Kameswari, 2015). This kind of rubbish is increasing rapidly particularly in metropolitan areas due to improved living standards and creation of stricter regulations making remodelling to occur. Renovation waste is sometimes very difficult to separate since it can contain hazardous substances such as adhesives, paints, solvents, and among others. It includes approximately 1520% of all construction and demolition waste, and it is significant to the models of the circular economy as the materials are in good condition and may be reused or recycled in the provided case of their recovery (Sakthibala et al., 2025).

2.1 Major Sources and Components of C&DW Waste

The recover and come up with good management plans, you need to know what the main parts of C&DW waste are (Fig. 1). These parts are usually grouped by the type of material they are made of because they are used in so many construction projects.

Concrete and Cementitious Materials

Concrete and other cement-based materials make up the biggest and heaviest parts of C&DW waste. In reports, they make up more than 50% of the weight. These materials are mostly broken concrete, reinforced concrete, cement blocks, mortar, bricks, and rubble. Their high bulk density makes them expensive to move and throw away, but it also means that they can be recycled in a big way. People often use crushed concrete as recycled aggregate in road sub-bases, backfilling, and sometimes even as a partial replacement in new concrete mixtures (Shajidha & Mortula, 2025). But if recycled aggregate is contaminated with reinforcing steel, gypsum, or adhesives, its quality can go down, and it may not be able to be used again. Recent studies stress the importance of better sorting technologies and pre-processing to raise recycling rates (Pineda-Martos et al., 2025).

Fig. 1 Circular flow of major sources, components, and recovery pathways of Construction and Demolition Waste (C&DW)

Metals and Alloys

Metals, especially steel, aluminium, and copper, make up a smaller part of the total weight but are worth a lot of money and can be recycled. Common types of metal waste are steel reinforcement bars (rebars), structural steel elements, wiring, and mechanical fixtures. It's especially important to keep metals separate because they can be recycled repeatedly without losing quality. This saves energy and raw materials compared to making new metal. In many places, more than 80% of the metal in

C&DW waste is recycled. This is because there is a high demand for recycled metal and well-established recycling supply chains (Bonifazi et al., 2025). Metals can be mixed in with or stuck to other types of waste, so it's important to have good separation technologies and on-site dismantling methods to get the most out of them (Pallewatta et al., 2023).

Wood, Glass, Ceramics, and Plastics

Form work, framing, flooring, and furniture disassembly all create wood waste. The quality of the wood varies from clean, untreated wood that can be reused or chipped to treated or painted wood that needs to be handled carefully. Windows, facades, and interior partitions are common sources of glass waste. It is often broken up and mixed with other trash, making it harder to recycle. Ceramics, like tiles, sanitary ware, and bricks, are more fragile, but they can be broken up and used again as road base or fill material (Ponnada & Kameswari, 2015). Plastic is used for things like packaging, pipes, insulation, and vapour barriers. It can be hard to work with because there are so many diverse types of polymers, and they can get dirty. Some plastics can be recycled mechanically, but many need to be down cycled or have their energy recovered because there isn't enough sorting infrastructure.

2.2 Physical and Chemical Characteristics of Typical Construction Waste

It is particularly essential to define C&DW waste physically and chemically to design the recycling process, anticipate the impact it will have on the environment, and ensure that they abide by the regulations in a circular economy. The construction waste materials vary in terms of their size, density, moisture content, and the contamination levels. In the case of concrete and masonry waste, the dimensions of particles may be either fine dust created during crushing of the substance or even large chunks bigger than 100 mm. The bulk densities of such wastes lie between 1, 200 and 2, 400 kg/m3, depending on the manner in which they are packed, as well as their composition. Metals are heavy and vary in elemental properties as per the type of alloy. The density of wood wastes is normally low (approximately 400–600 kg/m3) yet the moisture content is high thus making the wood wastes less easy to work with and more difficult to recycle (Baggio et al., 2024). Silica (SiO2), calcium oxide (CaO), aluminium oxide (Al2O3), and iron oxide (Fe2O3) is some of the oxides contained in construction waste. They contain hydrated calcium silicates and aluminates which are cementitious wastes. Should the materials, when reused in concrete or other building works, they react with other materials. They are complicated by such contaminants as asbestos fibres, heavy metals, plastic additives, paints, and solvents and require extraordinary care. Typical methods of figuring out what elements and minerals are in various types of waste include X-ray diffraction (XRD), X-ray fluorescence (XRF), thermogravimetric analysis (TGA), and scanning electron microscopy (SEM), and contribute to the development of the appropriate recycling method, such as thermal activation, chemical stabilisation, or mechanical separation, and the ability of people to adhere to environmental regulations (Laadila et al., 2022).

3 Challenges in Solid Waste Management in Construction

Efficient management of solid waste in construction faces multiple barriers that hinder the achievement of sustainable and circular economy goals. Despite growing awareness, construction and demolition (C&DW) waste management stays challenging due to systemic inefficiencies, regulatory gaps, economic constraints, infrastructure limitations, and lack of awareness.

3.1 Inefficient Segregation and Collection Systems

Separating waste at the source is one of the most effective ways of enhancing the material recovery rates as well as the recycling rates in various parts of the world. However, it is not that most construction sites have good rules of segregation as most construction sites lack space, training of workers, or time. An example is that statistics in India indicate that nearly 1% of construction and demolition wastes are sorted and recycled despite nearly 60% of these wastes having the ability to be reused or recycled in a clean condition. The issues of segregation dispose of mixed waste in landfills, where the material that can be recycled goes to waste, which complicates the establishment of a circular economy.

3.2 Absence of Regulatory Compliance and Enforcement

Although there are laws that have been enacted in many countries concerning the way to manage construction and demolition (C&DW) waste, it is still not necessarily the same in that enforcement. As an example, the Construction and Demolition Waste Management Rules (2016) in India state that waste has to be separated and recycled but it is not enforced due to the poor governance of the municipalities and absence of infrastructure. This has brought about a fair deal of illegal dumping. The same issues are faced in Brazil as over half of C&DW waste is disposed of illegally despite the laws promoting waste reduction. According to the Circular Economy Promotion Law in China (2008), waste should be separated, which is usually disregarded due to the rapid development of cities. Just under 10% of waste is recycled in certain provinces (Shajidha and Mortula, 2025) European Union Waste Framework Directive (2008/

98/EC) however targets 70% of recycling of non-hazardous C&DW waste. Recycling rates in Germany and Sweden are 80–90, whereas in Spain and Italy, it is less than 50. This is largely because they lack sufficient implementation. Environmental Protection Agency in the USA promotes using sustainable materials, however, since there are no federal requirements, then there are differences between states.

3.3 Economic and Logistical Obstacles

The construction industry is faced by economic constraints that are a major constraint of sustainable waste management. Recycling requires a lot of money in the initial stages in order to establish the structure of sorting, processing, and transportation. The first to be cut is the money saved, and hence the waste management budgets are subject to cut in construction companies. In 2024, the construction waste market amounted to 232.1 billion in the world. This demonstrates that it is large and costly to manage such materials in a manner that is eco-friendly. The cost of transporting heavy, large garbage is prohibitive in nature, more so in high-traffic cities, and therefore recycling industries have the challenge of having to compete with fluctuating prices on recycled materials since they are forced to compete with cheaper virgin aggregates (das Neves et al., 2025). Pay back these expenses, promote enterprises to adopt sustainable practices, new financial structures, such as subsidies, green bonds, and tax breaks are particularly valuable (Madirisha et al., 2024).

3.4 Insufficient Recycling and Reuse Infrastructure

Many places around the world don't have enough recycling facilities that can handle the amount of C&DW waste they produce. For instance, Tehran produces about 50,000 tonnes of construction and demolition (C&DW) waste every day. Because there aren't many places to recycle it, more than 30,000 tonnes of it goes straight to landfills. About 57% of this waste is classified as non-hazardous and recyclable, but because of problems with infrastructure, this potential is not being fully used. In the same way, developing countries have recycling plants that are not evenly spread out, with rural areas getting the least amount of service. Even in countries with advanced recycling technologies, dealing with hazardous wastes, mixed-material streams, and contaminants is still hard (Sakthibala et al., 2025). Modern solutions that use mechanised sorting, digital tracking, and higher material standards are still not widely used, which lowers the quality and efficiency of recovery. To make full circular cycles in construction waste management possible, facilities need to be upgraded, and their reach needs to be expanded.

3.5 Lack of Awareness, Training, and Data-Driven Assessment

One big problem with improving sustainable solid waste management in construction is that stakeholders don't know enough about it or how to do it. A lot of site managers and workers don't know how to properly separate things or what the rules are. For example, research from Kuwait shows that construction workers don't know enough to reduce waste effectively on sites. Also, the fact that waste generation data is broken up and there is no real-time monitoring makes it harder for governments and businesses to plan effective interventions (Karunasena et al., 2025). Digital tools like Building Information Modelling (BIM), IoT sensors for tracking waste, and artificial intelligence methods look promising for getting correct data and making waste audits easier. Still, not many people use them because they don't know how to use them, and they think integration is too hard. To improve compliance and cut down on waste, we need behavioural change programmes that focus on psychological and cultural barriers, as well as systematic training programmes (Ghobadi & Sepasgozar, 2025). Characterising C&DW waste physically and chemically is especially important for creating recycling processes, predicting how they will affect the environment, and making it easier for businesses to follow the rules in a circular economy. As an overview of key challenges, recycling performance, and influencing factors in Construction and Demolition Waste (C&DW) management across selected regions are listed in Table 1.

4 Strategies for Effective Solid Waste Management

A combination of waste reduction strategies, recycling and reuse technologies, and waste-to-energy and value addition strategies can greatly improve how construction manages solid waste.

4.1 Waste Minimisation Approaches

Design for Deconstruction and Modular Construction

Design for deconstruction (Dodd) is a construction technique that allows ease of dismantling of buildings when they are not required. This allows materials and parts to be recycled with minimal waste. This is further enhanced by modular construction which involves the creation of parts of buildings in regulated factory environments before handing them over. This reduces waste, error, and wasting at the workplace by an enormous percentage. It is estimated that modular construction can reduce construction waste by 23–100% as opposed to conventional methods (Shooshtarian

Table 1 Overview of key challenges, recycling performance, and influencing factors in Construction and Demolition Waste (C&DW) management across selected regions

Region/ country	Key waste management challenge	Real-world data/statistics	Notable factors/ observations	Ref.
India (Delhi, Noida, Greater Noida)	Inefficient segregation, weak enforcement	Only ~1% C&DW waste recycled; massive illegal dumping and landfill overuse	Fragmented municipal governance; lack of infrastructure and training	(Shajidha & Mortula, 2025)
Brazil	Regulatory non-compliance and illegal dumping	>50% of C&DW waste illegally dumped despite legislation	Informal construction sector; corruption; weak policy enforcement	
China	Low recycling rates due to rapid urbanisation	Some provinces <10% recycling rate for C&DW waste	Fast-paced construction outstrips recycling infrastructure capacity	
Europe (Germany, Sweden)	Successful policy enforcement and high recycling	Recycling rates between 80 and 90% for C&DW waste	Landfill bans; strong circular economy policies and incentives	(Ravish, 2025), (Jain et al., 2019)
Europe (Spain, Italy)	Poor monitoring, enforcement gaps	Recycling rates below 50%	Variability in institutional effectiveness and infrastructure quality	
Tehran, Iran	Insufficient recycling infrastructure	Generates ~50,000 tonnes/day C&DW waste; >30,000 tonnes landfilled daily	Limited recycling capacity; high landfill reliance despite recyclable waste	(Rouhi Broujeni et al., 2016), (Song & Zhao, 2024)

(continued)

Table 1 (continued)

Region/ country	Key waste management challenge	Real-world data/statistics	Notable factors/ observations	Ref.
Kuwait	Lack of awareness and workforce training	Knowledge gaps cause poor segregation and waste handling on construction sites	Training deficits; limited data-driven assessment	(Al-Raqeb et al., 2023)

et al., 2025). This is largely due to the fact that it is more precise in the utilisation of materials and the assembly process is standardised. The designs can be made better using digitalisation techniques such as Building Information Modelling (BIM) that allows the designer to consider how the project will be dismantled and how the material will be reclaimed at the onset of the project.(Song & Zhao, 2024)

Material Optimisation and Lean Construction

Material optimisation refers to good estimation and ordering of materials to ensure that there is less spoilage and waste. The principles of lean construction simplify work and reduce wastes through better planning, coordination of suppliers, and on-site management of materials. Collaborating with suppliers and appropriate quantity take-offs can contribute to reducing the packaging waste and excess stockpiling (Fig. 2). The different approaches to solid waste management in construction are illustrated below using Project Management Information Systems (PMIS) and data analytics, thus reducing the amount of waste, which are helpful in continuing to enhance it.

Substitution of High-Impact Materials with Sustainable Alternatives

This makes the environment easier since recycled or renewable materials are used in place of virgin concrete and wood which have a high effect on the environment. Some of them include the use of recycled aggregates, engineered wood products, and composite materials that are durable and can be recycled to ensure that the performance standards are maintained and the greatest environmental gains are achieved (Sakthibala et al., 2025). Lifecycle assessment aids in the selection of diverse materials with a lower carbon footprint and resource consumption.

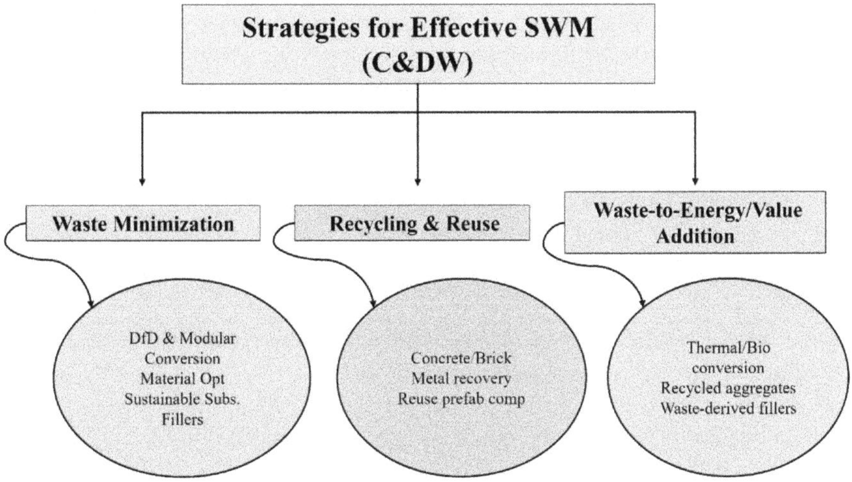

Fig. 2 Flowchart of strategies for effective solid waste management in construction

4.2 *Recycling and Reuse Technologies*

Recycling of Concrete, Asphalt, and Brick

To recycle hardened concrete and brick, you will have to crush and screen them to recycle aggregates which can be incorporated in new constructions as opposed to new materials. This reprocessing and milling of asphalt to recycle it is a common practice in road construction. Made mobile recycling plants allow you to recycle materials on the ground, which reduces the cost of transportation and handling of materials needed to be reused (Rahman et al., 2020). The quality control ensures that the recycled materials conform to the standards of their structure and durability, which are required to be used again. Such technologies assist in keeping much of the waste in the landfills.

Metal Recovery and Reuse

Some of the metals that can be recycled numerous times without considerable loss of quality are steel, aluminium, and copper. Mixed waste streams are easier to sort into metals by using eddy current sorting and magnetic separation. Having high levels of reclaimed metals which are recycled is beneficial to the economy and detrimental to the environment in terms of mining and refining new metals. The use of reclaimed and prefabricated components should be reused. Once these materials are checked and treated, reclaimed materials such as timber, bricks, and structural steel used in buildings that have been demolished can be exploited immediately in other new projects. Modular construction involves the use of prefabricated materials, which are reusable in other projects and this reduces waste. The reuse strategy maintains the high value of materials throughout the building cycle, which conserves energy, reduces emissions, and helps in the ideals of the circular economy.

4.3 Waste-to-Energy and Value Addition Strategies

Thermal Conversion Method and Biochemical Conversion

Waste-to-energy (WtE) is a set of processes that convert construction waste that cannot be re-used (such as in recycling) into useful energy, such as electricity, heat or biofuels. Thermal processes such as incineration, gasification, and pyrolysis decompose combustible materials that produce energy and also reduce their size. Biological processes (e.g. anaerobic digestion) convert organic matter (e.g. wood waste) into biogas (Alao et al., 2022). WtE does not only keep trash out of landfills, but also assists in the production of renewable energy that will substitute fossil fuels. The ashes that are left behind after the burning process can also be used as construction fillers and as such the chances of being left behind with a lot of waste are minimal.

The Recycling of Aggregates and Waste-Derived Fillers in Construction

Use of recycled aggregates in concrete production as a substitute for virgin aggregates in the production of concrete is on the rise. This is more economical of resources and has a lower effect on the environment. Concrete that is improved by using fillers created from waste such as ashes, slags or finely ground materials that have been demolished is better and consumes less cement which is environmentally friendly. These types of value addition render construction materials more circular and reduce the environmental impact of new buildings.

Use of Recycled Aggregates and Waste-Derived Fillers in Construction

Increasingly, recycled aggregates from C&DW waste are being used in concrete production as partial replacements for virgin aggregates. This makes better use of resources and has less of an impact on the environment. Fillers made from waste, like ashes, slags, or finely ground demolition materials, make concrete better and use less cement, which is good for the environment. These methods of adding value make construction materials more circular and lower the environmental impact of new buildings.

5 Integration of Circular Economy Principles in Construction

The application of the principles of Circular Economy (CE) in construction is a significant shift in the previous linear approach to doing business, where people extract resources, consume them, and discard them. CE on the other hand is preoccupied with resource optimisation, reuse, and sustainability. The change is needed due to the fact that the old construction technique exacerbates the depletion of resources, waste production, and deterioration of the environment. One of the significant elements of such change is the attempt to re-engineer the entire lifecycle of construction undertakings and make systemic changes that would incorporate CE elements at each stage. Making buildings with modular parts to be easier to take apart, change, and remake is the key to shifting to a circular approach and enables us to make the most of materials and reduce the amount of waste (Banihashemi et al., 2024). Design to deconstruct (DfD) is gaining more significance. It requires the construction of the parts to be disassembled at the expiry of their lifespan without causing any destruction so that materials can be reutilised and reclaimed. Besides, to move to circular models, it is necessary to modify the supply chain to enable closed-loop flows of materials. This implies the development of new business models such as product-as-a-service where builders retain ownership of the parts and materials and must ensure their durability, maintenance as well as recovery in the end. Such systems make manufacturers and contractors consider reusing and recycling materials during the planning stage. This makes economic interests in line with the environmental goals. One of the most important components of the implementation of CE in construction is its creation and utilisation of Material Circularity Indicators (MCIs). These indicators tell the efficiency of materials remaining in cyclical loops by examining such aspects as the recycling rate, the duration of materials, and their remanufacturability. MCIs enable companies and policymakers to make better decisions that benefit the environment, providing them with information, establishing standards and performance measurement. CE principles are even more formal and the system of sustainable procurement and green building certification (such as LEED, BREEAM, and GRIHA) introduces resource efficiency standards, waste management standards, and design decisions that make the circularity. Digital technologies will be needed to accelerate the transition to circular construction. Building Information Modelling (BIM) assists in the planning, pictorialisation, and management of materials during the life cycle of a building. This has enabled it to monitor everything in detail, dismantle them fast, and put them back together. IoT sensors provide real-time updates on the status of your assets and the surrounding, and thus it is easier to schedule maintenance and utilise resources in a more efficient way. Artificial Intelligence (AI) simplifies the process of data analysis, enhances the distribution of resources, and automates monitoring and decision-making, which significantly improves the practices of CE (Charef, 2022). By all stakeholders, such as governments, businesses, researchers, and residents, these strategies and modern technologies should be employed together. This comprehensive approach is required to transform the built environment into a regenerative system balancing economic feasibility with ecological soundness which will assist in meeting the global sustainability targets and initiatives to make the climate more resilient.

5.1 Transition from Linear to Circular Construction Models

A transition to the circular models of construction is a total shift in the values, management, and utilisation of resources in the lifecycle of the building. Construction has always been a straightforward process where raw materials were extracted, refined into a component, assembled into buildings and discarded when the latter were not in demand. This technique consuming lot resources and generating a lot of waste cannot be described as sustainable since the world resources are scarce and the pressures on the environment are increasing. The circular model promotes an interconnected model of systems approach, which promotes resource efficiency, maximum possible material life, and regeneration. This is a significant component of change that involves adopting design principles that enable taking things apart, repairing them, and creating new things. An example is the modular methods of construction, which allows parts to be swapped in and out, or reused. The concept of the Design for Disassembly (DfD) is to be able to disassemble a building in a simple way without damaging any useful material that will simplify the recycling and reusing of the materials. Another concept is adaptive design to last a long time, which implies that buildings are able to be re-used in other ways throughout their existence. This implies that they do not need to be demolished and constructed frequently. The change is rooted in lifecycle thinking, which is made possible through tools such as Building Information Modelling (BIM) that allow you to trace materials, energy consumption, and environmental impact of the cradle to cradle (Kirchherr et al., 2023). Governments and other policymakers are particularly significant as they introduce the rules that promote the use of CE practices, such as requiring the use of recycled materials or establishing building standards that are easy to tear down. Circular business models favouring renting over owning, establishing material banks, and industry formations could be some industry-led initiatives towards making the circular practice more standardised. Real-life evidence demonstrates that through the use of circular models, waste, emissions, and consumption of resources could be reduced significantly. This may help to save on money through material saving and making a business to shine in the market. Despite all these advantages, the reasons as to why they are not even more common among people are still numerous. Others include the contracts that are there, broken supply chains, and the exorbitant price of modern technologies. To overcome such issues we must change the whole system including promoting teamwork and formulating policies which support innovative thought. This will assist the construction sector to take steps towards the full-circle model (Bocken et al., 2025).

5.2 *Circular Business Models and Closed-Loop Material Flows*

The most important way of integrating the principles of CE in the construction is by having new business models that ensure the flow of materials and components in closed circle and retain value throughout the life of a building. Circular business models (CBM) differ with the traditional ones in that they are aimed at increasing the lifespan of resources, using them again, manufacturing them again, and reusing them. This is value addition as well as reducing wastage. A common CBM is the so-called Product-as-a-Service (PaaS) concept. The manufacturer in this model retains ownership of materials or parts and leases them to owners of buildings. This will ensure that they maintain, upgrade, and recover at the end of their life. PaaS reduces material usage and provides producers with a constant inflow of revenue, which maintains the flow of materials in a regenerative system (Zhang et al., 2025). One more model is to establish material banks or resource pools that are the central locations where reusable building materials and components are stored. This facilitates easy search and share of them in new projects. Such platforms render them open, faster, and cheaper to purchase. The modular construction supports this model in that the parts can be disassembled, reused, and reassembled in different ways. This will conserve raw materials and reduce wastes during demolition. Another approach that is of particular importance to enable closed loops would be Design for Deconstruction (DfD) and Adaptive Reuse strategies. Those plans allow the buildings to live longer or be utilised in a different way altogether. As an example, when old buildings are used, rather than demolishing them and constructing new ones, the impact on the environment is minimal. Supply chain integration is relevant to close loops. This implies that the whole team of people consisting of the designers, manufacturers, contractors, and end-users must collaborate with one another. This can be assisted by the digital platforms and information sharing. Certification regimes like the Cradle to Cradle (C2C) provide regulations for producing products with safe and reusable material, which facilitates reusability and creativity. Applying these models will require reconsidering the way you purchase goods, the way you enter into a contract, and the manner in which value chains operate. Government incentives and standards are also effective to make people use the technology to save a considerable amount of money, reduce their expenses, and discover new methods to earn money that are economically and environmentally friendly (Golinska et al., n.d.).

5.3 *Application of Material Circularity Indicators (MCIs)*

You require powerful, quantifiable measures to determine the success of the programmes and the progress achieved by the circularity programmes. This is done through Material Circularity Indicators (MCIs) which provide a full range of tools to quantify the circularity of materials and products in construction. The MCIs are used to determine the effectiveness of the material, part or system to maintain its value by reusing, recycling or remanufacturing, reducing linear flows of resources in the construction industry, identifying where linearity is an issue, and prioritising actions to be undertaken (Pilipenets et al., 2025). In order to implement MCIs, you must acquire data on the origin of the resources, their impact on the environment through their existence, and their possible recovery. Another concrete mix, which has much recycled aggregate and is simple to dismantle to be recycled later, e.g. would score higher on the MCI scale. The measure allows things to be more open, assists in certification and promotes future designs, which make things more circular. According to recent case studies, it is very important that MCIs are incorporated in Building Information Modelling (BIM), which allows one to test things in real time during the design and construction process. There is growing debate on the adoption of MCIs as possible rules and regulations, to support assertions of sustainability and ensure that projects align with the ambitions of a circular economy. Technologically, the innovation of digital dashboards with the inclusion of MCIs could demonstrate the circular value of materials, monitor the progress over time, and information could be used to make decisions (Waqar et al., 2023). Future studies are aimed at improving MCIs to include social, economic, and environmental issues, which will result in holistic tools of the transition towards fully circular construction methodologies.

5.4 Sustainable Procurement and Green Building Certification Systems

The primary means of ensuring that CE principles become a standard thing in construction projects is through sustainable procurement practices and green building certification systems. By introducing sustainability into all aspects such as the choice of materials, design solutions, and so on, these frameworks promote the development of environmentally friendly behaviours and ensure that the practices of industries align with global climate objectives. LEED (Leadership in Energy and Environmental Design), BREEAM (Building Research Establishment Environmental Assessment Method), GRIHA (Green Rating on Integrated Habitat Assessment), and other regional standards provide a set of steps that the project planners and builders can follow to ensure that their projects become more resource-efficient, waste less, and more circular. LEED emphasises maximising materials use through reusing, recycling, and examining their impact on life cycles. It awards points on recycled materials, materials locally sourced, deconstruction, and waste management strategies. BREEAM is a more thorough approach to assessment, considering such aspects as life cycle assessments (LCAs), occupant health, and biodiversity. The certifications influence decision-making processes in projects, purchasing of goods, and the stakeholder inclusion, which fosters the culture of sustainability. The rules you have to establish in order to use these tools include purchasing inputs that focus on recycled materials, greener materials, or regenerative materials, implementing product standards proven to be circular, and controlling your supply chain in a transparent and straightforward manner. They also contribute to marketing and client engagement by demonstrating your environmental concern-capacity building, collaboration with stakeholders and regulation-like compulsory waste diversion goals or recycled material acquisition quotas (Ji et al., 2017). The effectiveness of these certification systems has been confirmed by empirical studies which indicate that sustainable design certified projects have a better operational costs, occupant productivity, and resilience in comparison with their traditional counterparts.

5.5 Role of Digital Technologies in Circular Construction

The digital transformation is altering how CE is applied in the construction sector by providing the designers, operators, and managers with the access to the latest tools. One of the most vital technologies in this category is Building information Modelling (BIM). It allows the development of detailed building asset digital representations, and this is associated with lifecycle management, disassembly planning, material tracking, and optimal resource utilisation. Building Information Modelling (BIM) makes reuse, remanufacturing, and disassembly to be tested and compared in terms of design. BIM models have a sophisticated structure of data which is used to store information regarding material make-up, structure, and space relationships. People can use this information to make decisions that are circular. IoT appliances such as sensors installed on buildings and machines allow you to monitor all the things such as the weather, the wellbeing of the building as well as the amount of time you are utilising your resources. These sensors collect live data about occupancy, energy consumption and material wear. This data is further delivered to digital twins and analytics systems that assist in making maintenance planning, seeking means of resource saving, and justifying adaptive reuse solutions. The algorithms of Artificial Intelligence (AI) and Machine Learning (ML) are able to assist you with decision-making, minimise waste, and optimise processes by providing you with insights that are informed by data (Cano-Sunen et al., 2023). AI-based systems can ensure that projects run on time, that objects can be broken down in an optimal manner, and that resources and tasks can be organised. These allow conducting virtual inspections, collaborating remotely, and pre-planning the recovery of resources with high accuracy. Transparency and collaboration among stakeholders can also be encouraged by digital tools that maintain tracking of material provenance, waste creation, and cycling KPIs by relying on dashboards and data repositories. IoT infrastructure and digital twins are gaining popularity in the construction industry. This will take the industry a step closer to the fully connected, efficient and regenerative systems, which are based on the CE principles.

6 Policies, Regulations, and Standards in Circular Construction

Strong policies, regulatory frameworks, standards, and collaboration among stakeholders are all important for the successful use of circular economy (CE) principles in construction. To deal with waste, resource depletion, and sustainability in this field, we need comprehensive global and national frameworks, normative standards for recycled and green materials, and coordinated efforts between government, academia, and private industry. Also, financial tools and incentives are especially important for getting people to use CE in the built environment.

6.1 Global Policy Frameworks Supporting Circular Construction

The construction sector has become a major area of interest to policy frameworks across the world in a bid to achieve the objective of circularity and sustainability. The United Nations (the SDGs), the European Union (EU), and the World Economic Forum all indicate that circular construction is significant in climate action, a more efficient use of resources and a more resilient economy. The European Green Deal (2019) targets to reduce construction waste and promote reuse, refurbishment, and recycling specifically. The Circular Economy Action Plan by EU (2020) has legally binding targets of material recovery and waste reduction. The United Nations Environment Programme (UNEP) also assists the people across the world to exchange information and receive technical assistance to encourage the use of sustainable building practices that adhere to the CE. Lifecycle thinking metrics and material circularity are also included in international standards and policy debate to enable countries gauge their progress and make regulations to suit their demands. International processes are actively working to combine the measurement systems in place, such as Material Circularity Indicators (MCIs), and exchange the best practices in sustainable procurement, urban planning, and building regulations and certification systems (Norouzi et al., 2021).

6.2 National-Level Legislations and Waste Management Rules

The national policy landscapes vary widely but they are usually in line with the global trends by integrating the ideas of circularity in the laws that regulate the construction waste and management of resources. The existence of more progressive construction industries usually means that the countries have regulations necessitating the separation of construction and demolition waste (C&DWW), have targets to reach in terms of recycling, and restrict landfill. To cite an example, the Waste Framework Directive by the European Union states that by the year 2025, the member states will be required to recycle or reuse at least 70% of C&DW waste. This drives national policies and implementation. There are rapid changes in the laws in the developing economies. The Construction and Demolition Waste Management Rules (2016) of India state that waste has to be classified, establish recycling targets, and promote the use of recycled materials in construction. However, the complete adherence remains difficult due to the issues with the enforcement and infrastructure. Among the laws that advance the idea of a circular economy in China is the compulsion to reuse materials as well as the incentives of people to recycle. The USA has no federal legislation that mandates recycling of C&DW waste, but there are state regulations that promote such waste and green building policies that promote such waste. The importance of a well-managed construction in waste management is increasing as more nations belong to the group of countries that incorporate the goals of a circular economy into their national climate action plans (Caro et al., 2024). Waste management national regulations commonly collaborate with building codes and sustainability certifications to ensure that all aspects of procurement and construction have a circular theme. In order to bridge the policy-practice gap, effective implementation typically requires institutional alignment capacity, financing programmes, and community participation.

6.3 Standards for Recycled and Green Construction Materials

The use of circular construction methods is an issue that has been achieved through development and implementation of stringent guidelines on recycled and green materials. Other organisations such as the ISO, ASTM International, and EN (European Standards) establish standards concerning recycled aggregates, insulating material, composite products, and low-carbon cement alternatives all over the globe. These standards establish testing, classification, and application regulations in order to ensure that recycled materials address the demands of contemporary construction (Amarasinghe et al., 2025). National standards collaborate with the global work and can have eco-labels and product certifications that identify environmental characteristics. India, acting through the Bureau of Indian Standards (BIS), has created a guideline of environmentally friendly building materials and recycled aggregates that are compatible with the objectives of green construction. Examples of green certification tools that incorporate these standards in their procurement criteria are LEED, BREEAM, and GRIHA. It promotes their use by ensuring their safety and efficacy to overcome the scepticism of the industry and to establish the changes necessary to implement the use of recycled materials on a massive scale (Plakantonaki et al., 2023). To eliminate the uncertainty of the industry, it is also advisable to develop harmonised and science-based standards. The continuation of research and consultations with stakeholders allows maintaining these standards up-to-date and enhancing them to allow them to be able to operate with new circular materials and technologies (Huang et al., 2020).

6.4 Role of Government, Academia, and Private Sector in Implementation

Circular construction systems require governments, schools, and businesses to collaborate in a manner that each group has a distinct and yet important role to portray. Governments are in charge of formulating policies that can aid business to grow, including laws, standards, incentives, and methods of enforcing. They invest in infrastructure upgrades, promote information sharing, and establish regulations in public procurement prioritising circular products and practices. Circular construction and its benefits are demonstrated by setting flagship sustainable projects by their public authorities. Academia also has a part to play, in research and innovations, which offer the scientific platform and technological solutions that enable circularity. New materials are made by colleges and research centres, the lifecycle assessment models are advanced, and the economic and environmental impacts of the circular construction are examined. The knowledge to the future professionals and the ability to ensure that the principles of the circle become part of the daily existence is also provided by academic programmes. The private sector that implements the ideas of the circular economy includes developers, contractors, material manufacturers, and waste managers (Amarasinghe et al., 2024). Modular building, Material-as-a-Service models, and integrated waste management solutions are all the ways in which innovation in the industry facilitates progress. Individual firms are open and they have funds to test new ideas and nurture successful ones. Industry consortia, public–private partnerships (PPP), and innovation clusters can contribute to accelerating the diffusion and adoption of innovative ideas. A circular construction project will be ensured to achieve a considerable size and long-term outcome due to an ecosystem approach, emphasising shared responsibility, openness, and matched incentives.

6.5 Incentives and Financial Mechanisms Promoting Sustainable Construction

Economic incentives and monetary instruments are required to make people invest and utilise circular construction practices. These mechanisms are used to overcome the high initial costs, promote creative thinking, and give recognition to those who behave in a manner that is environmentally friendly. Common financial tools are grants and subsidies on recycling infrastructure and green materials, tax rebates or credits on projects based on the principles of the circular economy, low-interest loans and green bonds to sustain construction, and public procurement preferences favouring the projects based on the principles of the circular economy. There are locations where landfill charges or waste disposal fees are paid and in this case, it is more economical to reuse the materials rather than disposing of them. Bond-backing schemes on the construction of the parts motivate individuals to reuse, and get back the substances of demolition. Certification subsidies promote sustainability projects that are of high standards. One of the more sophisticated market-based mechanisms is carbon trading based on lesser embodied of materials in circular. Result-based financing models can focus on trading materials and establishing circular supply chains through platforms that enable them to make money by accomplishing circularity objectives, such as waste diversion rates (Kumar et al., 2025). Practical experience demonstrates that monetary rewards have a significant effect of increasing the utilisation of recycled materials, reducing wastefulness, and promoting the implementation of new ideas. Environmental objectives and economic instruments must collaborate to continue the process and transform construction markets into low-carbon and circular economy.

7 Future Perspectives and Research Directions

The construction industry is at an especially crucial point where the need for sustainability and modern technologies come together. This creates many chances to make circular economy principles more deeply and effectively. The future perspectives and research directions highlight the development of innovative sustainable materials, the enhancement of circularity metrics, the incorporation of advanced digital technologies, and the significance of interdisciplinary collaborations.

7.1 Emerging Sustainable Materials from Construction Waste

The opportunities of circular construction are also being transformed by the existence of eco-friendly building materials created using building waste that offer long-lasting and sustainable building materials as alternatives to building waste. The most environmentally friendly materials are geopolymers, recycled aggregates, and green composites which are also the most effective. Geopolymers refer to inorganic polymers prepared through the alkaline solution to activate industrial wastes or construction wastes rich in aluminosilicates such as fly ash, slag, or demolition rubble. These binders possess equal or superior mechanical strengths than conventional Portland cement (OPC) but they possess considerably lesser carbon footprints since they do not involve the manufacture of clinker. The current research trends have focussed on optimising the geopolymer formulations to include mixed waste streams, increase their workability, and longevity, making them potential candidates as low-carbon construction materials. Examples of recycled aggregates include crushed concrete, bricks, and masonry that is a construction and demolition (C&DW) waste. Aggregates are improved by new processing technologies in terms of physical and chemical characteristics. This implies that they may be utilised in both structural and non-structural concrete instead of natural aggregates with insignificant impact on performance. Continued studies are done on methods of treatment to reduce contaminants, enhance bonding performance, and check the durability over a long period of time under various climatic conditions. The growing supply of recycled aggregates is quite appropriate to the urban mining strategies since it will reduce the landfills and extractions of raw materials. Green composites are those lightweight building materials, which insulate and last long. They are composed of a mixture of natural fibres or agricultural waste with recycled polymers or cementitious binders. There are numerous applications of these materials which include: temperature regulation, reducing embodied energy, and being biodegradable. They are produced out of plastic waste and biomass waste. The products of these materials enable the technology to proceed even further besides resisting fire, making composites that can carry structural loads, and lasting longer in wet environments, which contributes to the emergence of the new market opportunities in green construction as well as supports the circular material flows (Nedeljkovic et al., 2021).

7.2 *Development of Circular Economy Indicators and Assessment Tools*

To help with policy, track progress, and encourage practices that close material loops, it's important to measure circularity using strong indicators and thorough assessment tools. Future research directions prioritise the enhancement and expansion of instruments such as Material Circularity Indicators (MCIs), Whole Building Circularity Indicators (WBCIs), and Sustainability Assessment frameworks. A circular economy indicators and assessment tools with their examples are listed in Table 2.

MCIs allow viewing the recycled content of each material, their service life, recovery possibilities, and the amount of waste that they produce. Recent studies seek to drive these indicators through the incorporation of ecological footprint measures, resource scarcity measures, and socio-economic aspects and come up with more inclusive assessments. Whole Building Circularity Indicators borrow this concept and extend it to entire buildings. They consider the process of movement of material in the building, easy dismantling, versatility, and reuse. Real-time application and decision support can be even more digitalised by being integrated with Building Information Modelling (BIM) systems in real time (Matos et al., 2023). Circular construction can be enhanced with the help of scenario modelling that involves economic, environmental, and social measures. The continuous evolution of these metrics offers a basis

Table 2 Circular economy indicators and assessment tools with their examples

Indicator/Tool	Key features	Operational mechanism	Application domains	Ref
Material circularity indicator (MCI)	– assesses recycled content, product lifespan, recovery potential, waste generation – captures degree of "closed loop"	– linear flow index (LFI): Ratio of virgin materials and unrecoverable waste to total material flow – utility factor: Adjusts for use duration/ intensity – final score 0 (linear) to 1 (circular)	Product/ material level	(Sazdovski et al., 2024) (Anuardo et al., 2022)
Product circular indicator (PCI)	– incorporates all MCI elements – adds quality of recycled materials and stage-specific waste – evaluates "tightness" of cycles	– weights losses during material/component/ feedstock stages – adjusts for down-cycling, multi-material products, and efficiency variations over lifecycle stages – considers mass and functional cycles	Supply chain/system level	
Whole building circularity indicator (WBCI/BCI)	– expands to building/ component/system level – measures modularity, demountability, reusability, adaptability - BIM integration for real-time assessment	– aggregates input/ output flows, technical/ service lifespans, design for disassembly – evaluates at material, product, system, and building scales – often index-based, adaptable via expert-defined KPIs	Whole buildings, portfolios	
Life cycle assessment (LCA), scenario modelling	– evaluates environmental (embodied carbon), economic, and social impacts – models recycling, reuse over multiple cycles with real data	– scenario modelling using system-wide data, includes quality of secondary materials, "embodied carbon" reductions, and circularity benefits	All scales, decision support	(Güngör et al., 2025)

to regulators, designers, and contractors to facilitate benchmarking, certification, and continuous improvement across the built environment.

7.3 Integration of Digital Twins and Blockchain for Traceability

The combination of digital twin and blockchain distributed ledger systems presents a strong argument in the need to make construction practices in a circular economy more open, accountable, and safe. Digital twins are virtual representations of physical properties which evolve with time to impart the details of performance, supply chains, and environmental conditions at any point in the lifecycle. When digital twins are tied to blockchain systems, they have the ability to maintain records of origin of materials, ownership, storing, and recycling which remain unchanged. This will help in easing the process of tracking materials and establishing their legitimacy. Such integration develops trust among the stakeholders since it enables identification of the content that has been recycled as well as adherence to regulations and verification of lifecycle assertions. Blockchain-based smart contracts are used to automate transactions in receiving materials, renting them, or recycling them. Circular business models are more user-friendly in this way. Currently, the current research is directed at the question of how to scale digital twin-blockchain platforms and how to integrate them with other platforms and how to interlink them with IoT sensor networks keeping an eye on the status of assets and their utilisation. Virtual audit of waste disposal may be one of the practical applications; and predictive maintenance would ensure that assets are not wasted prematurely and would make the introduction of the secondary material market more cost-effective and advanced in terms of the pace at which circular practices will be introduced to construction systems (Adu-Amankwa et al., 2023). Data transparency and trust are crucial in the operation of the circular economy, so enhancement of these technologies will be critical to make construction systems more resistant to the future and accelerate the transition to the circular practices.

7.4 Multidisciplinary Research and Stakeholder Collaboration Needs

The complexity of the implementation of a circular economy in the construction sector needs interdisciplinary research projects and collaboration between stakeholders. Engineering, environmental science, economics, policy, social sciences, and digital innovation must be combined in future studies to holistically address technical, economic, behavioural, and regulatory issues. Academia–industry–policymaker—community stakeholder collaboration platforms foster complete knowledge sharing and co-development of solutions. Transdisciplinary initiatives are needed to examine the circular products and processes in real-life contexts, determine and calculate multi-dimensional effects, and conduct socio-economic research to determine the barriers and drivers of different stakeholders. The money invested in programmes that develop skills, and knowledge produces a workforce familiar with how to make use of digital tools and adhere to the principles of a circle. Circular construction methods are becoming more significant because of climate change, resource scarcity, and urbanisation which are all global issues that demonstrate the importance of using integrated approaches. Innovative research methods, including the use of systems thinking, intersector partnerships, and new funding concepts will play a critical role in transitioning circular building projects to energy-saving industrywide change.

8 Conclusion

Solid waste management in the construction industry has ceased to be an environmental requirement but an important aspect of sustainable urban development. A shift towards a circular economy system, as opposed to a linear take-make-dispose can dramatically decrease environmental pressures, save resources, and create additional economic opportunities. Waste segregation at the source, reuse of materials and strengthening recycling infrastructure are sustainable measures to be made. The addition of the latest technologies, like Building Information Modelling (BIM), Artificial Intelligence (AI), and intelligent tracking software, also improves the efficiency and the traceability of materials. Nonetheless, to realise the mass transformation, strict policy implementation, industry cooperation, and awareness of the population are necessary. The industry can reduce the reliance on landfills, decrease carbon emissions, and build a regenerative system that will provide environmental protection and economic development by incorporating circular principles into the construction process.

References

Abedin Khan, Z., Balunaini, U., & Costa, S. (2024). Environmental feasibility and implications in using recycled construction and demolition waste aggregates in road construction based on leaching and life cycle assessment—A state-of-the-art review. *Cleaner Materials, 12*, 100239. https://doi.org/10.1016/j.clema.2024.100239

Adu-Amankwa, N. A. N., Pour Rahimian, F., Dawood, N., & Park, C. (2023). Digital twins and Blockchain technologies for building lifecycle management. *Automation in Construction, 155*, 105064. https://doi.org/10.1016/j.autcon.2023.105064

Alao, M. A., Popoola, O. M., & Ayodele, T. R. (2022). Waste-to-energy nexus: An overview of technologies and implementation for sustainable development. *Cleaner Energy Systems, 3*, 100034. https://doi.org/10.1016/j.cles.2022.100034

Al-Raqeb, H., Ghaffar, S. H., Al-Kheetan, M. J., & Chougan, M. (2023). Understanding the challenges of construction demolition waste management towards circular construction: Kuwait Stakeholder's perspective. *Cleaner Waste Systems, 4*, 100075. https://doi.org/10.1016/j.clwas.2023.100075

Amarasinghe, I., Hong, Y., & Stewart, R. A. (2024). Visualising a framework for enhancing material circularity in building construction projects: Drivers, barriers, and strategies. *Building and Environment, 253*, 111359. https://doi.org/10.1016/j.buildenv.2024.111359

Amarasinghe, I., Stewart, R. A., Sahin, O., & Liu, T. (2025). Enhancing construction material circularity: An integrated participatory systems model. *Sustainable Production and Consumption, 57*, 106–120. https://doi.org/10.1016/j.spc.2025.05.012

Anuardo, R. G., Espuny, M., Costa, A. C. F., & Oliveira, O. J. (2022). Toward a cleaner and more sustainable world: A framework to develop and improve waste management through organizations, governments and academia. *Heliyon, 8*, e09225. https://doi.org/10.1016/j.heliyon.2022.e09225

Baggio, T., Possan, E., & de Oliveira Andrade, J. J. (2024). Physical-chemical characterization of construction and demolition waste powder with thermomechanical activation for use as supplementary cementitious material. https://doi.org/10.2139/ssrn.4764143

Banihashemi, S., Meskin, S., Sheikhkhoshkar, M., Mohandes, S. R., Hajirasouli, A., & LeNguyen, K. (2024). Circular economy in construction: The digital transformation perspective. *Clean Eng Technol, 18*, 100715. https://doi.org/10.1016/j.clet.2023.100715

Bocken, N. M. P., Kimpimäki, J.-P., Ritala, P., & Konietzko, J. (2025). How circular are large corporations? Evidence from a large-scale survey with senior leaders. *Resources, Conservation and Recycling, 215*, 108151. https://doi.org/10.1016/j.resconrec.2025.108151

Bonifazi, G., Grosso, C., Palmieri, R., & Serranti, S. (2025). Current trends and challenges in construction and demolition waste recycling. *Curr Opin Green Sustain Chem, 53*, 101032. https://doi.org/10.1016/j.cogsc.2025.101032

Cano-Suñén, E., Martínez, I., Fernández, Á., Zalba, B., & Casas, R. (2023). Internet of things (IoT) in buildings: A learning factory. *Sustainability, 15*, 12219. https://doi.org/10.3390/su151612219

Caro, D., Lodato, C., Damgaard, A., Cristóbal, J., Foster, G., Flachenecker, F., & Tonini, D. (2024). Environmental and socio-economic effects of construction and demolition waste recycling in the European Union. *Science of the Total Environment, 908*, 168295. https://doi.org/10.1016/j.scitotenv.2023.168295

Charef, R. (2022). The use of building information modelling in the circular economy context: Several models and a new dimension of BIM (8D). *Clean Eng Technol, 7*, 100414. https://doi.org/10.1016/j.clet.2022.100414

das Neves, J. S., Yamane, L. H., Santos, R. P., & Siman, R. R. (2025). Challenges in commercial solid waste management: Generation indicators and willingness to pay for management. *Cleaner Waste Systems, 11*, 100250. https://doi.org/10.1016/j.clwas.2025.100250

Gamage, I., Senaratne, S., Perera, S., & Jin, X. (2024). Implementing circular economy throughout the construction project life cycle: A review on potential practices and relationships. *Buildings, 14*, 653. https://doi.org/10.3390/buildings14030653

Ghobadi, M., & Sepasgozar, S. M. E. (2025). Challenges and benefits of implementing AI in timber construction for circular economy goals. *Buildings, 15*, 1073. https://doi.org/10.3390/buildings15071073

Güngör, B., Agibayeva, A., Karaca, F., Askar, R., Giarma, C., Rosado, L., Pineda-Martos, R., Griffiths, P., & Bragança, L. (2025). Circularity tools and frameworks for new buildings, *431–458*. https://doi.org/10.1007/978-3-031-73490-8_16

Huang, B., Gao, X., Xu, X., Song, J., Geng, Y., Sarkis, J., Fishman, T., Kua, H., & Nakatani, J. (2020). A life cycle thinking framework to mitigate the environmental impact of building materials. *One Earth, 3*, 564–573. https://doi.org/10.1016/j.oneear.2020.10.010

Jain, S., Singhal, S., & Jain, N. K. (2021). Construction and demolition waste (C&DW) in India: Generation rate and implications of C&DW recycling. *International Journal of Construction Management, 21*, 261–270. https://doi.org/10.1080/15623599.2018.1523300

Jain, S., Singhal, S., & Jain, N. K. (2019). Construction and demolition waste generation in cities in India: An integrated approach. *International Journal of Sustainable Engineering, 12*, 333–340. https://doi.org/10.1080/19397038.2019.1612967

Ji, S., Banerjee, S., & Punekar, R. M. (2017). Assessment of GRIHA and LEED on the parameters of sustainable design and development of buildings, 179–188. https://doi.org/10.1007/978-981-10-3521-0_15

Karunasena, G., Gurmu, A., Shooshtarian, S., Udawatta, N., Ranthika Perera, C. S., & Maqsood, T. (2025). Effect of construction defects on construction and demolition waste management in building construction: A systematic literature review. *Integrated Environmental Assessment and Management, 21*, 233–244. https://doi.org/10.1093/inteam/vjae026

Kirchherr, J., Yang, N.-H. N., Schulze-Spüntrup, F., Heerink, M. J., & Hartley, K. (2023). Conceptualizing the circular economy (revisited): An analysis of 221 definitions. *Resources, Conservation and Recycling, 194*, 107001. https://doi.org/10.1016/j.resconrec.2023.107001

Kumar, B., Kumar, A., Sassanelli, C., & Kumar, L. (2025). Exploring the role of finance in driving circular economy and sustainable business practices. *Journal of Cleaner Production, 486*, 144480. https://doi.org/10.1016/j.jclepro.2024.144480

Laadila, M. A., LeBihan, Y., Caron, R.-F., & Vaneeckhaute, C. (2022). Physical and chemical characterization of construction, renovation and demolition waste in the Quebec province. *Cleaner Waste Systems, 1*, 100002. https://doi.org/10.1016/j.clwas.2022.100002

Madirisha, M. M., Dada, O. R., & Ikotun, B. D. (2024). Chemical fundamentals of geopolymers in sustainable construction. *Materials Today Sustainability, 27*, 100842. https://doi.org/10.1016/j.mtsust.2024.100842

Material waste in construction projects: Causes and effects. (2023). *Journal of Southwest Jiaotong University 58*. https://doi.org/10.35741/issn.0258-2724.58.3.10

Matos, J., Santos, S., Simões, C. L., Martins, C. I., & Simoes, R. (2023). Practical application of circularity micro-indicators to automotive plastic parts in an industrial context. *Sustainable Production and Consumption, 43*, 155–167. https://doi.org/10.1016/j.spc.2023.11.009

Nedeljković, M., Visser, J., Šavija, B., Valcke, S., & Schlangen, E. (2021). Use of fine recycled concrete aggregates in concrete: A critical review. *Journal of Building Engineering, 38*, 102196. https://doi.org/10.1016/j.jobe.2021.102196

Norouzi, M., Chàfer, M., Cabeza, L. F., Jiménez, L., & Boer, D. (2021). Circular economy in the building and construction sector: A scientific evolution analysis. *Journal of Building Engineering, 44*, 102704. https://doi.org/10.1016/j.jobe.2021.102704

Pallewatta, S., Weerasooriyagedara, M., Bordoloi, S., Sarmah, A. K., & Vithanage, M. (2023). Reprocessed construction and demolition waste as an adsorbent: An appraisal. *Science of the Total Environment, 882*, 163340. https://doi.org/10.1016/j.scitotenv.2023.163340

Patil, Y. R., Dakwale, V. A., & Ralegaonkar, R. V. (2024). Recycling construction and demolition waste in the sector of construction. *Advances in Civil Engineering, 2024*. https://doi.org/10.1155/2024/6234010

Pilipenets, O., Gunawardena, T., Hui, F. K. P., Mendis, P., & Aye, L. (2025). A novel circular economy framework: Assessing process circularity through resource flow and emissions analysis. *Resources, Conservation and Recycling, 215*, 108083. https://doi.org/10.1016/j.resconrec.2024.108083

Pineda-Martos, R., Askar, R., Karaca, F., De Simone, M., Borg, R. P., Malešev, M., Radonjanin, V., Bas, B., Çavdar, A. D., Cervantes Puma, G. C., Rosado, L., & Bragança, L. (2025). Circularity criteria and indicators at the construction material level, 299–333. https://doi.org/10.1007/978-3-031-73490-8_12

Plakantonaki, S., Kiskira, K., Zacharopoulos, N., Chronis, I., Coelho, F., Togiani, A., Kalkanis, K., & Priniotakis, G. (2023). A review of sustainability standards and Ecolabeling in the textile industry. *Sustainability, 15*, 11589. https://doi.org/10.3390/su151511589

Ponnada, M. R., & Kameswari, P. (2015). Construction and demolition waste management—A review. *International Journal of Advanced Science and Technology, 84*, 19–46. https://doi.org/10.14257/ijast.2015.84.03

Purchase, C. K., Al Zulayq, D. M., O'Brien, B. T., Kowalewski, M. J., Berenjian, A., Tarighaleslami, A. H., & Seifan, M. (2021). Circular economy of construction and demolition waste: A literature review on lessons, challenges, and benefits. *Materials, 15*, 76. https://doi.org/10.3390/ma15010076

Rahman, M. T., Mohajerani, A., & Giustozzi, F. (2020). Recycling of waste materials for asphalt concrete and bitumen: A review. *Materials, 13*, 1495. https://doi.org/10.3390/ma13071495

Ravish, P. (2025). Sustainable Management of Landfill Sites in India: Addressing environmental, health, and socioeconomic challenges. *Current World Environment, 20*, 19–34. https://doi.org/10.12944/CWE.20.1.3

Rouhi Broujeni, B., Omrani, G. A., Naghavi, R., & Afraseyabi, S. S. (2016). Construction and demolition waste management (Tehran case study). *Engineering, Technology & Applied Science Research, 6*, 1249–1252. https://doi.org/10.48084/etasr.812

Sakthibala, R. K., Vasanthi, P., Hariharasudhan, C., & Partheeban, P. (2025). A critical review on recycling and reuse of construction and demolition waste materials. *Cleaner Waste Systems, 12*, 100375. https://doi.org/10.1016/j.clwas.2025.100375

Sazdovski, I., Batlle-Bayer, L., Bala, A., Margallo, M., Azarkamand, S., Aldaco, R., & Fullana-i-Palmer, P. (2024). Comparative assessment of two circularity indicators for the case of reusable versus single-use secondary packages for fresh foods in Spain. *Heliyon, 10*, e27922. https://doi.org/10.1016/j.heliyon.2024.e27922

Shajidha, H., & Mortula, M. M. (2025). Sustainable waste management in the construction industry. *Frontiers in Sustainable Cities, 7*. https://doi.org/10.3389/frsc.2025.1582239

Shooshtarian, S., Wong, P. S. P., & Maqsood, T. (2025). Circular economy in modular construction: An Australian case study. *Journal of Building Engineering, 103*, 112182. https://doi.org/10.1016/j.jobe.2025.112182

Siddiqua, A., Hahladakis, J. N., & Al-Attiya, W. A. K. A. (2022). An overview of the environmental pollution and health effects associated with waste landfilling and open dumping. *Environmental Science and Pollution Research, 29*, 58514–58536. https://doi.org/10.1007/s11356-022-21578-z

Song, W., & Zhao, K. (2024). Navigating the innovation policy dilemma: How subnational governments balance expenditure competition pressures and long-term innovation goals. *Heliyon, 10*, e34787. https://doi.org/10.1016/j.heliyon.2024.e34787

Wang, B., Yan, L., Fu, Q., & Kasal, B. (2021). A comprehensive review on recycled aggregate and recycled aggregate concrete. *Resources, Conservation and Recycling, 171*, 105565. https://doi.org/10.1016/j.resconrec.2021.105565

Waqar, A., Othman, I., Hayat, S., Radu, D., Khan, M., Galatanu, T., Almujibah, H., Hadzima-Nyarko, M., & Benjeddou, O. (2023). Building information Modeling—Empowering construction projects with end-to-end life cycle management. *Buildings, 13*, 2041. https://doi.org/10.3390/buildings13082041

Zhang, B., Larsson, J., & Reim, W. (2025). Circular business models for construction companies: A literature review and future research directions. *Sustainability, 17*, 4688. https://doi.org/10.3390/su17104688

Solid Waste Materials for the Removal of Environmental Contaminants

Gurjinder Sing, Atul Kumar, Ashish Kumar, and Anuj Kumar

1 Introduction

For the sustainable development of our society, ensuring clean water and air is an issue of high priority. The increase in air and water pollution due to recent industrial development and urbanization is putting a huge stress on ecosystem and human health. There is thus urgent need to remove pollutants from both air and water, and this is attracting enormous attention from the scientific community (Ejiohuo et al., 2024; Gupta & Gupta, 2016, 2018; Singh et al., 2018, 2020, 2022). The efficient methods for treatment of wastewater includes adsorption, degradation under sunlight, oxidative degradation, etc. While air purification involves nanomaterial-based filters, photocatalysis, biofilterations, smart devices and coatings, smog towers and air purification units, etc. Although mechanism and process of operations are entirely different in these methods but all these require nanomaterials for removal of contaminants. Thus it is very clear that development of advanced nanomaterials is very important in efficient removal/degradation of pollutants from both air and water. In this scenario, creating advanced functional nanomaterials from solid waste can be good strategy (Munien et al., 2025).

G. Sing · A. Kumar (✉)
Department of Chemistry, Akal University, Talwandi Sabo, Bathinda, Punjab, India
e-mail: atul79prashar@gmail.com

G. Sing
e-mail: gsingh39308@gmail.com

A. Kumar · A. Kumar
Department of Chemistry, School of Applied and Life Sciences, Uttaranchal University, Dehradun, Uttarakhand, India
e-mail: ashishkumar.iict@gmail.com

H. Gupta et al. (eds.), *Advanced Technologies for Assessment, Recycling, and Management of Solid Waste*, Waste as a Resource,
https://doi.org/10.1007/978-3-032-24788-9_11

Solid waste is unwanted or unusable material in solid state. It consists of solid material generated from variety of sources such as households, industries, and agriculture activities. Industrial waste includes scrap metals, wood and wood products, plastics, coal ash, construction debris, and slag. There are various sources of agriculture solid waste which include animal, food, crop, horticulture, industrial agricultures, and chemicals. Household waste includes food scraps, paper, plastic packaging, glass bottles, metal cans, yard waste, old textiles, electronic waste (e-waste), batteries, and even human waste from toilets. Improper management of these waste materials has resulted in soil, water, and air contamination. It can also have harmful effect on human health and its environmental impact can contribute to climate change (Gebrekidan et al., 2024). Various methods which are used to manage the solid waste includes landfilling, incineration, recycling, reuse, composting, waste to energy, hazardous waste treatment, source reduction and biogas generation, etc. (Abdel-Shafy & Mansour, 2019). Among these recycling and reuse are the best strategies to conserve resources and manage environment. Recycling means converting the waste in to new products and main focus of reuse is on extending the life of the original item which may be for original purpose or through repurposing (Hajam et al., 2023). These solid waste materials can also be reused to convert them into valued nanomaterials after proper treatment which can be used for the removal of contaminates from both air and water. This principle of "waste to value" is gaining large attention from scientific community because it reduces the negative impact of solid waste, reduces waste volume, conserve metal resources and thus carries a great economic and environmental significance.

In this strategy, a lot of work is being done on reusing the solid waste for environment remediation (Shovon et al., 2024). A lot of nanomaterials which are based on carbon and transition metals have been developed from the solid materials using various chemical and physical treatment methods and used for water treatment. In this scenario, various types of municipal solid wastes, biomass wastes, industrial wastes, and waste from electronic industry have been converted in to variety of adsorbents, photocatalysts, electerocatalyts, etc., for removal of contaminants from water (Mishra et al., 2024). There has also been several reports for the removal of different pollutant gases from air by using different industrial solid wastes. This is being done in various ways which include using industrial solid waste directly due to presence of metal content in them, use after physical or chemical treatment, and synthesizing nanomaterials from the metal precursors extracted from these solid wastes. Different industrial sectors such as coal and steel plants, metal extraction industries, etc., are the sources of these raw solid wastes. These industries are also the sources of various pollutant gases, for example coal fired power plants produces CO_2, SO_2, NO_x, Hg^0, etc., which are among the major pollutant gases. There are several such reports which discusses about removing these gas pollutants from the sources using materials developed from solid wastes generated from these places only (Biswal et al., 2020).

In this chapter, we summaries the work done in case of three main industrial solid wastes namely blast furnace slag (BFS), red mud (RM), and coal fly ash (CFA) for removal of contaminants from air and water. There has been no recent report in literature where use of these solid waste materials for the both air and water pollution mitigation has been reviewed simultaneously and comprehensively.

2 Blast Furnace Slag (BFS)

It is nonferrous byproduct formed during production of iron from blast furnace process and mainly consists of calcium and silicon oxides along with oxides of magnesium and aluminum. Other elements such as Fe, Ti, and Mn are also present in it. The quality of iron ore and fluxes used during smelting process decides its actual composition. Its structure depends upon the cooling rate of the slag (Iluţiu-Varvara & Aciu, 2022). It is one of the most prominent waste which is produced in iron and steel industry. It is produced about 175–225 Mt. per year worldwide (Peys et al., 2022). So if properly used it can help to reduce energy consumption and environmental pollution. It is being explored for the waste water as well as air pollution remediation along with its main use in cement production and civil engineering work. Here in following section, we discuss about various recent reports where it has been used for the removal of heavy metal ions, dyes, nitrogen, and phosphorus compounds, pharmaceuticals and waste oils from water as well as gaseous contaminants such as CO_2 and NO_x from air.

2.1 BFS for Waste Water Remediation

BFS for Metal Ions Removal from Waste Water

Sadawy et al. (2025) modified BFS by a two-step process involving hydration with water and named it as hydrated nanoscale blast furnace slag (NBFS) very recently. XRD analysis confirmed that hydrated NBFS is mainly composed of crystalline phases of $CaSiO_4$, $Ca_2SiO_4.H_2O$, $Ca(OH)_2$, and Fe_2O_3 (Fig. 1a). Its removal efficiency for Fe, Cu, and Zn was found to be 99.4%, 96.2%, and 93.7%, respectively, at 70 °C and pH 6 within 90 min. This was attributed to increase in its reactivity and surface area (Fig. 1b).

Ngoc et al. (2021) synthesized Slag-Ox materials by treating BFS with different concentration of oxalic acid. Slag-Ox which is a mesoporous material showed an adsorption (576 $mg.g^{-1}$) of Co^{2+} ions which is considerably higher than original slag. For the removal of Co^{2+} ions, this is maximum adsorption capacity of any material obtained up to date. The pH range in which Slag-Ox works consistently was found to be between 5 and 9.Adsorption kinetics and isotherm can be described by quasi-second-order model and Sip model, respectively. Hydroxyapatite-zeolite composites (HAP-ZE) were synthesized by Li et al. by alkaline melting and hydrothermal treatment (Li et al., 2021). The exchange of Mn^{2+}, NH_4^+, and phosphate ions with Ca^{2+} ions present in their structure was found to be responsible for their high adsorption

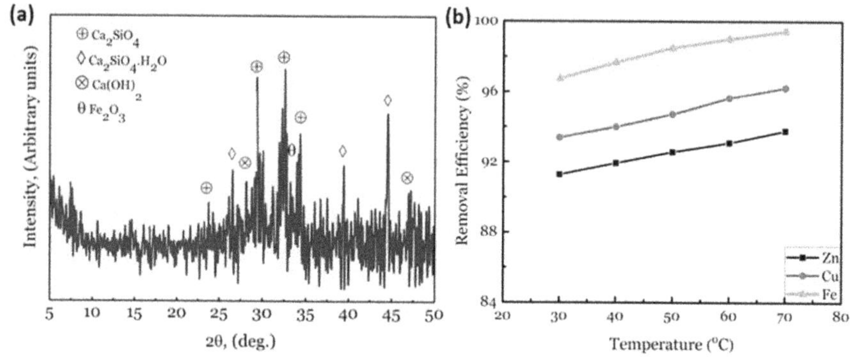

Fig. 1 **a** XRD pattern of adsorbent (hydrated NBFS). **b** Effect of temperature on the adsorption of Fe^{2+}, Cu^{2+}, and Zn^{2+} metallic ions. Reprinted from original open access article by Sadawy et al. (2025) under a Creative Commons CC BY license, http://creativecommons.org/licenses/by/4.0/

efficiency. HAP-ZE was found to be mesoporous with 44.22 $m^2\ g^{-1}$ surface area and presence of metal ions (magnesium, potassium, iron, etc.) in its structure resulted in its synthesis with in short time. Abdelbasir and Khalek (2022) used BFS for Co^{2+} and Pb^{2+} ion removal from water. Ion exchange was found to be the primary adsorption mechanism (Fig. 2). Adsorption of Co^{2+} and Pb^{2+} reached to its maximum of 43.8 and 30.2 mg g^{-1} respectively, within 1 h at pH 6. Adsorption data fit well to the Avrami isotherm and Freundlich Kinetic model.

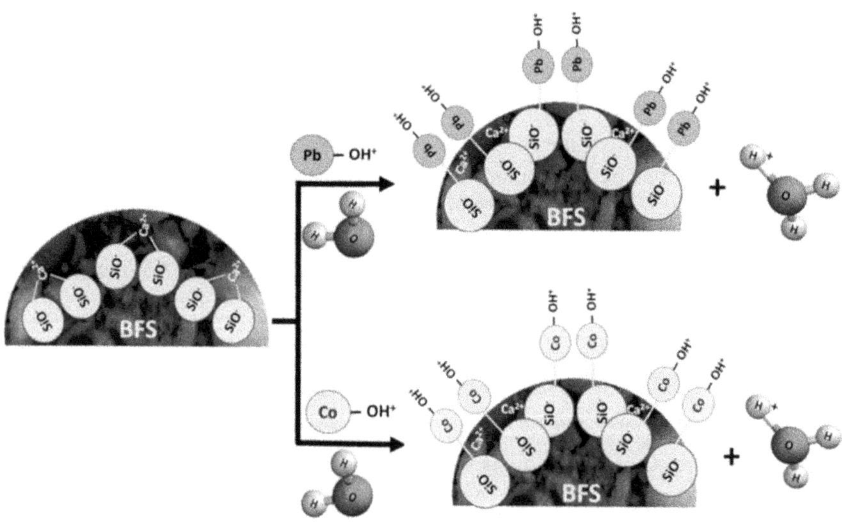

Fig. 2 Mechanism of adsorption of Pb and Co metal ions by BFS. Reprinted from original open access article by Abdelbasir and Khalek (2022) under a Creative Commons CC BY license, http://creativecommons.org/licenses/by/4.0/

BFS for Adsorption of Dye Molecules from Waste Water

To remove methylene from water Dhmees et al. (2018) prepared silica nanoparticles (NSBFS) form BFS and used both NSBFS and de-siliconized blast furnace slag (DBFS) for this purpose by varying key experimental parameters. The dye adsorption behavior of DBFS and NSBFS was systematically examined by varying key experimental parameters. Thermodynamic studies indicate that Langmuir isotherm model is the best fitted model for adsorption isotherm studies and maximum adsorption capacities (q_m) varied from 0.1370 to 0.3848 mg/g. The correlation coefficient (R^2) of the pseudo-second-order model was between 0.9986 and 0.9997, and it is the most fitted one according to kinetic studies. The surface area of NSBFS and DBFS was found to be 642.94 m^2/g and 40.05 m^2/g while pore volumes were 0.46 cm^3/g and 0.09 cm^3/g, respectively, from the BET measurements. The higher surface area and pore volume may be responsible for higher adsorption potential of NSBFS. The adsorbent's surface charge and the ionization level of adsorption sites were primarily influenced by the solution pH. DBFS exhibited a maximum dye removal efficiency of 85% at pH 10, whereas NSBFS achieved a higher removal rate of 99% at pH 8. Under identical conditions, q_m of NSBFS was 109.8 mg/g for methylene blue (MB), which was much higher than DBFS (80.8 mg/g).

Cheng et al. (2023) used titanium-bearing blast furnace slag as precursor and successfully synthesized a porous polymer which has adsorption as well as photocatalytic degradation capabilities using titanium-bearing blast furnace slag (TBBFS) as the precursor. The slag had high amount of amorphous phases and it dissolve easily in alkaline solution. Its maximum adsorption capacity for MB was found to be 97.18 mg/g. It showed monolayer chemisorption which was found to be spontaneous and exothermic. The adsorption process followed pseudo-second-order kinetics and Langmuir adsorption isotherm. The titanium containing blast furnace slag underwent selective crystallization at 1400 °C and resulted in the formation of a hybrid microstructure which consisted of crystalline $CaTiO_3$ phase embedded within an amorphous matrix. The resulting $CaTiO_3$-based mineral polymers exhibited dual functionalities—adsorption and photocatalytic degradation. MB molecules reached to the polymer surface through diffusion process before being chemically adsorbed. The electron-hole pairs generated on ionic surface of $CaTiO_3$ due to absorption of sunlight started the photocatalytic degradation process. The involvement of electrons in reduction reactions with oxidants and holes in oxidation reactions with methylene blue also prevented their recombination (Fig. 3).

Nitrogen and Phosphorus Removal from Waste Water

The presence of excess amount of nitrogen and phosphorus which are mainly present as ammonia nitrogen and phosphate, respectively, can also cause pollution. BFS can be used to remove these substances from water. Wu et al. (2022) studied the removal of phosphorus using BFS with different magnetic properties. They found that main adsorption pathway is monomolecular layer chemisorption which is indicated by the

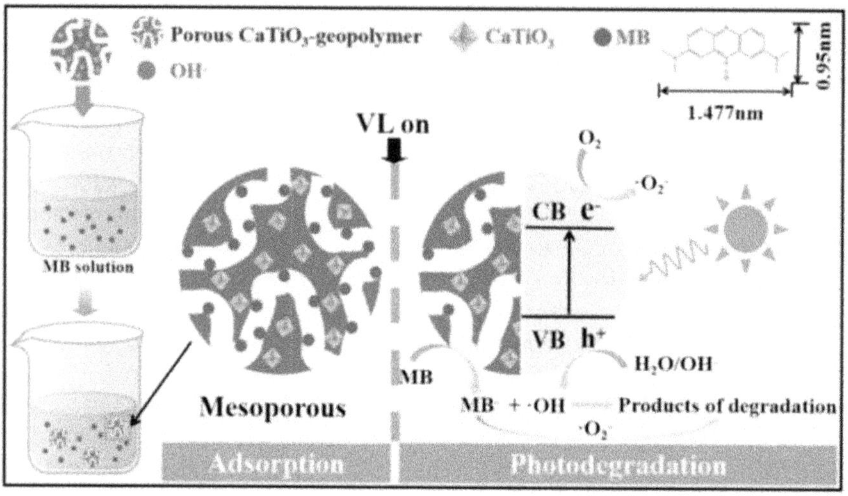

Fig. 3 Interactions of MB with porous $CaTiO_3$–geopolymer. Reprinted from original open access article by Cheng et al. (2023) under a Creative Commons CC BY license, https://creativecommons.org/licenses/by/4.0/

consistency of the data with the related isotherm and kinetic models. From X-Ray photoelectron spectroscopy (XPS), it was inferred that phosphate mainly existed as HPO_4^{2-} on the BFS surface. It is found that non-magnetic BFS with less iron content adsorbed more phosphate due to more preference of hydroxy active sites by phosphate and also on sulfuric acid elution more phosphate can be recovered from them. The variation curves of phosphate adsorption showed that non-magnetic BFS which has smaller particle size also have greater adsorption efficiency than with BFS having larger particle size.

Pharmaceutical's Removal from Waste Water

Rangappa et al. (2024) reported the tetracycline removal using oxalic acid-treated blast furnace slag (GGBS-Ox). The materials were characterized by techniques like Fourier transform infrared (FTIR) spectroscopy, X-Ray diffraction (XRD), scanning electron microscopy (SEM), XPS, Brunner Emmet Teller (BET), and dynamic light scattering (DLS). Adsorption experiments revealed that GGBS-Ox achieved a 68% removal efficiency for tetracycline at concentrations of 50 mg and 20 ppm for the adsorbent and contaminant, respectively.

Impressively, even five cycles of consecutive adsorption and desorption did not deteriorate its adsorption capacity. pH affected the ionization state and surface charge of the adsorbent and it showed point of zero charge at pH 5.30. The ionization state and surface charge of the adsorbent was pH-dependent and point of zero charge (PZC) was found to be at pH 5.30. GGBS-Ox exhibited a negatively charged surface above

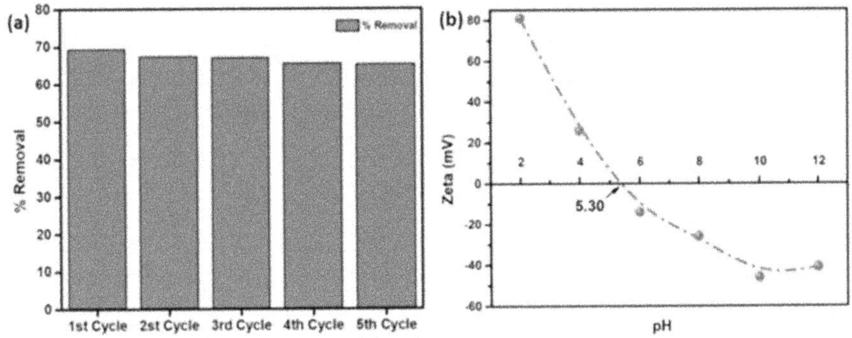

Fig. 4 **a** Regeneration of GGBS-Ox **b** zeta potential as a function of pH and PZC for GGBS−Ox. Reprinted from original open access article by Rangappa et al. (2024) under a Creative Commons CC BY license, https://creativecommons.org/licenses/by/4.0/

this pH and a positively charged surface below it. This indicates that at pH above 5.30, the surface of GGBS-Ox is negative which promoted tetracycline adsorption (Fig. 4). It was concluded that the adsorption mainly happened through filling of pores along with hydrogen bonding, and electrostatic interactions.

2.2 *BFS for Gaseous Pollutant Treatment*

The presence of oxides of magnesium, calcium, aluminum, etc., which are alkaline in nature makes BFS very reactive toward acidic gases such as CO_2, SO_2, and NO_X. Kuwahara et al. (2020) prepared mesoporous silica-calcium oxide composites (slag CaO-MS) by adding PEOPPO-PEO (P123) as pore-forming agent to the Mg and Ca extracted from BFS. It showed good CO_2 adsorption (18.8 wt %) under the optimal conditions and could be recycled at least 10 times. Yan et al. (2022) prepared material named BFSx-yTEPA by functionalizing BFS with Tetraethylenepentamine (TEPA) after acid treatment. Material with 50% loading of TEPA (BFS10.5-50TEPA) showed 5.64 mmol g^{-1} CO_2 absorption capacity at 80 °C. TEPA in the upper layer of the adsorbent acts as strong active sites which is responsible for this absorption as shown in Fig. 5, and these releases CO_2 at high temperature which is indicated by the prominent CO_2 desorption peak at higher temperature. The desorption peak at low temperature showed larger area indicating weak chemisorption and physio sorption at lower temperature. Ammonia selected catalytic reduction (NH_3-SCR) is a technology that is used for the reduction of NO_x in exhaust gases using NH_3 as reductant over catalyst surface.

NO_x gets converted to N_2 and water. Ti-Bearing blast furnace slag (Ti-BFS) is being studied for preparing catalyst for NH_3-SCR. Ti-BFS is a byproduct in iron making process from Vanadium bearing titano magnetite rocks. It is a complex material consisting mainly TiO_2 with small percentage of SiO_2, Al_2O_3, and MgO.

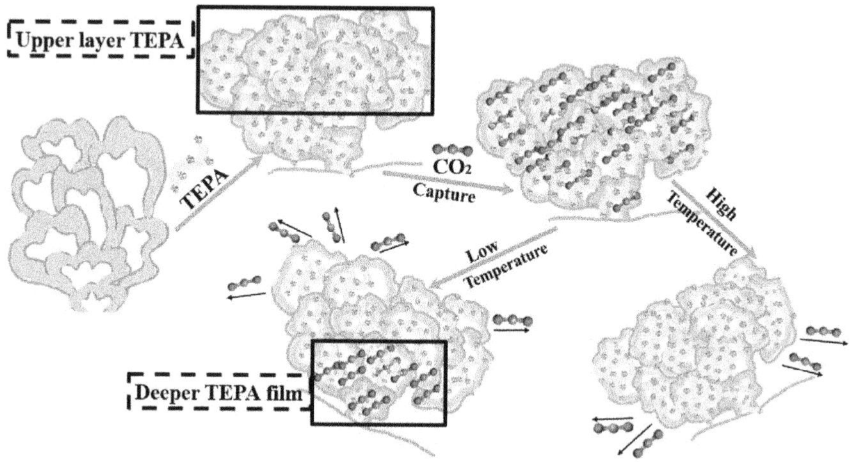

Fig. 5 CO_2 adsorption process for TEPA functionalized BFS. Reprinted from Yan et al. (2022) after taking permission from Elsevier

Due to presence of high amount of TiO_2, it can act as support for catalysts in the low temperature NH_3-SCR process (Bao et al., 2024). Xu et al. (2017) prepared Mn–Ce/Ti–BFS catalyst by wet impregnation of MnOx and CeO_2 on Ti-BFS support. This catalyst resulted in >95% conversion of NO within 180–260 °C temperature range. This could be attributed to better redox properties and greater dispersion of MnO_x due to interaction between CeO_2 and MnO_x. However, N_2 selectivity was decreased to 72.2% in the temperature range between 80 °C and 270 °C. Liu et al. (2017) found that by doping heteroatom into the Mn– Ce/Ti-BFS catalyst, its SCR temperature window can be broadened. The doping of metals like Co and Fe in raw Ti-BFS accelerated the electron transfer due to increased dispersion of loaded metal which happens because of increased lattice distortion of metal oxides.

3 Red Mud (RM)

It is solid waste which is produced during the production of alumina from Bauxite in Bayer process. Its annual production was 170 million tons in 2023.It is based on the calculation that for every ton of alumina 1.23 times red mud is produced for that year 142 million tons of alumina was produced (Silveira et al., 2021). Its main components are oxides of Fe, Al, Si, Ca, Na, and Ti, and their actual percentage depends upon the bauxite source and type of alumina production process. Due to excess alkalinity, it largely remains unused and it causes serious environmental concerns. The collaborating efforts of industry and science have found some potential applications of red mud which includes recovery of iron, cement production, synthesis of road materials, and source of rare earth elements. They are also being used as catalyst and as an adsorbent for removal of pollutant from water and air. RM needs to be activated by acidification, thermal, or hydrothermal treatment before being used as adsorbent (Wang et al., 2019).

3.1 RM for Wastewater Remediation

It is a good adsorbent for heavy metal ions, radioactive substances, compounds having phenolic groups, dyes, bacteria, and acid mine drainage from water. Milenkovic et al. (2016) used rinsed Bosnian red mud for the removal of Co^{2+} and Sr^{2+} ions and studied their adsorption mechanism. They found that adsorption of Co^{2+} ions increases with increase in pH and maximum adsorption happens at isoelectric point (pH 6.6) of red mud. Sr^{2+} adsorption increases on the negatively charged surfaces and only marginal adsorption is obtained at pH 6.9. The adsorption efficiency for the Co^{2+} ions was better in comparison to Sr^{2+} ions. Yang et al. found that simple heat treatment can enhance the toxic metal ion removal capacity of the red mud. The highest and fastest adsorption capacity (42.64 mg g^{-1}) of red mud for Cd^{2+} ions was obtained at 500 °C. The increased surface area along with the formation of inner-sphere complex (–OCdOH) and metal-metal ion exchange were found to be responsible for adsorption of metal ions (Yang et al., 2020) (Fig. 6).

Recently Tehrani et al. used red mud neutralized with sea water for the removal of Pb(II) ions from waste water obtained from battery industry. The mechanism for the

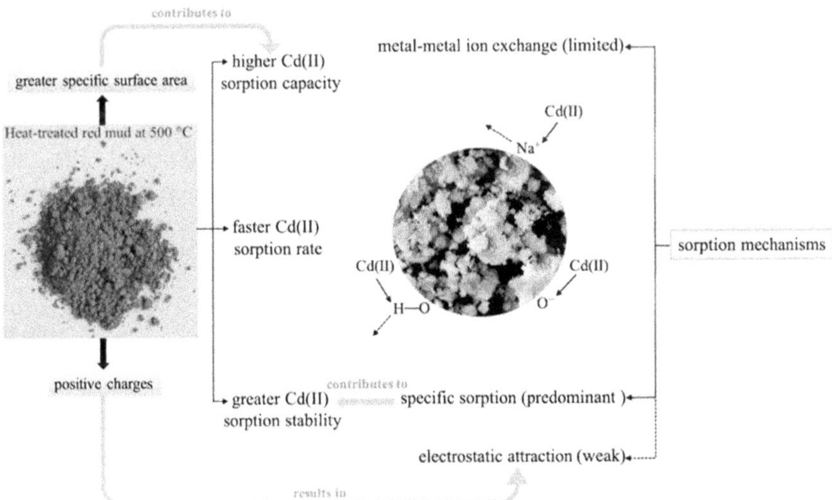

Fig. 6 Mechanism of adsorption of Cd^{2+} ions by heated 571 mud. Reprinted from Yang et al. (2020) after taking permission from Elsevier

removal of Pb(II) ions involved the formation of inner-sphere complex and precipitation. The removal percentage of Pb(II) ions was 91% with increase in pH of solution to 6 which was within permissible limits according to the Iran regulations (Tehrani et al., 2023). Hu et al. used digestion precipitation method for the activation of red mud. The adsorption capacity of 232 and 274 mg/g for MB was obtained for activated red mud and activated red mud-200, respectively, at pH 7. They demonstrated that adsorption happens by electrostatic interaction and experimental data fit well to pseudo-second-order kinetic model along with Langmuir, Freundlich, and Temkin isotherms (Hu et al., 2017).

3.2 *RM for Treatment of Gaseous Pollutants*

It is good support for medium-high SCR catalysts. Li et al. (2017) used ball milling and acid base neutralization to prepare Fe-based denitration catalyst from RM. This catalyst showed >90% NO conversion above 400 °C even in the presence of SO_2 and H_2O. It was better than commercial V–W–Ti catalyst. The characterization techniques revealed that absence of alkaline metals and high dispersion of RM particles are responsible for its high catalytic activity and it further get enhanced by SO_2 activation process. They demonstrated that its low cost, no toxicity, and high efficiency make it a potential De NO_x catalyst with industrial prospectus.

Niu et al. (2021) found that pore structures and surface features of activated carbon get improved when RM is used as additive. The surface area and pore volume of activated carbon prepared with 10% RM (named as 10RM/AC) was more than AC and it showed increased O/C ratio and π-π* groups. They used it at low temperature for the removal of SO_2 and NO_x. Sulphur capacity of 10RM/AC was 17.9% higher than AC. But high content of CaO/Na_2O in RM resulted in poisoning effect which decreased its nitrate capacity. They also found that SO_2 and NH_3 created new acid sites through sulfate species which promoted nitrate capacity of activated carbon and RM provided the inorganic components (Fe-, Ca- and Na-containing species) which catalyzed the SCR reaction (Fig. 7).

Due to higher iron content in RM, it also catalysis VOC degradation. Pande et al. (2021) modified RM by leaching with oxalic acid or hydrochloric acid which

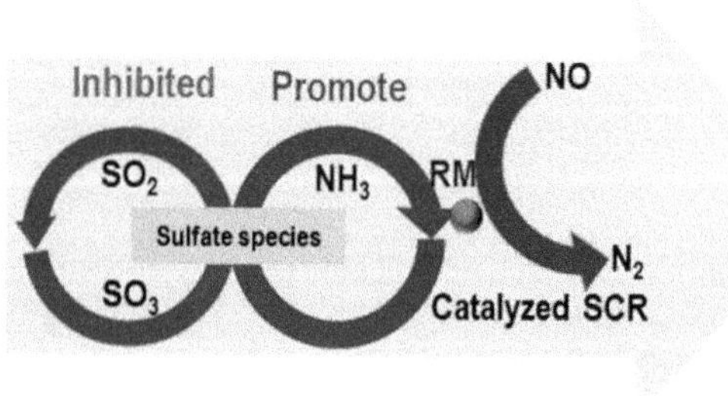

Fig. 7 Proposed mechanism of simultaneous SO_2 and NO removal along with enhanced NO adsorption. Reprinted from Niu et al. (2021) after taking permission from Elsevier

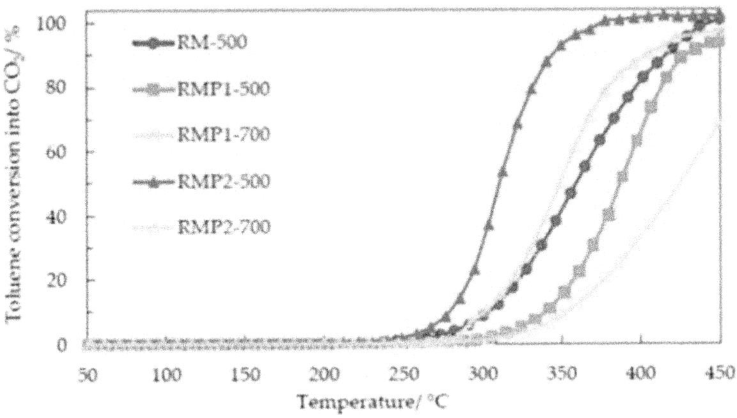

Fig. 8 Toluene conversion to CO_2 in the presence of the calcined and precipitated catalysts. Reprinted from original open access article by Pande et al. (2021) under a Creative Commons CC BY license, https://creativecommons.org/licenses/by/4.0/

was followed by precipitation or evaporation. These catalysts were used to oxidize toluene which is added in to unleaded gasoline and found in the exhaust of industrial processes. The presence of highly dispersed Fe oxides on catalyst surface led to the better performance of catalyst. The best performing catalyst (RMP2–500) showed light off effect at 310 °C (Fig. 8) and almost complete oxidation took place at 380 °C.

Red mud can also capture and store CO_2 through mineral carbonation. This happens through the chemical reaction of CO_2 with the alkaline component of RM. Very recently Suman and Tripathy (2024) mechanically activated red mud for long time and found that it improves its CO_2 adsorption capacity. They studied the effect of various factors like of mill ball size, flow rate of CO_2 and reaction time. The activated red mud showed CO_2 adsorption capacity of 18.28 mg/g in 5 h at ambient temperature and pressure. Labidi et al. (2024) found that loading organic amine on red mud also improves its CO_2 adsorption capacity. They tested different organic amines and found that MRM-20-TETA gave maximum adsorption capacity (23.94 mg/g) which is very high in comparison adsorption by pristine red mud (0.8 mg/g).

4 Coal Fly Ash (CFA)

It is one of the major byproduct of the coal power plants which are used for the production of power all over the world by burning coal. Fly ash constitutes 65–95% of the ash generated by these power plants. It generally consists of SiO_2, Al_2O_3 and oxides of alkali, alkaline, and transition metal oxides (Liu et al., 2016a). India contributes about 140 million tons in 780 million tons ash generated worldwide per year (Banerjee et al., 2014). Its health hazards include disruption of DNA repair mechanism through generation of reactive oxygen species which is caused by inhaled fine particulate matter (Matzenbacher et al., 2017). It causes smog and haze in the air and it also causes water and soil pollution (Liu et al., 2016a). As its disposal requires large space and is not environmentally friendly researchers are looking for its high value uses.

4.1 CFA in Wastewater Treatment

It is being extensively studied for environment remediation processes. Various waste water treatment methods in which fly ash can be used are highlighted in Fig. 9 (Mushtaq et al., 2019). These methods are explained with examples in following sections.

CFA in Adsorption

CFA can be used for the removal of contaminants such as heavy metal ions, dyes, phenols, pharmaceutical compounds, polychlorinated biphenyls, etc., from water in

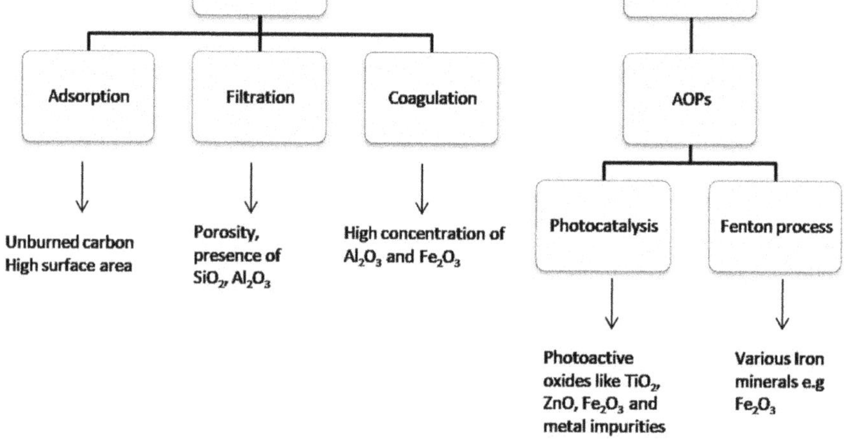

Fig. 9 CFA waste water treatment method layout. Reprinted from Mushtaq et al. (2019) after taking permission from Elsevier

one of the three ways: (i) directly, (ii) after modification, and (iii) after its transformation (Hower et al., 2017). CFA is alkaline in nature and its surface is negatively charged which by electrostatic interaction adsorb the ions. Various modification methods of fly ash includes hydrothermal treatment, functionalization of surface, and mixing with different metal oxide nanoparticles. The materials like zeolites, mesoporous silica, and geopolymers which can be derived from CFA has enhanced adsorption capacities. Table 1 gives the adsorption capacities of these materials for heavy metal ions. CFA-based materials also have potential for the adsorption of dyes. Modification of FA includes treatment with acid (Kurniasih et al., 2025), base (Thamaraiselvan et al., 2025), salt or surfactant (Mushtaq et al., 2022). It also includes microwave or sonochemical treatment (Hamidi et al., 2024). Most of these treatments improve the adsorption capacity of CFA for dyes due to increased pore size and surface area.

Fly Ash-Based Membrane Filters

Inorganic ceramic membranes are better than polymer membranes in many aspects. Due to high cost of metal oxides like Al_2O_3, SiO_2, TiO_2, ZrO_2, etc., present in them their applications are restricted (Sandhya Rani & Kumar, 2021).CFA is a suitable candidate for making porous inorganic membranes due to presence of higher percentage of SiO_2 and Al_2O_3 in it (Sawunyama et al., 2024). These membranes can be used for effluent treatment in large volumes. These membranes have high oil recovery efficiency. In filtration application pore, size is the major factor and various additives such as calcium carbonate (Kaur et al., 2015), starch (Lorente-Ayza et al., 2015), activated carbon (Das et al., 2016), and graphite (Huang et al., 2023) helps to control it. CFA can also play the same role of increasing the permeability and porosity of other ceramic membranes. It can act as permeable support for making membrane. It is due to the reason that in comparison to other metal oxide precursors such as mullite, alumina, and cordierite it has low sintering temperature and cost (Almandoz et al., 2015).

CFA Application in Coagulation

Aluminum and iron oxides are present in a suitable weight ratio required for making coagulants. The CFA-based coagulants improves their coagulation properties by absorbing pollutants on residual particles. CFA aided coagulation has advantages such as formation of larger and strengthen flocs due to increased settling rate and presence of activated silica in them. Their performance is similar to conventional and polymeric coagulants but along with that they are cheap also (Hu et al., 2015). The conversion of CFA oxides in to coagulants is mainly affected by leaching time and temperature (Hu et al., 2015). Zhang et al. treated wastewater containing suspended solid and high amount of oil with coagulants prepared from CFA. These coagulants mainly consisted of polymeric aluminum ferric chloride (Zhang et al.,

Table 1 Adsorption capacities of FA and modified FA adsorbents for various heavy metal ions

Heavy metal ion	Adsorbent	Adsorption capacity(mg/g)	References
Zn(II)	NaOH-modified CFA	26.98	Astuti et al., 2020
	Flyash(FA)-loaded nano-FeS composites	10.42	Guo et al., 2024
	Lime modified FA	33.13	Karanac et al., 2018
	Hydrothermally activated FA	7.89	Visa, 2016
Cu(II)	Fly ash-loaded nano-FeS composites	8.8	Guo et al., 2024
	CFA-red mud modified composite	135.96	Zhao et al., 2023
	FA-geopolymer	113.41	Duan et al., 2016
Cr(II)	Chitosan/FA	36.22	Adamczuk & Kołodyńska, 2015
Pb(II)	FA	48.03	Liu et al., 2016a, 2016b
	FAG	74.48	Liu et al., 2016a, 2016b
	NaOH-modified CFA CFA-RM modified composite	31.38 149.81	Astuti et al., 2020 Zhao et al., 2023
	CFA nano zeolite	196.24	Sivalingam & Sen, 2018
	Lime modified FA	26.06	Karanac et al., 2018
	AgeFe_3O_4/FA	526.5	Joshi et al., 2015
Cd(II)	CFA-RM modified composite	127.82	Zhao et al., 2023
	Zeolite synthesized from CFA and oil shale ash	236.41	Bai et al., 2025
As(IV)	FAM	19.14	Karanac et al., 2018
	Chitosan/FA	19.10	Adamczuk & Kołodyńska, 2015
	Lime modified FA	29.71	Karanac et al., 2018

2018). Gao et al. subjected coal fly ash to HCl-H_2SO_4 leaching process and prepared aluminum ferric chloride sulfate based polymeric coagulants (PAFCS). The biochemical effluent of cooking wastewater containing humic acid and fulvic acid-like matters got efficiently treated with these coagulants. Both recovery factor and strength factor were quite high in case PAFCS (Gao et al., 2023). Nguyen et al. prepared aluminum and iron salt containing complex coagulant by treating fly ash with H_2SO_4. These coagulants showed almost 83.6% efficiency for total suspended solid removal from a solution having initial TSS of 60 g/L. For the fly ash produced from coal power plants in huge amount this can considered as very useful approach (Nguyen et al., 2023).

Use of CFA in Photo Catalysis

CFA has been used in waste water remediation by photocatalytic process in one of following ways: (i) without any modification, (ii) as support, (iii) in composites, and (iv) by converting in to zeolites and geopolymers. CFA has all properties of a good photocatalytic support like high surface area and stability to oxidative radicals and light illuminations. It attaches strongly to the photocatalytic materials and increases their adsorption affinity and catalytic activity toward the contaminants (Srikanth et al., 2017). CFA has these properties due to presence of transparent materials such as glass, quartz, silica, and activated carbon in CFA. Metal or metal oxide nanoparticles (Fe, Ag, ZnO, and TiO_2) can be easily dispersed on surface of FA particles (Hlekelele et al., 2017; Thirumalai et al., 2016). Thirumalai et al. made photo catalyst by supporting ZnO on fly ash by hydrothermal decomposition method. ZnO particles had nanorod morphology and hexagonal wurtzite crystal lattice. Their reduced photoluminescent intensity indicated that recombination of electrons and holes gets reduced which resulted in enhanced degradation of azo dyes Reactive Orange 4, Rhodamine-B (RhB), and Trypan Blue. It also showed enhanced electrocatalytic oxidation of methanol than bare ZnO (Thirumalai et al., 2016). Cheng et al. prepared TiO_2/SiO_2 composite using fly ash acid sludge (FASS) and titanium sulfate as precursor (Cheng et al., 2018). FASS has amorphous silica as main component, and it is prepared by acid leaching of circulating fluidized bed fly ash. They found that 90% of RhB was degraded with in 4 h by TiO_2/SiO_2 composite which were prepared by keeping calcination temperature below 600 °C.

Babu et al. studied degradation of reactive dye in ultraviolet rays using fly ashes. They used four different ashes and from XRF analysis they found that these ashes contain different amounts of SiO_2, Al_2O_3, CaO, Fe_2O_3, and TiO_2. According to them, contribution of adsorption to degradation was almost 18%, and rest was due to oxidation processes. They concluded that in degradation process an important role is played by CaO and Fe_2O_3 along with small quantity of TiO_2 (Babu & Swaminathan, 2016).

Since fly ash itself contains oxides like Fe_2O_3, TiO_2, and CaO in small quantities doping it with photoactive metals improve its photocatalytic activity. Introduction of dopant like transition metals suppresses electron hole recombination and introduces an intermediatory level between valence band (VB) and conduction band (CB). Both these changes help to improve photocatalytic activity. Giribabu and Swaminathan (2016) doped lignite fly ash with cobalt and further used it for degrading reactive dye Acid Red 1 under UV-C radiation. It found to be more effective in degrading the dye in comparison to unmodified fly ash. Specifically, the study determined that cobalt-doped fly ash achieved 99% degradation of the dye within 60 minutes, while untreated fly ash achieved 95% degradation under the same conditions. The researchers also explored the reusability of the catalysts, finding that the cobalt-doped

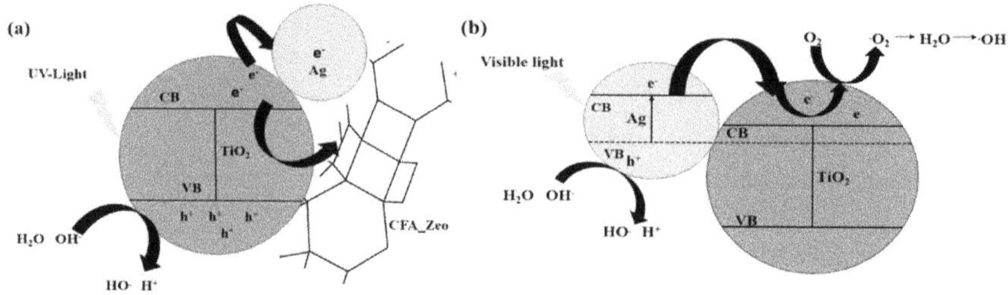

Fig. 10 A schematic representations of **a** charge separation by Ag and CFA_Zeo under UV-light and **b** SPR induced visible light driven photocatalysis. Reprinted from Hlekelele et al. (2018) after taking permission from royal society of chemistry

fly ash retained 93% of its efficiency after the first reuse, and the untreated fly ash retained 88% (Giribabu & Swaminathan, 2015).

Fly ash can also be converted to zeolite by alkali treatment which has porous structure and helps to improve the photocatalytic activity. Wang et al. prepared a composite material named as zeolite fly ash bead/TiO_2 (ZFABT) which on alkali treatment led to different morphologies. These materials gave higher adsorption capacity and photocatalytic efficiency than only TiO_2 for RhB removal. Their better efficiency was attributed to the increased adsorption as well as penetration of light to the pores (Yang et al., 2017).

Hlekelele et al. used alkali fusion hydrothermal method to prepare zeolitic material from CFA which was then composited with TiO_2 nanoparticles. These composites were then deposited with 1% Ag using deposition precipitation technique. UV-vis DRS spectra showed that band gap of TiO_2 is not affected in CFA_Zeo composites while it does in Ag containing composites. The photocatalytic activity of TiO_2 composites with and without Ag was better than uncomposited TiO_2. The better photocatalytic activity was due to the separation of electron and holes in UV-light while surface plasmon resonance by Ag was responsible in visible light (Hlekelele et al., 2018) (Fig. 10).

CFA Application in Fenton Process

CFA can be used as a catalyst for the Fenton process because it contains several active metals such as Fe, Mn, Cu, and Ni mostly in the form of their oxides'. Wang et al. converted CFA into three types of Fenton-like catalysts: MCFA powder, granular MCFA, and Fe_2O_3-supported granular MCFA. CFA showed varied catalytic activity depending on the modification method used. Acid treated, alkali treated, and thermally modified CFA achieved 64.3, 41.1, and 44.9% removal rate, respectively, for p-nitrophenol (p-NP). In forming granular MCFA, attapulgite performed better as a binder than kaoline, with removal efficiencies of 43.1% and 28.6%, respectively. For preparing granular MCFA, the optimum calcination temperature was found to

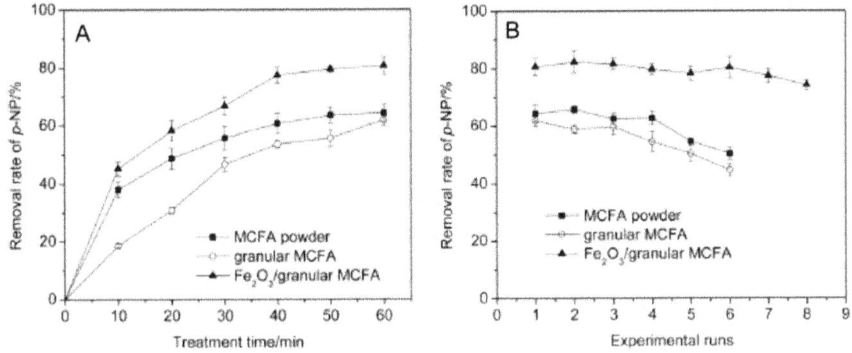

Fig. 11 Comparison of the three kinds of MCFA **a** catalytic performance and **b** stability. Reprinted from original open access article by Wang et al. (2017) under a Creative Commons CC BY-NC 3.0. https://creativecommons.org/licenses/by-nc/3.0/

be 723 K. Enhancing granular MCFA by supporting Fe_2O_3 on its surface significantly improved its catalytic activity (Fig. 11). Using $Fe(NO_3)_3$ as the precursor, a pre-adjusted pH with oxalate and a 12-h impregnation period yielded an effective Fe_2O_3/granular MCFA catalyst. Final calcination at 673 K for 2 h was essential for catalyst activation. All three MCFA catalysts demonstrated reusability for at least three cycles and achieved a 57% COD removal rate when applied to real secondary effluent. They also elucidate the mechanism of heterogeneous Fenton-like reaction after thoroughly characterizing the CFA-based catalysts (Wang et al., 2017).

Zhao et al. used magnetite nanoparticles of 40–400 nm extracted from CFA for mesotrione degradation. They used cathodic Fenton process which was much faster than microbial electrochemical degradation and traditional microbial fermentation and it degraded mesotrione at a rate of 1.6 mg/L/h. And also this magnetite material exhibited stronger degradation performance than the commercial magnetite (Zhao et al., 2023).

4.2 Fly Ash in Removal of Gaseous Pollutants

CFA holds potential as a medium–low temperature SCR catalyst as it is rich in silicon, aluminum, and iron. Enhancing its redox properties can be effectively achieved by loading metals onto its surface. Bakirov et al., (2024) evaluated FA-based catalysts supported with Cu, Fe, Ni, and V using wet impregnation methods for SCR applications. 10% Cu/FA catalyst exhibited the highest NO conversion efficiency among these in the temperature range of 270–310 °C. Calcination temperature during catalyst preparation was the main factor which affected the NO conversion efficiency. In contrast, nitric acid pre-treatment had minimal impact, suggesting that acid treatment alone is insufficient to significantly modify the surface characteristics of FA. Additionally, Wang and colleagues prepared CeO_x supported over fly ash-derived porous cordierite ceramics (Ce/FPSS) and explored it as SCR catalyst at low temperatures. The Ce/FPSS catalyst showed batter SCR activity in comparison to its commercial counterpart due to its increased oxygen exchange capabilities. This reaction happened through both the Eley-Rideal (E-R) and Langmuir-Hinshelwood (L-H) mechanisms (Wang et al., 2019). Zang et al. prepared FA supported Mn–Ce/FA catalysts for NO_n removal. They showed increased oxygen mobility and redox properties due to formation of Mn–O–Ce solid solutions (Zhang et al., 2021). However, these catalysts showed limited stability due to the depletion of oxygen vacancies during the reaction and maintained only 60% deNO_n efficiency for 18 min. Additionally, high silica zeolites such as Na–P1, and Zeolites *X* and *Y* can be synthesized from FA and these can be used for SO_2 adsorption applications (Czuma et al. 2018). Despite this, the adsorption capacity of these zeolites remained relatively low (3–3.5 mg S/g adsorbent). To address this, zeolite X were synthesized from FA by hydrothermal treatment with NaOH which was preceded by magnetic separation and this showed an improved sulfur adsorption capacity of 24 mg/g (Bao et al., 2024). These had also been several studies about the effect of chemical modifications on the SO_2/NO_n activity of CFA. Alkali or acid treatments also modify the mineral composition of FA and promote the catalytic decomposition of H_2O_2 into reactive free radicals, which facilitate the simultaneous removal of NO_n and SO_2 (Darmansyah et al. 2024; Ding et al. 2019; Meng et al., 2020). Asaoka et al. prepared granulated coal ash by treating CFA with blast furnace cement. It could remove H_2S in both aerobic and anaerobic conditions but removing capacity in anaerobic conditions was very low (5.1 mg-Sg^{-1}). This is because of the reason that manganese oxide which oxidizes H_2S to S could not be regenerated in anaerobic conditions. But the rate of H_2S removal was quite high in aerobic conditions because manganese oxide could be regenerated by dissolved oxygen (Asaoka et al., 2018). Belyaev et al. proposed a promising method which was combined with catalytic ozonation using wood ash—containing mullite ($3Al_2O_3{\cdot}2SiO_2$), magnetite (Fe_3O_4), and hematite (Fe_2O_3) for the efficient oxidation of H_2S, and other sulfur containing compounds at room temperature (Belyaev et al., 2012). The study showed that H_2S was oxidized to SO_2 at a rate of 1.3×10^{-9} $mol{\cdot}m^{-2}{\cdot}s^{-1}$ in the presence of ozone, while in an oxygen (O_2) environment, elemental sulfur was produced at a slower rate. On wood ash surfaces, O_3 operates via a different mechanism compared to O_2 which is confirmed by the fact that DMS was only oxidized to DMSO and $DMSO_2$ in the presence of ozone. The catalytic oxidation of reduced sulfur species proceeds either through radical formation or the L-H mechanism.

CFA has also emerged as a excellent raw material for zeolites aimed at volatile organic compounds (VOCs) adsorption. VOCs are organic compounds having boiling point between 50 and 250 °C, and these includes aliphatic hydrocarbons, aromatic, and oxygen-containing organic compounds. These compounds have unpleasant smell, high toxicity, and carcinogenicity along with their contribution to local ozone and PM 2.5 pollution which makes them harmful to human health. Liu et al. modified fly ash with MnO_2 using ball milling process and found that it has enhanced VOC

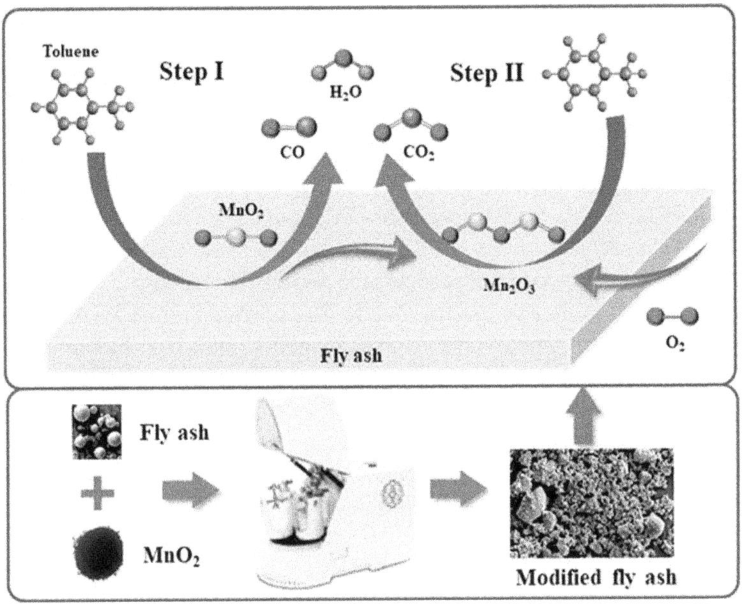

Fig. 12 Mechanism of MnO_2 modified ash for VOC oxidation. Reprinted from Liu et al. (2021) after taking permission from Elsevier

oxidation activity. They concluded that this simple modification of fly ash increased the adsorption and oxidation of toluene due to increased specific surface area and MnO_2.They proposed based on the valence state and reduction ability of MnO_2 that oxidation of adsorbed toluene in second step is preceded by the reduction of MnO_2 to Mn_2O_3in first step (Fig. 12) (Liu et al., 2021).

Various silicate-based materials can be produced from CFA with simple alkaline treatment which has adsorption capacities for common VOCs comparable to those of commercial activated carbons (ACs) (Etacheri et al., 2015).

For example, Mokrzycki et al. prepared MCM-41 from alkaline treatment of fly ash and using that they prepared series of Cu/Mn bimetallic catalyst which were investigated for the catalytic combustion of toluene. These catalyst have good thermal stability and they found that bimetallic catalyst with Mn/Cu ration of 2.5/2.5 gave the best results and it catalytic activity was comparable to the catalyst having three fold higher metal loading. It has temperature of 50% toluene conversion at 300 °C (Mokrzycki et al., 2024).

Weng et al. synthesized Y-type zeolite from CFA with 72.5% production cost when compared conventional method. They used alkali fusion hydrothermal method coupled with dealumination by ammonium hexafluorosilicate (AHFS) The adsorption capacity of optimized Y-0.4AHFS for toluene was 67.7 $mg \cdot g^{-1}$ under 80% relative humidity and low toluene concentration which was three times higher than pristine zeolite. The increased Si/Al ratio along with tailored framework and high micro porosity were found to be responsible for enhanced performance (Weng et al., 2025).

Coal fly ash and its derived materials also been considered as very suitable material for CO_2 mitigation. It helps in CO_2 removal through adsorption and catalytic methods (Abdelkader Labidi et al., 2024). These have been used as adsorbent for CO_2 either directly or after modification with organic or organic molecules. Islam et al. synthesized zeolite Na P1 (SynZ) from CFA and used these materials for CO_2 capturing and subsequent potential fire radiation and methylene blue removal. The synthesized zeolite showed high surface area and cation exchange capacity and its CO_2 adsorption capacity was 30 fold higher in comparison to CFA (Islam et al., 2024). Zhou et al. extracted silica from CFA and synthesized hierarchal porous SiO_2 which allowed them to effectively incorporate pentaethylenehexamine(PEHA) organic amine in to these lamellar-spherical stacking structures. Their hierarchal porous structure the pore blocking due to amine functionalization was minimum and they showed 4.815 mmol/g CO_2 adsorption capacity at 70 °C which could b maintained up to 10 consecutive cycles (Zhou et al., 2025).

CO_2 methanation is another approach in which fly ash-derived catalysts has been used. This process produces various chemicals such as methanol, methane, formic acid, etc. In one report, Czuma' s studied physicochemical properties of Ni/zeolite X synthesized from CFA and studied its chemical and structural properties using various techniques like XRD, XRF, etc. They studied CO_2 conversion in a tubular fixed-bed quartz reactor in a temperature between 250 and 470 °C and found that it resulted in almost 50% CO_2 conversion (Czuma et al., 2020). In another study, Dong et al. (2020) used co-impregnation to synthesize bimetallic Ni–Re/CCFA catalyst from CFA. They characterized the material using techniques like BET isotherm, XRD, TPD, SEM, and TEM (Fig. 13). They found that surface area of CCFA increased from 10.43 to 23.53 $m^2 . g^{-1}$ at the Ni dispersion 14.8% in Ni–Re/CCFA. They found that in the temperature range between 350 to 450 °C conversion of CO_2 was 94% while CH_4 selectivity was 50% (Dong et al., 2020).

5 Global Impact of Using Solid Wastes in Environmental Remediation

It is very clear from above studies that solid waste materials such as RM, BFS, and CFA have definite positive impact on solving waste accumulation and environmental pollution problem. These materials which are produced as solid waste worldwide and using them for waste water remediation and air pollution control can surly add value to the waste. But there are several challenges to the implementation of this strategy such as transportation cost, potential for secondary pollution, high processing cost, scale and consistency, technical and regulatory hurdles which need to be addressed carefully.

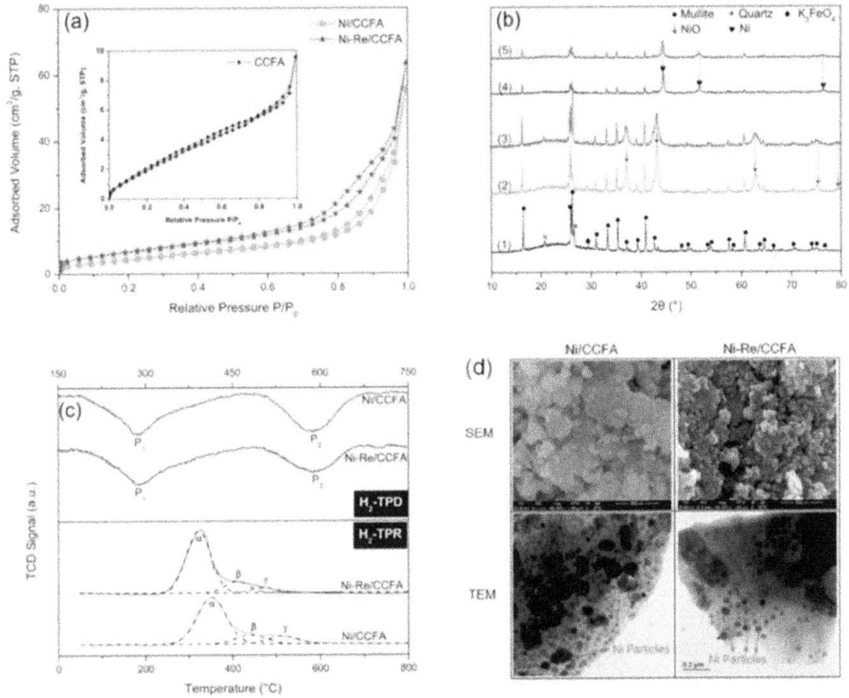

Fig. 13 **a** N_2 adsorption isotherms **b** XRD patterns in sequence 1–5 represents CCFA, as-prepared Ni/CCFA, as-prepared Ni-Re/CCFA, reduced Ni/CCFA and reduced Ni-Re/CCFA respectively **c** H_2-TPR plus H_2-TPD curves, and **d** SEM plus TEM morphologies of reduced catalysts. Reprinted from Dong et al. (2020) after taking permission from Elsevier

6 Conclusion

From the literature review discussed above, it must be concluded that very extensive studies are going across globe for the using solid waste materials especially industrial wastes like BFS, RM, and CFA in environmental applications. BFS which is rich in silica and calcium oxide after treatment with water, oxalic acid, etc., can be used for the elimination of metal ions, dyes, drugs, etc., from water. It can also be modified with other metal oxides like TiO_2, MnO_2, and CeO_2 to be effectively used for removal of gaseous pollutants like CO_2 and NOx. RM which is by product of Bayer process also found to be efficient for the removal of Cd(II), Pb(II) and other heavy metal ions from water after activation. It is also possible to remove gases like CO_2, SO_2, and NO_x from air using RM because of its metal oxide content. CFA has also found extensive uses as adsorbent, coagulant and in the synthesis of inorganic membranes. Materials derived from fly ash such as zeolites, mesoporous silica, and other materials can be put in various applications of contaminant remediation from both air and water especially in photocatalysis, Fenton process, VOCs degradation, H_2S and NO_x

removal, CO_2 capturing and CO_2 methanation. In conclusion, it can be said that these waste materials can act as two sided swords to solve the problems of solid waste accumulation and environmental pollution. But this approach of using solid waste material for environment remediation also faces challenges as mentioned in the previous paragraph which need to be addressed with great caution.

References

Abdel-Shafy, H. I., & Mansour, M. S. M. (2019). Solid waste issue: Sources, composition, disposal, recycling, and valorization. *Egyptian Journal of Petroleum, 27*(4), 1275–1290. https://doi.org/10.1016/j.ejpe.2018.07.003

Abdelbasir, S. M., & Khalek, M. A. (2022). From waste to waste: Iron blast furnace slag for heavy metal ions removal from aqueous system. *Environmental Science and Pollution Research, 29*(38), 57964–57979. https://doi.org/10.1007/s11356-022-19834-3

Labidi, A., Ren, H., Zhu, Q., Liang, X., Liang, J., Wang, H., Sial, A., Padervand, M., Lichtfouse, E., Rady, A., Allam, A. A., & Wang, C. (2024). Coal fly ash and bottom ash low-cost feedstocks for CO2 reduction using the adsorption and catalysis processes. *Science of the Total Environment, 912*(20), 169179–169179. https://doi.org/10.1016/j.scitotenv.2023.169179

Adamczuk, A., & Kołodyńska, D. (2015). Equilibrium, thermodynamic and kinetic studies on removal of chromium, copper, zinc and arsenic from aqueous solutions onto fly ash coated by chitosan. *Chemical Engineering Journal, 274*(15), 200–212. https://doi.org/10.1016/j.cej.2015.03.088

Asaoka, S., Jadoom, W. A., Ishidu, T., Okamura, H., Oikawa, T., & Nakamoto, K. (2018). Removal of hydrogen sulfide with granulated coal ash under aerobic and anaerobic conditions. *Journal of Environmental Chemical Engineering, 6*(4), 4665–4670. https://doi.org/10.1016/j.jece.2018.06.063

Bai, S., Chang, Z., Ren, Z., Zhao, Y., & Pang, L. (2025). Adsorption of Cd^{2+} ions onto zeolites synthesized from a mixture of coal fly ash and oil shale ash in aqueous media. *RSC Advances, 15*(15), 11293–11300. https://doi.org/10.1039/d5ra00752f

Bakirov, A., Zhunusov, A., Oleinikova, N., & Ramazanova, R. (2024). Studies of the sintering method of ash and slag waste for the production of alumina from self-disintegrating sinters. *Heliyon, 10*(21), e39117. https://doi.org/10.1016/j.heliyon.2024.e39117

Banerjee, S., Sharma, G. C., Chattopadhyaya, M. C., & Sharma, Y. C. (2014). Kinetic and equilibrium modeling for the adsorptive removal of methylene blue from aqueous solutions on of activated fly ash (AFSH). *Journal of Environmental Chemical Engineering, 2*(3), 1870–1880. https://doi.org/10.1016/j.jece.2014.06.020

Bao, J., Sun, X., Ning, P., Li, K., Yang, J., Wang, F., Shi, L., & Fan, M. (2024). Industrial solid wastes to environmental protection materials for removal of gaseous pollutants: A review. *Green Energy & Environment, 10*(1). https://doi.org/10.1016/j.gee.2024.01.006

Belyaev, A. V., Sozin, A. Y., Aleksandrov, Y. A., & Churbanov, M. F. (2012). Homogeneous oxidation of hydrogen selenide with ozone in an argon-oxygen medium. *Russian Journal of Applied Chemistry, 85*(12), 1919–1922. https://doi.org/10.1134/s107042721212021x

Biswal, S., Pahlevani, F., & Sahajwalla, V. (2020). Wastes as resources in steelmaking industry — Current trends. *Current Opinion in Green and Sustainable Chemistry, 26*, 100377. https://doi.org/10.1016/j.cogsc.2020.100377

Munien, C., Rathilal, S., & Tetteh, E. K. (2025). Challenges and prospects of TiO_2-based Photocatalysis for wastewater treatment: Keyword analysis. *Catalysts, 15*(9), 801–801. https://doi.org/10.3390/catal15090801

Cheng, Y., Luo, F., Jiang, Y., Li, F., & Wei, C. (2018). The effect of calcination temperature on the structure and activity of TiO_2/SiO_2 composite catalysts derived from titanium sulfate and fly ash acid sludge. *Colloids and Surfaces A: Physicochemical and Engineering Aspects, 554*(5), 81–85. https://doi.org/10.1016/j.colsurfa.2018.06.032

Cheng, Y., Wang, K., Li, P., Guo, H., Yan, B., Chen, D., & Zhao, W. (2023). A porous Geopolymer containing Ti-bearing blast furnace slag: Synthesis, characterization, and adsorption-Photodegradation studies towards methylene blue removal under visible light condition. *Molecules, 28*(9), 3673. https://doi.org/10.3390/molecules28093673

Czuma, N., Baran, P., Franus, W., Zabierowski, P., & Zarębska, K. (2018). Synthesis of zeolites from fly ash with the use of modified two-step hydrothermal method and preliminary SO_2 sorption tests. *Adsorption Science & Technology, 37*(1–2), 61–76. https://doi.org/10.1177/0263617418810607

Czuma, N., Zarębska, K., Motak, M., Gálvez, M. E., & Da Costa, P. (2020). Ni/zeolite X derived from fly ash as catalysts for CO_2 methanation. *Fuel, 267*(1), 117139. https://doi.org/10.1016/j.fuel.2020.117139

Darmansyah, D., You, S.-J., & Wang, Y.-F. (2024). Modification of recycled industrial solid waste incineration Fly ash as a low-cost catalyst for NOx removal. *Aerosol and Air Quality Research, 24*(3), 230247. https://doi.org/10.4209/aaqr.230247

Das, B., Chakrabarty, B., & Barkakati, P. (2016). Preparation and characterization of novel ceramic membranes for micro-filtration applications. *Ceramics International, 42*(13), 14326–14333. https://doi.org/10.1016/j.ceramint.2016.06.125

Dhmees, A. S., Khaleel, N. M., & Mahmoud, S. A. (2018). Synthesis of silica nanoparticles from blast furnace slag as cost-effective adsorbent for efficient azo-dye removal. *Egyptian Journal of Petroleum, 27*(4), 1113–1121. https://doi.org/10.1016/j.ejpe.2018.03.012

Ding, S., Hülsey, M. J., Pérez-Ramírez, J., & Yan, N. (2019). Transforming energy with single-atom catalysts. *Joule, 3*(12), 2897–2929. https://doi.org/10.1016/j.joule.2019.09.015

Dong, X., Jin, B., Cao, S., Meng, F., Chen, T., Ding, Q., & Tong, C. (2020). Facile use of coal combustion fly ash (CCFA) as Ni-re bimetallic catalyst support for high-performance CO_2 methanation. *Waste Management, 107*(15), 244–251. https://doi.org/10.1016/j.wasman.2020.04.014

Duan, P., Yan, C., Zhou, W., & Ren, D. (2016). Development of fly ash and iron ore tailing based porous geopolymer for removal of cu(II) from wastewater. *Ceramics International, 42*(12), 13507–13518. https://doi.org/10.1016/j.ceramint.2016.05.143

Ejiohuo, O., Onyeaka, H., Akinsemolu, A., Nwabor, O. F., Siyanbola, K. F., Tamasiga, P., & Al-Sharify, Z. T. (2024). Ensuring water purity: Mitigating environmental risks and safeguarding human health. *Water Biology and Security, [online], 4*(2), 100341. https://doi.org/10.1016/j.watbs.2024.100341

Etacheri, V., Di Valentin, C., Schneider, J., Bahnemann, D., & Pillai, S. C. (2015). Visible-light activation of TiO_2 photocatalysts: Advances in theory and experiments. *Journal of Photochemistry and Photobiology C: Photochemistry Reviews, 25*, 1–29. https://doi.org/10.1016/j.jphotochemrev.2015.08.003

Forghani Tehrani, G., Rubinos, D. A., Rahimi-Nia, A., Bagherian, G., & Goudarzi, N. (2023). Lead(II) removal from aqueous solutions and battery industry wastewater by sorption using seawater-neutralized red mud. *International journal of Environmental Science and Technology, 20*(4), 3713–3732. https://doi.org/10.1007/s13762-023-04801-3

Gao, J. M., Wang, B., Li, W., Li, C., Guo, L. Y., & Cheng, F. (2023). High-efficiency leaching of Al and Fe from fly ash for preparation of polymeric aluminum ferric chloride sulfate coagulant for wastewater treatment. *Separation and Purification Technology, 306*, 122545. https://doi.org/10.1016/j.seppur.2022.12254

Gebrekidan, T. K., Weldemariam, N. G., Hidru, H. D., Gebremedhin, G. G., & Weldemariam, A. K. (2024). Impact of improper municipal solid waste management on fostering one health approach in Ethiopia — Challenges and opportunities: A systematic review. *Science in One Health, [online], 3*, 100081–100081. https://doi.org/10.1016/j.soh.2024.100081

Giri, S., & Swaminathan, G. (2016). Comparative study on synergetic degradation of a reactive dye using different types of Fly ash in combined adsorption and Photocatalysis. *Journal of The*

Institution of Engineers (India) Series A, 97(3), 239–246. https://doi.org/10.1007/s40030-016-0166-y

Giribabu, P. V. S., & Swaminathan, G. (2015). Synergetic degradation of reactive dye acid red 1 by cobalt-doped lignite fly ash. *Desalination and Water Treatment, 57*(36), 1–8. https://doi.org/10.1080/19443994.2015.1082509

Giri Babu, P. V. S., & Swaminathan, G. (2016). Comparative Study on Synergetic Degradation of a Reactive Dye Using Different Types of Fly Ash in Combined Adsorption and Photocatalysis. *Journal of The Institution of Engineers (India): Series A, 97*(3), 239–246. https://doi.org/10.1007/s40030-016-0166-y

Guo, X., Fu, H., Gao, X., Zhao, Z., & Hu, Z. (2024). Study on the adsorption of Zn(II) and cu(II) in acid mine drainage by fly ash loaded nano-FeS. *Scientific Reports, 14*(1). https://doi.org/10.1038/s41598-024-58815-z

Gupta, H., & Gupta, B. (2016). Adsorption of polycyclic aromatic hydrocarbons on banana peel activated carbon. *Desalination and Water Treatment, 57*(20), 9498–9509.

Gupta, H., & Gupta, B. (2018). Vehicular tire as potential adsorbent for the removal of polycyclic aromatic hydrocarbons. *Polycyclic Aromatic Compounds, 38*(4), 354–368.

Hajam, Y. A., Kumar, R., & Kumar, A. (2023). Environmental waste management strategies and vermi transformation for sustainable development. *Environmental Challenges, [online], 13*, 100747. https://doi.org/10.1016/j.envc.2023.100747

Hamidi, A., Bastami, Y. S., Shakibania, S., Mahmoudi, A., Rashchi, F., & Vahidi, E. (2024). Fly ash treatment via conventional and microwave-assisted organic acid leaching: Kinetics and life cycle assessment. *Environmental Science and Pollution Research, 31*(20), 30039–30058. https://doi.org/10.1007/s11356-024-33208-x

Hlekelele, L., Franklyn, P. J., Djiike, F., & Durbach, S. H. (2018). Novel synthesis of Ag decorated TiO_2 anchored on zeolites derived from coal fly ash for the photodegradation of bisphenol-A. *New Journal of Chemistry, 42*(3), 1902-1912.https://doi.org/10.1039/C7NJ02885G

Hower, J. C., Groppo, J. G., Graham, U. M., Ward, C. R., Kostova, I. J., Maroto-Valer, M. M., & Dai, S. (2017). Coal-derived unburned carbons in fly ash: A review. *International Journal of Coal Geology, 179*(15), 11–27. https://doi.org/10.1016/j.coal.2017.05.007

Hu, K., Zhao, Q., Chen, W., Tang, F., & Xu, H. (2015). Preparation and performance of fly ash-based coagulants in chemically enhanced primary treatment of domestic wastewater. *Desalination and Water Treatment, 57*(10), 1–10. https://doi.org/10.1080/19443994.2014.996774

Hu, Z.-P., Gao, Z.-M., Liu, X., & Yuan, Z.-Y. (2017). High-surface-area activated red mud for efficient removal of methylene blue from wastewater. *Adsorption Science & Technology, 36*(1–2), 62–79. https://doi.org/10.1177/0263617416684348

Huang, J., Chen, H., Qi, R., Yang, J., Li, Z., & Zhang, H. (2023). Porous ceramic membranes from coal fly ash with addition of various pore-forming agents for oil-in-water emulsion separation. *Journal of Environmental Chemical Engineering, 11*(3), 109929–109929. https://doi.org/10.1016/j.jece.2023.109929

Iluțiu-Varvara, D.-A., & Aciu, C. (2022). Metallurgical wastes as resources for sustainability of the steel industry. *Sustainability, 14*(9), 5488. https://doi.org/10.3390/su14095488

Islam, M. R., Biswas, B., & Naidu, R. (2024). CO_2 capture using zeolite synthesized from coal Fly ash and its subsequent utilization for fire retardation and dye removal. *ACS Sustainable Resource Management, 1*(4), 799–809. https://doi.org/10.1021/acssusresmgt.4c00033

Karanac, M., Đolić, M., Veličković, Z., Kapidžić, A., Ivanovski, V., Mitrić, M., & Marinković, A. (2018). Efficient multistep arsenate removal onto magnetite modified fly ash. *Journal of Environmental Management, 224*, 263–276. https://doi.org/10.1016/j.jenvman.2018.07.051

Kaur, H., Bulasara, V. K., & Gupta, R. K. (2015). Preparation of kaolin-based low-cost porous ceramic supports using different amounts of carbonates. *Desalination and Water Treatment, 57*(32), 15154–15163. https://doi.org/10.1080/19443994.2015.1068226

Kuwahara, Y., Hanaki, A., & Yamashita, H. (2020). A direct conversion of blast furnace slag to a mesoporous silica–calcium oxide composite and its application in CO_2 captures. *Green Chemistry, 22*(12), 3759–3768. https://doi.org/10.1039/d0gc00722f

Hlekelele, L., Franklyn, P. J., Dziike, F., & Durbach, S. H. (2017). Novel synthesis of ag decorated TiO_2 anchored on zeolites derived from coal fly ash for the photodegradation of bisphenol-a. *New Journal of Chemistry, 42*(3), 1902–1912. https://doi.org/10.1039/c7nj02885g

Li, C., Li, X., Yu, Y., Zhang, Q., Li, L., Zhong, H., & Wang, S. (2021). A novel conversion for blast furnace slag (BFS) to the synthesis of hydroxyapatite-zeolite material and its evaluation of adsorption properties. *Journal of Industrial and Engineering Chemistry, 105*, 63–73. https://doi.org/10.1016/j.jiec.2021.08.017

Li, C., Zeng, H., Liu, P., Yu, J., Guo, F., Xu, G., & Zhang, Z. (2017). The recycle of red mud as excellent SCR catalyst for removal of NO_x. *RSC Advances, 7*(84), 53622–53630. https://doi.org/10.1039/c7ra10348d

Liu, J., Dong, Y., Dong, X., Hampshire, S., Zhu, L., Zhu, Z., & Li, L. (2016a). Feasible recycling of industrial waste coal fly ash for preparation of anorthite-cordierite based porous ceramic membrane supports with addition of dolomite. *Journal of the European Ceramic Society, 36*(4), 1059–1071. https://doi.org/10.1016/j.jeurceramsoc.2015.11.012

Liu, J., Shi, N., Wang, T., Yang, J., Liu, P.-F., Zhang, Y., & Pan, W.-P. (2021). Significant enhancement of VOCs conversion by facile mechanochemistry coupled MnO_2 modified fly ash: Mechanism and application. *Fuel, 304*, 121443–121443. https://doi.org/10.1016/j.fuel.2021.121443

Liu, Y., Yan, C., Zhang, Z., Wang, H., Zhou, S., & Zhou, W. (2016b). A comparative study on fly ash, geopolymer and faujasite block for Pb removal from aqueous solution. *Fuel, 185*(1), 181–189. https://doi.org/10.1016/j.fuel.2016.07.116

Liu, R., Xu, Y., Ye, F., Xia, F., & Xia, R. (2017). Influence of transition metal (Fe, Co, and Ag) doping on the MnO_x–CeO_2/Ti-bearing blast furnace slag catalyst for selective catalytic reduction of NO_x with NH_3 at low temperature. *New Journal of Chemistry, 41*(19),11299–11307. https://doi.org/10.1039/c7nj00775b

Lorente-Ayza, M.-M., Orts, M. J., Pérez-Herranz, V., & Mestre, S. (2015). Role of starch characteristics in the properties of low-cost ceramic membranes. *Journal of the European Ceramic Society, [online], 35*(8), 2333–2341. https://doi.org/10.1016/j.jeurceramsoc.2015.02.026

Almandoz, M. C., Pagliero, C., Ochoa, N. A., & Marchese, J. (2015). Composite ceramic membranes from natural aluminosilicates for microfiltration applications. *Ceramics International, 41*(4), 5621–5633. https://doi.org/10.1016/j.ceramint.2014.12.144

Joshi, M. K., Pant, H. R., Liao, N., Kim, J. S., Kim, H. J., Park, C. B., & Kim, C. S. (2015). In-situ deposition of silver–iron oxide nanoparticles on the surface of fly ash for water purification. *Journal of Colloid and Interface Science, 453*(1), 159–168. https://doi.org/10.1016/j.jcis.2015.04.044

Kurniasih, M., Aprilita, N. H., Roto, R., & Mudasir, M. (2025). Modification of coal Fly ash for high capacity adsorption of methylene blue. *Case Studies in Chemical and Environmental Engineering, 11*, 101101. https://doi.org/10.1016/j.cscee.2025.101101

Matzenbacher, C. A., Garcia, A. L. H., Dos Santos, M. S., Nicolau, C. C., Premoli, S., Corrêa, D. S., de Souza, C. T., Niekraszewicz, L., Dias, J. F., Delgado, T. V., Kalkreuth, W., Grivicich, I., & da Silva, J. (2017). DNA damage induced by coal dust, fly and bottom ash from coal combustion evaluated using the micronucleus test and comet assay in vitro. *Journal of Hazardous Materials, 324*(Pt B), 781–788. https://doi.org/10.1016/j.jhazmat.2016.11.062

Meng, Z., Wang, C., Wang, X., & Li, H. (2020). Efficient and stable catalyst of α-FeOOH for NO oxidation from coke oven flue gas by the catalytic decomposition of gaseous H_2O_2. *RSC Advances, 10*(14), 8207–8211. https://doi.org/10.1039/d0ra00533a

Milenković, A. S., Smičiklas, I. D., Šljivić-Ivanović, M. Z., Živković, L. S., & Vukelić, N. S. (2016). Effect of experimental variables onto $Co^{2+}$and Sr^{2+}sorption behavior in red mud-water suspensions. *Journal of Environmental Science and Health, Part A, 51*(8), 1–12. https://doi.org/10.1080/10934529.2016.1159884

Mokrzycki, J., Fedyna, M., Duraczyńska, D., Marzec, M., Panek, R., Franus, W., Bajda, T., & Karcz, R. (2024). Mesoporous silica MCM-41 from Fly ash as a support of bimetallic cu/Mn catalysts for toluene combustion. *Materials, 17*(3), 653–653. https://doi.org/10.3390/ma17030653

Mushtaq, F., Zahid, M., Bhatti, I. A., Nasir, S., & Hussain, T. (2019). Possible applications of coal fly ash in wastewater treatment. *Journal of Environmental Management, 240*(15), 27–46. https://doi.org/10.1016/j.jenvman.2019.03.054

Mushtaq, S., Aslam, Z., Ali, R., Aslam, U., Naseem, S., Ashraf, M., & Bello, M. M. (2022). Surfactant modified waste ash for the removal of chloro and nitro group substituted benzene from wastewater. *Water Science and Technology, 86*(8). https://doi.org/10.2166/wst.2022.324

Ngoc, T., Vivas, E. L., & Cho, K. (2021). Oxalated blast-furnace slag for the removal of cobalt(II) ions from aqueous solutions. *Journal of industrial and engineering chemistry/Journal of Industrial and Engineering Chemistry - Korean Society of Industrial and Engineering Chemistry, 95*, 57–65. https://doi.org/10.1016/j.jiec.2020.12.003

Nguyen, N. T., Trinh, Q. H., Nguyen, T. T. H., Tran, V. D., & Ho, T. T. H. (2023). Study on the modification of fly ash as a coagulant. *Ministry of Science and Technology, Vietnam, 65*(2), 20–25. https://doi.org/10.31276/vjste.65(2).20-25

Niu, J., Zhang, H., Li, L., & Guo, Y. (2021). Cost-effective activated carbon (AC) production from partial substitution of coal with red mud (RM) as additive for SO_2 and NO_x abatement at low temperature. *Fuel, 293*(1), 120448–120448. https://doi.org/10.1016/j.fuel.2021.120448

Pande, G., Selvakumar, S., Ciotonea, C., Giraudon, J.-M., Lamonier, J.-F., & Batra, V. S. (2021). Modified red mud catalyst for volatile organic compounds oxidation. *Catalysts, 11*(7), 838. https://doi.org/10.3390/catal11070838

Peys, A., Isteri, V., Yliniemi, J., Yorkshire, A. S., Lemougna, P. N., Utton, C., Provis, J. L., Snellings, R., & Hanein, T. (2022). Sustainable iron-rich cements: Raw material sources and binder types. *Cement and Concrete Research, [online], 157*, 106834. https://doi.org/10.1016/j.cemconres.2022.106834

Rangappa, H. S., Mon, P. P., Herath, I., Madras, G., Lin, C., & Subrahmanyam, C. (2024). Modeling tetracycline adsorption onto blast furnace slag using statistical and machine learning approaches. *Sustainability, 16*(1), 464. https://doi.org/10.3390/su16010464

Mishra, R. K., Singh, B., & Acharya, B. (2024). A comprehensive review on activated carbon from pyrolysis of lignocellulosic biomass: An application for energy and the environment. *Carbon Resources Conversion, 7*(4), 100228–100228. https://doi.org/10.1016/j.crcon.2024.100228

Sadawy, M. M., Fayed, Saad, M., Sharafedin, H. E., Salem, A. I., Hosny, A., & Li, S. (2025). Hydrated Nano-blast furnace slag: A green and efficient adsorbent for industrial wastewater treatment. *Journal of Metals*. https://doi.org/10.1007/s11837-025-07692-6

Sandhya Rani, S. L., & Kumar, R. V. (2021). Insights on applications of low-cost ceramic membranes in wastewater treatment: A mini-review. *Case Studies in Chemical and Environmental Engineering, 4*, 100149. https://doi.org/10.1016/j.cscee.2021.100149

Sawunyama, L., Olatunde, O. C., Oyewo, O. A., Bopape, M. F., & Onwudiwe, D. C. (2024). Application of coal fly ash based ceramic membranes in wastewater treatment: A sustainable alternative to commercial materials. *Heliyon, 10*(2), e24344–e24344. https://doi.org/10.1016/j.heliyon.2024.e24344

Shovon, S. M., Akash, F. A., Rahman, W., Rahman, M. A., Chakraborty, P., Hossain, H. M. Z., & Monir, M. U. (2024). Strategies of managing solid waste and energy recovery for a developing country—A review. *Heliyon, 10*(2), e24736. https://doi.org/10.1016/j.heliyon.2024.e24736

Silveira, N. C. G., Martins, M. L. F., Bezerra, A. C. S., & Araújo, F. G. S. (2021). Red mud from the aluminium industry: Production, characteristics, and alternative applications in construction materials—A review. *Sustainability, 13*(22), 12741. https://doi.org/10.3390/su132212741

Singh, S., Parveen, N., & Gupta, H. (2018). Adsorptive decontamination of rhodamine-B from water using banana peel powder: A biosorbent. *Environmental Technology and Innovation, 12*, 189–195.

Singh, S., Kumar, A., & Gupta, H. (2020). Activated banana peel carbon: A potential adsorbent for rhodamine B decontamination from aqueous system. *Applied Water Science, 10*(8), 185. https://doi.org/10.1007/s13201-020-01274-4

Singh, S., Gupta, H., Dhiman, S., & Sahu, N. K. (2022). Decontamination of cationic dye brilliant green from the aqueous media. *Applied Water Science, 12*(4), 61.

Sivalingam, S., & Sen, S. (2018). Rapid ultrasound assisted hydrothermal synthesis of highly pure nanozeolite X from fly ash for efficient treatment of industrial effluent. *Chemosphere, 210*, 816–823. https://doi.org/10.1016/j.chemosphere.2018.07.091

Srikanth, B., Goutham, R., Badri Narayan, R., Ramprasath, A., Gopinath, K. P., & Sankaranarayanan, A. R. (2017). Recent advancements in supporting materials for immobilised photocatalytic applications in waste water treatment. *Journal of Environmental Management, 200*(15), 60–78. https://doi.org/10.1016/j.jenvman.2017.05.063

Suman, N., Tripathy, A. (2024). CO_2 capture using red mud & mechanically-activated red mud and its kinetics under ambient conditions. *Chemical Engineering Journal, 498*(15), 155609–155609.https://doi.org/10.1016/j.cej.2024.155609

Thamaraiselvan, C., Wisdom, C., & Wickramasinghe, S. R. (2025). Efficient dye removal from wastewater using fly ash-based adsorbents. *Separation Science and Technology*, 1–12. https://doi.org/10.1080/01496395.2025.2567479

Thirumalai, K., Balachandran, S., & Swaminathan, M. (2016). Superior photocatalytic, electrocatalytic, and self-cleaning applications of Fly ash supported ZnO nanorods. *Materials Chemistry and Physics, 183*(1), 191–200. https://doi.org/10.1016/j.matchemphys.2016.08.018

Visa, M. (2016). Synthesis and characterization of new zeolite materials obtained from fly ash for heavy metals removal in advanced wastewater treatment. *Powder Technology, 294*, 338–347. https://doi.org/10.1016/j.powtec.2016.02.019

Wang, L., Sun, N., Tang, H., & Sun, W. (2019). A review on comprehensive utilization of red mud and Prospect analysis. *Minerals, 9*(6), 362. https://doi.org/10.3390/min9060362

Wang, N., Chen, J., Zhao, Q., & Xu, H. (2017). Study on preparation conditions of coal fly ash catalyst and catalytic mechanism in a heterogeneous Fenton-like process. *RSC Advances, 7*(83), 52524–52532. https://doi.org/10.1039/c7ra09925h

Weng, S., Ye, F., Jing, C., He, X., Zhong, X., Tian, J., Yang, C., Lei, K., Huang, H., & Fu, M. (2025). Coal Fly ash-based Y zeolite modified by ammonium Hexafluorosilicate and its enhanced toluene adsorption under high humidity conditions. *Journal of Environmental Chemical Engineering, 13*(5), 118356–118356. https://doi.org/10.1016/j.jece.2025.118356

Astuti, W., Chafidz, A., Al-Fatesh, A. S., & Fakeeha, A. H. (2020). Removal of lead (Pb(II)) and zinc (Zn(II)) from aqueous solution using coal fly ash (CFA) as a dual-sites adsorbent. *Chinese Journal of Chemical Engineering, 34*, 289–298. https://doi.org/10.1016/j.cjche.2020.08.046

Wu, X., Zhan, R., Liu, L., Lan, J., Zhao, N., & Wang, Z. (2022). Phosphorus adsorption on blast furnace slag with different magnetism and its potential for phosphorus recovery. *Water, 14*(16), 2452–2452. https://doi.org/10.3390/w14162452

Xu, Y., Liua, R., Yea, F., Jiad, F., & Ji, L. (2017). Manganese-cerium oxide (MnO_x-CeO_2) catalysts supported by titanium-bearing blast furnace slag for selective catalytic reduction of nitric oxide with ammonia at low temperature. *Journal of the air & waste management association, 67*(8), 899–909. https://doi.org/10.1080/10962247.2017.1302021

Yang, L., Wang, F., Hakki, A., Macphee, D. E., Liu, P., & Hu, S. (2017). The influence of zeolites fly ash bead/TiO_2 composite material surface morphologies on their adsorption and photocatalytic performance. *Applied surface science, 392*(15), 687–696. https://doi.org/10.1016/j.apsusc.2016.09.023

Yang, T., Wang, Y., Sheng, L., He, C., Sun, W., & He, Q. (2020). Enhancing cd(II) sorption by red mud with heat treatment: Performance and mechanisms of sorption. *Journal of Environmental Management, 255*(1), 109866. https://doi.org/10.1016/j.jenvman.2019.109866

Yan, H., Zhang, G., Wang, Y., Liu, J., Li, G., Zhao, Y., Xu, Y., & Lv, Y. (2022). A green synthesis strategy for low-cost multi-porous solid CO_2 adsorbent using blast furnace slag. *Fuel, 329*(1), 125380–125380. https://doi.org/10.1016/j.fuel.2022.125380

Zhang, Y., Li, M., Liu, D., Hou, X., Zou, J., Ma, X., Shang, F., & Wang, Z. (2018). Aluminum and iron leaching from power plant coal fly ash for preparation of polymeric aluminum ferric chloride. *Environmental Technology, 40*(12), 1568–1575. https://doi.org/10.1080/09593330.2018.1426639

Zhang, Z., Yang, Y., & Cheng, D. (2021). Fabrication of N-doped TiO2 Nanorods and its visible-light photocatalytic activity for resorcinol degradation in wastewater. *International Journal of Electrochemical Science, 16*(11), 21115. https://doi.org/10.20964/2021.11.01

Zhao, Y., Luan, H., Yang, B., Li, Z., Song, M., Li, B., & Tang, X. (2023). Adsorption of Pb, cu and cd from water on coal Fly ash-red mud modified composite material: Characterization and mechanism. *Water, 15*(4), 767–767. https://doi.org/10.3390/w15040767

Zhou, R., Yang, Z., Yang, X., Zhang, M., Gao, J., Hussain, S., & Li, Q. (2025). Design of amine-functionalized fly ash-based hierarchically porous SiO_2 adsorbents and mechanistic insights into CO_2 adsorption. *Journal of CO2 Utilization, 101*, 103213. https://doi.org/10.1016/j.jcou.2025.103213

Microorganisms for the Treatment of Toxic Solid Wastes for Environmental Sustainability

Samruddhi Mane, Naman Dosi, Mitali Mahajan, and Nidhi Gupta

1 Introduction

The rate of industrialisation and pollution, increasing population, has led to the release of various toxic chemicals and harmful pollutants. The waste from municipalities, biomedical waste, electronic waste and agricultural waste are directly associated with these pollutants that affect human health and well-being. Microorganisms create metabolites that break down and change solid waste-based pollution to various small molecules (Ghosh et al., 2023). The site can be restored with potential microbial communities which act toward solid waste valorisation, utilising agricultural solid waste, organic solid waste, plastic solid waste and various industrial solid waste. Municipal waste collection in India is about 42 million tonnes/year, and of this, only 80% gets processed or treated. As compared to 2016, the waste processing rate has increased by 18% (data from CPCB and Swachh Bharat Mission). In Nigeria, waste generation and disposal is a very serious problem. Even after the segregation of waste and having various disposal grades, the treatment of liquid waste remains a problem. The adaptation of specific use of microorganisms in waste management has been a very successful and proven solution in Nigeria (Ezeudu and Ugochukwu, 2024). Kazakhstan also has been facing challenges with industrial waste management. Here, it was found that strains of *Pseudomonas* sp. Such as KPA1, KPA2, KPA3, KPA4, KPA5 and *Bacillus* sp. isolated from the soils of the Shymkent-Kokys landfill can recycle domestic waste with a high content of cellulose and organic substances up to 95–97%. The findings can be used to develop more effective organic cellulosic waste management strategies and improve the environmental sustainability of various industries (Akhmet et al., 2024). Different groups of bacteria, their types,

Naman Dosi and Mitali Mahajan are equally contributed authors.

S. Mane · N. Dosi · M. Mahajan · N. Gupta (✉)
Department of Biotechnology, TERI School of Advanced Studies, Vasant Kunj, New Delhi, India
e-mail: nidhi.gupta2@terisas.ac.in; nidhimic@gmail.com

H. Gupta et al. (eds.), *Advanced Technologies for Assessment, Recycling, and Management of Solid Waste*, Waste as a Resource,
https://doi.org/10.1007/978-3-032-24788-9_12

mechanisms and other factors affect the microbial remediation. The two factors or properties due to which microbes are generally used, first, their ability to multiply rapidly and secondly, their ability to easily manipulate, which enhances the bioremediation properties. It is an easy and sustainable technology, as compared to traditional chemical and physical methods (Ebsa et al., 2024). We have discussed the role of different microorganisms in waste management, sewage treatment, oil and heavy metal contamination and other environmental applications. The different types of heavy metal released from industrial units associated with mercury, chromium and mercury, which are directly released in the environment, can be remediated using various consortia of microorganisms. An intensive study has been done on the same. The oil spills in the oceanic body can be treated by specific microbial enzymes, and many microbes' benefits are being researched; a few of them are discussed below. This chapter will focus on the role of microorganisms in solid waste composting and bioaccumulation of heavy metals, and on how microbial enzymes can act in the treatment of wastewater. The waste generated by pulp and paper industries is remediated in various ways (Madan et al., 2018; Shahzad et al., 2024).

2 Different Types of Solid Wastes

Solid waste includes a wide variety of waste that is often generated from different settings like households, industries or from agricultural and forest-based activities (Ziraba et al., 2016). When it happens to be dealing with waste, we need to properly understand and classify it so that appropriate technologies can be used to process and deal with it. Waste can be categorised according to its content, source, degree of hazard, biodegradability, etc.

Solid Waste in Municipalities: Municipal waste is the kind of waste that usually originates from houses, schools, institutions, restaurants or from cleaning of the public areas, roads, etc. It includes organic material like food waste, scraps or leaves, branches, etc. that were obtained while cleaning. It also includes inorganic waste like plastic, glass, paper and clothes. This kind of waste exists in the highest ratio and is owed to rapid urbanisation. It is predicted that the waste generated by 2050 will be reached according to the World Bank (2018), 3.4 billion tonnes. Most of this waste usually ends up in landfills, which can be very degrading toward the environment if not segregated and processed properly. The organic fraction of the waste degrades and leads to the release of the greenhouse gas methane and plays a major role toward global warming. While recycling and waste management are very efficient in high-end cities like Tokyo or San Francisco (Hoornweg & Bhada-Tata, 2012), it is very difficult in underdeveloped and developing countries, which lack a proper system for waste management (Wang et al., 2025). **Industrial Waste**: This kind of waste arises from the mining, manufacturing or processing of certain goods. It includes metal shavings, chemicals, sludge, plastic, etc. The form in which the waste varies from industry to industry. Like the textile industry produces dye and chemical-ridden waste, and metal-based industries produce chemicals, scales and scrapes, they are all very harmful to the environment if not taken care of while being discarded. It not only harms the ecosystem, but it is also highly detrimental toward plants, animals and humans. While there are laws in place to take care of and minimise the adverse effects in countries like the USA, called the Resource Conservation and Recovery Act (Kumar et al., 2021), waste minimisation and disposal may not be as controlled in other countries, but is gaining momentum. **Biomedical Waste**: This kind of waste is generated from hospitals, healthcare institutions, diagnostic centres or research-oriented institutions. It usually consists of syringes, gauze or other sharp waste which met humans or animal systems. It can also include expired medications, bodily fluids, etc. which may potentially be a biohazard. 15% or so of the total garbage produced in such cases is hazardous and requires special handling and treatment before disposal (WHO, 2018). This kind of waste is especially risky if discarded without treatment, as it can cause disease transmission and outbreaks, especially in serious diseases like HIV or hepatitis. These can be handled by segregating and treating with methods like autoclave, incineration, microwave treatment or even chemical disinfection, which is very prevalent in the modern healthcare waste management (Gupta & Boojh, 2006). **Electronic Waste**: All electronic devices and their components, like mobiles, computers, TVs, circuit boards, etc. comprise E-waste. High rates of technological advancement and distribution among the population have made this a rapidly developing form of waste. Over 59.4 million tonnes were generated in 2022 (Forti et al., 2020). Such waste contains rare earth metals, which can be both valuable and very toxic to the environment and to us. Such waste needs to be recycled to not only obtain and reuse these metals but also to prevent such elements from leaching into the surrounding environment and causing pollution. Recycling methods include mechanical separation, hydrometallurgical recovery and extended producer responsibility frameworks (EPR) (Tsydenova & Bengtsson, 2011). **Agricultural Waste**: This kind of waste includes all the biodegradable residues from farms, like crop stalks, leaves, husks or animal waste like manure, litter, etc. These kinds of wastes, although biodegradable and very helpful if properly degraded, also contain pesticide and herbicide contaminants. If these could be treated and removed, it would not only eliminate the environmental threat that pollution poses but also be used to produce clean energy. Anaerobic decomposition can produce biocompost, biogas. These are clean and renewable forms of energy. This also reduces pollution load, greenhouse gases and pathogens load from the environment (Holm-Nielsen et al., 2009).

The generation of food waste has significantly increased via various food and fermentation industries. In recent studies, it has been noted that the high-temperature-resistant strains, with high degradation capacities, were isolated and screened for a new approach to prepare microbial formulations (Liu et al., 2022). Other types of waste include trash from building and destruction that was produced during the construction of new buildings or during maintenance. It includes bricks, pipes, concrete, drywall, etc. This kind of waste is generally not very harmful to the environment, but it leads to space occupation in landfills. Hazardous waste is the kind of waste that can be toxic, flammable or corrosive to the environment. This includes batteries, solvents, chemicals, etc. and presents a significant risk to the environment. These can be managed by high temperature incineration, stabilisation, vitrification,

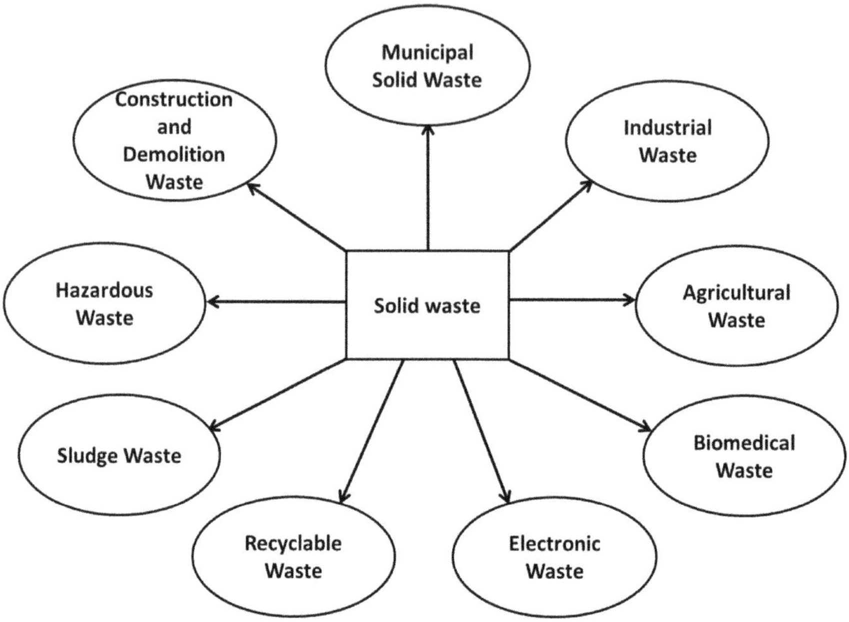

Fig. 1 Different types of solid waste

etc. Sludge is a waste byproduct generated when treating wastewater. It contains organic matter, nutrients, heavy metals and pathogens. If treated, this sludge can be either used as manure or safely disposed of in landfills (Fytili & Zabaniotou, 2008). Inert waste is the kind of waste which doesn't have any detrimental effects on the environment, but occupies space, which can pose logistical issues with dealing with other forms of waste. Recyclable waste is the kind of waste which is capable of being retrieved, processed and turned into new goods. These materials consist of copper, steel, aluminium, paper and glass. Recycling not only saves raw material but also reduces net waste generated, promotes conservation of energy in the finding and processing of the materials (Hamidinasab & Nabavi-Pelesaraei, 2025) (Fig. 1).

3 Microbes as Vital Additives for Solid Waste Recycling

The natural process of composting is where organic waste, such as food scraps, leaves, branches or agricultural waste, is broken down and converted into nutrient-rich soil-like material called compost. It has manure-like properties and is very useful in farming. Microorganisms play a major role in composting, or breaking down the complex organic matter into a simpler form using enzymes. Naturally, composting can occur with the help of fungi or microbes, but herein we can control the number of microbes by governing the amount of inoculum added for waste, which we want to decompose. This can increase the efficiency, safety, quality and production of compost, which is of use to us. Microbial composting is especially advantageous as it leads to the production of organic fertilisers, promotes sustainable agriculture and reduces the amount of organic waste that might end up in landfills, which in turn reduces the amount of GHG that is being produced and released uncontrollably (Rastogi et al., 2020).

Composting is a discontinuous exothermic process which involves different sets of microorganisms working at different time points in a composting cycle. The first stage is called the mesophilic phase, where the temperature ranges from 25–40 °C. Energy-dense substances like carbohydrates and proteins are prevalent during this stage and are broken down by primary decomposers. These include fungi, actinobacteria and bacteria from other genera like *Pseudomonas, Lactobacillus.* Insects like worms, mites and millipedes also help in the turnover and help in decomposition of living things. This also increases the temperature. Now the thermophilic phase starts, in which the decomposition of the substrate is accelerated. Temperature ranges from 35 to 65 °C. There, thermophilic fungi and bacteria take over the decomposition process. This can help in killing pathogenic bacteria, phage or other miscreants that might be harmful (Kausar et al., 2012). Insects turning the material helps to spread the temperature evenly and decompose organic matter such as cellulose, hemicellulose, fats and lignin. As the temperature decreases, most bacteria die off, allowing the remaining actinomycetes and fungi (such as Trichoderma and Aspergillus) to take over the decomposition. These organisms break down cellulose and lignin by secreting specialised enzymes (Ding et al., 2025). The addition of selected microbe cultures or a mixture of them can have a synergistic effect on decomposition. This can increase the degradation by enhancing certain biochemical pathways, which will reduce the composting time. This can be obtained by the addition of high-speed decomposers and condition optimisation, which can reduce the time required from months to just weeks (Goyal et al., 2005). The addition of microbe inoculum, which contains your property of interest, also improves the decomposition rate as well as overall decomposition. For example, fungi or bacteria producing cellulose and lignin peroxidase can help improve the degradation of plant or paper-based waste (Hassen et al., 2001). There are certain strains which lead to a decrease in bad odour produced during decomposition. The soil's nutritional quality can be improved by including certain bacteria, which can fix nitrogen into the soil (*azobacter, rhizobium*), hence enhancing the soil's fertility (Tiquia-Arashiro, 2014). Many different strains of microbes are used as additives. It includes *Bacillus subtilis, Lactobacillus* spp., *Pseudomonas fluorescens* and *Cellulomonas* spp., fungi like: *Trichoderma harzianum, Aspergillus niger, Penicillium chrysogenum* and *Phanerochaete chrysosporium* and *actinomyces like Streptomyces* (Pointing, 2001; Nakasaki et al., 1993; Haug, 1993). All these microbes can synergistically act and accelerate the process of breakdown. Due to the diversity of microbes in the culture, various biomolecules can be simultaneously broken down, increasing the efficiency as compared to a single microbe inoculum (Mehta et al., 2014).

3.1 Types of Composting Systems

There are several different composting methods that are used according to the kind of waste being processed, conditions and resources available via using microbes as additives (Fig. 2).

Windrow Composting

It is a method in which the organic material is arranged in long rows. It is an aerobic process in which the rows are regularly turned to facilitate air provision to the microbes present. This is a simple and cost-effective method for large amounts of waste, which can be from municipal solid waste or agricultural waste. The waste is continuously monitored and made to ensure that there is no leaching or odour. It has been studied that certain pesticides in the soil can be degraded by using this method (Ali et al., 2016). In India, landfilled waste soil-like materials can be processed by windrow composting. The impact assessment on using various impact categories has shown that higher environmental benefit (Cheela et al., 2023).

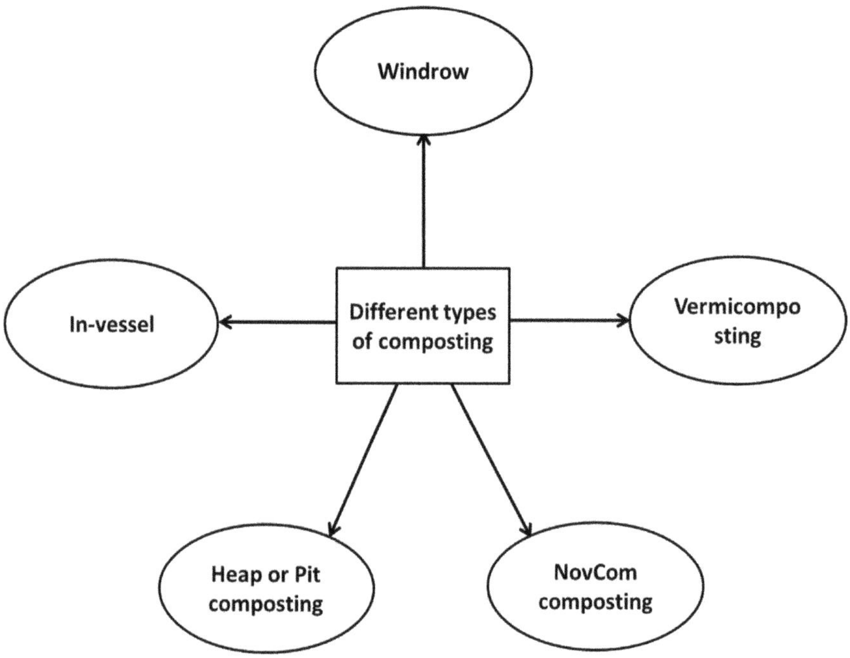

Fig. 2 Different types of composting techniques

In-Vessel Composting

It is a method that utilises containers in which the solid waste is added, the temperature is regulated with appropriate aeration and moisture content for the microbes to carry out decomposition. In such methods, there are minimal environmental effects on the process and leading to a much faster decomposition. The microbes in the mix are highly controlled and are useful for industrial use (Gajalakshmi & Abbasi, 2008). The silver nanoparticles (AgNPs) and titanium dioxide nanoparticles (Ag-TiO2NPs) found in municipal solid waste, under in-vessel composting, are used for microbial degradation and in the application of composting. The presence of nanoparticles does not inhibit the growth of microbes or their metabolic activity. The fluorescence excitation-emission matrix (EEM) spectroscopy is the tool used to evaluate the impact of composting. The results suggest that the humin concentration remains stable in the soil containing NPs during in-vessel composting and has no effect on microbes (Stamou & Antizar-Ladislao, 2016).

Vermicomposting

This is a method in which earthworms are used. This is one of the most common methods used, which requires minimal to no resources. Worms are deployed, which can break down organic matter into semi-digested matter, which is much easier to decompose for the microorganisms. It produces very high-quality compost, although it requires some temperature and moisture modulation to work effectively (Domínguez & Edwards, 2004). In urban residential area of Pune, vermicomposting for the treatment of solid organic waste using earthworm—*Eudrilus eugeniae* was studied. The results indicated that the speed of decomposting increases three times with almost 60 per cent reduction in waste volume (Chaudhari et al., 2011).

Heap or Pit Composting

It is also referred to as farmyard manure composting. It involves the layering of organic matter, like dung or leaves or branches, in piles or in fills. It requires minimal resources but a lot of time since there is no fixed microbe inoculum. Due to its ease, it is often used in rural settings or on farms (Bera et al., 2024). Inoculating the heap with thermophilic bacteria (Bacillus species) ensures the temperature of the pile (generally agricultural waste) remains high for longer times. Further, the actinomycetes increase the enzymatic activity for the breakdown of lignin and cellulose, which in turn enhances the decomposition speed of waste in the field (Rastogi et al., 2020).

Novcom Composting

It is an advanced method that uses a bio-enriched solution prepared from plant extracts like *Cynodon dactylon* (Doob grass), *Sida cordifolia* (Bel) and *Ocimum basilicum* (Basil). It uses a mix of multiple microbes in a certain amount and is regularly applied to the organic matter. This leads to rapid decomposition and gives high-quality compost. It is very cost-effective, time-saving and has highly reduced greenhouse gas release (Ali et al., 2023). The Novcom compositing has been studied to increase the soil nutrient value, post-soil application. The microflora has increased. The greenhouse gas emission level under this composting has found to be 17 times lower than that of other biodegradation techniques (Bera et al., 2022).

4 Microbes' Contribution to Heavy Metal Bioaccumulation in Municipal Solid Waste

Heavy metals like lead (Pb), cadmium (Cd), mercury (Hg), chromium (Cr), nickel (Ni) and arsenic (As) are hazardous, persistent and non-biodegradable; hence, heavy metal contamination in MSW poses a threat. Batteries, electronics, pigments, paints, plastics waste, etc., are the sources of these metals. Since they can seep into soil and water, if left unsupervised, they can accumulate in the food chain and be highly detrimental to the environment as well as human health if improperly treated (Alloway, 2013). Certain microorganisms, such as Pseudomonas, Flavobacterium, Bacillus and Nitrosomonas, have characterised pathways and the potential to treat pollutants (Sharma et al., 2022).

Microbial bioaccumulation, a type of bioremediation, has become a popular and sustainable way to remove heavy metals from contaminated trash and landfill leachates. Heavy metals can be taken in by microbes by active transport into the cell or through passive adsorption through the cell wall. Since there are several microorganisms that are known to evolve to tolerate the accumulation of toxic metals, these can be harnessed as a source of clean-up in the environment containing waste (Gadd, 2010).

4.1 Mechanism of Microbial Bioaccumulation

There are two main ways in which bacteria can be used for bioaccumulation of heavy metals: passive biosorption and active bioaccumulation.

Passive Biosorption: Biosorption is a method by which the metals bind to negatively charged groups like carboxyl, phosphate and hydroxyl on the surface of microbes like bacteria, algae and fungi. Such a metabolically independent method is known as biosorption. The speed and reversibility make it possible to reuse biosorbent materials and recover metals effectively by using both live and dead microbial biomass. However, dead biomass is preferred due to its stability and lack of nutritional needs (Volesky, 2007). Metal ions can be actively transported into microbial cells during bioaccumulation, where they are either bound with intracellular molecules such as phytochelatins and metallothioneins or sequestered in vacuoles. Active bioaccumulation offers long-term metal containment and is especially useful for low-concentration pollutants, despite being slower and more selective than biosorption (Gadd, 2010).

4.2 Microbial Groups Involved in Heavy Metal Accumulation

Several microbial taxa are renowned for their high metal ion affinity and tolerance. It is well-known that certain bacterial species of *Pseudomonas, Bacillus, Cupriavidus* and *Desulfovibrio* absorb heavy metals. For example, *Cupriavidus metallidurans* can accumulate Zn, Cd and Ni and carry plasmids encoding metal resistance (Mergeay et al., 2003). Several fungi species like *A. niger, P. chrysogenum* and *Rhizopus arrhizus* are examples of species that are effective biosorbents due to their large cell wall structures, which contain high concentrations of proteins and polysaccharides. To improve metal precipitation and immobilisation, fungi can also alter their surroundings, such as pH and redox conditions (Hu et al., 2025). Cyanobacteria and algae are also known to have bioaccumulation properties. Alginates, which bind metal ions, are abundant in the biomass of algae, particularly brown algae. They may bioaccumulate metals through both adsorption and active absorption modes. Cyanobacteria such as *Spirulina* and *Nostoc* are appropriate for use in leachate treatment systems and artificial wetlands (Chugh et al., 2022).

4.3 Application of Microbial Bioaccumulation in MSW Management

Bioaccumulation has several applications when it comes to the treatment of solid waste generated. Leachate from landfills contains a variety of pollutants, both organic and inorganic, such as heavy metals. Before discharge, heavy metal concentrations can be successfully decreased in bioreactors and constructed wetlands that have been inoculated with metal-resistant microorganisms. For example, metals are less mobile when precipitated as insoluble sulphides by sulphate-reducing bacteria such as *Desulfovibrio* spp., (Serrano et al., 2011). Trace metals may still be present in the final product of anaerobic digestion or composting MSW. Compost is safer for agricultural use by inoculating it with fungi or actinomycetes that can immobilise metals, which lowers metal bioavailability, hence stabilising the digestate that is obtained after composting (Park et al., 2011). Heavy metals can also be extracted from liquid or gaseous waste material using microbial consortia in engineered systems like packed-bed biofilters and membrane bioreactors. These methods enable great efficiency and continuous operation, especially when utilising microbial biofilms or immobilised cells (Sharma et al., 2022; Wang & Chen, 2009).

4.4 Factors Influencing Microbial Bioaccumulation

Several biological and environmental factors impact the efficiency of microbial metal accumulation by affecting the ability of microorganisms to interact with and bind heavy metals. Temperature and pH are important variables because most microorganisms absorb metals well in specific pH ranges. For example, mildly acidic environments, usually between pH 4.5 and 6.0, are generally more conducive to the process for fungal biomass to biosorb heavy metals, including lead and cadmium (Sharma et al., 2022). Additionally, temperature influences enzyme activity and microbial metabolism, both of which influence the efficiency of biosorption. The significance of metal concentration is twofold: extremely high concentrations can become toxic, preventing microbial growth and activity, while trace levels may not trigger the required absorption systems in microorganisms. Hence, to maximise microbial biosorption, a balanced concentration between metal ions and microbes must be maintained. Further, by occupying the same adsorption sites, competing ions like calcium, magnesium or sodium might obstruct the binding of target heavy metals, decreasing overall effectiveness. Lastly, biosorption capability is strongly influenced by the kind and density of biomass utilised and how efficiently metals are bound to the microbial surface depends on several factors, including surface area, the microbial cell wall's structure and the presence of functional groups (such as carboxyl, hydroxyl or amine groups) (Gupta & Diwan, 2017).

Recombinant organisms like bacteria and fungi with improved metal-binding abilities can be made with the help of recent advancements in genetic and metabolic engineering. To increase specificity and metal uptake capacity, genes producing metallothioneins, metal transporters or stress response proteins might be inserted into the right microbial hosts. For instance, metallothionein-expressing recombinant *Escherichia coli* has demonstrated a notable buildup of mercury and cadmium from wastewater. The microorganisms also tend to change the pathways under high-toxicity stress, which we studied, can be used for degradative properties (Rafeeq et al., 2023).

5 Mechanisms of Electroactive Microorganism Detoxification under Toxicity Stress

Electroactive microorganisms (EAMs) fill a special ecological niche by using extracellular electron transfer (EET) to connect metabolic activities to extracellular electron acceptors. This skill is a potent detoxifying tactic in high-toxicity situations, in addition to promoting energy conservation. This section includes the biochemical underpinnings of stress caused by pollutants, intracellular defences and constraints, the specific extracellular detoxification pathways mediated by EET, engineered improvements to support these pathways and practical applications demonstrating their effectiveness (Feng et al., 2022) (Fig. 3).

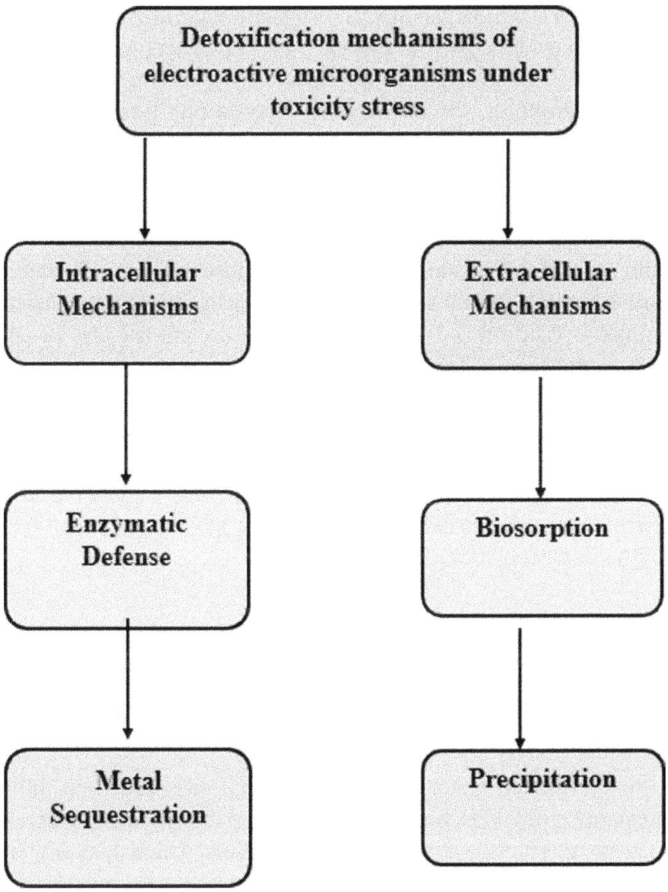

Fig. 3 Detoxication mechanism of electroactive microorganism under toxicity stress

This flowchart illustrates the primary detoxification strategies employed by electroactive microorganisms when exposed to environmental toxicity. These mechanisms are categorised into intracellular and extracellular pathways. Intracellular mechanisms involve enzymatic defences and metal sequestration, while extracellular mechanisms include biosorption and the precipitation of toxic metals and compounds, mitigating intracellular damage and enhancing survival (Feng et al., 2022).

5.1 Biochemical Foundations of Pollutant Toxicity

Environmental pollutants that cause covalent and oxidative damage include heavy metals (such as Cr (VI), Hg (II) and U(VI)), nitroaromatic chemicals and polycyclic aromatic hydrocarbons (PAHs). For instance, Cr (VI) enters cells through sulphate transporters and attaches to thiol groups, deactivating enzymes and causing DNA strand breaks. Electrophilic species establish long-lasting interactions with nucleophilic residues in proteins and nucleic acids (Feng et al., 2022). At the same time, reactive oxygen species (ROS) such as superoxide ($O_2 \cdot^-$), hydrogen peroxide (H_2O_2) and hydroxyl radicals ($\cdot OH$) are produced by the redox cycling of organic contaminants. Cellular integrity is compromised by these ROS because they break DNA, oxidise amino acid side chains and peroxidise membrane lipids (Zhao et al., 2021). Redox-sensitive transcriptional regulators are used by microbial sensors to identify these dangers. While Sox RS detects superoxide and triggers the production of superoxide dismutase (SodA/SodB) and efflux pumps, OxyR reacts to H_2O_2 by activating the genes for catalase (KatG) and alkyl hydroperoxide reductase (AhpCF) (Feng et al., 2022). To sequester or export metal ions, metal-responsive regulators like mercury (MerR) and chromate (ChrR) stimulate the expression of metallothionein and P-type ATPase (Zhao et al., 2021). Although they serve as the initial line of defence, these coordinated reactions are not always sufficient to counteract long-term high-dose exposures (Ta et al., 2023).

5.2 Intracellular Detoxification: Enzymatic Transformations and Redox Balancing

Enzymatic transformation occurs once contaminants are absorbed. Initiating the oxidative breakdown of PAHs, monooxygenases and dioxygenases insert molecular oxygen into aromatic rings to produce cis-dihydrodiols, which dehydrogenases then transform into catechols and cleave via ortho or meta-pathways into metabolites that are directed into central carbon metabolism (Feng et al., 2022). In anaerobic environments, chromate reductases change Cr (VI) into Cr (III), and nitroreductases change nitroaromatics into aromatic amines. Nevertheless, if antioxidant enzymes are not present to counteract electron "leakage" from these processes, semiquinone radicals may be produced, increasing the load of ROS (Zhao et al., 2021). Quinone pools, such as ubiquinone and menaquinone, are essential for electron flow because they transfer electrons from dehydrogenases to periplasmic c-type cytochromes, such as cytochrome c4 and c5. For the transformation of intracellular pollutants, these cytochromes transfer electrons to terminal reductases (Feng et al., 2022). Despite its adaptability, intracellular pathways have a lower detoxification efficacy when pollutant loads are large due to transporter saturation, the accumulation of hazardous intermediates, and the requirement to preserve vital redox cofactors (Zhao et al., 2021).

5.3 *Extracellular Electron Transfer Mediated Detoxification*

EAMs export electrons to extracellular acceptors, which minimises contaminants outside the cytoplasm, hence reducing intracellular overload. A sequence of small-molecule and protein components is followed by EET: **Inner-Membrane Oxidoreductases**: Enzymes such as *Shewanella* (CymA) and *Geobacter* (ImcH) oxidise quinols at the cytoplasmic membrane, transferring electrons to periplasmic partners (Shi et al., 2016). **Periplasmic Electron Relays**: Creating a bridge to the outer membrane, fumarate reductases (FccA) and small tetraheme cytochromes (STC) transfer electrons within the periplasm (Zhao et al., 2021). **Outer-Membrane Conduits**: To reduce insoluble Fe (III)/Mn (IV) oxides and soluble pollutants like Cr (VI), multiheme cytochromes MtrCAB and OmcA in *Shewanella* and OmcB and OmcZ in *Geobacter*—span the outer membrane and make direct contact with extracellular substrates (Feng et al., 2022). **Soluble Redox Mediators**: Endogenous phenazines and flavins (riboflavin, FMN) transfer electrons to contaminants that are not near cell surfaces by diffusing between cells and substrates (Shi et al., 2016). **Extracellular Polymeric Substances (EPS)**: By concentrating cytochromes and mediators, biofilm matrices produce microenvironments with decreased diffusion losses and increased electron flow (Zhao et al., 2021). This modular EET network enables EAMs to reduce and immobilise a diversity of toxic compounds at or beyond the cell periphery, preserving intracellular redox homeostasis and minimising direct pollutant contact (Rong et al., 2025).

5.4 *Engineering Enhancements for EET-Based Detoxification*

To adapt the natural EET pathways, researchers have applied genetic, electrochemical and material-science strategies to degrade various high-toxicity pollutants.

Genetic Overexpression of EET Pathways

When multi-copy mtrCAB and cymA operons were introduced into *S. oneidensis*, the outer-membrane cytochrome density rose, and the rates of Fe (III) reduction doubled in comparison to the wild-type (Zhao et al., 2021). Likewise, *G. sulfurreducens'* mediated Cr (VI) reduction rates were approximately doubled when ribADE flavin biosynthesis genes were expressed heterologous, increasing flavin secretion by 150% (Feng et al., 2022). Directed Evolution of Cytochromes: High-throughput screening of random mutagenesis libraries of multiheme cytochromes has revealed variations with enhanced binding affinity for nitroaromatic substrates, resulting in a 30% increase in nitrobenzene reduction kinetics (Shi et al., 2016).

5.5 *Field Applications and Pilot-Scale*

Remediation of Chromium (VI) in Groundwater: Within 24 h, flow-through bioreactors with *S. oneidensis* MR^{-1} biofilms were able to remove >95% of the Cr (VI) from 10 mg/L^{-1} influents; 80% of the reduction was due to extracellular routes (Zhao et al., 2021). Uranium (VI) Immobilisation: In contaminated aquifers, U(IV) was precipitated by in situ stimulation of native *Geobacter* spp. Through acetate amendments. Over the course of two weeks, soluble U(VI) decreased from 1 mg/L to <10 μg/L, exhibiting sustained extracellular reduction under subsurface conditions (Feng et al., 2022). The accumulation of partially reduced intermediates may necessitate downstream mineralisation steps; environmental factors (pH, ionic strength, competing acceptors) may reduce the effectiveness of EET; and scaling bioelectrochemical reactors necessitates striking a balance between mass transfer and reactor design, although EET-based detoxification is excellent at shielding cells from intracellular damage. To sustain long-term functionality, future research should investigate phototrophic EAM partnerships for solar-driven detoxification, use multi-omics to identify new EET pathways and design self-healing biofilms with programmed cell turnover (Feng et al., 2022; Shi et al., 2016).

6 Microbial Engineering Approaches for Sewage and Wastewater Treatment

Four key strategies are used in modern wastewater treatment: (1) consortium design and adaptive evolution; (2) genetic enhancements of functional strains; (3) incorporation of bioelectrochemical systems; and (4) hybrid process configurations coupled with real-time smart monitoring. These strategies combine microbial ecology, synthetic biology and bioelectrochemistry to address complex pollutant mixtures, reduce energy footprints and enable resource recovery (Awasthi et al., 2023) (Fig. 4).

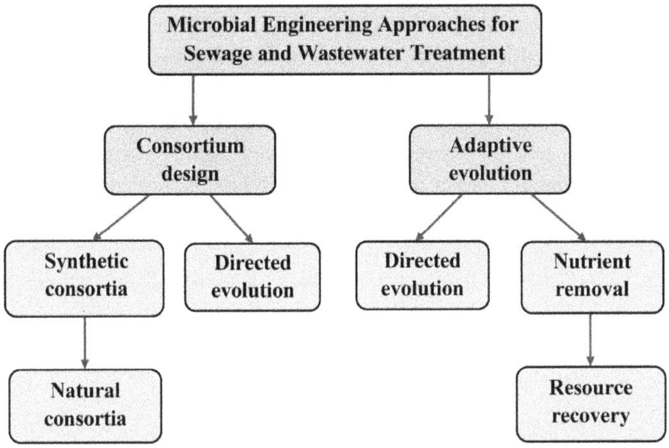

Fig. 4 Microbial engineering approaches for sewage and wastewater treatment are divided into two main strategies: consortium design and adaptive evolution. Consortium design includes creating synthetic consortia, which are based on natural consortia, and using directed evolution. Adaptive evolution involves directed evolution and nutrient removal, with nutrient removal leading to resource recovery. This flow chart shows how different microbial engineering methods connect to improve sewage and wastewater treatment by designing microbial groups or evolving them to remove nutrients and recover resources

The figure outlines key microbial strategies applied to improve sewage and wastewater treatment efficiency. It presents two primary approaches: **consortium design and adaptive evolution**. Under consortium design, synthetic and directed evolution methods lead to either natural or optimised microbial consortia. Adaptive evolution strategies focus on nutrient removal and resource recovery, reflecting the integration of microbial ecology and process engineering.

6.1 Consortium Design and Adaptive Evolution

Diverse metabolic capacities are necessary for microbial populations in complex wastewater influents. According to Sharma et al. (2023), tailored consortia combine nitrifiers (like *Nitrosomonas* and *Nitrobacter*), denitrifiers (like *Paracoccus* and *Pseudomonas*) and organotrophic degraders (like *Acinetobacter* and *Rhodococcus*) in specific ratios to guarantee the balanced removal of nitrogen, phosphorus and chemical oxygen demand (COD). Through a series of batch or continuous culture cycles, adaptive laboratory evolution (ALE) exposes consortia to progressively increasing pollutant loads, such as fluctuating ammonia concentrations or sporadic heavy metal surges, which improves community resilience. Under stress, ALE-selected communities show enriched stress-resistance genes, efflux pump expression and horizontal gene transfer events, and they maintain >90% COD removal efficiency (Waldrop & Firestone, 2006).

Co-cultivation with microalgae (e.g. *Chlorella vulgaris*) further improves consortia stability by reducing external aeration demands by up to 30% by absorbing CO_2 and surplus nutrients and providing oxygen for aerobic degraders through photosynthesis (Barker & Stuckey, 1999). **Genetic Enhancements of Key Strains**: Targeted integrations and knockouts are made possible by CRISPR-Cas9 systems that are tailored for wastewater bacteria that are not models. At high ammonia concentrations, the addition of *Nitrosomonas europaea's* ammonia monooxygenase (amoA) to *Pseudomonas putida* increased nitrification rates by 50%, lowering residual NH_4^+ to less than 1 mg L^1 in bench-scale reactors (Waldrop & Firestone, 2006). At higher temperatures (>40 °C), yeast strains modified to express fungal laccases (*P. chrysosporium*) detoxify phenolic substances with elimination efficiencies that are 40% greater than those of wild-type controls (Daims et al., 2006).

6.2 Integration of Bioelectrochemical Systems (BES)

Microbial fuel cells (MFCs) use exoelectrogenic microorganisms to produce power and remove organic materials at the same time. *Geobacter sulfurreducens* is enriched by anodic polarisation at −0.2 V, resulting in densities of up to 1.5 A m^2 and > 95% BOD elimination in continuous-flow MFCs. Pilot MFCs report hydrogen yields of 0.5 m^3 H_2 m^3 treated wastewater with concurrent 90% COD removal. Microbial electrolysis cells (MECs) convert influent organics to hydrogen gas at the cathode. Anaerobic digestion coupled with MFC/MEC units results in net-positive energy recovery, according to energy balance calculations (Awasthi et al., 2023).

6.3 Hybrid Process Configurations and Smart Monitoring

In hybrid trains, MFC/MEC modules are combined with built membrane bioreactors (MBRs), wetlands, or sequencing batch reactors (SBRs) to take advantage of complementary strengths. An MFC upstream of an MBR, for instance, lowers organic loads, lowering fouling rates and prolonging membrane life; downstream MBR cleaning guarantees effluent turbidity <10 NTU (Barker & Stuckey, 1999). Real-time biosensors, such as impedance spectroscopy arrays and redox-responsive reporter strains that express fluorescent flavoproteins, can be embedded to monitor microbial activity, pollutant concentrations and electrode performance continuously. Artificial intelligence (AI)-powered control systems modify variables including electrode potential, aeration and hydraulic retention time, increasing treatment efficiency by 25% while consuming less energy (Waldrop & Firestone, 2006).

7 Microbial Remediation of Agricultural Residues

The numerous lignocellulosic wastes from agriculture, such as straw, husks and stalks, can be remediated and made valuable by microorganisms. The following topics are covered in this section: (1) lignocellulose deconstruction techniques; (2) composting upgrades; (3) bioenergy production pathways; (4) residue upcycling into feed and food; and (5) future advances (Fig. 5).

This flowchart summarises the three principal microbial processes used to remediate and valorise agricultural residues: composting, anaerobic digestion and enzymatic hydrolysis. Each pathway leads to beneficial outcomes such as organic matter degradation, methane production, nutrient recovery, cellulose degradation and bioethanol production, demonstrating the potential of microbial technologies in sustainable agriculture and waste management.

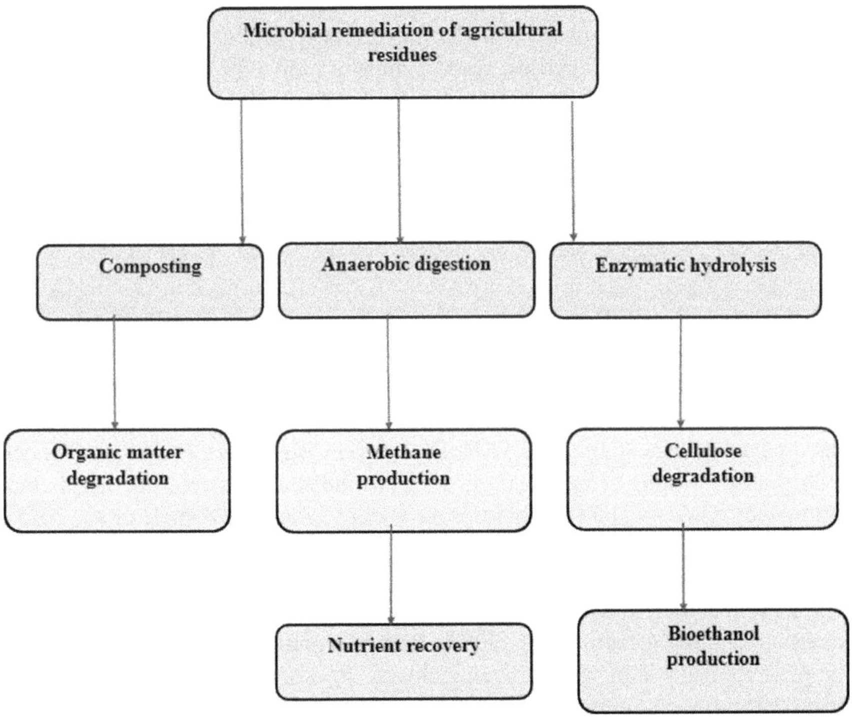

Fig. 5 Microbial remediation of agricultural residues involves three main processes: composting, anaerobic digestion, and enzymatic hydrolysis. Composting leads to organic matter degradation. Anaerobic digestion results in methane production, which further enables nutrient recovery. Enzymatic hydrolysis causes cellulose degradation, which supports bioethanol production. This flowchart highlights how different microbial methods break down agricultural waste into useful products like nutrients and biofuels

7.1 Lignocellulose Deconstruction: Enzymatic and Ecological Interplay

Lignocellulose is composed of cellulose (35–50%), hemicellulose (20–35%) and lignin (10–25%). Cooperation is necessary for effective remediation: According to (Sharma et al., 2023), hydrolytic bacteria and fungi, such as *Cellulomonas* and *Trichoderma reesei*, release endo- and exoglucanases to break down cellulose into cellobiose and glucose. Xylanases and acetylxylan esterases are produced by hemicellulolytic actinomycetes, such as *Streptomyces* spp., to break down hemicellulose into xylose and arabinose (Sharma et al., 2023). *P. chrysosporium* and *Pleurotus ostreatus*, two white-rot fungi produce lignin peroxidases and laccases, which oxidatively break down phenylpropanoid bonds and reveal carbohydrate polymers (Sharma et al., 2023). By inhibiting pathogens and accelerating residue stabilisation, these functional guilds sequentially colonise residues. Mesophiles start hydrolysis around 30–40 °C, whereas thermophiles drive deeper matrix breakdown during the thermophilic composting phase (50–70 °C). Effective residue turnover is ensured by optimising compost properties, such as moisture (50–60%), aeration rate (0.1–0.3 L air kg^{-1} dry matter min^{-1}), particle size (<10 cm) and C: N ratio (25–30:1). Bioaugmentation using certain microbial consortia improves soil health and carbon sequestration by reducing maturation time by 30%, increasing cellulase and laccase activities by 35–40% and producing compost with a greater humic acid content (Mahalle et al., 2025).

7.2 Bioenergy Production from Agricultural Waste

Bioethanol: *Saccharomyces cerevisiae* fermented pretreated hydrolysates and engineered pentose-fermenting yeasts (*Klumyveromyces marxianus*, *Pichia stipitis*) can produce ethanol titres of up to 80 g/L^{-1}. Consolidated bioprocessing strains that co-produce cellulases and ethanol improve energy balances (Mahalle et al., 2025). **Biobutanol**: Butanol titres of approximately 15 g/L^{-1} are produced by *Clostridium acetobutylicum* fermenting hydrolysates using Clostridial Acetone-Butanol-Ethanol (ABE). Product toxicity is decreased, and productivity is increased by in situ solvent extraction methods (Sharma et al., 2023). **Biohydrogen**: Alkaline and steam explosion pre-treatments improve hydrogen yields by 20%, while *Clostridium butyricum* and mixed thermophilic consortia dark ferment up to 2.5 mol H_2 mol^1 hexose (Kumar et al., 2023).

7.3 Upcycling Residues into Feed and Food

Animal Feed: Fungi (*A. niger*, *Pleurotus* spp.) enable solid-state fermentation, increasing the protein content of straw from less than 5% to 15–18% and improving in vitro digestibility by 30–40%. This makes residue-based ruminant feeds sustainable and viable (Sharma et al., 2023). Edible Mushrooms: Oyster mushroom (*P. ostreatus*) cultivation on pasteurised straw produces biological efficiencies of 20–25%, and the spent substrate retains nutrients that can be used as animal bedding or as a soil supplement. The objective of synthetic ecology advancements is to combine fermentation and deconstruction in a single bioreactor by creating minimal consortia with defined interspecies metabolite exchange. Next generation biorefineries will be powered by metabolic engineering of yeast and clostridial chassis for increased substrate ranges and better solvent tolerance. AI-driven process control and real-time monitoring using enzyme-based biosensors will allow for dynamic modifications to preserve reactor performance at its best and optimise resource recovery. Agricultural leftovers are transformed from environmental hazards to renewable feedstocks for electricity, feed and soil health using these integrated microbial engineering techniques, exemplifying the circular bioeconomy (Pekkoh et al., 2024) (Table 1).

Table 1 Simplified table explaining different microbes and their function studied above

Microorganism/microbial group	Specific application or function
Shewanella oneidensis	Enhancing Fe (III) reduction; pilot-scale Cr (VI) remediation in groundwater
Geobacter sulfurreducens	Enhancing Cr (VI) reduction (via flavin secretion); anode enrichment in microbial fuel cells (MFCs)
Native Geobacter spp.	Immobilisation of soluble U(VI) (uranium) in contaminated aquifers
Pseudomonas putida	Genetic enhancement for nitrification (NH_4^+ removal)
Chlorella vulgaris (microalgae)	Co-cultivation with microbial consortia to reduce external aeration demand in wastewater
Saccharomyces cerevisiae (and engineered yeasts)	Production of bioethanol
Clostridium acetobutylicum	Production of biobutanol
Clostridium butyricum	Production of biohydrogen via dark fermentation
Yeast strains (modified)	Expression of fungal laccases for phenolic substance detoxification

8 Microplastic Remediation

The small pieces of plastic which less than 5 mm in length are considered harmful to the environment, leading to plastic pollution and are called microplastics. These are present in synthetic clothing to cosmetics. Plastic bags as well as plastic bottles. They consist of hydrogen, simple carbon-carbon atoms, which are connected in the polymer chains. Chemical compounds which have highly present in microplastics are polybrominated diphenyl ethers (PBDEs), phthalates and tetrabromobisphenol A (TBBPA), which have adverse effects when they enter the environment. Microplastics are generally of two types: Primary, which includes microbeads used in industrial and synthetic textiles, and secondary microplastics are formed due to breakdown of high toxicity and higher molecular weight plastics through exposure to, for instance, wind abrasion, wave movement and solar UV exposure. The molecules are not biodegradable. Air pollution is also caused by microplastics, which occur in dust and airborne fibrous particles. Microplastics can absorb a variety of harmful environmental chemical pollutants and microorganisms. Currently, it poses a significant threat to human health. The filtration, absorption and coagulation are current highlight solutions in microplastic technology for remediation in wastewater and water treatment operations. One of the most popular techniques for effectively and sustainably removing microplastics is membrane filtration. The modern is integrated to remediate the degradation and recycling of the environment. Electrocoagulation is one of them; it is low-cost, environmental-friendly and used for removing contaminants from water bodies. The optimised electric field is applied so that the metal electrode produces the cations, which attach themselves to microplastics and cause them to congeal into bigger particles that can be filtered out. The efficacy of the method in treating microplastics in wastewater is evaluated by testing the impacts of wastewater pH, conductivity and current density (Martic et al., 2022). This technique of removal of the electrocoagulation process for microplastics is important for energy conservation. The microbial activities are also considered the foundation for the process by which trash is broken down by microorganisms, including bacteria, fungi and algae, in the natural environment. Green microbiology is the new approach to remediate microplastics from the environment. The microbial properties to degrade the substance work advantageously for waste degradation and promote environmental sustainability. There are three mechanisms in charge of the degradation of larger plastic material through physical, chemical and biological processes to create microplastics or nanoplastics. The oxidation of the plastic by UV rays, physical abrasion yields tiny plastic fragments. Such other factors, such as travel, aquaculture, fishing, both commercial and recreational and other maritime enterprises, can potentially result in the development of secondary microplastics, which are obviously linked to climate change. The above factors are also responsible for the travel of these tiny fragments to various ecosystems. Cosmetics, as well as the personal care industries in Asia alone, generate 70% of the product-oriented microplastic pollution (Akinsemolu & Onyeaka, 2024). The current study acknowledges various clean-up strategies to reduce the environmental impact and stop the environment from being contaminated by plastic. Filtration, sediment collection and sieving are basic techniques for detecting microplastics. With some restrictions, these materials can be further investigated in laboratories using methods like scanning electron microscopy and atomic force microscopy. Over the past decades, various microbes have been discovered with the capacity to break down microplastics. The categorisation of the microorganism depends on the environment they need to degrade microplastic. Various microbial combinations are also capable of degrading microplastics in the environment. According to their unique characteristics, bacteria like *Bacillus mycoides, Acinetobacter baumannii, P. fluorescens, Staphylococcus cohnii, Staphylococcus xylosus, Bacillus cereus, Micrococcus luteus, Micrococcus lylae, Rahnella aquatilis, Arthrobacter viscosus and Rhodococcus ruber,* as well as fungi like *Aspergillus clavatus, Aspergillus flavus, Zalerion maritimum, Aspergillus tubingensis and A. niger* can enzymatically break down microplastics in both terrestrial and aquatic environments. The enzymes and metabolic pathways are employed to degrade microplastics into small molecules, which are more environmentally friendly through anaerobic and aerobic processes. Every microorganism is different in terms of pace and condition. Understanding the various processes and activities involved in this process is crucial (Akinsemolu & Onyeaka, 2024) (Figs. 6 and 7).

This topic has been extensively researched and explored. The process occurs in four different stages, i.e. hydrolysis followed by acidogenesis, acetogenesis and methanogenesis. Hydrolysis includes the degradation of insoluble polymers of microplastics into soluble organic molecules, similar to biodeterioration in an aerobic environment. In the acidogenesis stage, carbon dioxide, alcohol, hydrogen and volatile fatty acids are produced from organic molecules. This is a very selective phase. Acetogenesis is the process of converting volatile fatty acids with extra hydrogen and carbon dioxide into acetate. The methanogenic archaea convert carbon dioxide, hydrogen and acetate into methane and water during methanogenesis. Methane, a renewable energy source, is produced because of this process. Environmental variables and polymer properties are the main determinants of the microbial biodegradation process (Khan & Majeed, 2019). However, the polymer type and the plastic material's surface area determine the overall rate of biodegradation. Because of their polymer chain topologies, high-molecular-weight synthetic polymers like polyethene and polypropylene are less hydrophilic and more biodegradable than polymers with low molecular weights (Kawai, 1995). Microplastics are depolymerised in this step, producing less complicated monomers that are easier to metabolise in the stages of breakdown that follow (Ru et al., 2020).

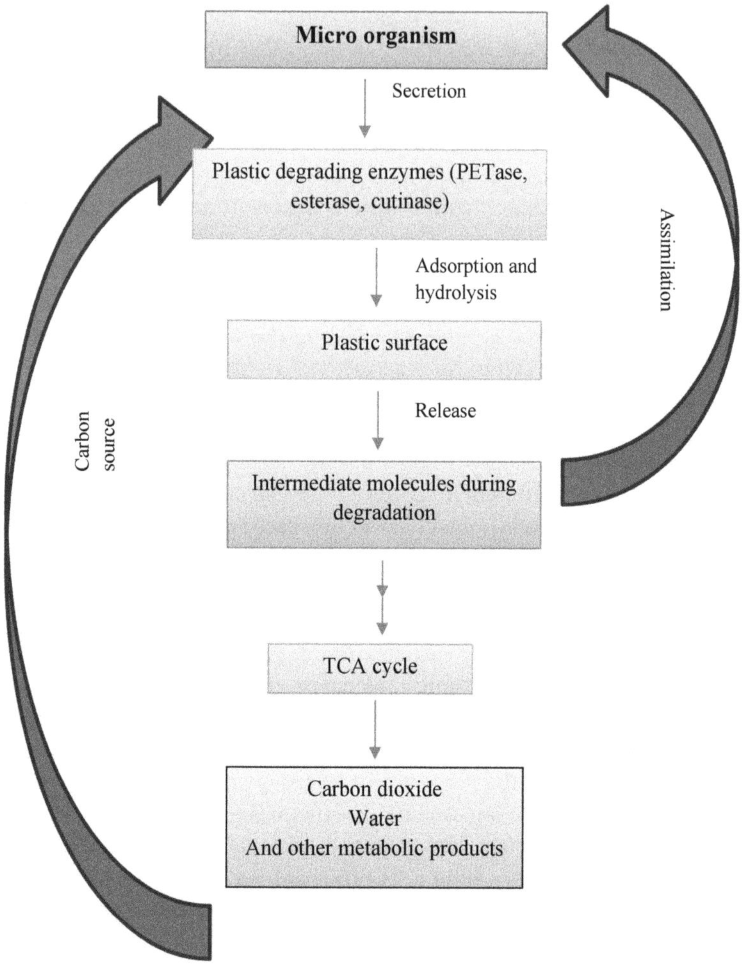

Fig. 6 Above schematic diagram shows the various processes of microplastic degradation through biotic and abiotic factors

9 Implementing Bioremediation to Clean Wastewater from the Paper and Pulp Industries

India's pulp and paper industry is not as advanced as those in Western Europe. Western European countries pulp and paper industries of Western Europe according to estimations, approximately 525 pulp and paper plants with a cumulative installed efficiency of 6.25 million tonnes annually and a roughly 67% increase in production. PPI uses a range of cellulosic-based materials to make paper in India; roughly 43% is forest lumber, 28% from agricultural and agricultural goods and 29% from recycled waste paper (Khumchai et al., 2022). PPME's features and wastewater generation

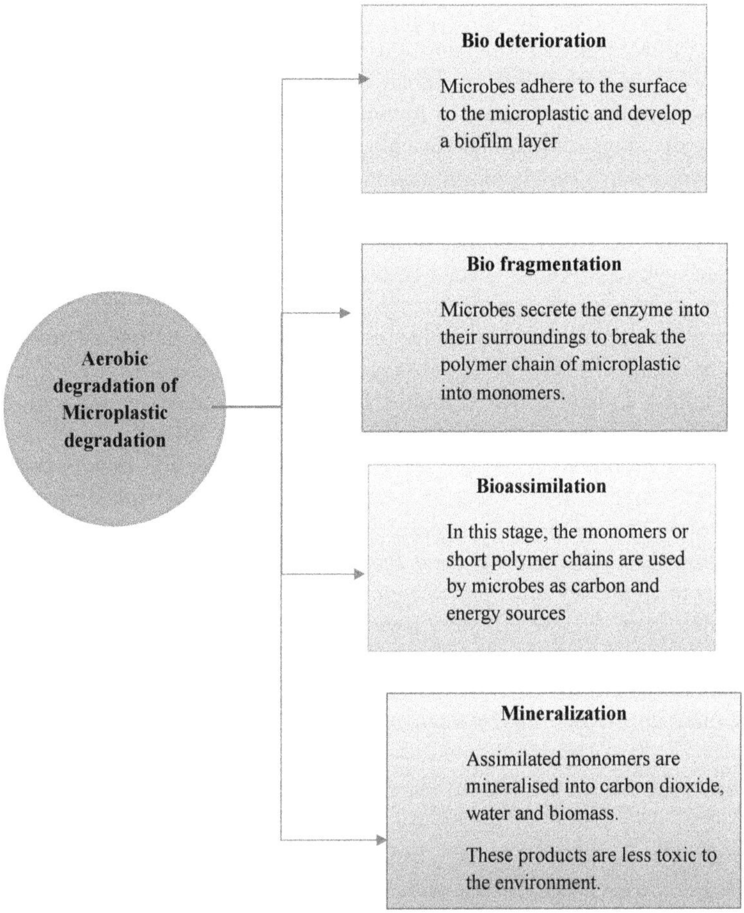

Fig. 7 Aerobic degradation of microplastics by microorganisms

are entirely dependent on the manufacturing process used and the amount of water reused in the facility. Characterised by factors like colour, absorbable organic halides (AOX) and related organic chemicals, kraft pulping effluent is extremely polluted. The primary source of colour in bleach plant effluent is lignin and its derivatives during the alkaline extraction step. Sludge buildup, eutrophication and related microbial growth were the most obvious impacts on incoming water, along with decreased oxygen levels. Receiving water, PPME and biota exposed to pulp and paper emissions have been shown to contain a variety of chlorinated organic chemicals. Phenolics, fatty acids, resin acids, dioxins and furans are among the most hazardous chlorinated chemicals. One of the most polluting businesses, the pulp and paper sector, releases every year 100 million kg of dangerous pollutants into the environment. 260 chemicals have been detected in the different stages of papermaking. These PPME absorb light and heat, which leads to an increase in oxygen level due to a high number of tannins, and it negatively affects the aquatic flora and fauna. The removal of contaminants from PPME has been a significant issue and research topic for many years. The primary issue with PPME is colour, which is mostly caused by lignin and its derivatives. A significant amount of lignin is generated by a number of procedures comprising divisions for pulping, bleaching and recovery. The colour of PPME has been removed using a variety of techniques (Kumar et al., 2022). Certain chlorinated lignins, colour, toxicity, suspended particles and COD are eliminated by physically and chemically affected procedures, which are comparatively costly. However, BOD and certain other compounds cannot be effectively removed. Colour and COD can be reduced more effectively with the biological technique, which also lowers BOD and other pollutants. A technique called bioremediation uses biological processes to catalyse or change different toxic substances into less dangerous forms. The paper and pulp industries have different multi-tolerant major indigenous bacterial species that can be used to remediate PPI effluents (Madan et al., 2018).

The use of microorganisms such as fungi, bacteria, algae and enzymes as a one-step treatment or in conjunction with other physical and or chemical treatments is known as biological treatment (Singhal & Thakur, 2009). One of such is *Streptomyces tuirus*, whose bioremediation process increases at 35 °C. It can produce protein-enriched content and lessen organic contaminants in PPI effluents (Khumchai et al., 2022). It has been discovered that *Trametes versicolor*, *P. chrysosporium*, *Schizophyllum commune* and *Tinctoria borbonica* aid in the breakdown and metabolism of lignin and carbohydrates. *Trichoderma* sp. and *A. niger* can also break down lignin and discolour hardwood pulp bleaching effluents (Dashtban et al., 2010). According to Tyagi et al. (2014), bacterial species such as *B. subtilis* and *M. luteus* can lower the BOD level by 87.2%, the COD by 94.7%, and the lignin content by 97% after nine days. A combined culture of Chlorella, Chlamydomonas, and Microcystis was used to eradicate AOX; in 30 days, there was an 80% colour reduction and a 70% AOX decrease. After analysing the alkaline extraction of algae biomass, numerous writers concluded that metabolism, not adsorption, was the primary mechanism for colour removal (Preeti Sharma et al., 2014). There are a few enzymes that can both remove organic debris from PPME and enhance effluent quality, including ligninase, cellulase and peroxidase. The most significant enzymes among them include peroxidase, which is utilised to remove colour from bleaching effluents. For the decolourisation process, it is also feasible to combine enzymes with unique bacteria that often do not have strong enzyme activity. Peroxidase and other extracellular enzymes are produced by certain fungi. It appears that this enzyme bleaches wastewater by oxidising the chromophores and removing their colour. The inclusion of peroxidase improved the colour removal from neutral pH effluents caused by hydrogen peroxide (H_2O_2) in trace amounts (Gao et al., 2013; Raj et al., 2014). As the demand for paper increases, treating effluents becomes one of the most pressing environmental protection challenges. Since bioremediation is more effective and cost-effective than chemical treatment, it is currently viewed as a desirable alternative for lowering the pollution load from contaminated water. The native dominating method could be used to treat large pulp and paper mill effluents of bacterial and fungal isolates. The comparison of decolourisation by various organisms revealed that certain bacteria

(*Pseudomonas ovalis, Pseudomonas aeruginosa,* and *B. cereus*) and fungi (*A. niger, P. chrysosporium* and *T. versicolor*) are appropriate for the effluent degradation of recalcitrant chromophore material in paper mill effluent. Bacterial and fungal isolates with the proper enzyme activity and optimal physical conditions are essential to the bioremediation process (Madan et al., 2018).

Global Impact and Benefit: The microbial remediation has benefitted at global scale. The detoxification of the contaminated sites by enzymatic activity of microbes, which converts pollutants into less toxic substances. Certain traditional methods may as well contribute to GHGs. The microbial treatment can capture the methane into a usable product (Tang et al., 2024). The microbial treatments are low cost and less energy intensive, as well do not require a high or very modified infrastructures. They generate value-added products like bioenergy, like the production of bioethanol, biobutanol and biohydrogen mentioned above (Dash et al., 2025; Dixit et al., 2015). Certain microbes are also used in the extraction of valuable metals like copper, gold and nickel; the process is known as bioleaching (Dash et al., 2025). The use of genetically modified organisms like (*S. oneidensis or P. putida)* significantly increases the rate and efficiency of pollutant degradation, accelerating global clean-up efforts. By using engineered biological systems for large-scale, sustainable industrial production and environmental control worldwide, the research paves the way for a transition from a petroleum based economy to a bio-based economy (Ojuederie & Babalola, 2017).

10 Conclusion and Future Prospective

A lot of research has been done on the use of microbial enzymes for waste bioremediation, yet improvement in the process is required for a safe and more sustainable environment. The novel microbe and their enzymes have efficient and steady properties to bioremediate the heavy pollutants. The above studies and different techniques are certain unique and well-known methods to use microbes in the cleaning up process. It would be preferable to use a variety of microbial consortia rather than just one for bioremediation, since this would result in the presence of many organisms that use diverse substrates, which would speed up microbial biodegradation.

In nature, the different degradation methods of different compounds are carried out at either higher or extremely lower temperatures, as microbes are very temperature-specific. More research and fieldwork are needed to determine the appropriate temperature for the survival of different microbes. The potential of microbes is yet to be determined. They are effective against pollutants, being it is plastic, paper, oil, biological waste or human remains, yet there are after-effects on the environmental and microbial system that must be followed up by research. There are high chances that mutations within the microorganism can be used in favour of the environment.

Lastly, if microbes or enzymes are introduced directly to the soil, they may die or lose their effectiveness before the remediation process starts; consequently, their activity may be increased by combining them with additional agents, like nanoparticles. The public and governments will be better able to use microbial degradation if there is greater awareness of its adoption. Many people who are not aware of this process may choose to use the conventional method that is accessible, which may not be as safe and efficient as microbial biodegradation.

Acknowledgement Prof. Chaithanya Madhurantakam, Principal Investigator-SMBL, Department of Biotechnology, TERI SAS, Vasant Kunj New Delhi is acknowledged for his scientific insights and laboratory facility.

CRediT Statement S.M. is involved in writing-review and editing, writing-original draft preparation, validation. N.D. is involved in writing the original draft and reviewing. M.M. is involved in writing the original draft and reviewing. N.G. involved in sketching, writing-review and editing, validation and proof reading.

References

Akhmet, A., Karimova, S., Zhumadullayeva, A., Ashirbayeva, S., Tlegenova, K., Alikhan, A., Issayeva, A., Tleukeyeva, A., & Kedelbayev, B. (2024). Microorganisms of solid waste as an opportunity for waste disposal and increasing environmental sustainability in the south of Kazakhstan. *Journal of Infrastructure, Policy and Development, 8*, 7880. https://doi.org/10.24294/jipd7880

Akinsemolu, A., & Onyeaka, H. (2024). Application of green microbiology for microplastic remediation: Current progress and future perspectives. *Environmental Advances, 17*, 100569. https://doi.org/10.1016/j.envadv.2024.100569

Ali, M., Gani, K. M., Kazmi, A. A., & Ahmed, N. (2016). Degradation of aldrin and endosulfan in rotary drum and windrow composting. *Journal of Environmental Science and Health B, 51*, 278–286.

Alloway, B. J. (2013). *Heavy metals in soils: Trace metals and metalloids in soils and their bioavailability*. Springer.

Awasthi, M. K., Ganeshan, P., Gohil, N., Kumar, V., Singh, V., Rajendran, K., Harirchi, S., Solanki, M. K., Sindhu, R., Binod, P., Zhang, Z., & Taherzadeh, M. J. (2023). Advanced approaches for resource recovery from wastewater and activated sludge: A review. *Bioresource Technology, 384*, 129250. https://doi.org/10.1016/j.biortech.2023.129250

Barker, D. J., & Stuckey, D. C. (1999). A review of soluble microbial products (SMP) in wastewater treatment systems. *Water Research, 33*(14), 3063–3082.

Bera, R., Datta, A., Bose, S., Dutta, S., Mukhopadhyay, K., Goswami, K. K., Debnath, M., Barik, A. K., Ganguli, M., Narasimhan, V. L., Quah, E., Bhattacharya, P., & Seal, A. (2022). Recycling of waste utilizing Novcom composting technology towards GHG abatement from source. *International journal of environment and climate change*, 1605–1629. https://doi.org/10.9734/ijecc/2022/v12i121604

Bera, R., Saha, S., Datta, A., Mukhopadhyay, K., Barik, A. K., Bhattacharya, P., Mallick, R., Ganguli, M., Narasimhan, V. L., & Quah, E. (2024). Review of different composting methods towards end product quality, cost of production, process adoptability potential, post soil application efficacy and GHG mitigation potential. *Asian Journal of Environment & Ecology, 23*(12), 23–46.

Chaudhari, R. D., Datar, M. T., & Babookani, M. R. (2011). Municipal solid waste management through vermicomposting employing exotic species of earthworm *Eudrilus eugeniae. Journal of Environmental Science & Engineering, 53*, 129–134.

Cheela, V. R. S., John, M., Biswas, W., & Dubey, B. (2023). Environmental impact evaluation of landfill mining of legacy waste with on-site sorting using life cycle assessment. *Environmental*

Science and Pollution Research International, 30, 30033–30047. https://doi.org/10.1007/s11356-022-24210-2

Chugh, M., Kumar, L., Shah, M. P., & Bharadvaja, N. (2022). Algal bioremediation of heavy metals: An insight into removal mechanisms, recovery of by-products, challenges, and future opportunities. *Energy Nexus, 7*, 100129. https://doi.org/10.1016/j.nexus.2022.100129

Daims, H., Taylor, M. W., & Wagner, M. (2006). Wastewater treatment: A model system for microbial ecology. *Trends Biotechnology, 24*, 483–489. https://doi.org/10.1016/j.tibtech.2006.09.002

Dash, J., Ojha, R., & Pradhan, D. (2025). Progress in bioleaching and its mechanism: A short review. *Discover Environment,* 3, 238. https://doi.org/10.1007/s44274-025-00454-w

Dashtban, M., Schraft, H., Syed, T. A., & Qin, W. (2010). Fungal biodegradation and enzymatic modification of lignin. *International Journal of Biochemistry and Molecular Biology, 1*, 36–50.

Ding, S., Zhong, J., Du, S., Liu, X., Yao, A., Xu, X., & Wu, D. (2025). Exploring the function of key species in different composting stages for effective waste biotransformation. *Journal of Environmental Management, 381*, 125234. https://doi.org/10.1016/j.jenvman.2025.125234

Dixit, R., Wasiullah, X., Malaviya, D., Pandiyan, K., Singh, U. B., Sahu, A., Shukla, R., Singh, B. P., Rai, J. P., Sharma, P. K., Lade, H., & Paul, D. (2015). Bioremediation of heavy metals from soil and aquatic environment: An overview of principles and criteria of fundamental processes. *Sustainability, 7*, 2189–2212. https://doi.org/10.3390/su7022189

Domínguez, J., & Edwards, C. A. (2004). Vermicomposting organic wastes: A review. In C. A. Edwards (Ed.), *Earthworm ecology* (pp. 401–424). CRC Press.

Ebsa, G., Gizaw, B., Admassie, M., Degu, T., & Alemu, T. (2024). The role and mechanisms of microbes in dichlorodiphenyltrichloroethane (DDT) and its residues bioremediation. *Biotechnology Reports, 42*, e00835. https://doi.org/10.1016/j.btre.2024.e00835

Ezeudu, O. B., & Ugochukwu, U. C. (2024). Financing mechanism for solid waste management in Anambra, Nigeria: Analyses of emerging challenges and implication for circular economy. *Environmental Science and Pollution Research International, 31*, 27634–27652. https://doi.org/10.1007/s11356-024-33178-0

Feng, H., Xu, L., Chen, R., Ma, X., Qiao, H., Zhao, N., Ding, Y., & Wu, D. (2022). Detoxification mechanisms of electroactive microorganisms under toxicity stress: A review. *Frontiers in Microbiology, 13*, 1084530. https://doi.org/10.3389/fmicb.2022.1084530

Forti, V., Baldé, C. P., Kuehr, R., & Bel, G. (2020). *The Global E-waste Monitor 2020*. United Nations University.

Fytili, D., & Zabaniotou, A. (2008). Utilization of sewage sludge in EU application of old and new methods—A review. *Renewable and Sustainable Energy Reviews, 12*(1), 16–140.

Gadd, G. M. (2010). Metals, minerals and microbes: Geomicrobiology and bioremediation. *Microbiology, 156*(3), 609–643.

Gajalakshmi, S., & Abbasi, S. A. (2008). Solid waste management by composting: State of the art. *Critical Reviews in Environmental Science and Technology, 38*, 311–400. https://doi.org/10.1080/10643380701413633

Gao, J., Xu, G., Qian, H., Liu, P., Zhao, P., & Hu, Y. (2013). Effects of nano-TiO_2 on photosynthetic characteristics of Ulmus elongata seedlings. *Environmental Pollution, 176*, 63–70. https://doi.org/10.1016/j.envpol.2013.01.027

Ghosh, D., Ghorai, P., Sarkar, S., Maiti, K. S., Hansda, S. R., & Das, P. (2023). Microbial assemblage for solid waste bioremediation and valorization with an essence of bioengineering. *Environmental Science and Pollution Research International, 30*, 16797–16816. https://doi.org/10.1007/s11356-022-24849-x

Goyal, S., Dhull, S. K., & Kapoor, K. K. (2005). Chemical and biological changes during composting of different organic wastes and assessment of compost maturity. *Bioresource Technology, 96*(14), 1584–1591.

Gupta, S., & Boojh, R. (2006). Report: Biomedical waste management practices in India. *Waste Management, 26*(6), 579–591.

Gupta, R., & Diwan, B. (2017). Bacterial exopolysaccharide mediated heavy metal removal: A review on biosynthesis, mechanism and remediation strategies. *Biotechnology Reports, 13*, 58–71.

Hamidinasab, B., & Nabavi-Pelesaraei, A. (2025). Systematic review on environmental impact assessment of incineration technologies. *Energy Conversion and Management: X, 26*, 101039. https://doi.org/10.1016/j.ecmx.2025.101039

Hassen, A. et al. (2001). Microbial characterization during composting of municipal solid waste. *Bioresource Technology, 80*(3), 217–225.

Haug, R. T. (1993). *The practical handbook of compost engineering*. Lewis Publishers.

Hoornweg, D., & Bhada-Tata, P. (2012). *What a waste: A global review of solid waste management*. World Bank.

Holm-Nielsen, J. B., Al Seadi, T., & Oleskowicz-Popiel, P. (2009). The future of anaerobic digestion and biogas utilization. *Bioresource Technology, 100*(22), 5478–5484.

Hu, W., Duan, R., Dai, Q., Yang, H., Zhang, Y., Meng, F., & Lin, Y. (2025). Effects of biochar immobilization *Bacillus subtilis* on heavy metal accumulation, rhizosphere microorganisms, and metabolism in rice. *Journal of Applied Microbiology, 136*, lxaf083. https://doi.org/10.1093/jambio/lxaf083

Kausar, H., Shahzad, S. M., & Aslam, M. (2012). Nitrogen-fixing Bacteria as Biofertilizer. *Soil and Environment, 31*(2), 127–136.

Kawai, F. (1995). Breakdown of plastics and polymers by microorganisms. In M. Hiroto, H. Nishimura, Y. Kodera, F. Kawai, P. Rusin, H. L. Ehrlich, S. Y. Lee, H. N. Chang, S. A. Markov, M. J. Bazin, D. O. Hall, A. L. Gutman, M. Shapira, Y. Inada, & A. Matsushima (Eds.), *Microbial and Eznymatic bioproducts* (pp. 151–194). Springer. https://doi.org/10.1007/BFb0102319

Khan, A., & Majeed, T. (2019). Biodegradation of synthetic and natural plastics by microorganisms: A mini review, *1*, 180–184.

Khumchai, J., Wongchai, A., On-uma, R., Sabour, A., Alshiekheid, M., Narayanan, M., Karuppusamy, I., Pugazhendhi, A., Brindhadevi, K., & Lan Chi, N. T. (2022). A viable bioremediation strategy for treating paper and pulp industry effluents and assessing the prospect of resulted bacterial biomass as single cell protein (SCP) using indigenous bacterial species. *Chemosphere, 304*, 135246. https://doi.org/10.1016/j.chemosphere.2022.135246

Kumar, J. A., Sathish, S., Prabu, D., Renita, A. A., Saravanan, A., Deivayanai, V. C., Anish, M., Jayaprabakar, J., Baigenzhenov, O., & Hosseini-Bandegharaei, A. (2023). Agricultural waste biomass for sustainable bioenergy production: Feedstock, characterization and pretreatment methodologies. *Chemosphere, 331*, 138680. https://doi.org/10.1016/j.chemosphere.2023.138680

Kumar, R., Singh, A., Maurya, A., Yadav, P., Yadav, A., Chowdhary, P., & Raj, A. (2022). Effective bioremediation of pulp and paper mill wastewater using Bacillus cereus as a possible kraft lignin-degrading bacterium. *Bioresource Technology, 352*, 127076. https://doi.org/10.1016/j.biortech.2022.127076

Liu, D., Ma, X., Huang, J., Shu, Z., Chu, X., Li, Y., & Jin, Y. (2022). Study on personalized microbial formulation during high-temperature aerobic fermentation of different types of food wastes. *Sci Total Environ, 814*, 152561. https://doi.org/10.1016/j.scitotenv.2021.152561

Madan, S., Sachan, P., & Singh, U. (2018). A review on bioremediation of pulp and paper mill effluent—An alternative to conventional remedial technologies. *Journal of Applied and Natural Science, 10*, 367–374. https://doi.org/10.31018/jans.v10i1.1632

Mahalle, S., Bhende, R. S., Bokade, P., Bajaj, A., & Dafale, N. A. (2025). Emerging microbial remediation methods for rejuvenation of pesticide-contaminated sites. *Total Environment Microbiology, 1*, 100026. https://doi.org/10.1016/j.temicr.2025.100026

Martic, S., Tabobondung, M., Gao, S., & Lewis, T. (2022). Emerging electrochemical tools for microplastics remediation and sensing. *Frontiers in Sensors, 3*. https://doi.org/10.3389/fsens.2022.958633

Mehta, C. M. et al. (2014). Compost microbiology: Contribution of microorganisms in organic waste management. In D. K. Choudhary et al. (Eds.), *Microbial inoculants in sustainable agricultural productivity* (pp. 221–245). Springer.
Mergeay, M. et al. (2003). The genetic basis of metal resistance in cupriavidus metallidurans. *FEMS Microbiology Reviews, 27*(2–3), 385–410.
Nakasaki, K., Akiyama, T., & Nagata, H. (1993). Composting of organic waste stabilized by thermophilic actinomycetes. *Journal of Fermentation and Bioengineering, 76*(3), 191–193.
Ojuederie, O. B., & Babalola, O. O. (2017). Microbial and plant-assisted bioremediation of heavy metal polluted environments: A review. *International Journal of Environmental Research and Public Health, 14*, 1504.
Park, J. H. et al. (2011). Effect of biochar on heavy metal sorption and bioavailability in soil. *Journal of Hazardous Materials, 185*(2–3), 829–834.
Pekkoh, J., Thurakit, T., Ruangrit, K., Chaichana, C., Phinyo, K., Lomakool, S., Wichaphian, A., Cheirsilp, B., & Srinuanpan, S. (2024). Co-bioaugmentation with microalgae and probiotic bacteria: Sustainable solutions for upcycling of aquaculture wastewater and agricultural residues into microbial-rice bran complexes. *Environmental Research, 261*, 119760. https://doi.org/10.1016/j.envres.2024.119760
Preeti Sharma, P.S., Aksha Sharma, A.S., Jasuja, N.D., Joshi, S.C., 2014. Organophosphorous compounds and oxidative stress: A review.
Pointing, S. B. (2001). Feasibility of bioremediation by white-rot fungi. *Applied Microbiology and Biotechnology, 57*(1–2), 20–33.
Rafeeq, H., Afsheen, N., Rafique, S., Arshad, A., Intisar, M., Hussain, A., Bilal, M., & Iqbal, H. M. N. (2023). Genetically engineered microorganisms for environmental remediation. *Chemosphere, 310*, 136751. https://doi.org/10.1016/j.chemosphere.2022.136751
Raj, A., Kumar, S., Haq, I., & Singh, S. K. (2014). Bioremediation and toxicity reduction in pulp and paper mill effluent by newly isolated ligninolytic *Paenibacillus* sp. *Ecological Engineering, 71*, 355–362. https://doi.org/10.1016/j.ecoleng.2014.07.002
Rastogi, M., Nandal, M., & Khosla, B. (2020). Microbes as vital additives for solid waste composting. *Heliyon, 6*, e03343. https://doi.org/10.1016/j.heliyon.2020.e03343
Rong, Q., Zhang, C., Ling, C., Lu, D., & Jiang, L. (2025). Mechanism of extracellular electron transport and reactive oxygen mediated Sb(III) oxidation by Klebsiella aerogenes HC10. *Journal of Environmental Sciences (China), 147*, 11–21. https://doi.org/10.1016/j.jes.2023.09.013
Ru, J., Huo, Y., & Yang, Y. (2020). Microbial degradation and valorization of plastic wastes. *Frontiers in Microbiology, 11*. https://doi.org/10.3389/fmicb.2020.00442
Serrano, A. et al. (2011). Bioremediation of heavy metals in acid mine drainage by sulfate-reducing bacteria immobilized on autoclaved and raw sludge. *Bioresource Technology, 102*(15), 7075–7082.
Shahzad, A., Zahra, A., Li, H. Y., Qin, M., Wu, H., Wen, M. Q., Ali, M., Iqbal, Y., Xie, S. H., Sattar, S., & Zafar, S. (2024). Modern perspectives of heavy metals alleviation from oil contaminated soil: A review. *Ecotoxicology and Environmental Safety, 282*, 116698. https://doi.org/10.1016/j.ecoenv.2024.116698
Sharma, P., Dutta, D., Udayan, A., Nadda, A. K., Lam, S. S., & Kumar, S. (2022). Role of microbes in bioaccumulation of heavy metals in municipal solid waste: Impacts on plant and human being. *Environmental Pollution, 305*, 119248. https://doi.org/10.1016/j.envpol.2022.119248
Sharma, P., Sangwan, S., Kaur, H., Patra, A., & Mehta, S. (2023). Microbial remediation of agricultural residues, 325–358. https://doi.org/10.1007/978-3-031-24181-9_13
Sharma, P., Singh, S. P., Parakh, S. K., & Tong, Y. W. (2022). Health hazards of hexavalent chromium (Cr (VI)) and its microbial reduction. *Bioengineered, 13*, 4923–4938. https://doi.org/10.1080/21655979.2022.2037273
Shi, L., Dong, H., Reguera, G., Beyenal, H., Lu, A., Liu, J., Yu, H.-Q., & Fredrickson, J. K. (2016). Extracellular electron transfer mechanisms between microorganisms and minerals. *Nature Reviews Microbiology, 14*, 651–662. https://doi.org/10.1038/nrmicro.2016.93

Singhal, A., & Thakur, I. (2009). Decolourisation and detoxification of pulp and paper mill effluent by Cryptococcus sp. *Biochemical Engineering Journal, 46*, 21–27. https://doi.org/10.1016/j.bej.2009.04.007

Stamou, I., & Antizar-Ladislao, B. (2016). The impact of silver and titanium dioxide nanoparticles on the in-vessel composting of municipal solid waste. *Waste Management, 56*, 71–78. https://doi.org/10.1016/j.wasman.2016.07.008

Swachh Bharat Mission—SBM Urban 2.0 [WWW Document] (n.d.). Accessed February 11 2025. URL https://sbmurban.org/

Ta, M., Wei, J., Ye, S., Zhang, J., Song, T., & Li, M. (2023). Investigation of dissolved organic matter's influence on the toxicity of cadmium to the cyanobacterium Microcystis aeruginosa by biochemical and molecular assays. *Environmental Science and Pollution Research International, 30*, 94790–94802. https://doi.org/10.1007/s11356-023-29000-y

Tang, H., Xiang, G., Xiao, W., Yang, Z., & Zhao, B. (2024). Microbial mediated remediation of heavy metals toxicity: mechanisms and future prospects. *Frontiers in Plant Science, 15*, 1420408. https://doi.org/10.3389/fpls.2024.1420408

Tiquia-Arashiro, S. M. (2014). *Microbial ecology of composting*. CRC Press.

Tsydenova, O., & Bengtsson, M. (2011). Chemical hazards associated with treatment of waste electrical and electronic equipment. *Waste Management, 31*(1), 45–58.

Tyagi, S., Garg, N., & Paudel, R. (2014). Environmental degradation: Causes and consequences. *European Researcher, 81*, 1491. https://doi.org/10.13187/er.2014.81.1491

Waldrop, M. P., & Firestone, M. K. (2006). Seasonal dynamics of microbial community composition and function in oak canopy and open grassland soils. *Microbial Ecology, 52*, 470–479. https://doi.org/10.1007/s00248-006-9100-6

Wang, J., & Chen, C. (2009). Biosorbents for heavy metals removal and their future. *Biotechnology Advances, 27*(2), 195–226.

Wang, Y.-N., Li, Y., Yang, W., Wang, W., Wang, H., Sun, Y., Xie, J., Zheng, Y., Cui, X., Li, W., & Bian, R. (2025). Effects of different types of municipal solid waste incineration slag on landfill stabilization and related microbiological mechanism. *Waste Management, 202*, 114833. https://doi.org/10.1016/j.wasman.2025.114833

WHO. (2018). Health-care waste. Available at: World Bank. (2018). *What a waste 2.0: A global snapshot of solid waste management to 2050*. World Bank.

World Development Report. (2018). Learning to realize education's promise [WWW Document], n.d. World Bank. URL https://www.worldbank.org/en/publication/wdr2018 (accessed 6.15.26)

Zhao, J., Li, F., Cao, Y., Zhang, X., Chen, T., Song, H., & Wang, Z. (2021). Microbial extracellular electron transfer and strategies for engineering electroactive microorganisms. *Biotechnology Advances, 53*, 107682.

Ziraba, A. K., Haregu, T. N., & Mberu, B. (2016). A review and framework for understanding the potential impact of poor solid waste management on health in developing countries. *Archives of Public Health, 74*, 55. https://doi.org/10.1186/s13690-016-0166-4

Strategies for the Minimisation of Hazardous Solid Waste for Environmental Protection

Meghna Rawat, Anjali, Srushti Gadiyaram, and Pawan Kumar

1 Introduction

The quick rise of industrialisation, urbanisation and increased consumption has significantly raised the volume and diversity of waste, mainly solid waste, much of which contains lethal materials. The worldwide generation of solid waste is estimated to reach nearly 3.4 billion metric tons per annum by 2050 (Kaza et al., 2018). Waste belongs to materials that are no longer useful (Sutherland et al., 2013). It can be classified as agricultural, biomedical, municipal solid and industrial waste, with the majority of it being hazardous.

Hazardous waste is defined as unused materials that hold radioactivity, toxicity, corrosivity, reactivity and infectiousness as one of the intrinsic physical properties, which are hazardous for the atmosphere and public well-being as reported by the Environmental Protection Agency (EPA) in 2016. Wastewater from dye industries contains toxic metals such as chromium, antimony and mercury-like substances (Methneni et al., 2021) and contaminants such as aluminium, zinc, cadmium and copper are used in agricultural fields in the form of fertilisers and pesticides (Ayilara & Babalola, 2023; Ayilara et al., 2020).

Conferring to the United Nations Environment Programme (UNEP), nearly 400 million tons of waste are produced globally annually, with an average of about 60 kg per person (UNEP, The World Counts Hazardous Waste Statistics). In 2005, nearly 11.62 million tons of industrial hazardous waste (IHW) were produced. Of this

M. Rawat · Anjali · P. Kumar (✉)
Department of Chemistry, Hemvati Nandan Bahuguna Garhwal University (A Central University), Srinagar Garhwal, Uttarakhand 246174, India
e-mail: Pawankumar@hnbgu.ac.in

S. Gadiyaram
Key Laboratory of Clean Chemistry Technology of Guangdong Regular Higher Education Institutions, School of Chemical Engineering and Light Industry, Guangdong University of Technology, Guangzhou 510006, China

H. Gupta et al. (eds.), *Advanced Technologies for Assessment, Recycling, and Management of Solid Waste*, Waste as a Resource,
https://doi.org/10.1007/978-3-032-24788-9_13

total, about recycled 43.4%, stored 33% and disposed of 23%, resulting in approximately 70,000 tons being discharged into the environment (Duan et al., 2008). This background requires a key strategy for sustainable development, like waste minimisation.

As stated by Davidson, waste management (WM) is a time-consuming procedure which involves collection, transportation, treatment and disposal managed under strict supervision following regulatory laws, various technologies and economic mechanisms (Davidson et al., 2011). Effective land disposal, recycling, waste minimisation, waste processing and waste transformation are all components of proper waste management. Saxena and Jotshi (1996) emphasised WM as a popular topic in urban planning and development. Promoting a circular economy requires an efficient waste management strategy that incorporates: refuse, reduce, reuse, repair, repurpose, recover and recycle practices, the 7R waste (Mohajan et al., 2020).

Hazardous waste (HW) represents a multifaceted challenge with widespread influence on atmospheric changes on a global scale. So, its disposal requires proper management as improperly discarded HW penetrates the ecosystem, thus contaminating soil, air and water resources. Various health impacts are caused by exposure to HW, ranging from respiratory ailments to neurological disorders, and even carcinogenic effects. Moreover, environmental challenges, such as climate change and biodiversity loss, can be worsened with the improper disposal of HW. The disruption of ecosystems due to pollutants can disturb delicate ecological balances (Abubakar et al., 2022).

According to a report in 2013, over 400 kids in Zamfara state were intoxicated by lead as a consequence of mining. Similarly, due to lead poisoning, 28 children were found dead in Niger state in Nigeria in the area of Rafin Local Government. A study on artisanal gold mining in Zimbabwe and Tanzania (Nweke et al., 2009) revealed that for each gram of gold mined, 1.46 g of mercury are dispersed, with 70–80% of that going to the atmosphere. Another study (Dresselhaus et al., 2001) on 248 patients in Malaysia found that exposure to water contaminated with arsenic caused certain high degrees of toxicity (Mazumder et al., 2003). Additional symptoms comprised burning sensations in the legs swelling, liver fibrosis, chronic lung disease, eyes and skin cancer. A report from the World Health Organisation (WHO) indicated that by 2010, an estimated 33,800 new HIV infections, 1.7 million hepatitis B infections and 315,000 hepatitis C infections were attributed to improper infection practices (Akpan & Olukanni, 2020; Pepin et al., 2014). Therefore, effective management of hazardous waste is crucial in today's world, focusing on waste reduction, proper disposal and safe storage. This chapter explores the various strategies employed globally to minimise hazardous solid waste, emphasising environmental protection and sustainability and aims to explore various strategies for minimisation of HWs rather than their treatment or disposal, i.e., management as minimisation aims to reduce generation of HW at source, which directly aims at the core issue and is more sustainable in the long run. It also signifies a transition from a linear economic model characterised by "take-make-dispose" to a circular economy approach, where materials are retained in use for extended periods and waste is eliminated from systems whenever possible.

2 Sources and Types of Hazardous Waste

2.1 Sources of Hazardous Wastes

Any solid material that is discarded and poses a significant risk to the environment and public health is considered hazardous solid waste, based on its physicochemical or biological properties. Household Hazardous Waste was defined as "leftover household products that can catch fire, react, or explode under certain circumstances, or that are corrosive or toxic" (Household Hazardous Waste, US EPA). The United Nations Environmental Programme (UNEP) has categorised waste as HW based on the characteristics, which result in the following consequences (Saeidi-Mobarakeh et al., 2020):

(i) Corrosive when exposed to air or some typical environment **(Corrosivity)**
(ii) Fires that occur during management **(Ignitability)**
(iii) Long-term effects on ecology, geology and environment **(Toxicity)**
(iv) Reactions that release toxic gases **(reactivity)**.

Hazardous waste may originate from a variety of sources, as illustrated in Fig. 1, including e-waste, biomedical waste and waste from personal computers (PCs). From 1994 to 2003, approximately 500 million PCs were reported to have concluded their service lives and these PCs reportedly contained tons of plastic. Apparently, in a 2005 report, their lifespan was limited to two years on average, a reduction of six years from 1997. According to reports, they typically contain 1363 tons of cadmium, 718,000 tons of lead and 2,872,000 tons of plastics (Widmer et al., 2005; Culver, 2024). China, the USA and Russia are often cited as top producers of hazardous waste. Africa is frequently portrayed as a dumping ground for obsolete electronic materials discarded by developed countries, highlighting the pressing issue of e-waste exploitation in the region. According to reports, Nigeria produces 2,469,000 tons of garbage per annum, which is more than any other country in Africa. Egypt comes in second with roughly 1,440,000 tons per annum. Reports indicate that Benin Republic and Morocco have the highest HW per capita, at 65 and 33 kg per person annually, respectively. Estonia has the highest per capita generation of hazardous six waste, at 4774 kg per person annually. From these data, we can conclude that waste management has become crucial for the survival of all living beings.

Hazardous solid waste arises from various sources:

Industrial Sources

Industries generate a diverse range of waste during the manufacturing process, which can be categorised as either hazardous or non-hazardous waste. They are the largest contributors to HW, particularly sectors such as chemical sectors (e.g. spent solvents, production residues), electronics (e.g. printed circuit board, heavy metals), textile (e.g. toxic dyes, wastewater sludge) and mining operations.

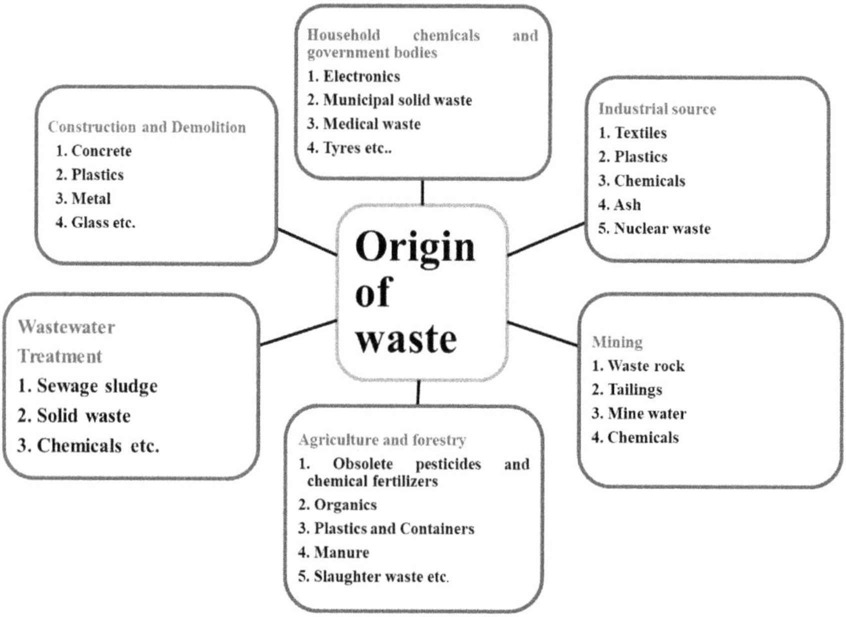

Fig. 1 Origin of waste from different sources

Healthcare and Biomedical Sources

Hospitals, labs and clinics generate wastes (Pasupathi et al., 2011) like needles, scalpels, human tissues and organs, expired or unused drugs, etc. According to Edzamba and Franck, approximately 85% of hospital waste is classified as safe, while 10% is considered contagious. The remaining 5% consists of non-contagious waste (Edzamba et al., 2024). Inopportune disposal of these can initiate infections or chemical exposure risks.

Agricultural Sources

They include pesticide and insecticide residues, chemical containers and more. These pose threats to both farmworkers and the environment, especially groundwater contamination.

Municipal and Domestic Sources

Municipal sources include household and commercial wastes. Households collectively generate large quantities of HW like chemicals (like bleach, ammonia), batteries, paints, electronics and small appliances. Their disposal is often inadequate due to a lack of public awareness.

Construction Sources

Construction wastes are generated during repair, remodelling, renovation and maintenance of various infrastructures like buildings, roads, airports, etc. (Hemalatha et al., 2008). They also generate HW containing asbestos, lead-based paints or treated wood products.

Radioactive Sources

Waste, consisting of radioactive substances (radionuclides), can harm living beings through radiation exposure and poses considerable risks to the environment. Nuclear technologies (such as nuclear reactors), nuclear fuel, contaminated materials, uranium mining waste or scrap metals from nuclear disposal are all capable of producing radioactive waste.

2.2 *Classification of Hazardous Waste*

It is indispensable to classify and identify the HW to secure their effective management, and every country has its own procedural methods for it. The USA enforces Subtitle C of the Resource Conservation and Recovery Act (RCRA) to delineate hazardous waste (USEPA 1976). China employs GB (Guobiao Standards) 5085.7 identification standards, while India uses the Hazardous and Other Waste Regulations—2016 (HWM rules-2016, 2019).

The United States Environmental Protection Agency (USEPA) notifies that the solid waste can be hazardous if it obeys one of the four groups (F, K, P and U) stated in the CFR or Code of Federal Regulations.

1. The F list categorises waste produced by traditional engineering and industrial activities as hazardous. For instance, residues from used solvents and substances utilised for wood treatments.
2. The K list classifies the hazardous waste from various industries and manufacturing sectors as source-specific waste. Example: production of iron and steel, wood preservation and pesticide production etc.
3. The P and U lists categorise hazardous waste derived from unadulterated and saleable grade formulations of specific waste types that need disposal (USEPA). They can also be categorised according to the toxicity that they initiate. They are categorised into high waste, intermediate risk waste and low–risk wastes (Misra et al., 2005). High-risk waste is described as the highly toxic, mobile

substances that can bioaccumulate, such as cyanide waste, PCB (polychrome biphenyl) and chlorinated solvents. Intermediate risk waste includes sludge from metal hydroxide. Whereas low–risk wastes are wastes that contain a significant amount of putrescible waste and low-hazard waste.

2.3 Characterisation and Quantification

According to the EPA 2016, hazardous waste refers to waste that has more than one inherent physiochemical characteristics, including ignitability, reactivity, toxicity and corrosivity. HW can probably be applied as a substitute for fossil fuel (Sadala et al., 2019) since it contains both sustainable and non-sustainable compounds with significantly increased calorific values (Karpan et al., 2021). The most effective methods of turning HW into cleaner, sustainable energy (Okolie et al., 2019) are thermal processing (pyrolysis, gasification, incineration) (Dhyani et al., 2018) and hydrothermal gasification and liquefaction (Kumar A et al., 2023). HW contains many components, like organic wastes, polythene, plastic and paper waste.

Organic wastes account for the majority of all waste streams. They therefore bear the greatest burden of disposal expenses and pose the greatest risk of greenhouse gas emission. It results in releasing landfill gases (carbon dioxide and methane) along with toxic trace elements, which are harmful to living beings. There are approximately three primary ways to repurpose organic waste, such as generating energy, composting and soil amendment. In developing countries, 60–70% of municipal solid waste (MSW) is organic content. Polythene also makes an important contribution to the composition of hazardous waste.

Plastics, especially high or low-density plastics, are durable but pose a significant environmental threat. The majority of polymers have great potential for repurposing by institutions and the local populations. Separating them at the source is one of the best methods to gather them. However, a large amount goes undetected when local customers pick them up or purchase them straight from the store owners. The quantity of plastic waste generated worldwide has increased since 2018–2020 due to increased usage during the COVID-19 pandemic (Choudhary et al., 2023). In which 46% of the usage was particularly from Asia (Peng et al., 2021). Compared to 2018 (4902.8 tonnes/day), China's medical waste production increased significantly after the epidemic, reaching 6,606.8 tonnes per day, while in the USA, it rose from 2.5 to 5.0 metric tons per month (Ilyas et al., 2021). Paper waste is less hazardous compared to other factors, as the large universities and cleaners resell it to the recycling agent directly (Fig. 2).

Due to industrialisation, urbanisation and economic development, the amount of waste generated varies significantly from country to country. More developed countries have higher incomes and economies, resulting in more waste per capita. The usual amount of MSW is 2.01 billion metric tons annually worldwide. This figure is expected to grow by 2.59 billion metric tons annually by 2030 and 3.40 billion metric tons annually by 2050 (Kaza et al., 2018). The common MSW includes metals

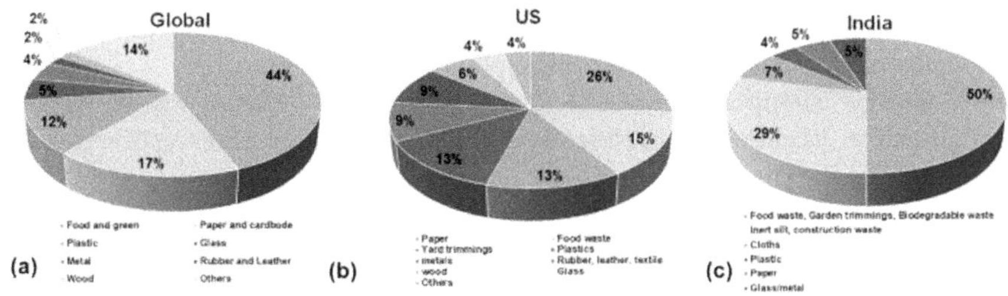

Fig. 2 Comparison of municipal solid waste composition: **a** global, **b** USA and **c** India (Reprinted from original article by Kumar et al. (2023) after taking permission from Springer Nature)

(1.1–2.2%), rubber (0.1–1.4%), glass (0.5–1.1%) and polyethylene (2.8–4.3%). Food waste (68.3–81.1%), textile (1.3–2.2%), paper (7.2–10.7%) and others (4.5–10.4%) are other non-hazardous fractions in MSW. India generates nearly 51.1 MMT of waste annually, with about 7.46 MMT of hazardous waste produced by 43,936 industries. Approximately 3.41 MMT (46%) is dumped in landfills, 0.69 MMT (9%) is combusted, and 3.35 MMT (45%) is recycled. Gujarat is the largest producer of hazardous waste, generating nearly 7751 tons and accounting for 28.76% of the country's total waste.

Approximately one-third of the 2.01 billion tons of municipal solid waste (MSW) produced globally each year is not disposed of in an environmentally friendly way. In developing countries, waste typically has a high organic content (approximately 60–70%), low calorific value and high moisture levels (20–30%). Due to its poor calorific value, MSW does not burn properly in developing nations. It is estimated that by 2050, the world will generate 27 billion tons of waste annually, with Asia accounting for approximately one-third of this total. The generation of MSW varied with the factors, such as urbanisation, geography, culture and financial conditions.

2.4 *Status and Trends in Global Municipal Solid Waste Management*

Global waste production is projected to reach 27 billion tons annually by 2050. Among Asian countries, India produces 0.5–0.9 kg of waste/person daily, whereas 0.44–4.3 kg of waste is generated in China per person. In contrast, Gulf Cooperation Council (GCC) nations- Saudi Arabia, Bahrain, UAE, Qatar, Oman and Kuwait produce each day a higher per capita waste amount compared to the global average of 1.2 kg. Some European countries have even succeeded in bridging the gap between domestically generated and treated hazardous waste. According to the World Bank and UNEP, HW generated globally exceeds 400 million metric tons annually, with the following trends:

(a) More waste is generated by the high-income countries per capita due to more industrial processes and high consumption rates.
(b) Countries comprising low- and middle-income are experiencing the fastest growth in hazardous waste due to expanding industrial bases and a lack of proper management.
(c) Fastest-growing e-waste stream globally due to its use as a daily necessity, much of which is hazardous due to lead, mercury and other toxic elements.

India mainly encounters issues in HWM due to an inadequate constitution and an absence of expertise (Kumar et al., 2020). As per estimates by the Department of Economic Affairs (Government of India), India's population would reach 1823 million by 2051, producing around 300 million tonnes of MSW annually, requiring 1450 km^2 of land for its proper disposal. In regions with populations over two crores, the daily waste generation is 0.43 kg per individual, while in areas with populations below 1 lakh, it is 0.36 kg per individual.

According to the 2014 Planning Commission Report, if the waste generation rate remains at 0.45 kg per person per day, urban centers are expected to generate approximately 165 million tonnes of garbage annually by 2031, potentially increasing to 436 million tonnes by 2050. This would result in a land requirement of 1175 ha each year by 2031, which could rise to 43,000 hectares annually by 2050. A report from the National Environmental Engineering Research Institute (NEERI) found that the 59 cities in India selected produced 39,031 tonnes per day (TPD) of MSW during 2004–2005, which grew to 50,592 TPD in 2009–2010. By 2011, India generated 127,486 TPD of MSW, but merely 89,334 TPD was collected, and just 15,881 TPD was processed (CPCB, 2013). As per the Central Pollution Control Board (**CPCB**) data, the amount of MSW produced in Indian cities increased exponentially between 2001 and 2018; this pace of growth is predicted to continue to accelerate (Rajak et al., 2022).

3 Strategies for Minimisation of Hazardous Solid Wastes

Sustainable waste management is a socially and environmentally responsible practice that is achieved through community involvement, public–private partnerships, fiscal incentives, institutional capacity building, strategic planning and technologically and economically feasible technology.

The U.S. Congress stated in the Hazardous and Solid Waste Amendments of 1984 that the national policy aims to minimise or eliminate hazardous waste (HW) generation whenever possible. It also emphasises managing waste through treatment, storage, or disposal methods that lower current and future risks to the environment and public health. HWM techniques can be established in a hierarchy depending on their preferences. The most prevalent approach is HW reduction at the source; subsequently, various approaches like reuse, recycling and composting are in the later preference (Wang et al., 2018). Minimisation focuses on avoiding or reducing waste at the point of generation, thereby lowering treatment costs, environmental hazards and public health risks.

The HWM can be achieved by identifying the nature and characteristics of waste, such as plastic waste, nuclear waste, medical waste and biowaste, among others and classifying suitable treatment facilities based on waste type. Waste minimisation encompasses the first four categories of EPA's preferred hazardous waste management strategy, as outlined below:

1. *Reducing Waste*—Involves limiting the amount generated at its source.
2. *Waste Separation and Concentration*—Isolating wastes from their natural mixtures.
3. *Interchange of Waste*—Allocating the waste for reprocessing.
4. *Energy and Material Recovery*—Involves reusing and recycling waste to generate energy or reclaim materials.
5. *Incineration/Treatment*—Break down, detoxify and neutralise waste into less harmful substances.
6. *Secure Land Disposal*Involves depositing waste across different sites, utilizing methods like volume reduction, encapsulation, monitoring and controlled release of air and surface contaminants.

HWM's primary objective is to encourage industries and communities to enhance their performance while reducing the use of raw materials. The secondary strategies include reuse and recycling, which allow for saving time and energy on new materials. When reuse and recycling options are the preferred approach, which commonly uses strategies like pyrolysis (Chew et al., 2021), gasification (Hameed et al., 2021) and incineration (Wajda et al., 2022). Recently, electronic wastes have also been reported for recycling in a number of ways. When waste-to-energy conversion is not feasible, physical, chemical and biological methods are employed. Disposal approaches like landfilling are the last method in the hierarchy as they are the least preferred option due to their unsustainable nature.

This study provides a detailed description of several existing and new advancements and multidisciplinary approaches in the effective minimisation of hazardous waste.

3.1 Source Reduction or Reduce

Source reduction involves strategies that aim to avert the creation of hazardous waste at its origin. This is often considered the most desirable form of waste minimisation because it addresses the problem before it is caused. This can be done by:

1. Process Optimisation: Optimising the industrial processes to use fewer hazardous materials or generate less waste
2. Equipment Upgrade: Replacing outdated machines with new and efficient systems

3. Inventory Control: Buying and using chemicals and materials in precise amounts to avoid overuse and wastage
4. Green Chemistry: To reduce the generation of hazardous substances through designing products and processes.

Source reduction in electroplating can be done by replacing cyanide-based electroplating with a non-toxic alternative, such as alkaline-based electroplating or metal-based electroplating, which reduces hazardous waste output by 85%. But this strategy also faces some problems like capital cost leading to resistance to change, lack of awareness or technical expertise or unclear return on investment.

3.2 Recycling

Recycling is a regenerative method of managing HW that involves converting it into functional forms by reprocessing or reusing materials of great economic value (plastics), which helps in saving energy and resources (Kong et al., 2020). Recycling is commonly achieved through these methods, including breaking down waste into useful products (Savoldelli et al., 2017), recovering resources (Sadala et al., 2019) and extracting useful products from waste (llyas et al., 2021).

Studies indicate that recycling 10 kg of aluminum saves 90% of energy and prevents the release of 20 kg of CO_2, 0.11 kg of SO_2 and 13 kg of bauxite waste (Kumar et al., 2017). In a similar vein, recycling iron and steel leads to energy savings of 74%, an 86% reduction in air pollution, a 40% decrease in water consumption, a 76% drop in water contamination, a 97% reduction in mining waste and a 90% decrease in the reliance on raw materials (Kumar et al., 2017; Cui et al., 2003).

Researchers have developed new constituents (bitumen for construction purposes) using recycled plastics, glass, etc. Recently, many researchers have explored various technologies to turn WCO (Waste Cooking Oil) into biodiesel. Some applications of these technologies, such as the use of cork biochar as a heterogeneous catalyst (Bhatia et al., 2020) and MM-CaO catalysts derived from eggshells (Nadeem et al., 2021), are examples of newly developed technologies. To summarise, recycling is a hygienic but underdeveloped waste reduction method that demands high investment; thus, it must be thoroughly evaluated before being deployed.

3.3 Waste-to-Energy Conversion

MSW releases energy during the breakdown of chemical bonds and can be used as a source of fuel through biochemical and thermochemical conversions. The conversion of waste-to-energy refers to the conversion of waste into energy or valuable products (like biochar, methane, syngas, bio-oil) as a substitute for fossil fuels (Sadala et al., 2019; Shovon et al., 2024). Thermal treatment, which generates

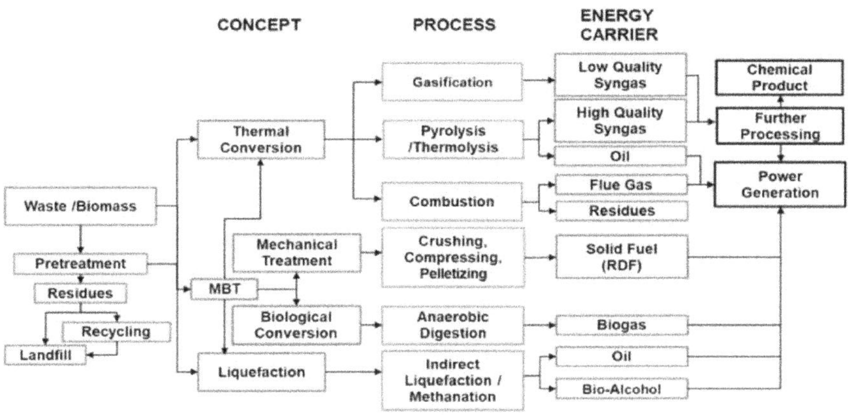

Fig. 3 Various methods for waste-to-energy treatment and their end products (Reprinted from original article by Ugwu, C.O et al. (2021) after taking permission from Elsevier)

cleaner, renewable energy, is considered most advantageous (Okolie et al., 2019). Figure 3 is shown, demonstrates some waste-to-energy conversion methods along with processes involved and their end products.

Thermochemical Conversion

Thermochemical conversion is a method of waste breakdown through heat (Chandraratne & Daful, 2022; Monir et al., 2020, 2021). It usually converts waste into biofuels. This method generally does not require any chemical add-ons and thus can be applied to a range of stocks throughout time. Thermochemical conversion follows techniques such as incineration, pyrolysis and gasification.

(i) **Pyrolysis**

Pyrolysis is a technique that decomposes waste through anaerobic thermal degradation, yielding three main products: a liquid (bio-oil), non-condensable gases and solid carbon residue (biochar) (Kumar et al. 2022). The product's production depends upon many aspects such as the design of the reactor, the composition and the type of feedstock. Pyrolysis is currently regarded as one of the most promising technologies because it efficiently produces bio-oil in high yields at relatively moderate temperatures within a brief time frame. After various analyses (Chabbra et al. 2021), when the moisture content is less than 20%, pyrolysis is energy sustainable.

During a study, pyrolysis using dolomite as a catalyst was used to manufacture syngas (Santagata et al., 2021). Producing syngas with 47 to 67 mol% of dolomites significantly affects syngas output (He et al., 2010), and it was found that yield increases with temperature.

(ii) **Gasification**

Gasification involves a thermochemical process that breaks down carbon-based materials into a gaseous syngas (Chanthakett et al., 2021). In this process, waste is kept under high temperature (1073–1473 K) along with a gasifying agent to produce syngas (Bourtsalas et al., 2023), and the composition depends on the gasifying agent used. This process includes steps such as drying to remove moisture and pyrolysis to release volatile components (Jeong et al., 2022). A study on the catalytic gasification of MSW found that adding the Ni-CaO-TiO_2 catalyst (with around 20% Ni) increased the H_2 concentration from 35.1% to 57.7%, lowered the tar content from 9.38 to 2.55% and increased the dry gas output from 0.75 to 1.74 Nm^3/kg.

As of right now, 114 gasification projects are operational, 27 are on hold and 13 are being built (Munir et al., 2019). Of these, 106 are dedicated to the production of electricity, 24 to the manufacture of liquid fuel, 8 to gaseous fuel and 7 to chemicals (Molino et al., 2018).

In comparison to pyrolysis and incineration, the gasification method is said to show better efficiency, due to its many benefits, like controlling the product's composition or the versatile use of its products. However, it also faces challenges like complex operation, small-scale capacity and less flexibility (Palmer et al., 2021).

(iii) **Incineration**

Incineration is a widely used technique to convert waste into heat energy (Lino et al., 2017). It burns (at temperatures above 800 °C) toxic organic components of HW and reduces the size of waste, and is well-known for material recovery or energy production (Block et al., 2015). HW incinerators come in a variety of forms (Trinh et al., 2019), including fluidised bed units (Azam et al., 2019), liquid injection units (Anufriev, 2021) and rotary kilns (Jiang et al., 2019). The incineration process produces byproducts like fly ash, bottom ash, etc. and its efficiency ranges from 60 to 90%. India (Gupta et al., 2021), Vietnam (Nguyen et al., 2021) and Thailand generally use this technique for waste management.

This method has several advantages, including energy recovery and preventing groundwater pollution, but it also has several disadvantages, such as toxic pollutant emissions and higher plant costs.

Biochemical Conversion

Biochemical conversion is a technique involves the biomass utilisation with the help of biological organisms and often produces biomass sludge as a by-product. This process is temperature sensitive and can even be negatively affected by sunlight exposure and thus offers limited productivity. It can be done in the following ways:

(i) **Composting**

The term 'compost' has derived from the Latin word 'componere', meaning 'to put together'. Compost is a natural biotransformation process that uses bacteria, fungi and actinomycetes to turn organic waste into a nutrient-rich, non-toxic fertiliser in a regulated environment with optimum amounts of moisture, aeration and temperature.

It requires a carbon to nitrogen ratio of 30:1 and a moisture level between 40 and 60%. The composting process can be categorised into aerobic and anaerobic types. Aerobic composting occurs with oxygen, resulting in compounds such as CO_2 and NH_3, as well as water and heat. Contrarily, anaerobic processes occur in environments devoid of oxygen, resulting in various by-products, including CO_2, CH_4, NH_3 and other trace gases, along with organic acids. This metabolic process is significant in various ecological and industrial contexts, facilitating the decomposition of organic matter while contributing to nutrient recycling in anaerobic ecosystems.

The composting approach for dealing with waste products from a variety of sources, such as yard waste, industrial waste, sewage, municipal solid waste, agricultural waste and animal manure and with lowest concentrations of hazardous substances (Sayara et al., 2020; Wang et al., 2018). Composting is a popular method for recycling organic waste that can be used in agricultural activities, as it recovers essential minerals such as phosphorus, nitrogen and potassium (Ahmad et al., 2019; Mupambwa et al., 2018). Vermicomposting is a popular composting strategy that involves the decomposition of organic waste by earthworms into a humus-like substance, increasing microbial diversity and accelerating nutrient mineralisation rates (Zucco et al., 2015). It is significant as a sustainable waste management practice and can be assessed in terms of soil enrichment, waste reduction and carbon sequestration, among others.

(ii) **Dark Fermentation**

Dark fermentation, as the name suggests, occurs in either sunlight or the lack of it (Mishra et al., 2019). This process uses bioreactors to convert biological energy into other forms (Sarangi et al., 2020). Dark fermentation utilises organic matter from wastewater to produce hydrogen with the assistance of living organisms (Pal et al., 2022). Dark fermentation offers several benefits like utilisation of diverse feedstocks, versatile end-products, hydrogen production from organic waste and fewer pollutants, thus considered environment-friendly. However, its main disadvantage is its lower capacity per unit capital expenditure compared to alternative technologies and limited scalability (Rathi et al., 2022).

(iii) **Anaerobic Digestion**

Anaerobic digestion (AD) is a biological process where organic material breakdown by microorganisms in anaerobic resulting in methane-rich biogas and usable digestate. The traditional AD is not considered efficient due to its long digestion period and large energy consumption (Mao et al., 2015), so nowadays, several pretreatment techniques are being used to increase methane production and energy output (Ali et al., 2018). According to Fan and Klemes (Fan et al., 2018), using microwaves to treat MSW limits energy consumption and lowers carbon emissions. AD has numerous advantages, such as less consumption of operating energy, producing fertiliser and electricity generation (Kumar et al., 2020). But it also has its own complications like technical complexity, a long and high capital start-up and limited feedstock.

3.4 Physical Remediation

Physicochemical and biological methods are considered when waste-to-energy conversion is not feasible. In this sorbent material, skimmers and booms are used for the physical remediation.

(i) The material-based physical barrier boom that accumulates oil pollutants and avoids them from spreading until further remediation is carried out (Vocciante et al., 2019).
(ii) Sorbents and Skimmers are used after booms for further absorption of pollutants (Kumari et al., 2019).

Physical remediation also includes methods like filtration and separation, which is the method of separating solid particles from liquid with the help of a porous medium. It also includes evaporation, which involves the conversion of a liquid from a slurry or solution into vapour form, with evaporation mainly being used as a pre-treatment method to reduce waste quantities.

This remediation faces a major challenge because it relies heavily on buoyancy and roll response, which rotate the bloom from its upright position through torque. The buoyant boom floats and remains on the surface of water, resulting in higher remediation due to increased roll response (Dhaka et al., 2021).

3.5 Chemical Remediation

In this remediation method, chemicals like biochar, phosphate, clay minerals, sulphide and aluminium salts are added to eliminate heavy metals from the ecosystem. This process involves mechanisms such as complexation, precipitation, adsorption, redox reaction and ion exchange (Xu et al., 2022). While chemical remediation offers a quick, simple and straightforward solution, it's important to remember that the chemicals used can also contribute to pollution concerns environment.

3.6 Bioremediation

Bioremediation is a safe, affordable and ecologically acceptable method of cleaning up hazardous waste using bacteria and plants (Patel et al., 2022). This section mainly deals with the use of microbes, as plants favour them due to their rapid and easy growing, easy manipulation and ability to survive in abnormal environments (Perera & Hemamali, 2022).

The Breakdown of toxic materials into less harmful substances such as carbon dioxide, water and other compounds with the help of microorganisms like bacteria, fungi and algae, which are further broken down by other microbes through mineralisation (Mahmoud et al., 2021; Kumar et al., 2022). Similarly, immobilisation transforms substances into a form that renders them unavailable in a given environment, such as the conversion of nitrate into organic nitrogen (Pratush et al., 2018), and is also employed for heavy metals, especially in highly polluted areas. Some microorganisms, along with their preferred bioremediation method, are listed in Table 1.

(i) **Enzymatic Oxidation**

Enzymatic oxidation is a process in which the oxidoreductase enzyme oxidises pollutant molecules from a higher to a lower oxidation state, resulting in the heavy metal becoming less hazardous due to electron loss. This method is effective for removing phenol, dyes and other pollutants that are challenging for microorganisms to break down (Unuofin et al., 2019). The oxidoreductase enzyme is laccase, which is primarily used for the oxidation of aromatic amines. *Pycnoporus* sp. and *Leptosphaerulina* sp. have been reported to produce laccase, which was found to be capable of degrading heavy metals (Copete-Pertuz et al., 2018; Tian et al., 2020).

(ii) **Bioaugmentation**

Adding microorganisms to contaminated locations so they can consume contaminants is known as bioaugmentation, and it is a quick, efficient and economical kind of bioremediation (Mahmoud et al., 2021). Generally, microbes are introduced to contaminated sites from external sources, but in some cases, microbes native to the site are isolated and genetically modified for use modification. For example, at about 30 °C and a slightly acidic pH, Burkholderia sp. Fenitrothion Degrading Strain-1 (FDS-1) has been shown to transform nitrophenolic compounds in pesticide-contaminated soil into less harmful forms when introduced to a polluted site (Goswami et al., 2018).

Table 1 Bioremediation mechanism in various microorganisms

Microorganism	Pollutant remediated	Mechanism of remediation
Cupriavidus sp.	Cadmium	Bioprecipitation
Oudemansiella canarii	Congo red dye	Enzymatic reduction
Bacillus subtilis	Lead	Bioimmobilisation
Pseudomonas aeruginosa	Cadmium	Biosorption
Pseudomonas aeruginosa and *Bacillus cereus*	Lead and cadmium	Bioaugmentation
Desolfovibrio desulfuricans	Copper, zinc and cadmium	Extracellular sequestration
Bacillus sp., Lysinibacillus sp. and *Rhodococcus sp.*	Aluminium, lead, cadmium and copper	Bioaugmentation
Bacillus sp.	Nickel	Biosorption
Lysinibacillus sphaericus	Azo dyes	Enzymatic reductase
Pseudomonas sp.	Copper and lead	Bioattenuation
Sulfolobus solfataricus	Copper	Intracellular sequestration

(iii) **Biostimulation**

Biostimulation involves adding nutrients, electron donors, metabolites, biosurfactants, enzymes and other substances to soil to enhance the activity of existing microbes and speed up the remediation process (Ojuederie & Babalola, 2017). Bioaugmentation is preferred because of the use of indigenous microbes, which are more competitive than introduced ones (Sayed et al., 2021), and it also maintains the natural biodiversity balance. Microbes such as *Bacillus* sp., *Rhodococcus* sp., *Staphylococcus* sp., and *Klebsiella* sp. have been reported to be effective in bioremediating heavy metals (Nivetha et al., 2023). Biostimulation, though effective, can cause environmental issues such as eutrophication and might contribute to further pollution.

(iv) **Bioleaching**

In the bioleaching process, the acidophilic microbes are employed to enhance the solubilisation of heavy metals, mainly iron or sulphur-containing pollutants (Sun et al., 2021). For example, *A. thiooxidans*, *Aspergillus* sp., *Mucor* sp., *Penicillium* sp., *Cladosporium* sp. and *Rhizopus* sp. are among the species of Fe-S oxidising bacteria that are commonly used (Medfu Tarekegn et al., 2020).

(v) **Biosorption**

The removal of heavy metals from contaminants by means of complexation, chelation, proton displacement and interaction with electrostatic forces is known as biosorption (Mahmoud et al., 2021). It aids in the recovery of useful industrial metals, such as gold and zinc. Some microorganisms like *Bacillus anthracis, Rhodococcus erythropolis* and *Streptomyces* sp. K11 is effective in bioremediation through the biosorption process (Baltazar et al., 2019). E.g., Cyanobacteria have been reported to be effective for bioremediating complex compounds like polythene.

(vi) **Bioaccumulation**

In bioaccumulation, a lethal stockpile of chemicals is found in the microbes intracellular region due to the rate of absorption being more than the rate of consumption. Many organisms like *Rhizopus arrhizus* (bioremediates mercury), *Pseudomonas putida* (for cadmium) and *Aspergillus niger* (thorium) have been found to be highly active in bio-accumulation for heavy metals (Sharma et al., 2022).

(vii) **Precipitation**

Precipitation is the process by which heavy metals are converted into precipitates or crystals, thereby reducing their toxicity levels. For example, under alkaline pH (above 7), sulfate-reducing bacteria can convert organophosphate into orthophosphate (Pratush et al., 2018). Similarly, *Bacillus subtilis* has been linked to environmental heavy metal precipitation (Maity et al., 2019).

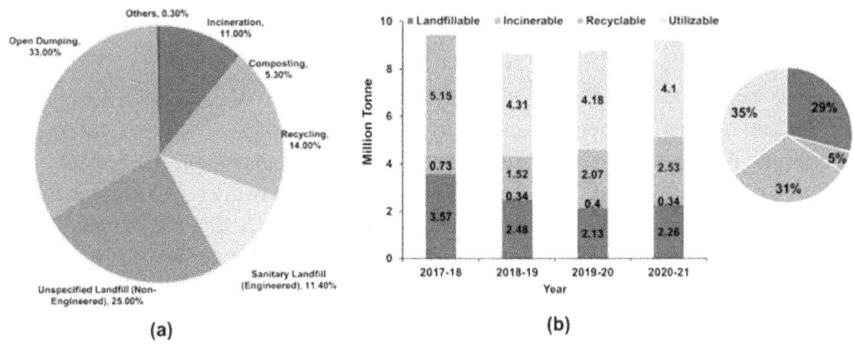

Fig. 4 **a** Global MSW management status. **b** India's category-wise HW data generated from 2017 to 2021. (Reprinted from original article by Kumar A et al. (2023) after taking permission from Springer Nature)

3.7 Landfilling and Open Dumping

Landfilling is a widely used but unsustainable method of non-liquid HWM. To minimise environmental damage, the HW is separated before being dumped in landfills, where it goes through several physicochemical and biological changes. Landfilling is an inexpensive, direct, labour-efficient, large-capacity method that works best with waste that has less organic and moisture content. According to Misra and Pandey (Misra et al., 2005), landfills can be categorised into common, approved and secured landfills. They can deteriorate into open dumps if not properly treated (Nemathaga et al., 2008). Figure 4a displays the global status of MSW management worldwide, while Fig. 4b presents category-wise statistics of HW generated in India from 2017 to 2021 (CPCB I inventory 2022). These data showed that most of the HW generated is landfillable.

3.8 Treatment of Radioactive Waste

Radioactive waste comprises radioactive elements or radioactive contaminants that can damage living biological organisms and the environment when they come into contact with them. Due to the presence of radionuclides, radioactive wastes are managed separately from other MSW. The nuclear industry has implemented many technologies for permanent disposal (nearly 90%) (IAEA, 2009). Radioactive waste management involves collecting the waste, treating it as needed, and then coming up with a long-term plan that involves storing, disposing of, or transforming it into a non-toxic form. All the while, maintaining the waste's safety and ensuring that it poses no unnecessary burden to present or future generations (Ojovan & Lee, 2005). Some treatment methods include incineration, compaction, vitrification and cementation.

4 Conclusion

The improper management, storage, treatment and disposal of hazardous waste have become a major environmental issue affecting all living things. Therefore, having an appropriate management system for hazardous waste management is necessary. Its production is closely related to a nation's socioeconomic progress and income level. So, focus should be on suitable waste management strategies that are sustainable in design but also efficient, budget-friendly and eco-friendly. The reduction of waste at the source or through prevention is considered the most preferred strategy, followed by waste minimisation (via reuse, recycling and composting), recovery or waste-to-energy processes, treatment and material recovery or waste-to-energy conversion, treatment and disposal. Besides environmental restoration, these treatment technologies can also generate value-added products such as chemicals and energy, based on the types and characteristics of hazardous and non-hazardous materials waste.

Acknowledgements The authors would like to thank Hemvati Nandan Bahuguna Garhwal University, Srinagar Garhwal (Uttarakhand), and Key Laboratory of Clean Chemistry Technology of Guangdong Regular Higher Education Institutions, School of Chemical Engineering and Light Industry, Guangdong University of Technology, Guangzhou, China.

References

Abubakar, I. R., Maniruzzaman, K. M., Dano, U. L., AlShihri, F. S., AlShammari, M. S., Ahmed, S. M. S., Al-Gehlani, W. A. G., & Alrawaf, T. I. (2022). Environmental sustainability impacts of solid waste management practices in the global South. *International Journal of Environmental Research and Public Health, 19*(19), 12717.

Ahmad, G., Nishat, Y., Haris, M., Danish, M., & Hussain, T. (2019). Efficiency of soil, plants and microbes for healthy plant immunity and sustainable agricultural system. *Plant Microbe Interface* (pp. 325–346). Springer International Publishing.

Akpan, V. E., & Olukanni, D. O. (2020). Hazardous waste management: An African overview. *Recycling, 5*(3), 15.

Ali, M., Zhang, J., Raga, R., Lavagnolo, M. C., Pivato, A., Wang, X., Zhang, Y., Cossu, R., & Yue, D. (2018). Effectiveness of aerobic pretreatment of municipal solid waste for accelerating biogas generation during simulated landfilling. *Frontiers of Environmental Science & Engineering, 12*(3), 5.

Anufriev, I. S. (2021). Review of water/steam addition in liquid-fuel combustion systems for NOx reduction: Waste-to-energy trends. *Renewable and Sustainable Energy Reviews, 138*, Article 110665.

Ayilara, M. S., & Babalola, O. O. (2023). Bioremediation of environmental wastes: The role of microorganisms. *Frontiers in Agronomy, 5*, 1183691.

Ayilara, M. S., Olanrewaju, O. S., Babalola, O. O., & Odeyemi, O. (2020). Waste management through composting: Challenges and potentials. *Sustainability, 12*(11), 4456.

Azam, M., Setoodeh Jahromy, S., Raza, W., Wesenauer, F., Schwendtner, K., & Winter, F. (2019). Comparison of the characteristics of fly ash generated from bio and municipal waste: Fluidized bed incinerators. *Materials, 12*(17), 2664.

Baltazar, M. D. P. G., Gracioso, L. H., Avanzi, I. R., Karolski, B., Tenório, J. A. S., do Nascimento, C. A. O., & Perpetuo, E. A. (2019). Copper biosorption by Rhodococcus erythropolis isolated from the Sossego Mine–PA–Brazil. *Journal of Materials Research and Technology*, *8*(1), 475–483.
Bhatia, S. K., Gurav, R., Choi, T. R., Kim, H. J., Yang, S. Y., Song, H. S., Park, J. Y., Park, Y. L., Han, Y. H., Choi, Y. K., & Kim, S. H. (2020). Conversion of waste cooking oil into biodiesel using heterogenous catalyst derived from cork biochar. *Bioresource Technology, 302*, Article 122872.
Block, C., Van Caneghem, J., Van Brecht, A., Wauters, G., & Vandecasteele, C. (2015). Incineration of hazardous waste: A sustainable process? *Waste and Biomass Valorization, 6*(2), 137–145.
Bourtsalas, A. C. (2023). Energy recovery from solid wastes in China and a Green-BRI mechanism for advancing sustainable waste management of the global South. *Waste Disposal & Sustainable Energy, 5*(3), 309–321.
Chandraratne, M. R., & Daful, A. G. (2022). Recent advances in thermochemical conversion. In: *Recent perspectives in pyrolysis research* (p. 145).
Chanthakett, A., Arif, M. T., Khan, M. M. K., & Oo, A. M. (2021). Performance assessment of gasification reactors for sustainable management of municipal solid waste. *Journal of Environmental Management, 291*, Article 112661.
Chew, K. W., Chia, S. R., Chia, W. Y., Cheah, W. Y., Munawaroh, H. S. H., & Ong, W. J. (2021). Abatement of hazardous materials and biomass waste via pyrolysis and co-pyrolysis for environmental sustainability and circular economy. *Environmental Pollution, 278*, Article 116836.
Chhabra, V., Parashar, A., Shastri, Y., & Bhattacharya, S. (2021). Techno-economic and life cycle assessment of pyrolysis of unsegregated urban municipal solid waste in India. *Industrial & Engineering Chemistry Research, 60*(3), 1473–1482.
Choudhary, R., Mukhija, A., Sharma, S., Choudhary, R., Chand, A., Dewangan, A. K., Gaurav, G. K., & Klemeš, J. J. (2023). Energy-saving COVID–19 biomedical plastic waste treatment using the thermal-Catalytic pyrolysis. *Energy, 264*, Article 126096.
Copete-Pertuz, L. S., Plácido, J., Serna-Galvis, E. A., Torres-Palma, R. A., & Mora, A. (2018). Elimination of Isoxazolyl-Penicillins antibiotics in waters by the ligninolytic native Colombian strain Leptosphaerulina sp. considerations on biodegradation process and antimicrobial activity removal. *Science of the Total Environment, 630*, 1195–1204.
Cui, J., & Forssberg, E. (2003). Mechanical recycling of waste electric and electronic equipment: A review. *Journal of Hazardous Materials, 99*(3), 243–263.
Culver, J. (2024). *CPU history-the CPU museum-life cycle of the CPU* [online].
Davidson, G. (2011). *Waste management practices: Literature review* (pp. 1–59). Dalhousie University.
Dhaka, A., & Chattopadhyay, P. (2021). A review on physical remediation techniques for treatment of marine oil spills. *Journal of Environmental Management, 288*, Article 112428.
Dhyani, V., & Bhaskar, T. (2018). A comprehensive review on the pyrolysis of lignocellulosic biomass. *Renewable Energy, 129*, 695–716.
Dresselhaus, M. S., & Thomas, I. L. (2001). Alternative energy technologies. *Nature, 414*(6861), 332–337.
Duan, H., Huang, Q., Wang, Q., Zhou, B., & Li, J. (2008). Hazardous waste generation and management in China: A review. *Journal of Hazardous Materials, 158*(2–3), 221–227.
Edzamba, M. F., Cheuyem, F. Z. L., Sime, P. S. D., Minlo, F. E. N., Sani, B. N., Nkongo, F. K., & Mossus, T. (2024). Hospital waste management: An overview of the situation in health districts outside Yaounde in the centre region, Cameroon. *medRxiv* 2024–2012.
Goswami, M., Chakraborty, P., Mukherjee, K., Mitra, G., Bhattacharyya, P., Dey, S., & Tribedi, P. (2018). Bioaugmentation and biostimulation: A potential strategy for environmental remediation. *Journal of Microbiology & Experimentation, 6*(5), 223–231.
Guha Mazumder, D. N. (2003). Chronic arsenic toxicity: Clinical features, epidemiology, and treatment: Experience in West Bengal. *Journal of Environmental Science and Health, Part A, 38*(1), 141–163.

Gupta, G., Datta, M., Ramana, G. V., Alappat, B. J., & Bishnoi, S. (2021). Contaminants of concern (CoCs) pivotal in assessing the fate of MSW incineration bottom ash (MIBA): First results from India and analogy between several countries. *Waste Management, 135*, 167–181.

Hameed, Z., Aslam, M., Khan, Z., Maqsood, K., Atabani, A. E., Ghauri, M., Khurram, M. S., Rehan, M., & Nizami, A. S. (2021). Gasification of municipal solid waste blends with biomass for energy production and resources recovery: Current status, hybrid technologies and innovative prospects. *Renewable and Sustainable Energy Reviews, 136*, Article 110375.

He, M., Xiao, B., Liu, S., Hu, Z., Guo, X., Luo, S., & Yang, F. (2010). Syngas production from pyrolysis of municipal solid waste (MSW) with dolomite as downstream catalysts. *Journal of Analytical and Applied Pyrolysis, 87*(2), 181–187.

Hemalatha, B. R., Prasad, N., & Subramanya, B. V. (2008). Construction and demolition waste recycling for sustainable growth and development. *Journal of Environmental Research and Development, 2*(4), 759–765.

Ilyas, S., Srivastava, R. R., & Kim, H. (2021). O2-enriched microbial activity with pH-sensitive solvo-chemical and electro-chlorination strategy to reclaim critical metals from the hazardous waste printed circuit boards. *Journal of Hazardous Materials, 416*, Article 125769.

Jeong, Y. S., Mun, T. Y., & Kim, J. S. (2022). Two-stage gasification of dried sewage sludge: Effects of gasifying agent, bed material, gas cleaning system, and Ni-coated distributor on product gas quality. *Renewable Energy, 185*, 208–216.

Jiang, X., Li, Y., & Yan, J. (2019). Hazardous waste incineration in a rotary kiln: A review. *Waste Disposal & Sustainable Energy, 1*(1), 3–37.

Karpan, B., Raman, A. A. A., & Aroua, M. K. T. (2021). Waste-to-energy: Coal-like refuse-derived fuel from hazardous waste and biomass mixture. *Process Safety and Environmental Protection, 149*, 655–664.

Kaza, S., Yao, L., Bhada-Tata, P., & Van Woerden, F. (2018). *What a waste 2.0: A global snapshot of solid waste management to 2050.* World Bank Publications.

Kong, F., Ren, H. Y., Pavlostathis, S. G., Nan, J., Ren, N. Q., & Wang, A. (2020). Overview of value-added products bioelectrosynthesized from waste materials in microbial electrosynthesis systems. *Renewable and Sustainable Energy Reviews, 125*, Article 109816.

Kumar, A., & Agrawal, A. (2020). Recent trends in solid waste management status, challenges, and potential for the future Indian cities–A review. *Current Research in Environmental Sustainability, 2*, Article 100011.

Kumar, A., & Reddy, S. N. (2022). Study the catalytic effect on pyrolytic behavior, thermal kinetic and thermodynamic parameters of Ni/Ru/Fe-impregnated sugarcane bagasse via thermogravimetric analysis. *Industrial Crops and Products, 178*, Article 114564.

Kumar, A., & Samadder, S. R. (2020). Performance evaluation of anaerobic digestion technology for energy recovery from organic fraction of municipal solid waste: A review. *Energy, 197*, Article 117253.

Kumar, A., Holuszko, M., & Espinosa, D. C. R. (2017). E-waste: An overview on generation, collection, legislation and recycling practices. *Resources, Conservation and Recycling, 122*, 32–42.

Kumar, A., Thakur, A. K., Gaurav, G. K., Klemeš, J. J., Sandhwar, V. K., Pant, K. K., & Kumar, R. (2023). A critical review on sustainable hazardous waste management strategies: A step towards a circular economy. *Environmental Science and Pollution Research, 30*(48), 105030–105055.

Kumari, A., Kaur, R., & Kaur, R. (2019). A review on fate and remediation techniques of oil spills. *International Journal of Pharmaceutical Sciences Review and Research, 10*(1), 111–116.

Lino, F. A. M., & Ismail, K. A. R. (2017). Incineration and recycling for MSW treatment: Case study of Campinas, Brazil. *Sustainable Cities and Society, 35*, 752–757.

Mahmoud, G. A. E. (2021). Microbial scavenging of heavy metals using bioremediation strategies. In *Rhizobiont in bioremediation of hazardous waste* (pp. 265–289). Springer Singapore.

Maity, J. P., Chen, G. S., Huang, Y. H., Sun, A. C., & Chen, C. Y. (2019). Ecofriendly heavy metal stabilization: Microbial induced mineral precipitation (MIMP) and biomineralization for heavy metals within the contaminated soil by indigenous bacteria. *Geomicrobiology Journal, 36*(7), 612–623.

Mao, C., Feng, Y., Wang, X., & Ren, G. (2015). Review on research achievements of biogas from anaerobic digestion. *Renewable and Sustainable Energy Reviews, 45*, 540–555.

Medfu Tarekegn, M., Zewdu Salilih, F., & Ishetu, A. I. (2020). Microbes used as a tool for bioremediation of heavy metal from the environment. *Cogent Food & Agriculture, 6*(1), 1783174.

Methneni, N., Morales-González, J. A., Jaziri, A., Mansour, H. B., & Fernandez-Serrano, M. (2021). Persistent organic and inorganic pollutants in the effluents from the textile dyeing industries: Ecotoxicology appraisal via a battery of biotests. *Environmental Research, 196*, Article 110956.

Mishra, P., Krishnan, S., Rana, S., Singh, L., Sakinah, M., & Ab Wahid, Z. (2019). Outlook of fermentative hydrogen production techniques: An overview of dark, photo and integrated dark-photo fermentative approach to biomass. *Energy Strategy Reviews, 24*, 27–37.

Misra, V., & Pandey, S. D. (2005). Hazardous waste, impact on health and environment for development of better waste management strategies in future in India. *Environment International, 31*(3), 417–431.

Mohajan, H. K. (2020). Circular economy can provide a sustainable global society. *Journal of Economic Development, Environment and People, 9*(3), 38–62.

Molino, A., Larocca, V., Chianese, S., & Musmarra, D. (2018). Biofuels production by biomass gasification: A review. *Energies, 11*(4), 811.

Monir, M. U., Abd Aziz, A., Khatun, F., Vo, D. V. N., Mokhtar, N. M. (2021). Lignocellulosic biomass conversion to syngas through Co-gasification approach. In *Progressive thermochemical biorefining technologies* (pp. 125–142). CRC Press.

Monir, M. U., Khatun, F., Abd Aziz, A., & Vo, D. V. N. (2020). Thermal treatment of tar generated during co-gasification of coconut shell and charcoal. *Journal of Cleaner Production, 256*, Article 120305.

Munir, M. T., Mardon, I., Al-Zuhair, S., Shawabkeh, A., & Saqib, N. U. (2019). Plasma gasification of municipal solid waste for waste-to-value processing. *Renewable and Sustainable Energy Reviews, 116*, Article 109461.

Mupambwa, H. A., & Mnkeni, P. N. S. (2018). Optimizing the vermicomposting of organic wastes amended with inorganic materials for production of nutrient-rich organic fertilizers: A review. *Environmental Science and Pollution Research, 25*(11), 10577–10595.

Nadeem, F., Bhatti, I. A., Ashar, A., Yousaf, M., Iqbal, M., Mohsin, M., Nisar, J., Tamam, N., & Alwadai, N. (2021). Eco-benign biodiesel production from waste cooking oil using eggshell derived MM-CaO catalyst and condition optimization using RSM approach. *Arabian Journal of Chemistry, 14*(8), Article 103263.

Nemathaga, F., Maringa, S., & Chimuka, L. (2008). Hospital solid waste management practices in Limpopo Province, South Africa: A case study of two hospitals. *Waste Management, 28*(7), 1236–1245.

Nguyen, T. D., Kawai, K., & Nakakubo, T. (2021). Drivers and constraints of waste-to-energy incineration for sustainable municipal solid waste management in developing countries. *Journal of Material Cycles and Waste Management, 23*(4), 1688–1697.

Nivetha, N., Srivarshine, B., Sowmya, B., Rajendiran, M., Saravanan, P., Rajeshkannan, R., Rajasimman, M., Pham, T. H. T., Shanmugam, V., & Dragoi, E. N. (2023). A comprehensive review on bio-stimulation and bio-enhancement towards remediation of heavy metals degeneration. *Chemosphere, 312*, Article 137099.

Nweke, O. C., & Sanders, W. H., III. (2009). Modern environmental health hazards: A public health issue of increasing significance in Africa. *Environmental Health Perspectives, 117*(6), 863–870.

Ojovan, M. I., & Lee, W. E. (2005). Immobilisation of radioactive wastes in glass. *An introduction to nuclear waste immobilisation* (pp. 213–249).

Ojuederie, O. B., & Babalola, O. O. (2017). Microbial and plant-assisted bioremediation of heavy metal polluted environments: A review. *International Journal of Environmental Research and Public Health, 14*(12), 1504.

Okolie, J. A., Rana, R., Nanda, S., Dalai, A. K., & Kozinski, J. A. (2019). Supercritical water gasification of biomass: A state-of-the-art review of process parameters, reaction mechanisms and catalysis. *Sustainable Energy & Fuels, 3*(3), 578–598.

Pal, D. B., Singh, A., & Bhatnagar, A. (2022). A review on biomass based hydrogen production technologies. *International Journal of Hydrogen Energy, 47*(3), 1461–1480.

Palmer, G., Roberts, A., Hoadley, A., Dargaville, R., & Honnery, D. (2021). Life-cycle greenhouse gas emissions and net energy assessment of large-scale hydrogen production via electrolysis and solar PV. *Energy & Environmental Science, 14*(10), 5113–5131.

Pasupathi, P., Sindhu, S., Ponnusha, B. S., & Ambika, A. (2011). Biomedical waste management for health care industry. *International Journal of Biological and Medical Research, 2*(1), 472–486.

Patel, A. K., Singhania, R. R., Albarico, F. P. J. B., Pandey, A., Chen, C. W., & Dong, C. D. (2022). Organic wastes bioremediation and its changing prospects. *Science of the Total Environment, 824*, Article 153889.

Peng, Y., Wu, P., Schartup, A. T., & Zhang, Y. (2021). Plastic waste release caused by COVID-19 and its fate in the global ocean. *Proceedings of the National Academy of Sciences, 118*(47), Article e2111530118.

Pepin, J., Abou Chakra, C. N., Pepin, E., Nault, V., & Valiquette, L. (2014). Evolution of the global burden of viral infections from unsafe medical injections, 2000–2010. *PLoS ONE, 9*(6), Article e99677.

Perera, I. C., & Hemamali, E. H. (2022). Genetically modified organisms for bioremediation: current research and advancements. In: *Bioremediation of environmental pollutants: Emerging trends and strategies* (pp. 163–186).

Pratush, A., Kumar, A., & Hu, Z. (2018). Adverse effect of heavy metals (As, Pb, Hg, and Cr) on health and their bioremediation strategies: A review. *International Microbiology, 21*(3), 97–106.

Rajak, R., Mahto, R. K., Prasad, J., & Chattopadhyay, A. (2022). Assessment of bio-medical waste before and during the emergency of novel Coronavirus disease pandemic in India: A gap analysis. *Waste Management & Research, 40*(4), 470–481.

Rathi, B. S., Kumar, P. S., Rangasamy, G., & Rajendran, S. (2022). A critical review on Biohydrogen generation from biomass. *International Journal of Hydrogen Energy, 52*, 115–138.

Sadala, S., Dutta, S., Raghava, R., Jyothsna, T. S., Chakradhar, B., & Ghosh, S. K. (2019). Resource recovery as alternative fuel and raw material from hazardous waste. *Waste Management & Research, 37*(11), 1063–1076.

Saeidi-Mobarakeh, Z., Tavakkoli-Moghaddam, R., Navabakhsh, M., & Amoozad-Khalili, H. (2020). A bi-level and robust optimization-based framework for a hazardous waste management problem: A real-world application. *Journal of Cleaner Production, 252*, Article 119830.

Santagata, R., Ripa, M., Genovese, A., & Ulgiati, S. (2021). Food waste recovery pathways: Challenges and opportunities for an emerging bio-based circular economy. A systematic review and an assessment. *Journal of Cleaner Production, 286*, 125490.

Sarangi, P. K., & Nanda, S. (2020). Biohydrogen production through dark fermentation. *Chemical Engineering & Technology, 43*(4), 601–612.

Savoldelli, J., Tomback, D., & Savoldelli, H. (2017). Breaking down polystyrene through the application of a two-step thermal degradation and bacterial method to produce usable byproducts. *Waste Management, 60*, 123–126.

Saxena, S. C., & Jotshi, C. K. (1996). Management and combustion of hazardous wastes. *Progress in Energy and Combustion Science, 22*(5), 401–425.

Sayara, T., Basheer-Salimia, R., Hawamde, F., & Sánchez, A. (2020). Recycling of organic wastes through composting: Process performance and compost application in agriculture. *Agronomy, 10*(11), 1838.

Sayed, K., Baloo, L., & Sharma, N. K. (2021). Bioremediation of total petroleum hydrocarbons (TPH) by bioaugmentation and biostimulation in water with floating oil spill containment booms as bioreactor basin. *International Journal of Environmental Research and Public Health, 18*(5), 2226.

Sharma, A., Aloysius, V., & Visvanathan, C. (2019). Recovery of plastics from dumpsites and landfills to prevent marine plastic pollution in Thailand. *Waste Disposal & Sustainable Energy, 1*(4), 237–249.

Sharma, P., Dutta, D., Udayan, A., Nadda, A. K., Lam, S. S., & Kumar, S. (2022). Role of microbes in bioaccumulation of heavy metals in municipal solid waste: Impacts on plant and human being. *Environmental Pollution, 305*, Article 119248.

Shovon, S. M., Akash, F. A., Rahman, W., Rahman, M. A., Chakraborty, P., Hossain, H. Z., & Monir, M. U. (2024). Strategies of managing solid waste and energy recovery for a developing country–A review. *Heliyon, 10*(2).

Sun, W., Cheng, K., Sun, K. Y., & Ma, X. (2021). Microbially mediated remediation of contaminated sediments by heavy metals: A critical review. *Current Pollution Reports, 7*(2), 201–212.

Sutherland, L. A. (2013). Can organic farmers be 'good farmers'? Adding the 'taste of necessity'to the conventionalization debate. *Agriculture and Human Values, 30*(3), 429–441.

Tian, Q., Dou, X., Huang, L., Wang, L., Meng, D., Zhai, L., Shen, Y., You, C., Guan, Z., & Liao, X. (2020). Characterization of a robust cold-adapted and thermostable laccase from *Pycnoporus sp.* SYBC-L10 with a strong ability for the degradation of tetracycline and oxytetracycline by laccase-mediated oxidation. *Journal of Hazardous Materials, 382*, 121084.

Trinh, M. M., Kuo, C. H., & Chang, M. B. (2019). Characterization of PCDD/Fs and dl-PCBs emission from combustion of PCB-containing oil in a fluidized-bed incinerator. *Chemosphere, 225*, 35–42.

Unuofin, J. O., Okoh, A. I., & Nwodo, U. U. (2019). Aptitude of oxidative enzymes for treatment of wastewater pollutants: A laccase perspective. *Molecules, 24*(11), 2064.

Van Fan, Y., Klemeš, J. J., Lee, C. T., & Perry, S. (2018). Anaerobic digestion of municipal solid waste: Energy and carbon emission footprint. *Journal of Environmental Management, 223*, 888–897.

Vocciante, M., Finocchi, A., De Folly D′ Auris, A., Conte, A., Tonziello, J., Pola, A., & Reverberi, A. P. (2019). Enhanced oil spill remediation by adsorption with interlinked multilayered graphene. *Materials, 12*(14), 2231.

Wajda, A., Brociek, R., & Pleszczyński, M. (2022). Optimization of energy recovery from hazardous waste in a waste incineration plant with the use of an application. *Processes, 10*(3), 462.

Wang, Q., Awasthi, M. K., Ren, X., Zhao, J., Wang, M., Chen, H., & Zhang, Z. (2018). Recent advances in composting of organic and hazardous waste: a road map to safer environment. *Biosynthetic Technology and Environmental Challenges*, pp. 307–329.

Widmer, R., Oswald-Krapf, H., Sinha-Khetriwal, D., Schnellmann, M., & Böni, H. (2005). Global perspectives on e-waste. *Environmental Impact Assessment Review, 25*(5), 436–458.

Xu, Q., Wu, B., & Chai, X. (2022). In situ remediation technology for heavy metal contaminated sediment: A review. *International Journal of Environmental Research and Public Health, 19*(24), 16767.

Yu, K. H., Zhang, Y., Li, D., Montenegro-Marin, C. E., & Kumar, P. M. (2021). Environmental planning based on reduce, reuse, recycle and recover using artificial intelligence. *Environmental Impact Assessment Review, 86*, Article 106492.

Zakaria, N. M., Hassan, M. K., Ibrahim, A. N. H., Rosyidi, S. A. P., Yusoff, N. I. M., Mohamed, A. A., & Hassan, N. (2018). The use of mixed waste recycled plastic and glass as an aggregate replacement in asphalt mixtures. *Jurnal Teknologi (Sciences & Engineering), 80*(1).

Zucco, M. A., Walters, S. A., Chong, S. K., Klubek, B. P., & Masabni, J. G. (2015). Effect of soil type and vermicompost applications on tomato growth. *International Journal of Recycling of Organic Waste in Agriculture, 4*(2), 135–141.

Index

H. Gupta et al. (eds.), *Advanced Technologies for Assessment, Recycling, and Management of Solid Waste*, Waste as a Resource,
https://doi.org/10.1007/978-3-032-24788-9

www.ingramcontent.com/pod-product-compliance
Ingram Content Group UK Ltd.
Pitfield, Milton Keynes, MK11 3LW, UK
UKHW021615260926
475829UK00010B/68
* 9 7 8 3 0 3 2 2 4 7 8 7 2 *